Probability and Random Processes for Electrical Engineers

Communications and Signal Processing

Senior Consulting Editor

Stephen W. Director, University of Michigan, Ann Arbor

Antoniou: *Digital Filters: Analyses and Design*
Auñón/Chandrasekar: *Introduction to Probability and Random Processes*
Candy: *Signal Processing: The Model-Based Approach*
Candy: *Signal Processing: The Modern Approach*
Carlson: *Communications Systems: An Introduction to Signal and Noise in Electrical Communication*
Cherin: *An Introduction to Optical Fibers*
Collin: *Antennas and Radiowave Propagation*
Collin: *Foundations for Microwave Engineering*
Cooper and McGillem: *Modern Communications and Spread Spectrum*
Davenport: *Probability and Random Processes: An Introduction for Applied Scientists and Engineers*
Drake: *Fundamentals of Applied Probability Theory*
Huelsman and Allen: *Introduction to the Theory and Design of Active Filters*
Jong: *Method of Discrete Signal and System Analysis*
Keiser: *Local Area Networks*
Keiser: *Optical Fiber Communications*
Kershenbaum: *Telecommunications Network Design Algorithms*
Kraus: *Antennas*
Kuc: *Introduction to Digital Signal Processing*
Mitra: *Digital Signal Processing: A Computer-Based Approach*
Papoulis: *Probability, Random Variables, and Stochastic Principles*
Papoulis: *Signal Analysis*
Papoulis: *The Fourier Integral and Its Applications*
Proakis: *Digital Communications*
Schwartz: *Information Transmission, Modulation, and Noise*
Schwartz and Shaw: *Signal Processing*
Siebert: *Circuits, Signals, and Systems*
Smith: *Modern Communication Circuits*
Taub and Schilling: *Principles of Signals and Systems*
Taylor: *Principles of Signals and Systems*

Also Available from McGraw-Hill

**Schaum's Outline Series in Electronics
& Electrical Engineering**

Most outlines include basic theory, definitions, and hundreds of example problems solved in step-by-step detail, and supplementary problems with answers.

Related titles on the current list include:

Analog & Digital Communications
Basic Circuit Analysis
Basic Electrical Engineering
Basic Electricity
Basic Mathematics for Electricity & Electronics
Digital Principles
Electric Circuits
Electric Machines & Electromechanics
Electric Power Systems
Electromagnetics
Electronics Communication
Electronic Devices & Circuits
Feedback & Control Systems
Introduction to Digital Systems
Microprocessor Fundamentals
Signals & Systems

Schaum's Electronic Tutors

A Schaum's Outline plus the power of Mathcad software. Use your computer to learn the theory and solve problems. Every number, formula, and graph can be changed and calculated on screen.

Related titles on the current list include:

Electric Circuits
Feedback & Control Systems
Electromagnetics
College Physics

Available at most college bookstores, or for a complete list of titles and prices, write to:
Schaum's

11 West 19th Street
New York NY 1001-4285
(212-337-4097)

PROBABILITY AND RANDOM PROCESSES FOR ELECTRICAL ENGINEERS

YANNIS VINIOTIS

NORTH CAROLINA STATE UNIVERSITY

WCB McGraw-Hill

BOSTON   BURR RIDGE, IL   DUBUQUE, IA   MADISON, WI   NEW YORK   SAN FRANCISCO   ST. LOUIS
BANGKOK   BOGATÁ   CARACAS   LISBON   LONDON   MADRID
MEXICO CITY   MILAN   NEW DELHI   SEOUL   SINGAPORE   SYDNEY   TAIPEI   TORONTO

WCB/McGraw-Hill
   A Division of The McGraw-Hill Companies

Probability and Random Processes for Electrical Engineers

This book is printed on acid-free paper.

1 2 3 4 5 6 7 8 9 0   DOC DOC   9 0 0 9 8 7

ISBN 0-07-067491-4

Publisher: Tom Casson
Sponsoring editor: Lynn B. Cox
Developmental editor: Michael Morales
Marketing manager: John Wannemacher
Project manager: Eva Marie Strock
Production supervisor: Richard DeVitto
Jacket designer: Nicole Leong
Printer: R. R. Donnelley & Sons

Library of Congress Cataloging-in-Publication Data

Viniotis, Yannis.
   Probability and random processes for electrical engineering /
Yannis Viniotis.
      p.   cm.
   Includes bibliographical references and index.
   ISBN 0-07-067491-4
   1. Probabilities. 2. Stochastic processes.   3. Electric
engineering—Mathematics. I. Title.
   QA273.V55 1997
   519.2'024'62—dc21   97-37609
                                                              CIP

http//www.mhhe.com

To my wife Maria

and

in the memory of my father Socrates

About the Author

Yannis Viniotis received his Ph.D in Electrical Engineering from the University of Maryland. He is Associate Professor of Electrical and Computer Engineering at North Carolina State University, where he has worked since 1988. He has served as a Principal Investigator on projects for the Army, IBM, and the NSF, and he continues to consult for industry. In 1992 and 1993, Dr. Viniotis chaired two international conferences on networking. In 1995 he was a guest editor for the *Performance Evaluation Journal's* special issue on high-speed networks. His primary area of research is the design and analysis of computer communication systems, with an emphasis on quality of service, multicasting, and adaptive network control algorithms. The author of more than forty articles, Dr. Viniotis remains an active researcher and lecturer in his field.

CONTENTS

PREFACE   xv

1   INTRODUCTION   1
    1.1   Why This Course?   2
    1.2   Why an Abstract Theory?   6
    1.3   Why Probability Theory?   7
    1.4   What Can Probability Theory Do?   7
    1.5   What Is the Theory in a Nutshell?   7
    1.6   Modeling Tools?   9
    1.7   Main Models?   9
    1.8   Main Application Areas   10
    1.9   A Few Detailed Examples   11
    1.10 Did I Learn the Theory?   18
    1.11 Where Do I Learn More?   18
    1.12 Summary of Main Points   19
    1.13 Problems   20

2   BASIC CONCEPTS   23
    2.1   Random Experiments   24
    2.2   Sample Spaces and Outcomes   26
    2.3   Events   27
    2.4   Probability Axioms and the Sets of Interest   30
    2.5   Elementary Theorems   33
    2.6   Conditional Probabilities   35
    2.7   Independent Events   40
    2.8   How Do We Assign Probabilities to Events?   42
    2.9   Probabilistic Modeling   44
    2.10 Difficulties   46
    2.11 Summary of Main Points   46

2.12 Checklist of Important Tools   47
2.13 Problems   47

3   CONCEPT OF A RANDOM VARIABLE   67
3.1   Sets of the Form $(-\infty, x]$ 68
3.2   Concept of a Random Variable   70
3.3   Cumulative Distribution Function   76
3.4   Probability Density Function   82
3.5   Probabilistic Model Revisited   95
3.6   Useful Random Variable Models   95
3.7   Histograms   109
3.8   Transformations of a Random Variable   115
3.9   Moments of Random Variables   121
3.10 Reliability and Failure Rates   129
3.11 Transforms of Probability Density Functions   134
3.12 Tail Inequalities   139
3.13 Generation of Values of a Random Variable   145
3.14 Summary of Main Points   150
3.15 Checklist of Important Tools   151
3.16 Problems   152

4   A VECTOR RANDOM VARIABLE   195
4.1   Experiments with More Than One Measurement   196
4.2   The Sets $(-\infty, x] \times (-\infty, y]$ 198
4.3   Joint Cumulative Distribution Function   204
4.4   Joint Probability Density Function   212
4.5   Probabilistic Model Revisited   221
4.6   Conditional Probabilities and Densities   223
4.7   Independence   230
4.8   Transformations of a Random Vector   236
4.9   Expectation, Covariance, and Correlation Coefficient   256
4.10 Useful Joint Distributions   260
4.11 More Than Two Random Variables   264
4.12 Generation of Values of a Random Vector   270
4.13 Summary of Main Points   272
4.14 Checklist of Important Tools   273
4.15 Problems   273

5 INTRODUCTION TO ESTIMATION 303

5.1 Criteria to Consider 304

5.2 MMSE Estimation, Single Measurement 307

5.3 Linear Prediction, Multiple Measurements 315

5.4 Dow Jones Example 318

5.5 Maximum-Likelihood Estimation 322

5.6 An Historical Remark 324

5.7 Summary of Main Points 327

5.8 Checklist of Important Tools 327

5.9 Problems 327

6 SEQUENCES OF (IID) RANDOM VARIABLES 333

6.1 Experiments with an Unbounded Number of Measurements 333

6.2 IID Random Variables 334

6.3 Sums of IID Random Variables 335

6.4 Random Sums of IID Random Variables 340

6.5 Weak Law of Large Numbers 344

6.6 Strong Law of Large Numbers 347

6.7 Central Limit Theorem 350

6.8 Convergence of Sequences of Random Variables 355

6.9 Borel-Cantelli Lemmas 366

6.10 Summary of Main Points 368

6.11 Checklist of Important Tools 369

6.12 Problems 369

7 RANDOM PROCESSES 387

7.1 Definition of a Random Process and Examples 388

7.2 Joint CDF and PDF 393

7.3 Expectation, Autocovariance, and Correlation Functions 398

7.4 Some Important Special Cases 399

7.5 Useful Random Process Models 406

7.6 Continuity, Derivatives, and Integrals 409

7.7 Ergodicity 413

7.8 Karhunen-Loève Expansions 417

7.9 Generation of Values of a Random Process 420

7.10 Summary of Main Points   421
7.11 Checklist of Important Tools   422
7.12 Problems   422

**8   POISSON AND GAUSSIAN RANDOM
       PROCESSES   439**
8.1   Poisson Process   439
8.2   Gaussian Random Process   458
8.3   Summary of Main Points   463
8.4   Checklist of Important Tools   464
8.5   Problems   464

9   PROCESSING OF RANDOM PROCESSES   475
9.1   Introduction   475
9.2   Power Spectral Density Function   477
9.3   Response of Linear Systems to Random Processes   484
9.4   Optimal Linear Estimation   492
9.5   Kalman Filter   503
9.6   Periodograms   510
9.7   Summary of Main Points   513
9.8   Checklist of Important Tools   513
9.9   Problems   514

10   MARKOV CHAINS   527
10.1 Definition and Classification   528
10.2 Discrete-Time Markov Chains   535
10.3 Steady State of Markov Chains   541
10.4 Drifts and Ergodicity   557
10.5 Continuous-Time Markov Chains   561
10.6 Application to Ethernet LANs   571
10.7 Generation of Values of a Markov Chain   574
10.8 Summary of Main Points   576
10.9 Checklist of Important Tools   576
10.10 Problems   576

11 CASE STUDY: A BUS-BASED SWITCH ARCHITECTURE 595

11.1 Switch Architecture and Operation 596

11.2 Transition to a Stochastic Model 598

11.3 System Model 599

11.4 Performance Measures 601

11.5 Input Data Description 603

11.6 Discussion of Results 604

11.7 Description of the Simulator 610

A SET THEORY PRIMER 611

A.1 Sets and Subsets 611

A.2 Operations on Sets 618

A.3 Partitions of a Set 624

A.4 Limits of Sequences of Sets 624

A.5 Algebras of Sets 625

A.6 Problems 626

B COUNTING METHODS 633

B.1 Ordering of Elements 633

B.2 Sampling with Ordering and Without Replacement of Elements 634

B.3 Sampling Without Ordering and Without Replacement of Elements 635

B.4 Sampling Without Ordering and with Replacement of Elements 637

B.5 Problems 638

C HISTORICAL DEVELOPMENT OF THE THEORY 643

C.1 A Brief History 643

C.2 Alternative Axioms 645

C.3 Some of the Early Problems 647

D MODELING OF RANDOMNESS IN ENGINEERING SYSTEMS: A SUMMARY 651

D.1 Elementary Concepts 651

D.2 Probabilistic Models 651

D.3 What Models Have We Developed?   653
D.4 What Tools Have We Developed?   653
D.5 Mathematical Subtleties   654
D.6 What Can We Do with a Model?   655

REFERENCES   657

INDEX   663

PREFACE

My Writing Philosophy

In the summer of 1992, I was attending the American Society for Engineering Education conference in Toledo, Ohio; there I heard the following anecdote, from the dean of an engineering school, that influenced the philosophy of this book in a major way. I am freely quoting:

> . . . Once upon a time, in the Washington, D.C., area, there was a famous football coach named Goe Jibbs. One summer, coach Jibbs decided to hold summer training camps for elementary school students. Coach thought the whole process out: He would *first* explain the theory and plays and strategy and all to the students; he would *then* show them video clips in slow motion and everything; he would draw *all* the necessary details on fancy chalkboards; he would even repeat the main points and tell a few jokes, to spice up the whole thing and make the students digest it more easily. Finally, at the end of the training period, he would give them an exam to make sure every concept was *properly* absorbed.
>
> So, it came as a big surprise to coach Jibbs when, after the first week, half the kids did not show up. He was worried, but did not say a thing. When he lost another quarter of the class, he decided to ask the inevitable question: What was wrong? To which the remaining students replied:
>
> –Coach, when are we going to play football?

It was my intention and goal to let the readers of this book play probability ball since day 1 of the training camp (well, after Chapter 2). I hope you will find the ball and necessary gear in the Web site (http://www.mhhe.com/viniotis). Toward the same goal, in the text I have made extensive use of MATLAB graphics to provide realistic and easy-to-grasp examples; I have incorporated numerous examples derived from *actual* experiments. Nowadays, with the proliferation of computer systems and campus networks, one can easily find realistic engineering systems in the students' surrounding environment, on which meaningful, tractable probabilistic experiments can be designed. A number of such experiments are included in the accompanying Web site. With the ubiquitous presence of the Internet, the world is just an (rlogin) call away.

Probability Theory in Engineering

Randomness and uncertainty are elements that electrical and computer engineering students are facing at increased rates, both in their everyday life and in technical environments. From traditional communication channels with noise; to performance evaluation of protocols, networks, and computers; to job searches; to boyfriend/girlfriend behavior; to stock market investment; engineering students must deal everyday with systems that seem to evolve in an unpredictable manner. Probability theory and statistics provide the mathematical language, the necessary concepts and formalizations, and the appropriate tools to describe, analyze, and eventually control such systems.

From my experience so far, *both as a student and as a teacher,* it is very difficult for the engineering student to make the connection between probability theory and engineering practice. There are good reasons for that:

- *Lack of realistic examples with true engineering relevance.* Most books focus on the mathematical technicalities of the theory, leaving the tasks of modeling and relating to the real world to the teacher and student. Examples involving dice, cards, and urns don't help much either.

- *Most of the engineering systems are dynamic.* The natural probabilistic model for such a dynamic system would be that of a random process (in continuous or discrete time). Unfortunately, this topic comes only late in the course, after seemingly irrelevant topics that can hardly be motivating.

- *There is little hands-on experience.* Engineering students are more likely to be of the do-it-yourself, hands-on experience type. It would be great to give them the opportunity to apply the theory *while taking the course,* not after a year or two (or five), when they have a job already or are in graduate school.

- *The what-if approach is not routinely encouraged in mathematical books,* which try to build up a rigorous theory from the bottom up, with as little distraction as possible. Yet to stray from the beaten path, posit questions, and solicit alternative approaches is an engineering attitude we try to endow our students with.

- *Pressure to cover as much theory as possible.* This leaves little time to spend on, say, description of an engineering system that could be used as a project or case study.

This book addresses these issues head-on. Realistic examples have been used throughout all chapters. A case study at the end shows how a lot of pieces of the theory are tied together to study a realistic, complex engineering system. The what-if approach is encouraged, both in the text and especially in the problem sections. Hands-on experience can be gained through the data sets in the Web site.

Theory/Intuition Balance

Throughout the book, we strive to achieve a balance between theory and intuition, and rigor and applicability of results. As a result, the book is not a detailed treatise of the subject (as are, for example, references [13], [5], and [25]). However, it does go a bit more deeply into mathematical details than the level usually taught at the undergraduate level.

Suggested Syllabi

I have taught parts of the book material at both the undergraduate and the graduate level. Chapters 1–4, along with selected sections from Chapters 5 and 6, can be used as an introductory-level, one-semester undergraduate course. Chapters 1–4, at a faster pace, along with selected sections from the rest of the chapters can be used as an introductory-level, one-semester graduate course. If emphasis on signal processing aspects is desired, Chapters 5, 7 (the Gauss part of) 8, and 9 can be used. If emphasis on performance analysis is desired, then Chapters 6, 7 (the Poisson part of) 8, and 10 can be used. There is plenty of material so that the instructor can tailor a course to specific needs and directions.

Problems

There are more than 1,250 problems in the book. They are logically grouped in four categories per chapter: mathematical drills, theory, programming, and "do-it-yourself experimental." The mathematical ones make sure that the technicalities of the theory are mastered and the analytical skills are developed. The theory-type questions are mostly what-if problems, where the presented theory is challenged and/or extended. This is a good way for the student to really understand the importance of a definition or a condition in a proof. Most of the programming problems are easily solved with MATLAB. In the accompanying Web site, I have provided the source code for plotting in two and three dimensions and calculating cumulative distribution functions, densities, histograms, etc. Finally, the experimental-type problems are easy to do in any multiuser operating system with networking facilities. The Unix-like commands for most of the experiments are provided in the text. Relevant data have been provided in the Web site, for students who lack the time to collect them. I found that my students hated it at first, but *really liked* this type of problem, since it was a very strong proof by example that the theory *does* pertain to engineering life.

Acknowledgments

I owe much to my undergraduate teachers G. Roussas, a statistician, and N. Tzannes, an engineer, at the University of Patras, Greece. The first taught me the merits of rigor and mathematical thinking. The second taught me the beauty of the what-if, let's-challenge-the-theory approach of engineering thinking. Getting a deserved B in a graduate-school random processes course definitely provided me with the motivation to really learn this stuff.

I owe much more to numerous authors, whose books have influenced my presentation plans and the balance between rigor and applicability. In particular, I would like to mention the books by K. L. Chung, R. M. Gray and L. D. Davisson, A. Leon Garcia, and A. Papoulis. I am very grateful to the students and anonymous reviewers whose valuable comments on early drafts of this book improved its style and presentation. For the same reason, I would like to thank in particular the following reviewers: Scott Budge, Utah State University; Shih-Chun Chang, George Mason University; Venkat Devarajan, University of Texas at Arlington; Monson H. Hayes, Georgia Institute of Technology; Subhash Kak, Louisiana State University; Philipp Kornreich, Syracuse University; and Jitendra K. Tugnait, Auburn University.

Thanks to NSF for a grant through the SUCCEED coalition, which made starting of this book possible. Thanks to the Electrical and Computer Engineering Department at North Carolina State University, for a sabbatical semester during which the book was finalized. Very special thanks to Prof. Serge Fdida, for providing a warm, conducive environment at Université de Paris 5; Chapters 3 and 4 were finalized there, despite the infinite distractions Paris has to offer. Also, very special thanks to Dr. Leo Georgiadis, of IBM T. J. Watson Research Center, for introducing me to the wonderful world of Markov Chains and for instilling in me a keen interest in the subject.

Finally, all my love to the "CNN" network (Candice, Nora, and Natalie) and my wife Maria; the latter for her patience and support and the former for their total lack of it. This made writing the book much longer but kept the proper balance.

Yannis Viniotis
Raleigh, NC

Abbreviation	Meaning
ARMA	autoregressive moving average
cdf	cumulative distribution function
CLT	central limit theorem
CTMC	continuous time Markov Chain
DTMC	discrete time Markov Chain
ftp	file transfer protocol
IID	independent, identically distributed
i.o.	infinitely often
IPP	interrupted Poisson process
jcdf	joint cumulative distribution function
jpdf	joint probability density function
jpmf	joint probability mass function
l.i.m.	limit in the mean
MC	Markov Chain
MLE	maximum likelihood estimation
MMPP	Markov modulated Poisson process
MMSE	minimum mean square estimation
OP	orthogonality principle
PASTA	Poisson Arrivals See Time Averages
pdf	probability density function
pmf	probability mass function
psd	power spectral density
SLLN	Strong Law of Large Numbers
WLLN	Weak Law of Large Numbers
WSS	wide sense stationary

Symbol	Meaning
R	set of real numbers
$\mathcal{B}(R)$	Borel field on R
$\text{Im}(x)$	imaginary part of the complex number x
$\text{Re}(x)$	real part of the complex number x
S	sample space
Q	set of events
$P[A]$	probability of an event
$P[A|B]$	conditional probability of an event
$f_X(x)$	probability density function
$F_X(x)$	cumulative distribution function
$F_{XY}(x,y)$	joint cumulative distribution function
$f_{XY}(x,y)$	joint probability density function
$f_X(x|y)$	conditional probability density function
$F_X(x|y)$	conditional cumulative distribution function
$R_L(t)$	reliability function
$r_L(t)$	failure rate function
$\Phi_X(\omega)$	characteristic function
$L_X(\omega)$	Laplace transform
$G_X(z)$	Z transform
$R_X(t,s), R_X(\tau)$	autocorrelation function
$C_X(t,s), C_X(\tau)$	autocovariance function
$m_X(t)$	mean function
$S_X(f)$	power spectral density
$R_{XY}(\tau)$	cross-correlation function
$S_{XY}(f)$	cross-power spectral density
S_X	range of random variable X
EX	mean of random variable X
σ_X^2	variance of random variable X
$cov(X,Y)$	covariance of random variables X, Y
$\rho_{X,Y}$	correlation coefficient
$X_n \xrightarrow{\text{a.s.}} X$	almost sure convergence
$X_n \xrightarrow{P} X$	convergence in probability
$X_n \xrightarrow{D} X$	convergence in distribution
$X_n \xrightarrow{\text{m.s.}} X$	convergence in mean square

1

INTRODUCTION

As an engineering student, the first questions that came to my mind, as I was about to take a class, and especially a mathematics class, always were these:

- Why this course? How does it relate to the other courses? How does it relate to my future job?

- Why this theory?

- Why this book?

- Why this instructor?

I felt it was crucial for my grasp of the material to have an answer to these questions as soon as possible. It gave me the feeling of obtaining the "global picture," of putting the pieces of the education puzzle together. Today, as an instructor and author of a book, I feel the same need. I firmly believe that understanding the *relevance, power, limitations, and tools* of a mathematical theory is crucial for an engineering student who will be asked to apply this theory to solve concrete, realistic problems. Moreover, having had a chance to apply the theory *while* studying it, as opposed to applying it during another semester, is *equally* crucial.

In this chapter we provide an answer to the first three questions and give a high-level, "executive summarylike" discussion of the relevance, power, limitations, and tools aspect. (The answer to the last question we cannot provide.) Students should read this chapter again, at the end of the course. If the "global picture" is clear by that time, the student, instructor, and author were successful.

1.1   WHY THIS COURSE?

Perhaps it is better to answer this question with a concrete, realistic example. The example draws from the *ftp* (*f*ile *transfer protocol*) network protocol which is available in most platforms (e.g., Unix, AIX, Mac).

The *ftp* is a protocol used to transfer files between two computers. During transfer of a file of a certain *size* (measured in kilobytes), a *delay* is incurred (measured in seconds); the transfer is done at a certain speed, or *through-put* (measured in kilobytes per second). Standard *ftp* output provides both delay and throughput measures. The following is a verbatim copy of the *ftp* output. We tried three separate connections between the author's machine (ececho.ncsu.edu, IP address 152.1.59.96) and the following machines:

- ftp.std.com (IP address 192.74.137.5) in Boston, United States (where the font styles for writing this book came from)

- descartes.univ-paris5.fr (IP address 193.48.200.13) in Paris, France

- ariadne.ics.forth.gr (IP address 139.91.1.1) in the island of Crete, Greece

```
From descartes.univ-paris5.fr:

150 ASCII data connection for tar.book (132.227.61.5,3192)
(2555904 bytes).
226 ASCII Transfer complete.
local: tar.book remote: tar.book
2611082 bytes received in 36 seconds (71 Kbytes/s)


From ftp.std.com:

226 Transfer complete.
59 bytes received in 0.012 seconds (4.7 Kbytes/s)
150 Opening ASCII mode data connection for INDEX (240 bytes).
226 Transfer complete.
local: INDEX remote: INDEX
245 bytes received in 0.47 seconds (0.51 Kbytes/s)


150 Opening ASCII mode data connection for README (1486 bytes).
226 Transfer complete.
local: README remote: README
1526 bytes received in 0.37 seconds (4 Kbytes/s)


150 Opening ASCII mode data connection for booksamp.tex
```

```
(12768 bytes).
226 Transfer complete.
local: booksamp.tex remote: booksamp.tex
13197 bytes received in 25 seconds (0.52 Kbytes/s)

150 Opening ASCII mode data connection for edbksamp.tex
(13084 bytes).
226 Transfer complete.
local: edbksamp.tex remote: edbksamp.tex
13540 bytes received in 19 seconds (0.71 Kbytes/s)

150 Opening ASCII mode data connection for editedbk.doc
(72379 bytes).
226 Transfer complete.
local: editedbk.doc remote: editedbk.doc
75121 bytes received in 44 seconds (1.7 Kbytes/s)

150 Opening ASCII mode data connection for editedbk.sty
(8140 bytes).
226 Transfer complete.
local: editedbk.sty remote: editedbk.sty
8427 bytes received in 0.12 seconds (70 Kbytes/s)


From ariadne.ics.forth.gr:

150 Opening ASCII mode data connection for README
(1018 bytes).
226 Transfer complete.
local: README remote: README
1046 bytes received in 0.93 seconds (1.1 Kbytes/s)

150 Opening ASCII mode data connection for README_FORTHnet
(6079 bytes).
226 Transfer complete.
local: README_FORTHnet remote: README_FORTHnet
6190 bytes received in 3.7 seconds (1.6 Kbytes/s)
```

As this output clearly shows, the throughput is a quantity that varies with the connection. This might be expected, given the relative distances of the three connections. However, even within the same connection, with the *same file* transfer, throughput varies. As we say, it appears that the throughput is *random*.

Generalizing this example, we can say that the electrical and computer engineering (ECE) student of the late twentieth century, and surely[1] that of the future, will be faced with systems that

1. Have *random* inputs (in the sense we describe shortly)

2. Have deterministic inputs but (seem to) behave *randomly*

3. Are too complex to be accurately described

4. Are a combination of the above

We mention briefly examples of such systems here. Some we study in greater detail in later chapters. A telephone switch, a central processing unit (CPU), a file server, a computer accepting packets transferred through the *ftp* protocol, and a signal processing filter are simple examples of type 1 systems. Calls arrive at the switch at time instants that cannot be predetermined (e.g., they arrive in random time instants); moreover, their duration is not known in advance. Programs are sent to a CPU (with multiple users) in random instants; their runtimes cannot be predicted in advance either. File retrieval and storage requests arrive at a file server in an unpredictable manner. Figure 1.1 is a realistic plot of the delay of packets transferred into machine descartes.univ-paris5.fr, during one of our experiments in Paris.

Typically, input signals to digital or analog filters have some unknown distortion in them, caused by noise or imperfect channels. A communication line with noise behaves as a system of type 2 because, for example, its output due to a particular input signal is not always the same. Figure 1.2 is an example of a simulated communication channel with noise.

The motion of gas molecules or electrons is an example of a type 3 system. A computer network is an example of a type 4 system. Your boyfriend or girlfriend or spouse may be another type 4 system. It is instructive to think about other examples, both technical and from everyday life (see Problems 1.1–1.9 at the end of this chapter).

This course and book will

- Help you *understand* the operation of systems with randomness; you will be provided with the appropriate *concepts* (such as random variables, random processes, averages, correlations, stochastic convergence).

- Help you *analyze and design* such systems; you will be provided with the appropriate *theorems* (such as the central limit theorem and the Strong Law of Large Numbers), *models* (e.g., the Gaussian and Poisson distribution functions) and *techniques* (e.g., estimation and prediction).

---

[1] After studying Chapter 6, read that as *almost* surely.

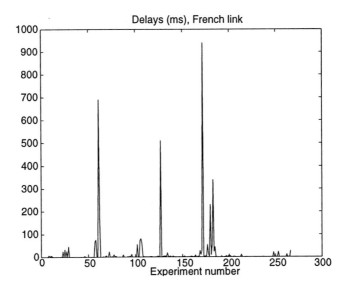

Figure 1.1 The *ftp* transferred packets.

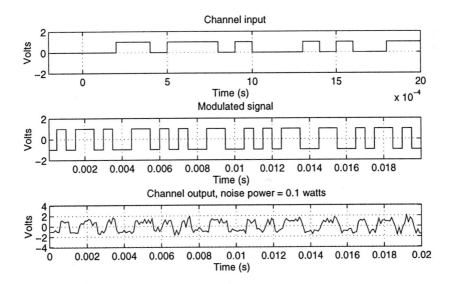

Figure 1.2 Channels corrupted by noise.

Probability theory is directly related to a number of engineering topics (and therefore courses). Probability concepts are as necessary as basic calculus if one is to understand communication systems (especially noise, signal recovery, protocols, and network performance) and signal processing (especially filtering, estimation, detection, and prediction). They are also applied in power distribution (especially load prediction and control) and semiconductor fabrication (especially implant diffusions, theoretical electron studies).

1.2   WHY AN ABSTRACT THEORY?

On one hand, the diversity of the example systems in Section 1.1 is superb for motivation; on the other hand, this diversity might lead to the hasty conclusion that such systems must be studied separately. There must be, however, crucial elements common to all these examples. An *abstract theory* is the only way to extract such common elements and provide further insights to them all. This approach has been vindicated historically many times. Markov Chains, e.g., were first used as a means to study linguistics, in 1907. About 60 years later, they became an invaluable tool in studying performance of computer systems and networks.

An *abstract* theory has numerous advantages:

- It is easier to teach since it does not require exact knowledge of any particular system.

- It provides economy since it can be taught on a chalkboard in a classroom environment. There is no need for expensive laboratories or time-consuming experiments (as you will notice later).

- It has wider applicability. A theory can be applied, anytime after it has been developed, to systems not even anticipated at the time of its creation. Democritus' theory of the atom was applied a couple of millennia after its inception.

- It is intellectually appealing.

If you still have doubts about an abstract theory being the best thing, can you provide a few arguments for an alternative approach?

Whether or not the merits of an abstract theory materialize depends largely on a simple fact: Will the student have the opportunity to *apply it, while still studying it*, to a realistic engineering example of his or her choice and interest? We hope that this book and the accompanying software will give you many chances and motivation while making it easy for you to work out a realistic case.

1.3   WHY PROBABILITY THEORY?

We can use humor to deal with a boyfriend or girlfriend; however, we must be more serious when it comes to *ftp* delays, etc. The only other alternative would be to use a *deterministic* theory, e.g., differential equations, but that would be too difficult, even impossible. Consider the movement of atoms and electrons around their nuclei. A few trillion differential equations could provide exact answers, but that would be "a bit difficult." Consider next the file transfer example in Section 1.1. We *could* determine the exact delay of the transfer if we knew who else was using the CPU and communication line(s) at the given time interval during which our transfer took place. But that would require future knowledge about users of the lines in the given time interval, so that is impossible.

It is instructive at this point to see how difficult (if possible at all) it is to model any of the systems we mentioned in Section 1.1, using any other theory you know. Problems 1.10–1.11 reinforce this point.

1.4   WHAT CAN PROBABILITY THEORY DO?

This is a very crucial question; the answer will be delivered *in pieces*, throughout the entire book. It is very easy for a student who learns a new theory to get absorbed in the "trees" and "miss the forest." Moreover, another issue that obscures the answer to this question is that *much effort* must be spent mastering (probably) new mathematical tools (e.g., set theory, integrals over two-dimensional spaces, transforms). Thus, I strongly suggest that students ask this question *again and again* as they go through the chapters. The case study in Chapter 11 provides a "proof by example" of what the theory can do. In short, the theory

- Is the natural tool to *describe and analyze* the inherent randomness of a variety of engineering systems

- Can be used to *design and control* such systems

1.5   WHAT IS THE THEORY IN A NUTSHELL?

This summarylike information is useful for attacking a realistic problem. One knows immediately which concepts and theorems can be applied in a given situation, similar to knowing whether to use a screwdriver or a wrench when repairing a car.

Here is the structure of the theory; complexity increases as you move down the list.

- Events (Chapter 2)

- One random variable (Chapter 3)

- A vector random variable (Chapter 4)

- A random process (Chapter 7)

The chapter layout follows this structure for the obvious reason: We can build more and more complex concepts only when we master the simpler ones. This natural selection, however, has a rather severe disadvantage: It is easier for the (impatient) student to get distracted and miss the forest for the trees.

Events are the simplest structural elements of the theory; they arise naturally when describing simple experiments. They are used to build a solid foundation of the theory. They draw mathematical tools from *set theory*. The notion of a **random variable** and its associated **probability density function** allows us to bring into the picture more powerful tools from *calculus and analysis*. A **vector random variable** is a natural extension of the concept of a single random variable. Both concepts are very common in engineering experiments; they are used much more than events.

None of the first three concepts deals directly with *time*. Therefore, they cannot be directly used to model dynamic systems.[2] The majority of the applications that an ECE student draws experiences from are time-evolving, dynamic systems, which makes the subject of probability theory quite difficult to motivate since its relevance becomes apparent only at the very end. A **random process** is the marriage of random variables with the notion of time. From the theoretical point of view, it is a vector random variable with infinite dimensionality.

The theory we build does not deal with time until almost the very end (Chapter 7), which may create an implicit doubt about the relevance of this theory to the ECE world.[3] This is the price we have to pay for building the theory in a bottom-up approach: There are *an awful lot* of (mathematical) details we have to take care of before we can go to the topics of Chapters 7, 8, 9, and 10.

---

[2] But see Chapters 3 and 4 for some limited handling of time and dynamic systems.

[3] The vast majority of books in the field indirectly fuel this doubt by providing lots of examples that seem irrelevant: dice throwing, card shuffling, genetic population studies, urns and balls, lightbulb lifetimes, etc.

So exercise patience throughout these beginning chapters. Before we can play "probability ball," we must spend time learning some basic rules (the axioms and some basic theorems) and "safety precautions" (the set Q, introduced for mathematical rigor).

1.6   MODELING TOOLS?

Of course, deciding that a wrench is the right tool to use solves about half the problem. The next question is, What size wrench to use? When one is confronted with a real problem (see, e.g, the case study chapter), it helps to know what the available tools are *exactly*. In a nutshell, the mainstream[4] tools that probability theory provides for modeling of engineering systems are

- Probability density and distribution functions, for single and vector random variables

- Infinite sequences of *independent and identically distributed* (*IID*) random variables

- Markov Chains

- Martingales

The first tool is applicable to static environments (or to the steady-state portion of dynamic environments). The last three tools are applicable to dynamic environments, or systems that evolve with time. Time can, of course, be discrete or continuous. Subtle mathematical problems arise in the continuous time case, as we will see in Chapter 7. The last three tools provide for increasing complexity in the system dynamics, as will become apparent later. The IID random variables do not have memory, but Markov Chains and martingales do. As expected, the latter tools become more involved. Martingales are beyond the scope of this introductory book. For an insightful introduction to the subject, see [38]. For more advanced exposition, see [11] and [7].

1.7   MAIN MODELS?

We must point out right away that there is a difference between modeling tools and models. The difference will become apparent immediately after you study random variables. For now, suffice it to say that we view the concept

---

[4] For more advanced tools, see [19].

of a random variable as a modeling tool (e.g., a wrench), while a Gaussian probability density function is a model (e.g., a size 14 wrench).

A tremendous number of detailed models were developed in the past. Since the seventeenth century, when Pascal studied and solved the first recorded problem in probability, hundreds of models have been suggested and are still being suggested even today, for a number of reasons: (1) A lot of mathematicians (such as Gauss, Rayleigh, Poisson, Markov) have been working on the subject, and (2) new applications create the need for new models. For example, in the late 1980s, the emergence of high-speed computer networks, in which traffic is expected to be bursty and correlated, sparked a big interest in the Interrupted Poisson Process (IPP) (Section 8.1.5) and the Markov Modulated Poisson Process (MMPP) model (Section 8.1.5).

The question of which model is the best choice for a given real problem is a tough one. The choice is affected at least by (1) the (in)experience of the modeler, (2) analytical tractability of the model, and (3) degree of desired accuracy. If I had to choose only a few of the hundreds of models that have appeared so far, my bet would be

- Gaussian random variable, the most widely used noise model in communication theory and the centerpiece of statistical models. As a probability historian said, the Gaussian curve [see Figure 3.25(b)] is as characteristic of statistics as the hexagon is of organic chemistry or the parabola of ballistics.

- Poisson random variable, the most widely used arrival model in computer network theory.

- Exponential random variable, the most widely used service time model in queuing theory.

- χ^2 random variable, very widely used in statistical testing,

- Student's t random variable, very widely used in statistical testing.

- Gaussian random process, the most widely used noise process model.

- Poisson random process, the most widely used arrival process model.

1.8   MAIN APPLICATION AREAS

Within ECE topics, three major areas have benefited the most from probability theory and in return have stimulated advances in the theory itself: (1) communication systems with noise, (2) digital signal processing (DSP), and

(3) performance evaluation and design of computer systems and communication networks. Topics in the first and second areas include (1) detection, (2) estimation, and (3) filtering. Topics in the third area include CPU and job scheduler performance, algorithm analysis, file server design, protocol design, and protocol performance.

1.9   A FEW DETAILED EXAMPLES

Before we embark on our theoretical adventures, let's look more closely at some examples. We use some of these examples later, in discussions, problems, and projects. The idea here is to investigate, from the engineering student's perspective, what is worth studying about them. The examples are drawn from everyday life and scientific and engineering and computer science environments.

1.9.1   Everyday Life

Example 1.1. *Lock Breaking and How Much Protection a Combination Lock Provides.* Around campus, you have seen bicycles with combination locks.[5] How much safety does such a lock buy you as a bike owner? Should you buy a lock with three digits or four? To ask the same question from a different perspective, as a thief, can you devise a clever strategy[6] to break the combination within a few tries?

Example 1.2. *Your Exam Score as an Indicator of How Much You Learned in a Course.* Your exam performance depends on a variety of factors, some of which you cannot even put your finger on. Among the easier to identify are the amount of study hours, the number of examples or homework problems you solve, and the quality of the instructor. Can we say, then, that an A in the course implies that you learned the material? Or, to reverse the question, does a grade of C imply that you did not learn it?

Example 1.3. *Your Grade Point Average as an Indicator of Your Job Potential.* Many recruiters use the grade point average (GPA) as a test to screen out applicants. How good is your GPA as an indicator (estimate) of your job potential?

Example 1.4. *The Amount of Rainfall in a Given Geographical Area and the "Poor" Weatherperson in the Local News.* Every self-respecting local newscast includes a weather center and reports that analyze radar and satellite images to tell you that tomorrow "there is a 20 percent chance of rain." Where do

---

[5] Unless you study in an extremely safe campus, where you see no locks, or in an extremely rich one, where you see no bicycles.

[6] Warning: the author and publisher strongly advise against *actual experimentation with those strategy ideas.*

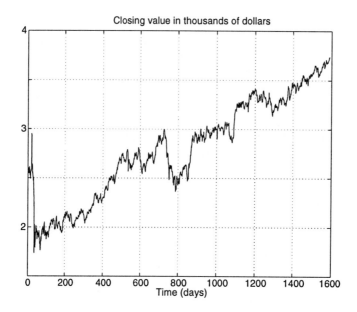

Figure 1.3 Dow Jones index, closing values.

these people get those numbers? And why is it that it pours when they said it would be a sunny day? Why hasn't someone sued them yet?

Example 1.5. *The Dow Jones Industrial Average Index.* The Dow Jones index measures the stock value of 30 companies whose shares are traded on the New York Stock Exchange. The index is monitored every business day by hundreds of thousands of investors. The actual values of the index at the end of a business day (its closing value) are shown in Figure 1.3, for a 5-year period starting in 1987. (You may find more values in the Web site.) The steep drop from about 2200 to 1700 at the beginning of the graph is the notorious Black Friday (October 17, 1987), which caused one confirmed death and countless headaches.

The Dow Jones index appears to be a "random" quantity. Yet there is a discernible trend that should be apparent to anyone, economically challenged or not: The index is going up, if viewed from a macroscopic point of view. There are "minor" perturbations (both up and down), but clearly, index values in 1992 are higher than those in 1987.

Here is a game, then, for the potential investor in you: How would you invest your future earnings or pension funds in order to maximize potential profits? All mutual fund managers are playing this game everyday.

Example 1.6. *Accident Rates and Car Insurance.* You may remember how your parents' car insurance premiums skyrocketed when you got your license at age 16. If you ask any insurance salesperson why this is the case, the answer is that young drivers are inexperienced and have (as a group) higher accident rates. Does it make sense, then, to charge them more?

Example 1.7. *The Behavior of Your Girlfriend or Boyfriend.* The behavior of your girlfriend or boyfriend is definitely random. She or he does not react to the same stimulus by giving the same response, as you may have noticed or heard. How should you judge such behavior? Is it reasonable to expect good behavior "all the time" or "on the average?"

Example 1.8. *The Behavior of a Child.* Here is another classic example of a random phenomenon, as every parent would promptly attest. You may remember that, too. So what is a good parenting strategy? Is it possible and is it worthwhile to guide that behavior toward a certain desired goal?

Example 1.9. *Traffic Jams on Highways.* Jams during rush hours are an almost everyday experience, awaiting you after your dreamed-about graduation day, provided you opt for the big city job. What is a good strategy to avoid taking part in such a mess?

Example 1.10. *Lotteries and Jackpots.* Is a lottery "good" for the state sponsoring it? Is it good for its gambling citizens? What is the meaning of the sentence "The odds of winning are 3,000,000 to 1?"

Example 1.11. *Dice and Card Playing in Casinos.* The whole subject of probability as a science began with the desire (or need) to study games of chance scientifically. In 1654, Pascal solved the following problem [40], laying the foundation of probability theory:

The Problem of Points and the Start of Probability Theory

Two players A and B play a game in which stakes are made. The first player to score N points wins. The players wish to stop the game *before it is finished*, while their scores are n_A and n_B. How should they divide the present stakes?

The problem was posed to Pascal by Chevalier de Méré, a well-known gambler who decided to enlist the help of science in his endeavors. For about two centuries, the main application of the theory was in the analysis of dice and card games. Conditional probabilities and sequences of random variables were first used in attacking such problems.

Philosophical or ethical questions notwithstanding, can you devise good strategies for winning in a card game?

Example 1.12. *Meeting the Right Person.* Meeting the right person is another game of chance, whose outcome affects someone for a lifetime. Can you describe the factors that cause randomness in this game? Given the inherent randomness, is it realistic to wait for the princess or prince, or should you compromise?

Example 1.13. *Our Expectations of Other People.* As the future manager of a group of newly hired, recently graduated engineers, would or should you expect a performance that is above average all the time?

Example 1.14. *Airline and Grocery Counters.* Waiting in line at a counter is a common experience. When designing such systems, designers need to calculate delays. They also must answer questions such as, One or multiple counters? Common waiting area for all counters or not?

Example 1.15. *NFL and NBA.*[7] The world of NFL and NBA is a haven for statistics—quarterback ratings, completion percentages, odds for winning the Super Bowl, the chances the Raiders have of ever winning on a Monday night, the strategies a team can come up with during the draft. The list goes on.

Example 1.16. *National Budget.* You have heard about the national deficit. Each year, taxes are collected, and a budget plan is put in place. Does it make sense to expect that in any given year, taxes collected equal expenses planned? Is a deficit unavoidable?

1.9.2   Scientific Environments

Example 1.17. *Electron Movement Around the Nucleus.* Until the start of the twentieth century, physicists believed that electrons moved around the nucleus in circular or elliptical orbits, with *well-defined* radii. The famous Heisenberg principle and the work of Schrödinger, however, established a different view. Physicists now believe that an electron is moving around the nucleus in an imprecise fashion: Only the probability of an orbit can be deduced, not its exact value.

Example 1.18. *Gas Molecule Movement.* The famous Einstein, Boltzmann, and Dirac statistics are models widely and successfully used in studying the way gases move in confined spaces. One of the early successes of the Markov Chain theory was, as a matter of fact, related to a controversial claim of Boltzmann's kinetic theory of gases [7].[8]

Example 1.19. *Chemical Reaction Yields.* Chemistry theory says that when we burn carbon, we produce carbon dioxide:

$$C + O_2 \rightarrow CO_2.$$

---

[7]I do not like baseball, so there are (almost) no American League or National League references in this book.

[8]The famous Ehrenfest Markov Chain model was used to predict that "on average," starting from an equilibrium state, gas molecules would take about 10^{6000} years to return to "unstable" states. For all practical purposes, then, they could be considered in "eternal equilibrium," static, as the disputed thermodynamics theory claimed.

In practice, we never find absolutely pure ingredients. When we put the ingredients in the laboratory tubes, a lot of chemical reactions take place; they produce new materials that may affect the actual yields of the designed reactions. Clearly, one must understand the probabilistic nature of the yields for better design, control, and pricing.

Example 1.20. *Implant Diffusion in* SiO_2. Electronic chips are manufactured by implanting various ions in silicon dioxide (SiO_2). The implanting process has a probabilistic nature. Some ions will go deeper than originally planned, others not. Therefore, the electrical characteristics of the chips (resistances, capacitances, etc.) must also be interpreted in a probabilistic sense.

In addition, the manufacturing process is not a perfect one. Some chips will be defective, due to mechanical inaccuracies in the machines, imperfect ion doping, etc. The price of the final product must of course reflect the difficulties of the manufacturing process. What is a fair and competitive pricing scheme?

1.9.3 Technical Environments

Example 1.21. *ftp or Any Communication Protocol (TCP, Kermit, HDLC, ATM, Appletalk)* [60]. The throughput and delays we achieve and experience with such protocols change with every try, even when the file sizes are kept the same. What, then, is a fair comparison of such protocols?

Example 1.22. *Customer Arrivals at an Automated Teller Machine (ATM).* Customers who want to access an ATM arrive at unpredictable times and at unpredictable rates. They also want to withdraw random amounts of money. These considerations must all be taken into account when one is designing the software that controls an ATM.

Example 1.23. *Packet Arrivals at a Computer Network Node.* Packets (e.g., files from a file transfer) arrive at a computer node at unpredictable times. This fact must be taken into account when the buffers are designed for such a node since when there is no buffer space available, the packet must be dropped and a retransmission must occur.

In modern communication switches, such buffers are a relatively scarce resource. Since these switches operate at speeds of hundreds or thousands of megabits per second, it is imperative that we understand their operating environment extremely well in order to design good buffer management schemes.

Example 1.24. *Terminal Log-Ins.* In a campuswide computer network, students share terminals and workstations. The duration of a student log-in session is random. How many terminals should a system administrator buy to satisfy student needs at low cost?

Example 1.25. *Disk Usage in a Multiuser System.* Large workstations provide better capabilities than smaller personal computers (PCs), at a greater

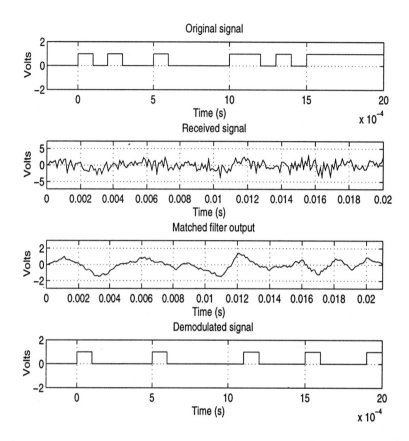

Figure 1.4   Errors in file transmission.

cost. Sharing them among multiple users is an efficient way to reduce per-user costs. Sharing introduces some disk-usage problems. Do we share the available disk space statically or on an as-needed basis? Do we compact files when we store them?

Example 1.26.   *Errors in a Protocol.*   When bits are sent over noisy communication channels, errors are introduced because of channel noise and filter imperfections. Some of the received bits differ from the ones originally sent. Figure 1.4 shows a scenario in which 0 and 1 bits are coded as 1-V and 0-V pulses, each 1 ms long.

In this example, the bits that start in times $t = 2, 10, 13, 16, 17, 18$ ms are all received in error. Error-checking codes, like the standard cyclic redundancy

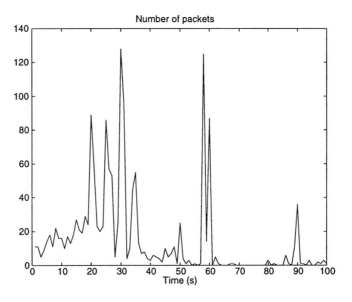

Figure 1.5   Ethernet traffic measurements, NCSU.

code (CRC) [60], are the primary way of checking for the presence of errors. In case an error is detected, retransmission of the erroneous data is requested, in the hope that the subsequent retransmission will be error-free. CRC codes introduce overhead; the optimum balance between overhead and error recovery capability can be studied only through probability theory [48].

Example 1.27. *A Router.* A router is a complex and fairly expensive device that manages traffic in a computer network. Its primary function is to (1) choose the path between a source that transmits packets and a destination node that receives them and (2) maintain smooth flow of packets, in case of errors, link failures, and congestion. Since actual packet arrivals at the router are unpredictable, the router must provide buffers. The performance of the router (in terms of throughput and delays) is an issue of importance to network users and designers alike.

Example 1.28. *Ethernet or Internet Traffic.* Figure 1.5 is a realistic pattern of traffic on an Ethernet local-area network (LAN). It represents the amount of packets flowing in an Ethernet during a given time interval. (It was obtained by using the Unix *netstat* facility.) As can be seen, the packet flow is random. Its analysis therefore can be done only through probability theory. Similar patterns can be observed in Internet traffic as well.

Example 1.29. *Airport Traffic Control.* Due to unforeseen delays such as late passengers and unexpected weather conditions, airplanes do not arrive at their destinations at times known in advance. This condition creates the need for air traffic control in the presence of uncertainties.

Example 1.30. *Radar Measurements and Decision Making.* Radars work on the principle of electromagnetic signal reflections. A signal of known frequency is sent to the aircraft; the reflected signal is analyzed, and a decision is made regarding characteristics of the airplane (e.g., speed, position, type). Errors are introduced, however, due to electromagnetic noise and reflections from wrong targets. A decision mechanism must be put in place in order to minimize the effect of such errors.

Example 1.31. *Yield of a Chip Manufacturing Process.* As discussed previously, because of imprecisions in the ion-implanting process, a chip manufacturing process will not yield 100 percent usable chips. To control this process better and to eventually produce cheaper chips, one must understand fully the nature of such imprecisions.

More examples follow. You can add your own interesting examples here and use them as checkpoints throughout the theory development process.

1.10   DID I LEARN THE THEORY?

Getting an A in a (mathematics) course does not necessarily mean that the theory has been mastered.[9] The litmus test that any students serious about their knowledge should subject themselves to is more or less the following: Undertake a project on (simulation or analytical) modeling of an engineering system similar to those in Chapter 11 or in the projects suggested in the problem sections.

1.11   WHERE DO I LEARN MORE?

Finally, for extremely serious students who want to pursue this study more deeply, suggestions are given throughout each chapter. The suggestions point to other books or papers that specialize in a particular subject.

Natural extensions of the basic theory we present here can be found in more advanced courses in

---

[9] In terms of the theory presented later in this book, your knowledge of the subject and your grade in the course are two random variables that are *strongly correlated* but not equal; see Chapter 4.

- *Queuing theory.* This subject deals with performance evaluation topics (such as delay and loss evaluation in communication systems). The seeds of the theory were planted by Erlang in the 1920s, with his seminal work on modeling telephone switches. The theory really flourished in the 1960s. The classical reference on the subject is [41]. Another good reference, with emphasis on discrete time systems, is [10].

- *Estimation and detection.* This subject deals with advanced filtering and random signal processing techniques, such as Kalman estimators. There is a plethora of good books on the subject: [63], [58], [61].

- *Measure theory.* This subject provides a rigorous foundation to the entire theory of probability. A very good introduction to the subject is [5]. More advanced treatises include [50] and [8].

- *Martingale theory.* This subject provides a useful, rigorous foundation for convergence problems (see [38] and [11]). The theory was systematically studied in the mid-1940s by Doob [17].

- *Stochastic differential equations.* This subject provides a rigorous treatment of noise models and dynamics of systems with random behavior. Stochastic differential equations appeared for the first time in the 1930s [6], and stochastic integrals in the 1940s [30]. Classical references on the subject are [2], [4], and [31].

1.12 SUMMARY OF MAIN POINTS

- Probability theory is useful in **understanding, describing, designing, and controlling** systems with **randomness**.

- The basic constructs of the theory are **events, random variables, random vectors,** and **random processes. Probability density functions** are associated with random variables, random vectors, and random processes; they are convenient means for calculating probabilities.

- Some basic models of the theory are the **Gaussian, Poisson, exponential, chi-squared,** and **Student's** t random variables; the **IID sequence** and **Markov Chains** in discrete time; and the **Gaussian** and **Poisson** random processes in continuous time.

1.13   PROBLEMS

1.1 Run *ftp* or any other file-transfer program from your machine to another one; transfer about 50–100 different files.
(a) Plot the file sizes you get.
(b) Transfer the same file, over and over, about 50–100 times. Plot the delays you get.
(c) Plot the throughput values, in both transfer cases. Are they the same? Can you explain why?

1.2 Certain workstations with window environments have a facility (called *xload* in Sun systems) that lets you see the load of the machine as a function of time.
(a) Run *xload* and see how the load changes as a function of time.
(b) Run other programs, such as *ftp* and *netstat*, in separate windows, and observe the jumps in the load as these programs start and terminate.
(c) Explain why the jumps occur in the graphs, and try to find out what they represent.

1.3 Run the Unix *netstat* command (or something equivalent) in two different environments. First, when no other program is running, you will see that packets are still transferred into and out of your machine. These are various network control packets. Second, run the *ftp* command, in a separate window. Observe the differences in the amount of traffic both into and out of your machine. Try to explain them.

1.4 Run a program on your PC or workstation that writes dummy data in a disk file. Measure the execution times of the program as the data size varies.
(a) Do you get a linear curve?
(b) If not, can you explain the discrepancies by, say, claiming randomness in the disk arm movements?

1.5 Survey your classmates. Ask them about their height, weight, and—if they are willing to disclose it—their GPA.
(a) Plot graphs for all three variables.
(b) Can you say whether height and weight are "related," in the sense that as height increases, so does weight? Why?
(c) Are height and GPA related in the same sense? Explain.

1.6 Throw a die N times (or simulate die throwing in your computer; in the book Web site you will find a MATLAB die simulator). The *relative frequency* of the throw showing number i, $i = 1, 2, \ldots, 6$, is defined intuitively as n_i/N, where n_i counts the number of throws that show face i.
(a) Calculate the relative frequencies for the six faces.
(b) Plot the relative frequency for face 6 as a function of N.
(c) Can you explain what you observe?

1.7 Consider a lottery game in your state (or the closest one, if lottery is not allowed in yours). On the back of some tickets you will find the *odds of winning* the lottery.
(a) Discuss how the lottery officials could have calculated those odds.
(b) Discuss any philosophical issues about lotteries.

1.8 In the book Web site you will find the Dow Jones data for the closing, opening, and high values of the index.
(a) Plot those values in chunks of 2 months, so you can see the detailed behavior of the index.
(b) Discuss any factors that you think cause randomness in the data.

1.9 Consider the boyfriend/girlfriend behavior example. At any given day n, this behavior, call it B_n, is one of, say, {good, bad, OK}.
(a) You *would like* to have B_n = good for all n. Is that humanly possible?
(b) Let's associate now +1 with good, −1 with bad, and 0 with OK. Would you be satisfied if

$$\frac{1}{n} \sum_{i=1}^{n} B_i \to 0$$

for large n, for example, $n > 2 \cdot 365$? As we say in time-varying systems, for values of n this large, you *go into steady state* (or you *go steady*). We will see what we mean by *satisfied* a lot more clearly in Chapter 6.

1.10 The face on which a die lands can be predicted *exactly* through a system of differential equations that describe the motion of the die through the air. Discuss the difficulties of such an approach in as much detail as possible.

1.11 Can the transfer delay that a file experiences be predicted exactly? Discuss what factors should be known for that prediction, in as much detail as possible.

2

BASIC CONCEPTS

In this chapter we develop the fundamental concepts of the theory. *A theory necessarily deals with concepts—not reality.* It uses mathematical *abstractions* that discard all the "unnecessary details," for clarity and economy. So even though we still use *ftp* jargon in this chapter, to illustrate the concepts and definitions, *the emphasis is on mathematical completeness and rigor.* This approach is necessary to avoid logical gaps and inconsistencies. We build the theory from the bottom up; as you will see, we posit three axioms only. The entire theory of the subsequent chapters is based on these three axioms—think of them as the "Constitution" from which all laws are derived.

Before we proceed, we stress that the subject of probability theory is *mass phenomena,*[1] i.e., phenomena that are repeated (or can be repeated) many times. It is a good idea to check all the examples in Section 1.9 and see how often each one can be repeated sequentially or concurrently. The theory does not apply to phenomena that cannot be repeated. Some misleading questions are commonly asked that can obscure the applicability of the theory; see Problems 2.1–2.7.

There are only a few tools developed in this chapter (total probability theorem and Bayes' rule) that the student can directly use in applications. Most of the tools come later, when random variables and random processes are introduced. We also introduce fundamental notions such as conditional probabilities and independence, which are used in later chapters (e.g., Chapters 4, 5, and 6) to develop fancier tools.

So there is no chance to play probability ball yet. We are still in the locker room and about to learn the ground rules of the game. Make sure that you

---

[1] R. E. von Mises, 1883–1953, an Austrian mathematician, and his school of thought (see Appendix C) were the first to clearly recognize and stress this point in the beginning of the twentieth century, despite the fact that the theory had been applied for more than 200 years. In Appendix C we present another approach, namely, subjective probabilities, which deals with nonmass phenomena. However, we do not follow this approach in this book.

understand the *transition process* that will bring us from *netstat*, *ftp*, NFL, and boyfriends or girlfriends into a probability space (S, Q, P). Also, after each theorem is presented, it is instructive to see how it applies to a concrete example, say, one of those presented in Section 1.9 (e.g., what does $A \cup B$ represent in the *ftp* environment?). Once again, a warning: The alert student will quickly observe that the concept of time is not present in this chapter. This concept will naturally come in Chapters 6 and 7.

Moreover, in anticipation of the material in Chapters 3 and 4, I strongly advise students to "criticize" the theory in the following respect: How powerful are the tools the theory offers me? How easy to use? For a deeper understanding of the theory, it is quite important that you come up with your own criticism. Only then will the powerful concepts of random variables and random processes appear as natural *needs* instead of "out-of-the-blue" mathematical "tricks." (Our criticism is offered in Section 2.10.)

In the next section, we introduce random experiments, the abstraction that unifies all the examples in Chapter 1. Sections 2.2 and 2.3 deal with the first two pieces of the probabilistic model (S, Q, P). The axioms are then introduced in Section 2.4. We develop a few basic theorems as immediate consequences of the axioms in Section 2.5. Two fundamental concepts are introduced in the next two sections. In Section 2.8 we discuss a question of practical concern, namely, how to assign probabilities to the events of an experiment. Modeling, the crucial transition from experiments to theory, is discussed in Section 2.9. Finally, in Section 2.10 we point out some difficulties we must deal with when modeling a random experiment; thus we pave the way for probability density functions, the most important tool of the theory, discussed in Chapter 3. Good knowledge of set theory is necessary to follow the subject of this chapter. We provided a fairly detailed appendix on set theory, separate from this chapter, so as not to disrupt the flow of probabilistic concepts. You should feel comfortable with the material in Appendix A, however, before you read this chapter.

2.1   RANDOM EXPERIMENTS

The examples in Section 1.9 are quite diversified. The basic concept that abstracts them and unifies their presentation is that of the *random experiment.*

What is an experiment? We all know of chemical, physical, and electrical experiments. We loosely define the term *experiment* as any process or procedure that transforms an input to an output. Figure 2.1 represents this idea pictorially.

Three elements are essential to an experiment: the input, the transformation process (labeled *processing* in the box in Figure 2.1), and the observed output. The key idea in a nonrandom experiment is that when the experi-

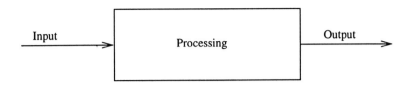

Figure 2.1   The notion of an experiment.

ment is repeated, with the same input and under identical conditions, the same output will be observed.

In contrast, in a *random experiment,* the output may not necessarily be the same, even in repetitions of the experiment with the same input and under identical conditions. For example, in the experiment in Example 1.28, the output can be the number of packets observed in a 1-s interval. In the girlfriend/boyfriend experiment, the output is *observed behavior.* In the NFL experiment, the output can be "the number of season games the Redskins win" (pretty low in the 93–94 and 94–95 seasons). So, loosely speaking, we can define a random experiment as follows:[2]

Definition: A *random experiment* is one in which the output cannot be uniquely identified from knowledge of the input.

You can see this aspect of nonunique outputs in the *ftp* example, when you repeatedly transfer the same file. You will also definitely observe it in the boyfriend/girlfriend scenario.

Two issues about random experiments are important:

- *They are repeatable many times* (in theory, an infinite number of times). If an experiment were to be performed only once, then it would have one and only one output.

- *We are unable or unwilling to predict the exact output* of the experiment.

Instead of being interested in an exact output, we focus on the *relative occurrence* of this output. For example, consider the Ethernet experiment in Example 1.28, in which we take measurements of packets for, say, 10,000 individual 1-s intervals. It is not possible to predict that the next measurement will

---

[2]The definition is vague on purpose. Consider the experiment of throwing a die in the air. The input to that experiment is the force and momentum exercised on the die by the player's fingers and the air. The conditions of the experiment are the initial position of the die in the player's hand and the initial air conditions. Then all repetitions of the experiment with the same input and under similar conditions will result in the same output. Throwing dice would not be considered a random experiment! A clearer definition of a random experiment that overcomes such apparent confusions is given in the next section.

be exactly equal to 3 packets. However, it is reasonable to expect that, in those 10,000 measurements, say, 100 measurements will be equal to 3. Probabilities are abstractions of those relative frequencies.

Notice that the same physical system may give rise to *multiple* random experiments. In the *ftp* example, we may observe transfer delays as the output of one experiment and transfer throughputs as the output of another. In Example 1.28, we may observe the number of packets as the output of one experiment and packet errors as the output of another. In the NFL example, one experiment might be to observe the wins and losses of the Washington Redskins; yet another one might be to observe the touchdown/interception ratio for all the quarterbacks together.

How can we describe such diverse experiments in a *common mathematical language?* A moment's thought should convince you that *set theory* is the most natural (or only) choice. Indeed, the output of a random experiment can be considered as an element of a set that contains all such outputs. Then repeating the random experiment once more is simply equivalent to selecting again an element from that set. In the next section we formalize these notions.

2.2   SAMPLE SPACES AND OUTCOMES

Definition: The set S of all possible outputs of a random experiment is called the *sample space* of the random experiment.

We can therefore abstract a random experiment as a *set*, the elements of which can be considered as the outputs of the random experiment. A keyword in the definition that can cause some confusion is *possible*. As we see later, the sample space may contain elements that have zero probability.

Definition: The elements of the sample space are called *outcomes*.

The key characteristic of an outcome is *indecomposability*—an outcome cannot be subdivided into more elementary outcomes. In the *ftp* example, with delay considered as the output of the experiment, the indecomposable outcomes are of the form "delay is equal to x s." In this definition, the notion of outcome is a *primitive* one; you may parallel it to the notion of a point in geometry.

Remark: A sample space is always associated with a given random experiment; different random experiments may have different sample spaces. Sample spaces are not unique. In abstracting a random experiment, one may come up with more than one sample space. That may cause a minor conceptual difficulty; however, as we see later, this is no problem as far as developing a theory is concerned.

Let's see now some sample spaces for the examples in Section 1.9. (We will see more in Chapters 3 and 4.)

Example 2.1. In the lock-breaking example (see Example 1.1), suppose that the experiment describes the thief's choice of a combination. Then, assuming a three-digit lock, we can take $S = \{000, 001, \ldots, 999\}$.

Example 2.2. In the silicon implant example (see Example 1.20), one experiment may describe how deep ions go into the silicon substrate. Then, measuring penetration in micrometers, we can take $S = [0, 20]$.

Example 2.3. In the automated teller machine example (see Example 1.22) suppose the experiment is to measure the amount of dollars a customer will withdraw. Then, assuming that the machine dispenses amounts in multiples of \$5, we can take $S = \{0, 5, 10, \ldots, 500\}$. (We made an implicit assumption here.)

Sample spaces can be classified as **discrete** and **continuous.** A discrete sample space has a *finite* or *countably infinite* number of elements. A continuous sample space has an *uncountably infinite* number of elements. The reason for this classification is mathematical. Discrete sample spaces are easier to work with, as we will see. In the *ftp* example, when delays are observed as outputs, the sample space is in principle continuous since delays can be any positive real number. In reality, however, the sample space will be discrete since any measuring device will have finite-precision arithmetic.

A particular form of sample spaces is that of *Cartesian products.* In Appendix A, Section A.2.6, we describe Cartesian products in greater detail. Here, suffice it to say that such sample spaces play a key role in independence, a very fundamental concept in probability theory.

2.3 EVENTS

2.3.1 Collections of Outcomes

The outcomes of a random experiment are basic, but not necessarily the only items of interest in a random experiment. In the *ftp* example, for instance, a user may be interested in how "large" her or his delay can be; as a matter of fact, the single outcome "delay is exactly equal to x s" may be of no interest at all to any user. We see therefore that *collections* of outcomes, i.e., *subsets* of S, rather than single outcomes of an experiment, are also of interest.

Are all subsets of S of interest?

Certainly there are a lot of them when S is "large," and especially when S is continuous. In Appendix A we show that when S has n elements, it has 2^n subsets. (Try to *enumerate* all the subsets of the set $[0, 1]$.)

Suppose that we choose some collection $\mathcal{A}$ of subsets as the subsets of interest, to which we will assign probabilities; suppose that $\mathcal{A}$ is *not* the collection of all subsets of S. Of course, we can develop a theory for these sets. But what happens if we change our mind in the future and, say, become interested in additional subsets? We must define probabilities for the new subsets from scratch.

It behooves us then to declare *all* subsets of the sample space as subsets of interest, even if this means we have to specify probabilities for uninteresting subsets as well. Now of course the obvious problem is: *How do we assign probabilities to them all?* And the more subtle, mathematical problem is: *Can we assign probabilities to them all?*

2.3.2   Special Collections of Subsets

Unfortunately, the answer to the latter question is a *qualified yes*. If S is a discrete set, we can consider all subsets of S without any problem. (See Section 2.8 for a suitable probability assignment.) However, if S is continuous, work by many mathematicians has shown that there is no universal way to assign probabilities to all subsets of S, even when S is as "simple" as $[0, 1]$. Some of those subsets are simply too "nasty."[3]

We must therefore consider only a *smaller* collection of subsets when S is continuous. *What kind of collection?* This collection must have a special structure,[4] namely, that of a σ algebra (also known as a σ field or Borel field). We discuss the structure of such collections in greater detail in Appendix A. See Remark 5 in the next section for an inutitive justification of this mathematical subtlety.

Definition: A collection $\mathcal{Q}$ of subsets of S that is a σ algebra, and to which we will assign probabilities, is called the *collection of events*.

Naturally, then,

Definition: An **event** A is an element of $\mathcal{Q}$.

When a random experiment is performed and an outcome ζ is observed we say that *event A has occurred* if $\zeta \in A$.

Example 2.4. Let's see now some events from Example 1.28 in Section 1.9. Consider the specific experiment E, where we measure, say, the number of packets received by the computer station in a 1-s interval. The outcomes of this experiment are integer numbers that can take any value 0, 1, 2, ...; the maximum value an outcome can have is of course determined by the speed of

---

[3] But also quite difficult to describe; see [8].

[4] This is rule 1, before we play our probability ball. We discuss the reason for this rule in the next section.

the connecting communication link. It is convenient to consider this maximum as ∞. What are events of interest here? To me, some are

$$A = \{10, 11, 12, \ldots\}, \quad B = \{\text{less than 20 packets}\}, \quad C = \{0\}.$$

When outcome 10 is observed, events A and B occur (simultaneously).

Remark: The empty set (see Appendix A) is always a subset of any set. In probabilistic jargon, it is called the *impossible* event. Any set is a subset of itself, and thus the sample space is a subset of itself. Again, in probabilistic jargon, we call the sample space the *certain,* or *universal,* event. Finally, the outcomes are alternatively called the *elementary* events.

2.3.3   Assigning Probabilities to Events

So we cannot assign probabilities to any arbitrary subset of S. *How* do we *assign* probabilities to those we can? For example, in the *netstat* experiment, how do we assign the probability of the event "less than 20 packets have been received in the last 1-s interval"? How do we calculate the probability of winning the lottery? The probability of the lock combination being broken?

Clearly, we have only two choices: assign probabilities to *all* events of interest or to *some* of them. If we choose the former choice, then what do we need a theory for? Such a theory would have nothing new to say. Therefore, we must go with the latter choice and assign probabilities to only a "few" elements of Q; then the theory should give us the tools to generate probabilities for the rest of them. This leaves open two questions:

- *What events* do we select for the initial assignment? Ideally, the initial selection should be large enough that no events of interest are left out and small enough that we do not have to do a lot of unnecessary work to experimentally compute those probabilities.

- On the basis of *what rules* do we generate the probabilities in question from knowledge of probabilities for the initial assignment?

Unfortunately, there is no universal answer to the first question for arbitrary S and Q. We address this question partly, in the examples in Section 2.8. We also provide an answer for the case of $S = R$, the real line, in the next chapter. We address the second question in the next section. The situation we are faced with here is similar to a familiar, simple problem from geometry: How do we represent any point A in the two-dimensional plane? A convenient, economical way in rectangular coordinates is to use two unit vectors $\imath$ and $\jmath$; then any point A can be represented by using two real numbers x and y such that $A = x\imath + y\jmath$. The rules of representation are of course addition of vectors and multiplication of a vector by a scalar.

2.4   PROBABILITY AXIOMS AND THE SETS OF INTEREST

In 1900, in a conference in Paris, the famous mathematician Hilbert[5] (of the Hilbert spaces fame) posed a series of 23 grand challenges to mathematics as a science in the twentieth century. In his own words, "Wir mussen wissen, wir werden wissen" (We must know, we will know). The sixth challenge dealt with the requirement to put the "applied subject" of probability into a rigorous framework, i.e., into an "axiomatic" form. This was the case with algebra, geometry, and set theory, which were axiomatized a little while ago. The rules (axioms) for generating probabilities should be *as few as possible*.[6] Needless to say, the axioms should be (1) *complete*, in the sense that any statement involving probabilities should be shown to be true or false based on these axioms only, (2) *consistent*, in the sense that no contradictions should result; and (3) *independent of each other*, in the sense that no axiom should be derivable from another one.

In 1933, Andrei Kolmogorov[7] responded to Hilbert's challenge by introducing the following Axioms 1–3. For the rest of the section, we assume that a fixed random experiment E is considered. The sample space S (discrete or continuous) is given, and the set of events of interest Q is somehow chosen. For a given event A, let $P[A]$ denote its assigned probability.

Definition: The probability assignment is a function from Q on R, the set of real numbers, such that Axioms 1–3 are satisfied.

Axiom 1.   Probabilities are nonnegative, i.e.,

$$P[A] \geq 0, \quad \text{for every } A \in Q. \tag{2.1}$$

Axiom 2.   The sample space has probability 1, that is,

$$P[S] = 1. \tag{2.2}$$

Axiom 3a.   Consider two disjoint events $A, B \in Q$, that is, $A \cap B = \emptyset$. Then

$$P[A \cup B] = P[A] + P[B]. \tag{2.3}$$

---

[5]David Hilbert, 1862–1943, a prominent German mathematician and the father of axiomatic geometry.

[6]Why? Mathematicians are known for their zeal to be as economical as possible.

[7]Andrei Nikolaevich Kolmogorov, April 25, 1903–October 20, 1987, a great Russian mathematician and the father of modern probability theory. For a complete account of this man's contributions, see the memorial articles in the *Annals of Probability*, vol. 17, nos. 3–4, 1989.

Axiom 3b (alternate). Let $A_1, A_2, \ldots$ be a sequence of events that are pairwise disjoint (that is, $A_i \cap A_j = \emptyset$, $\forall i \neq j$); then

$$P[\cup_{i=1}^{\infty} A_i] = \sum_{i=1}^{\infty} P[A_i]. \tag{2.4}$$

Axioms 1–3 are the *only* rules we impose on the probability ballgame besides the rule on Q. Note that Axiom 3b can be used to *derive* Axiom 3a as a special case. (Put $A_1 = A, A_2 = B, \emptyset = A_3 = A_4 = \ldots$.)

Remark 1: Axioms 1, 2, and 3a were used in the past (before 1933) in attempts to formalize probability theory. All such attempts have proved problematic or at least incomplete since limits of sets (like $\cup_{i=1}^{\infty} A_i$) cannot be handled by Axiom 3a. (For an interesting account of alternative approaches to defining or interpreting the theory, see Appendix C and references there.) Much hard work had been done and much experience gained by scientists before the theory took its final form in 1933. So do not feel intimidated if the axiomatic approach looks a little like magic.

Remark 2: Observe also that the only axiom we can use as a rule to generate probabilities of more events is Axiom 3a (or 3b) since the other two involve a single set only. Indeed, one way to interpret this axiom is the following: Given the probabilities of events A and B, we can find the probability of the *new* event $A \cup B$ by simply adding the probabilities $P[A]$ and $P[B]$.

Remark 3: Observe that the only set operation that appears in the axioms is union. So it must be that this operation suffices to generate all other events of interest in a random experiment. Indeed, as we can see in Appendix A, when Q is selected as a (σ) field, intersection is not necessary.

Remark 4: Observe that *addition* of probabilities is used in Axioms 3a and 3b. Why not subtraction or multiplication or division? We investigate some "what-if" possibilities regarding the form of axioms in Problems 2.46–2.49.

Remark 5: We need to discuss one mathematical subtlety here before we proceed. By writing Axiom 3a or 3b as we did, we *implicitly* assume that given that A and B are events of interest, $A \cup B$ is *also* an event of interest (i.e., it belongs to the collection Q). Without this assumption, Axioms 3a and 3b do not make mathematical sense. This is the reason for requiring that Q be a σ field. A σ field is closed under (finite or infinite) union, so there is no problem there.

2.4.1   A Bit of History

Axioms 1, 2, and 3b form the rigorous basis of modern probability theory. They were introduced by Andrei Kolmogorov as a response to Hilbert's challenge,[8] in [45], in a book of historical significance that can be thought of as a "little Bible of the theory."[9] We quote from page 1 of [45]:

> ...The theory of probability, as a mathematical discipline, can and should be developed from axioms in exactly the same way as Geometry and Algebra. This means that after we have defined the elements to be studied and their basic relations, and have stated the axioms by which these relations are to be governed, all further exposition must be based exclusively on these axioms, independent of the usual concrete meaning of these elements and their relations.

> ...The postulational basis of the theory of probability can be established by different methods in respect to the selection of axioms as well as in the selection of basic concepts and relations. However, if our aim is to achieve the utmost simplicity both in the system of axioms and in the further development of the theory, then the postulational concepts of a random event and its probability seem the most suitable.

Kolmogorov recognized up front that assigning probabilities to any arbitrary subset of the sample space would not be possible. He wrote on page 16:

> ...Only in the case of *Borel fields* of probability do we obtain full freedom of action, without danger of the occurrence of events having no probability.

Even though Kolmogorov's selection of σ fields was based on mathematical rigor, we can intuitively argue that such a selection is also a natural one. If A is an event of interest, is A^c, its complement with respect to S, an event of interest as well? It would be hard to argue not.[10] So naturally we make A^c a member of class Q. It is equally easy to argue that if A, B are events of interest, A *or* B is also of interest, and thus $A \cup B$ should belong to Q. Why the *infinite* union $\cup_i A_i$ should belong to Q is the hardest to argue for, on a purely intuitive basis. For some fierce arguments against Kolmogorov's selection, see Chapter 5 in [15].

---

[8] See [51] for a brief presentation of two other sets of different, unsuccessful axioms. S. N. Bernstein, 1880–1968, another Russian mathematician, introduced a similar set of axioms in 1917. His set was not as successful as Kolmogorov's.

[9] And probably a bargain since it cost only $2.12.

[10] If you are interested in the event "delay is less than 3 s," you are *also* interested in the event "delay is larger than 3 s." If you are interested in the event "Redskins win," are you not interested in the event "Redskins lose?"

Then we make Q a class of sets such that unions and complements of sets in the class are also members of the class! That collection of sets is precisely what set theorists call a *field*, if unions involve a finite number of sets (as in Axiom 3a), or a σ *field*, if an infinite number of sets are involved (as in Axiom 3b).[11]

To summarize this discussion, when the sample space S is a discrete set, we always take its powerset $2^{|S|}$ as the collection of events Q; when S is a continuous set, we take a σ field as Q. (See Appendix A for more details on powersets and σ fields.) *With this selection in mind, we are guaranteed that a probability assignment will be free of logical inconsistencies.*

2.5 ELEMENTARY THEOREMS

Let's see now some immediate results we can get when we put the axioms and the structure of Q together.

Recall that, by definition, for any $A \in Q$, $A^c \in Q$ as well. But A and A^c are disjoint sets, so from Axioms 3a and 2 we get

$$1 = P[S] = P[A \cup A^c] = P[A] + P[A^c]. \tag{2.5}$$

From Equation 2.5 we can immediately get

$$P[A^c] = 1 - P[A].$$

This equation is our first tool since it can be interpreted as a means to generate more probabilities! Let's state it as our first theorem.

Theorem 2.1 *For any $A \in Q$,*

$$P[A^c] = 1 - P[A]. \tag{2.6}$$

An immediate corollary can be obtained from this theorem and Axiom 2:

Corollary 2.1 *The probability of the impossible event is 0, that is,*

$$P[\emptyset] = 0.$$

In general, there are events other than $\emptyset$ whose probabilities are also 0. Such events are called *null* events.

---

[11] The two major alternative axiom systems, which we discuss in Appendix C, were based on arbitrary collections Q; logical inconsistencies were soon discovered and led to their fate.

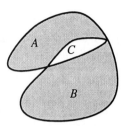

Figure 2.2   Venn diagram for overlapping sets A, B.

If $A \in \mathcal{Q}$, what about the subsets of A? Are they also in $\mathcal{Q}$? Yes if S is discrete and $\mathcal{Q} = 2^{|S|}$, but not necessarily when S is continuous. However, when a subset A' of A does belong to $\mathcal{Q}$, we can say something about its probability. Observe that $B \triangleq A - A' = A \cap (A')^c$, so $B \in \mathcal{Q}$ as well. Of course, $A = A' \cup B$, A' and B are disjoint, and thus

$$P[A] = P[A' \cup B] = P[A'] + P[B] \geq P[A'].$$

We used Axiom 3a in the second equality and Axiom 1 for the inequality in the last step. Let's state this result as a theorem.

Theorem 2.2 *For any $A \in \mathcal{Q}$ and any $A' \subseteq A$, also in $\mathcal{Q}$, we have*

$$P[A'] \leq P[A].$$

Now since $A \subseteq S$, for any $A \in \mathcal{Q}$, from Axiom 2 and Theorem 2.2 we get Corollary 2.2.

Corollary 2.2 *For any $A \in \mathcal{Q}$, we have $P[A] \leq 1$.*

The following theorem is an immediate consequence of Axiom 3a.

Theorem 2.3 *For any $A, B \in \mathcal{Q}$,*

$$P[A \cup B] = P[A] + P[B] - P[A \cap B]. \tag{2.7}$$

Proof: We use the Venn diagram in Figure 2.2 to illustrate the basic idea. From Corollary 2.1, if $A \cap B = \emptyset$, there is nothing to prove. So suppose that $C \triangleq A \cap B \neq \emptyset$.

Let $D \triangleq B - (A \cap B)$, so that B can be expressed as a union of disjoint sets $B = D \cup (A \cap B)$. Of course, $A \cup B = A \cup D$, where A, D are disjoint. Therefore, from Axiom 3a we have

$$P[A \cup B] = P[A \cup D] = P[A] + P[D].$$

Now, also from Axiom 3a,

$$P[B] = P[D \cup (A \cap B)] = P[D] + P[A \cap B],$$

and the theorem follows. □

Remark: Theorem 2.3 (and its extension to the case of three events, Theorem 2.4 below) was known to de Moivre and Lagrange,[12] who proved it in their studies of annuity problems in the seventeenth and eighteenth centuries.

The next theorem is an extension of Theorem 2.3. Its proof is easy once we associate the set $B \cup C$ in Equation 2.8 with the generic set B in Equation 2.7; therefore, it is left as an exercise. Problems 2.66–2.70 are relevant extensions to $n < \infty$ sets.

Theorem 2.4 *For any* $A, B, C \in \mathcal{Q}$,

$$P[A \cup B \cup C] = P[A] + P[B] + P[C] - P[A \cap B] - P[B \cap C] - P[C \cap A] + P[A \cap B \cap C] \tag{2.8}$$

The above theorems and corollaries are not earthshaking. They do present reason for celebration, though: They are the first, timid steps toward more exciting results. Note that only Axioms 1, 2, and 3a and elementary algebra and set relations were used. These theorems are as basic as it gets; you should compare them to the basic rules for algebra, such as $a - b = -(b - a)$. They are the bread and butter of the theory, even though at this stage (of the ballgame) we may not recognize this fact.

It is obvious, however, that for a rich theory, we need more tools; these basic theorems alone are not enough. The next two sections provide more tools.

2.6   CONDITIONAL PROBABILITIES

Set theory did provide the means for calculating probabilities of related events (such as unions, intersections, and complements), but the story does not end here. There is another form of relationship between events that set theory does not handle directly. Loosely speaking, the occurrence of an event A does convey some information about the possible occurrence of another event B. At least when we know that event A has occurred, event A^c *cannot* occur! Moreover, when event A has occurred, all events that contain A as a subset have necessarily occurred as well! Let's see a few examples of this new relationship.

---

[12]Abraham de Moivre, 1667–1754, and Joseph Louis Lagrange, 1736–1813, prominent French mathematicians.

Example 2.5. In the *ftp* example, it is intuitively clear that "large" delays are related to "low" throughputs. It is also clear that the event "the starting quarterback is healthy" is related to the event "the team loses." As a matter of fact, we can say that occurrence of the former event decreases the probability of the latter.

Example 2.6. The event "the Dow Jones index is above 7200 on Friday" is related to the event "the Dow Jones index is above 7000 on Monday." We can say that occurrence of the former event increases the probability of the latter.

A moment's thought about all these examples shows that occurrence of an event A in fact changes the notion of the sample space; in some sense, the sample space must be narrowed down from S to A once it becomes known that event A has been observed because now we are sure that outcomes in the set A^c cannot occur. How then should we recalculate the probability of an event $B \in Q$, given that another event $A \in Q$ has occurred?

Let's call this new probability the *conditional probability* of event B, given that event A has occurred; let's denote it as $P[B|A]$. How do we define it precisely?

Let's look at some of the intuitive properties of conditional probability, which we can already deduce from the above discussion. (A Venn diagram outlining the various cases might help.) First, if $A \cap B = \emptyset$, then it must be that $P[B|A] = 0$; for example, given that A has occurred, A^c cannot occur. If $A \cap B = A$, then $P[B|A] = 1$ (since A is a subset of B). Finally, if $A \cap B = B$, then $P[B|A] \leq 1$ but $P[B|A] \geq P[B]$.

One way to define $P[B|A]$ and capture all the above intuitive properties is the following.[13] Consider a fixed experiment E. Let $A, B \in Q$ be two arbitrary events. Let $P[\cdot]$ denote the probability assignment rule.

Definition: The **conditional probability** of event B, given that event A has occurred, is defined as

$$P[B|A] = \frac{P[B \cap A]}{P[A]}. \tag{2.9}$$

We must apologize to the mathematically alert student for a confusing notation here. On the right-hand side of Equation 2.9, the symbol P denotes the function that assigns a probability (a real number) to an event. Therefore, P is a function with a single set as its (only) argument. On the left-hand side of Equation 2.9, however, we are using the same symbol P in a different way, with two arguments. This notation has been used for ages, and we also use it here, with this remark in mind.

Remark 1: Of course, there is no problem in the definition when $P[A] > 0$; however, when $P[A] = 0$, we have an indeterminate form of $0/0$ since $B \cap A \subseteq A$,

---

[13] After you gain some experience, can you think of any other? See Problem 2.76.

and from Theorem 2.2 of the previous section, the numerator in Equation 2.9 is also 0. In this case, we will arbitrarily define $P[B|A]$ as 0, in order to make Equation 2.9 as general as possible.

Remark 2: Equation 2.9 was introduced by Bayes[14]; it first appeared in print in 1764. Bayes did not use the symbol $P[B|A]$; he used the term *compound event* instead of "conditional."

Let's show now that the definition in Equation 2.9 is indeed a probability that satisfies Axioms 1, 2, and 3 for any $A, B \in Q$.

Theorem 2.5 *For any $A, B \in Q$, with $P[A] > 0$, the expression $P[B|A]$ is a valid probability assignment.*

Proof: Axiom 1 is satisfied since both the numerator and the denominator in Equation 2.9 are nonnegative numbers. Axiom 2 is satisfied since

$$P[S|A] = \frac{P[S \cap A]}{P[A]} = \frac{P[A]}{P[A]} = 1.$$

Finally, observe that when events B and C are disjoint, we also have that $B \cap A$ and $C \cap A$ are disjoint; thus

$$
\begin{aligned}
P[B \cup C|A] &= \frac{P[(B \cup C) \cap A]}{P[A]} = \frac{P[(B \cap A) \cup (C \cap A)]}{P[A]} \\
&= \frac{P[B \cap A] + P[C \cap A]}{P[A]} = P[B|A] + P[C|A].
\end{aligned}
$$

Axiom 3a is therefore satisfied as well. Axiom 3b is shown in a similar fashion. $\square$

Theorem 2.5 has an important implication. It means that any theorems that we prove for "regular" probabilities will also hold for conditional probabilities; therefore, we do not have to state or prove them separately. Thus it gives us significant savings in notation and presentation.

The theorems in the following two subsections show that we can use conditional probabilities as a convenient means to calculate probabilities of *more* events.[15]

2.6.1   Total Probability Theorem

Consider any event $A \in Q$; consider a *partition* $\{B_k\}_{k=1}^n$ of S (see Appendix A for the definition and properties of a partition).

---

[14] Thomas Bayes, 1702–1761, an English pastor and mathematician.
[15] We see how convenient a means this is in Chapter 8.

Theorem 2.6 *Let $A \in \mathcal{Q}$ and let $\{B_k\}_{k=1}^n \in \mathcal{Q}$ be a partition of S. Then*

$$P[A] = \sum_{k=1}^n P[A|B_k] \cdot P[B_k]. \qquad (2.10)$$

Proof: Since the sets $\{B_k\}$ are disjoint, it follows that the sets $\{C_k\}$, where $C_k \triangleq A \cap B_k$, are also disjoint. Indeed, for $k \neq l$, we have

$$C_k \cap C_l = A \cap B_k \cap A \cap B_l = A \cap (B_k \cap B_l) = \emptyset.$$

It is clear that $A = \cup_{k=1}^n C_k$. From Axiom 3a or 3b (depending on whether n is finite or infinite), we have

$$P[A] = P[\cup_{k=1}^n C_k] = \sum_{k=1}^n P[C_k].$$

From Equation 2.9, we have $P[C_k] = P[A|B_k] \cdot P[B_k]$, and Equation 2.10 follows.   □

Theorem 2.6 is known as the **total probability theorem**. You can interpret it as a means to calculate the probability of event A when the conditional probabilities $P[A|B_k]$ and the probabilities of events B_k are known. We have ample opportunity to apply this powerful tool in the rest of the book (e.g., Chapter 8), so we do not present any examples here. Its power should come as no surprise once we observe that conditioning on an event changes the structure of the underlying sample space to something simpler and removes some of the randomness of the random experiment.

2.6.2   Bayes' Rule

Consider an event A and a *partition* of the sample space $\{B_k\}_{k=1}^n$. Suppose now we want to *reverse* the conditioning events and make A the conditioning one. Then what is $P[B_k|A]$?

By definition,

$$P[B_k|A] \triangleq \frac{P[B_k \cap A]}{P[A]}. \qquad (2.11)$$

From Equations 2.9 and 2.10, we get the following:

Theorem 2.7 *Let $A \in \mathcal{Q}$ and let $\{B_k\}_{k=1}^n$ be a partition of S. Then*

$$P[B_k|A] = \frac{P[A|B_k]P[B_k]}{\sum_{k=1}^n P[A|B_k]P[B_k]}. \qquad (2.12)$$

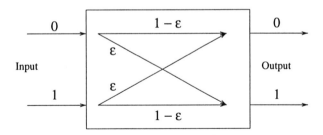

Figure 2.3   A simplified model for a communication link.

Equation 2.12 is known as **Bayes' rule**. It was published in 1764 (in a different form since the notation $P[A|B]$ was not known then) and was a landmark in the development of the theory of probability. At the time, the subject of probability had been around for 110 years. Most of the results obtained prior to then applied to *specific* problems, such as dice throwing, card games, mortality and life insurance. Bayes' rule, in contrast, applied to *arbitrary* events. It was the *second* theorem of the century-old theory.[16] Bayes' rule has been widely used in genetics over the past centuries. In engineering it is widely used in hypothesis testing and detection [63]. Let's look at an example of where it is useful.

Example 2.7.   Consider the following simplified model of a communication link between two computer network nodes, shown in Figure 2.3. The input can take logical values 0 and 1 only; due to transmission errors, the output may not be the same as the input. The probability of an error is ϵ, while the probability of sending a 0 or a 1 is 0.5. We wish to find the probability of a sending a 1 when the received bit has a value of 0. Define the events $A_i, B_i, i = 0, 1$, as follows:

$$A_i \triangleq \{\text{sent bit is equal to } i\}, \quad B_i \triangleq \{\text{received bit is equal to } i\}.$$

Then, from Bayes' rule, we have

$$
\begin{aligned}
P[A_1|B_0] &= \frac{P[B_0|A_1]P[A_1]}{P[B_0|A_0]P[A_0] + P[B_0|A_1]P[A_1]} \\
&= \frac{\epsilon \cdot \frac{1}{2}}{(1 - \epsilon) \cdot \frac{1}{2} + \epsilon \cdot \frac{1}{2}} = \epsilon,
\end{aligned}
$$

since by definition $P[B_0|A_1] = \epsilon = 1 - P[B_0|A_0]$.

---

[16]The first was the Weak Law of Large Numbers (Theorem 6.4, Section 6.5), a theorem developed by James Bernoulli in the late seventeenth century.

2.7   INDEPENDENT EVENTS

Independence is perhaps the single most important concept in the subject of probability theory.[17] To quote from the father of modern probability theory, [45], page 8,

> ...Historically, the independence of experiments and random variables represents the very mathematical concept that has given the theory of probability its peculiar stamp.

Probabilities of events may increase, decrease, or remain the same as other events take place. The case where $P[B|A] = P[B]$, is special; it means that the occurrence of event A does not alter the probability of event B. Moreover, we have

$$P[A|B] = \frac{P[A \cap B]}{P[B]} = \frac{P[B|A] \cdot P[A]}{P[B]} = \frac{P[B] \cdot P[A]}{P[B]} = P[A].$$

Thus, if the occurrence of event A does not alter the probability of event B, then the occurrence of event B does not alter the probability of event A either. In other words, events A and B do not *depend* on each other; i.e., they are independent. It is also easy to see that

$$P[A \cap B] = P[B|A] \cdot P[A] = P[A] \cdot P[B].$$

This last relationship is symmetric in A and B. We use it in the following definition. Consider again a fixed experiment E. Let $A, B \in Q$ be two arbitrary events. Let $P[\cdot]$ denote the probability assignment rule.

Definition: Two events $A, B \in Q$ are **independent** if and only if

$$P[A \cap B] = P[A] \cdot P[B]. \tag{2.13}$$

Definition 2.13 can be used to calculate $P[A \cap B]$ when we know that events A and B are independent. We see later (in Chapter 4, when we discuss "unrelated" experiments and Cartesian product spaces) that it is very easy to define independent events in a lot of experiments. Equation 2.13 is very useful in calculating probabilities in such cases.

Independence of events A and B implies something about events A^c and B. Observe first that $B = B \cap S = B \cap (A \cup A^c) = (B \cap A) \cup (B \cap A^c)$. Moreover, the events $B \cap A$ and $B \cap A^c$ are disjoint. Then

$$P[A^c \cap B] = P[B] - P[A \cap B] = P[B] - P[A]P[B] = (1 - P[A])P[B] = P[A^c]P[B],$$

---

[17] Its importance will become apparent later, throughout the development of the theory, e.g., in Chapter 6 when we discuss IID sequences, one of the most useful tools for modeling dynamic systems.

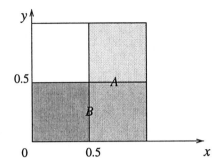

Figure 2.4 Two independent events.

and the events A^c and B are also independent! Similarly, events A^c and B^c can be shown to be independent. The implication of the above equation is that when A, B are independent events, knowledge of whether A has or has not occurred cannot alter the probability of B's occurring or not.

Example 2.8. Consider a sample space which contains (among others) two outcomes c and d. Are the elementary events $\{c\}$ and $\{d\}$ independent? Clearly not, since their intersection is the empty set, and $P[\{c\}]$ and $P[\{d\}]$ need not be zero, in general.

Example 2.9. Consider an experiment where S is the unit square, i.e., $S = [0,1] \times [0,1]$. Define the probability of a subset of S to be equal to its area. (Why is this a valid probability assignment?) Define the events

$$A \triangleq \{(x,y) \in S : x > 0.5\}, \quad B \triangleq \{(x,y) \in S : y \le 0.5\}.$$

Figure 2.4 shows these events. We can easily check that

$$A \cap B = \{(x,y) \in S : x > 0.5 \text{ and } y \le 0.5\},$$

and thus

$$P[A \cap B] = 0.25 = P[A]P[B].$$

Therefore, the two events are independent.

Given $(S, \mathcal{Q}, P)$, how do we *construct* two independent events? It is very tough to do so for an arbitrary sample space S. On the other hand, it is very easy (and we take advantage of it later, in Chapters 3 and 4) to do so when the sample space has a product or Cartesian form (see Appendix A).

So far, we have been talking about two events only. The definition of independence extends to three or more events. Consider again a fixed experiment E. Let $A_1, A_2, A_3 \in \mathcal{Q}$ be three arbitrary events.

Definition: The events $A_1, A_2, A_3 \in Q$ are called **independent** if they are pairwise independent, i.e., if

$$P[A_i \cap A_j] = P[A_i]P[A_j], \quad \forall i \neq j,$$

and

$$P[A_1 \cap A_2 \cap A_3] = P[A_1]P[A_2]P[A_3].$$

Independence of $n \geq 3$ events can now be defined by induction.

Definition: The events $A_1, A_2, \ldots, A_n, n \geq 3$, are called **independent** if any $n - 1$ of them are independent and

$$P[A_1 \cap A_2 \cap \cdots \cap A_n] = P[A_1]P[A_2] \cdots P[A_n].$$

2.8   HOW DO WE ASSIGN PROBABILITIES TO EVENTS?

So far, we have investigated the *properties* that *all* probabilities must satisfy; we have seen some tools to generate probabilities of more events, given the probabilities of other events. Now let's see who *gives* us those probabilities. This is an implementation question: Once a random experiment like the *netstat* and *ftp* examples is defined, how can probabilities of desired events be evaluated?

There are three possible ways to do this:

1. We can **measure** them as *relative frequencies.*

2. We can **compute** them from other known probabilities, using the theorems and corollaries we have developed.

3. We can **hypothesize** about them.

We discuss in Section 3.7 the intricacies of Method 1. This method was first suggested by D'Alembert.[18] It is the "brute force" way. It is very simple conceptually and can be applied to almost every experiment. The obvious disadvantage is that it can be time-consuming or expensive; moreover, in experiments with an infinite number of outcomes, its accuracy can become questionable since practically only a finite number of measurements can be taken. These are some of the major concerns with this method: How many measurements should we take before we are certain that probabilities are measured

---

[18] Jean le Rond D'Alembert, 1717–1783, a French mathematician and physicist. He was the first scientist to criticize the new theory, thus helping put it on a sounder basis. His major criticism of the new theory was about an infinite mean value in the then-famous St. Petersburg game (see Appendix C).

with adequate accuracy? How do we deal with imperfect measurements? The subject of *statistical experiment design* (see, for example, [22]) deals with such questions in a scientifically sound way.

Examples of expensive experiments from Section 1.9 are *ftp* (even though you may not directly pay for it), measuring rainfall in a given area, and air traffic-related ones. Relevant to this method is also Laplace's definition of probabilities, which we now discuss.

Example 2.10. Consider a discrete sample space S with $n < \infty$ elements; for any subset $A \subseteq S$, define n_A as the *cardinality* of A. Laplace defined $p[A]$, the probability of event A, as the *relative frequency*

$$p[A] = \frac{n_A}{n}.$$

We can easily check that Axioms 1, 2, and 3a are satisfied. Take $Q = 2^{|S|}$. Since n_A is a nonnegative integer,

$$p[A] \geq 0, \quad \forall A \in Q,$$

and Axiom 1 is satisfied. Since $n_S = n$, $p[S] = 1$, satisfying Axiom 2. For Axiom 3a, let A, B be two disjoint sets, with cardinalities n_A, n_B. We can easily see that $n_{A \cup B} = n_A + n_B$, and thus

$$p[A \cup B] = \frac{n_{A \cup B}}{n} = \frac{n_A}{n} + \frac{n_B}{n} = p[A] + p[B],$$

satisfying Axiom 3a.

Example 2.11. Let ζ_i denote the ith element of the sample space S. Observe that, with Laplace's definition of probabilities,

$$p[\{\zeta_i\}] = \frac{1}{n}, \quad i = 1, 2, \ldots, n,$$

i.e., all outcomes of the sample space must be equiprobable. Laplace's definition served its purpose at the time it was introduced since many problems of interest at the time (throwing dice, drawing cards from decks, balls from urns, etc.) *did* have finite sample spaces with equiprobable outcomes. It is possible to extend the definition for *countable* sample spaces, with outcomes that are not equiprobable. To that extent, consider a discrete sample space S with $n \leq \infty$ elements; to each element $\zeta_i \in S, i = 1, 2, \ldots, n$, assign a nonnegative number a_i such that $\sum_{i=1}^{n} a_i = 1$. Take $Q = 2^{|S|}$. For any given subset $A \in Q$, define the following function:

$$I_A(\zeta) = \begin{cases} 1, & \zeta \in A, \\ 0, & \zeta \notin A. \end{cases} \tag{2.14}$$

The function $I_A(\zeta)$ is called the **indicator function** of the event A. To each event A, assign now a number $p[A]$, calculated as follows:

$$p[A] = \sum_{i=1}^{n} I_A(\zeta_i)a_i. \tag{2.15}$$

This assignment satisfies all the axioms of probability. [Check that; the key observation for Axiom 3b is that

$$n_{\cup_i A_i} = \sum_i n_{A_i}$$

for any (countable) union of disjoint sets A_i.]

   In Section 2.3 we asked, *What* events do we select for the initial assignment? It is obvious now that we need only calculate the probabilities of the elementary outcomes ζ_i, in order to answer this question. Once these values (i.e., the a_i) are calculated, the probability of any event can be calculated via Equation 2.15.

Example 2.12. It is not possible to extend the definition for spaces more "complex" than countable sets, i.e., for $S = [0,1]$. The theory in Chapter 3 will be handy for such cases!

Example 2.13. An example for Method 2 is given in Section 3.6, where the binomial probabilities are computed from the Bernoulli probabilities. Method 2 does not involve any *direct* measurements; it relies on knowledge of some basic probabilities. We present in detail in Appendix B various enumeration and counting techniques that are very useful with this method. The method has an obvious disadvantage—its scope is rather limited.

   Examples for Method 3 are all the models in Section 3.6. In many cases, Method 3 is the least accurate, but surprisingly enough, it is the most widely used, mainly because often it is the only viable alternative. In practice, Methods 1 and 3 are often combined. A preliminary set of probabilities is measured, and then Method 3 is extensively used, in order to save time and money.

2.9   PROBABILISTIC MODELING

By now it should be clear that the elements needed in providing a probabilistic model of an engineering system (or, in general, of a random experiment) are specified by the triplet (S, Q, P).

Definition: The *probabilistic model* of a random experiment E is specified by the triplet (S, Q, P).

   Once a description of the actual system is given, the sample space S usually can be defined in a straightforward manner.

Example 2.14. Consider the *netstat* example, with the number of packets in a 1-s interval being measured. The smallest possible value we can measure is obviously 0; the largest one depends on the capacity of the connecting communication line and the size of the packets. Since these parameters vary from system to system, it is a convenient generalizing assumption to make this number ∞ and take $S = \{0, 1, 2, \ldots, \infty\}$. For any given system for which we know that these values cannot exceed, say, a maximum of 1000, we can make the probabilities of anything above 1000 identically equal to 0. Make sure you have no trouble with this suggestion. If you don't, then the alternative suggestions $S = [0, \infty)$ and $S = (-\infty, \infty)$ should be equally acceptable.

Example 2.15. In the girlfriend/boyfriend example, a suitable choice for S might be $S = \{$good, bad, OK, exceptional, unbearable$\}$.

Based on our discussions in Sections 2.3 and 2.4, Q can be taken as the collection of all subsets of S when S is discrete. When S is continuous, then Q can be taken as the collection of all Borel subsets of S. So the choice of Q is fairly straightforward, too, even though the mathematical nature of Q is much more subtle than that of S.

Specifying P, the last component of a probabilistic model, is the most difficult part of the modeling process.[19] We looked at three approaches for doing so, namely, measurements, computation, and hypothesis. There is no obvious choice among the three, as all have their respective advantages and disadvantages.

One rather subtle point here is that P *must be specified on all elements of* Q. Of course, it would be nice if one could assign probabilities to a limited set of events and then use theorems to *compute* the probabilities of the remaining events. Moreover, if we are given a candidate assignment P, we should check whether the three axioms are satisfied. This can be a formidable task, especially when the cardinality of S is "large."

The characteristics of a *good model* are **mathematical tractability** and **relevance**. After all, a model (probabilistic or not) that cannot be manipulated mathematically is of no use at all. A model remotely relevant to the (engineering) system at hand is of no use either. Typically, one characteristic comes at the expense of the other. As a model becomes simpler and thus more tractable, its relevance decreases. Therefore, the *real question* in the modeling process involves where to draw the line between tractability and relevance. Fortunately or unfortunately, the answer comes only with experience, and often after a few trials and errors.

At any rate, once we have specified a "good" model (S, Q, P), what can we do with it? We can mainly use it to: (1) understand a system better, (2) design a system, and (3) compare various designs.

---

[19] As we see in Chapters 3 and 4, histograms and density functions make this step easier.

Probabilistic models can be classified as **analytic** and **simulation** models. Analytic ones can be manipulated mathematically, and the desired results about the actual system can be obtained as solutions to (algebraic, integral, or differential) equations. Simulation models, on the other hand, do not involve mathematical handling; they are typically implemented in software, even though hardware implementations are also used. The results about the modeled physical system are obtained as outputs of a computer run. The main advantages of an analytic model are speed and low cost since the only need of the model is equation solving. The main advantage of a simulation model is accuracy; since no mathematical equations are formed and solved, a simulation model can capture more details of the actual physical system. As with all engineering systems, typically there is a tradeoff among speed, cost, and accuracy. We see in Chapter 11 a simulation model of a communication switch in detail.

2.10   DIFFICULTIES

So far we have used the "language" of sets to express the concepts of probability theory. This language gave us the means to found the theory in an axiomatic, rigorous way. It also gave us the means to develop a few tools (such as the total probability theorem and Bayes' rule). Unfortunately, it seems that we have run out of steam as far as tool development goes. The mathematically experienced student may have noticed the absence of integrals, derivatives, and limits so far. These tools are more widely known and much easier to use than sets.

How do we select the events in Q for the initial assignment of probabilities? We have not said much about that because there is no universal way of doing so for an arbitrary S and Q. Finally, how do we check whether a given assignment P satisfies the three axioms, especially Axiom 3b? The examples in Section 2.8 provide an answer, but there are difficulties when S is uncountable. See also Problem 2.61.

We see in the next chapter how these difficulties can be overcome through the concept of a *random variable*.

2.11   SUMMARY OF MAIN POINTS

- A **random experiment** E is a convenient mathematical abstraction for describing physical systems with uncertainties. A **probabilistic model** for such an experiment is specified by the triplet (S, Q, P), where S, the **sample space**, contains all possible **outcomes** of the experiment, Q is the set of all **events** of interest, to which we can assign probabilities, and P, the **probability assignment**, is a function from Q on $[0, 1]$ that specifies how events in Q are assigned probabilities.

- A valid assignment P must satisfy Axioms 1, 2, and 3b (or 3a).

- For an assignment P to make mathematical sense, Q must be a σ field.

- Assigning probabilities to events of a random experiment has been no fun. Conditional probabilities, independence, the theorem of total probability, and Equations 2.6 and 2.12 can be used to generate probabilities of "more complex" events, once probabilities of "simpler" events are known.

- There are three ways to assign probabilities to events: measurement, computation, and hypothesis.

2.12　CHECKLIST OF IMPORTANT TOOLS

- Total probability theorem, Equation 2.10

- Bayes' rule, Equation 2.12

- Equation 2.6

- Equation 2.7

- Equation 2.13

2.13　PROBLEMS

Mass Phenomena

2.1 Consider the Wheel of Fortune game. Which of the following statements can be answered with probability theory?
(a) If *you* play the game, what are *your* chances of winning?
(b) Is the probability of a *player's* winning equal to 1/3?

2.2 Which questions are valid ones, that the theory in this book can answer?
(a) What is the probability that a meteorite will hit the earth?
(b) What is the probability that there is life on Mars?
(c) What is the probability that there is life on a planet?

2.3 Which of the following statements are absurd?
(a) The chances of the space shuttle's colliding with an airplane are small.

(b) The chances of the space shuttle's colliding with a meteorite are small.

(c) The chances of an airplane's being hit by lightning are medium.

2.4 Discuss the exact meaning you may attach to the following statement: "I am 80 percent sure that my GPA will be above 3.5."

2.5 Is it fair to say that the "probability that *you* will find the right person in your life" is 50/50? Explain.

2.6 Is it (politically or otherwise) correct to say that the Democrats will win the next presidential election with high (or low) probability?

2.7 Consider this year's Super Bowl.
(a) What is the probability that the AFC team will win it?
(b) If you place a bet that the score is tied at the end of the first half, what is the probability that you will lose your bet?

Random Experiments

In this subsection's problems, you must specify the input/output of the experiment; you should also discuss what will introduce randomness.

2.8 Specify two random experiments from your everyday life.

2.9 Specify at least two random experiments from your electrical engineering-related experience.

2.10 Specify at least two random experiments from your computer engineering-related experience.

2.11 Specify at least three random experiments from your physics courses.

2.12 Specify at least two random experiments related to your humanities courses.

2.13 Specify a random experiment from an example in Section 1.9.3.

2.14 Specify 5 to 10 random experiments from NFL, NBA, NL, AL, or your school's conference (ACC, being the best conference, is preferred).

2.15 Specify two random experiments with respect to rolling dice.

2.16 Specify a random experiment with respect to playing cards.

2.17 Specify a random experiment related to lottery or casino games.

2.18 Is "John Doe throws a die, *and* Jane Doe selects a card" a random experiment?

2.19 Is "John Doe transfers a file, *and* rain falls in Minnesota" a random experiment?

Sample Spaces

2.20 Specify the sample space S in some of the Problems 2.8–2.19. For which problems is the sample space S discrete? Uncountably infinite?

2.21 Specify at least two choices for the sample space S in some of the examples in Section 1.9.3.

2.22 Let $N \leq \infty$ be a given positive integer number. Can you specify a random experiment with a sample space S that has exactly N elements for any given N? Explain.

2.23 Specify a random experiment with a sample space S that has an uncountably infinite number of elements.

2.24 Specify a random experiment with a sample space S whose elements are three-dimensional vectors.

2.25 Let S be any given set. Can you specify a random experiment with S as the sample space? Explain.

2.26 Suppose that the sample space S of a given random experiment is $S = [0, 1] \cup [2, 10]$. Someone suggests that it would "look nicer" to have $S = [0, 10]$ (i.e., no gaps). Is this artificial sample space augmentation a problem? Explain.

2.27 Let $S \triangleq \{1, 2, 3, 4, 5, 6\} \times \{1, 2, 3, 4, 5, 6\}$ (a Cartesian product). Is $\{4\}$ an outcome? What are the outcomes of the random experiment with such a sample space?

2.28 Consider the set $S = \{a, b, c, \ldots, z, A, B, \ldots, Z, 0, 1, 2, \ldots, 9\}$. Specify a random experiment related to computer security for which S would be a suitable sample space.

2.29 Let S be the set of all ASCII characters. Specify a random experiment for which this S would be a suitable sample space.

2.30 Consider a multiprocessing multiuser CPU that employs priorities for selecting the program to be run next. A lower-priority program can be interrupted when a higher-priority one arrives. A low-priority program cannot be executing while higher-priority ones are present in the system. Programs that are ready for execution are placed in one of the READY queues, shown in Figure 2.5, where 1 denotes the highest priority and N the lowest. The CPU *scheduler* selects the highest-priority, nonempty queue and forwards the first-in-line program to the CPU. For simplicity, assume that a program will run to completion if it is not interrupted (i.e., no round-robin scheduling). Suppose that at time $t = 0$, a low-priority program (call it P) is running on the CPU.

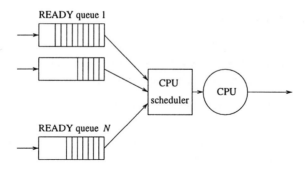

Figure 2.5   CPU scheduling.

(a) For simplicity, assume that only two priority levels are implemented. Suppose we observe the number of programs of higher priority that are executed before P. Specify a suitable sample space for this experiment.

(b) Now assume that $N > 2$ priority levels are implemented. Repeat part (a) for programs of every priority level. State any further assumptions you need to make.

Events

2.31 Specify two to three events of interest in some of the random experiments in Problems 2.8–2.19.

2.32 Specify two to three events of interest in some of the examples in Section 1.9.3.

2.33 How many events are there in the *ftp* random experiment? Are all of them of interest?

2.34 How many events are there in the *netstat* random experiment? Are all of them of interest?

2.35 Define events of interest in the lottery example in Section 1.9.

2.36 How many events can you associate with a random experiment that involves a deck of 52 cards? Explain.

2.37 Consider this random experiment: A coin is flipped twice; the faces showing up are noted.
(a) What is the sample space?
(b) How many events can you define on this random experiment?
(c) Can you define an event with exactly five elements? Explain.

2.38 Consider this random experiment: A die is tossed; the number of dots facing up is counted.
(a) What is the sample space S?
(b) Find two events with equal numbers of outcomes. Express them as subsets of S; then express them with an English statement.

2.39 Consider this random experiment: A card is drawn from a deck of 52 cards. Its suit and number are observed.
(a) Find an event that involves 26 outcomes; express it in words.
(b) How many events involve hearts?
(c) How many events involve hearts *and* diamonds?
(d) Find two events that involve hearts and diamonds *only*.

2.40 You have undoubtedly filled out a course evaluation form; define a few events of interest regarding this random experiment.

2.41 Suppose that you fill out a teacher evaluation form for a class.
(a) Define two events related to the teacher's performance.
(b) Define two events related to the students of the class.
(c) Define two disjoint events.
(d) Define two events that have a nonempty intersection.

2.42 The lifetime of a PC is defined as the time it takes for the PC to break down.
(a) Define a random experiment that involves the lifetime of the PC.
(b) What is the sample space?
(c) Define a few events.

2.43 Consider this random experiment: The time until a PC breaks down is observed.
(a) Define a sample space S for this experiment.
(b) Enumerate, if possible, all its subsets.
(c) Define two events that are disjoint.
(d) Define two events that have a nonempty intersection.
(e) Consider the set $S = \{$population in the United States$\}$. Would you consider S a suitable sample space for this experiment?
(f) Now consider the set $S = \{$all PCs in the world$\}$. Is S a suitable sample space for this experiment? Explain.

Axioms and Theorems

2.44 Discuss what problems might arise if the set of axioms is (a) incomplete, (b) inconsistent, (c) dependent.

2.45 List the axioms of algebra. (The point here is to compare them to the axioms of probability.)

2.46 *What If?* Al Steinein is an eccentric engineer and self-proclaimed mathematician who likes to try out things, just for the fun of it; we will deal with him a lot in the problems. Al suggests that we replace Axiom 3a by the following axiom, call it Axiom 3c.

$$P[A \cup B] = P[A] + P[B] - P[A \cap B], \quad A, B \in \mathcal{Q}. \tag{3c}$$

(The sets A and B are *not* necessarily disjoint.) Is Al's Axiom 3c equivalent to Kolmogorov's Axiom 3a? Explain.

2.47 Now Al says replace Axiom 3a by the following axiom, call it 3d.

$$P[A - B] = P[A] - P[B], \quad \text{when } A \supseteq B. \tag{3d}$$

(See Appendix A for the definition of the difference of two sets.) Show that Axiom 3d implies Axiom 3a.

2.48 Al wants to experiment with "negative" probabilities $P'[A]$. He suggests that we do it in one of the following ways:
(a) "Shift" all probabilities by -1 (i.e., set $P'[A] = P[A] - 1$, so that now $-1 \leq P'[A]$ for any A and $P'[S] = 0$).
(b) "Scale" all probabilities by -1 (i.e., set $P'[A] = -P[A]$, so that now $P'[A] \leq 0$ for any A and $P'[S] = -1$).
Do any of Al's suggestions make sense?

2.49 Al suggests a "scaling" of all probabilities, $P'[A] = 10P[A]$, so that now $0 \leq P'[A] \leq 10$. Do you agree with him?

The Third Axiom

2.50 Would you need Axiom 3b in the *ftp* example in Section 1.9.3?

2.51 Identify (if possible) an example in Section 1.9 where you do need Axiom 3b.

2.52 Explain why Axiom 3b is necessary, using an example of your own choice.

2.53 The following "continuity" axiom was Kolmogorov's original suggestion, instead of Axiom 3b. Consider a sequence of events $\{A_k\}_{k=1}^{\infty}$ such that

$$A_1 \supseteq A_2 \supseteq \cdots \supseteq A_k \supseteq \cdots$$

and

$$\cap_{k=1}^{\infty} A_k = \emptyset.$$

Then the axiom of continuity requires that the following equation hold:

$$\lim_{n \to \infty} P[A_n] = P[\lim_{n \to \infty} A_n] = 0.$$

Show that the axiom of continuity is equivalent to Axiom 3b.

Some Simple Problems

2.54 (a) Explicitly enumerate Q for the sample space $S = \{1, 2, 3\}$.
(b) Explicitly enumerate Q for the random experiment "a die is thrown once, and the face value is observed."
(c) Explicitly enumerate Q for the card-drawing experiment.
(d) Explicitly enumerate Q for an example from Section 1.9.3.

2.55 Consider the random experiment of tossing a single die once and counting the number of dots facing up. Assume that all faces have equal probability of showing up.
(a) Find the probability of the events $A = \{2, 4, 6\}$, $B = \{2, 4\}$, and $C = \{1, 3, 4, 5\}$. State explicitly which axioms you have used in your calculations.
(b) Find the probability of the event $A \cap B^c$.
(c) Calculate $P[A \cup (B \cap C)]$ three ways.

2.56 Repeat Problem 2.55 when the die is loaded, so that $P[\{6\}] = 0.3$ and all other faces are equiprobable.

Some Tricky Problems

2.57 The outcome of a certain random experiment is a positive integer number k, that is, $S = \{1, 2, 3, \ldots\}$. Measurements suggest that p_k, the probability of the elementary event $\{k\}$, is proportional to 0.5^k, that is,

$$p_k = c \cdot 0.5^k, \quad k = 1, 2, \ldots,$$

where c is a real number. Based on this information, can you check if this is a valid probability assignment?

2.58 The outcome of another random experiment is again an integer k, but this time $S = \{k : -10 \le k \le 10\}$. Measurements suggest that p_k, the probability of the elementary event $\{k\}$, is proportional to 0.5^k; that is, $p_k = c \cdot 0.5^k, k = -10, -9, \ldots, 9, 10$. Can this be a valid probability assignment?

2.59 The outcome of yet another random experiment is a positive integer k. Measurements suggest that p_k, the probability of the elementary event $\{k\}$, is proportional to $1/k$, that is, $p_k = c/k, k = 1, 2, \ldots$. Can this be a valid probability assignment?

Simple Mathematical Skills

2.60 The outcome of a random experiment is a real number in the interval $(0, 1]$. Measurements suggest that events of the form $(a, b]$, where a and b are real numbers with $0 < a < b \leq 1$, have probabilities proportional to the length of the interval.
(a) Determine the constant of proportionality.
(b) Determine the probability of the outcomes of S.
(c) Determine the probability of the event $[a, 1)$.
(d) Determine the probability of the event $A = \{$all rational numbers in $(0, 1)\}$.
(e) Make up an event of your own and determine its probability.
(f) Find two events A, B such that $P[A] = P[B] = 0.3$.

2.61 Let $S = \{a, b, \ldots, z\}$. Consider a set function $\tilde{P}$, defined on subsets of S, such that

$$\tilde{P}(A) = \begin{cases} 1, & \text{if } a \in A \text{ and } b \in A, \\ 0.5, & \text{if either } a \in A \text{ or } b \in A, \\ 0, & \text{otherwise.} \end{cases}$$

Does the function $\tilde{P}$ satisfy the axioms of probability?

2.62 *Bizarre.* Consider two *separate* random experiments E_1 and E_2. Consider two events A_1 and A_2 that refer to the two experiments, respectively. Let each event have probability equal to 0.6. Since $P[A_1] + P[A_2] = 1.2 > 1$, can we deduce the following?
(a) The two sets A_1 and A_2 are *definitely* not disjoint.
(b) The whole thing is nonsense.
(c) None of the above.

2.63 Consider an experiment E and two events A_1 and A_2 in Q. Show that the statement "only one of A_1 and A_2 occurs" defines another event $A \in Q$. Show that

$$P[A] = P[A_1] + P[A_2] - 2P[A_1 \cap A_2].$$

Can you generalize for three or more events?

2.64 Evaluate $P[A^c \cup B^c]$ and $P[A^c \cap B^c]$ in terms of $P[A], P[B]$, and $P[A \cap B]$.

2.65 Calculate the following probabilities in terms of events that do not contain complements.
(a) $P[A_1 \cap A_2 \cap A_3^c]$,
(b) $P[A_1 \cap A_2 \cap A_3^c \cap A_4^c]$.

2.66 Prove Theorem 2.4; can you generalize to n events?

2.67 Show that $P[A \cup B \cup C] \leq P[A] + P[B] + P[C]$; then show by induction on m that

$$P[\cup_{n=1}^{m} A_n] \leq \sum_{n=1}^{m} P[A_n].$$

This inequality is known as the *union bound*.

2.68 Consider m events $A_1, \ldots, A_m$. Let

$$P_1 = \sum_i P[A_i], \quad P_2 = \sum_{i<j} P[A_i \cap A_j], \quad P_3 = \sum_{i<j<k} P[A_i \cap A_j \cap A_k],$$

and so on. Show that

$$P[\cup_{n=1}^{m} A_n] \leq P_1, \quad P[\cup_{n=1}^{m} A_n] \geq P_1 - P_2, \quad P[\cup_{n=1}^{m} A_n] \leq P_1 - P_2 + P_3,$$

etc. Can you generalize the pattern?

2.69 Consider m events $A_1, \ldots, A_m$. Show that

$$P[\cap_{n=1}^{m} A_n] \geq 1 - \sum_{n=1}^{m} P[A_n^c].$$

2.70 Consider two sequences of events $\{A_n\}$ and $\{B_n\}$, where $n \geq 1$; set $A_0 = \emptyset$. Show that

$$P[\cup_{n=1}^{\infty}(A_n \cap B_n)] \geq \inf_{n \geq 1}\{P[B_n]\} \cdot P[\cup_{n=1}^{\infty} A_n]$$

if for all $n \geq 1$,

$$P[B_n \cap (A_n \cap A_{n-1}^c \cap \cdots \cap A_0^c)] = P[B_n]P[A_n \cap A_{n-1}^c \cap \cdots \cap A_0^c].$$

2.71 Consider two events A and B. Suppose that $P[A] - P[B] = 1$. What does this tell you about $P[A]$ and $P[B]$?

2.72 Consider a sample space S and an event A. Does $P[A] = 0$ imply that $A = \emptyset$ when
(a) S is finite?
(b) S is discrete but infinite?
(c) S is continuous?

2.73 Consider the symmetric difference of two events A, B (see Appendix A for its definition). Show that $P[A \triangle B] \geq |P[A] - P[B]|$.

2.74 Show that for any events A, B, C the *triangle inequality* holds:

$$P[A \triangle B] \le P[A \triangle C] + P[C \triangle B].$$

Under what conditions on A, B, C do we have equality?

2.75 Consider two events A, B and a partition $\{C_i\}$ of event A. Show that

$$P[A \cap B] = \sum_i P[B \cap C_i].$$

Conditional Probabilities; Clarify Concepts

2.76 *What If?* Al Steinein proposes two alternative definitions of conditional probability:
(a) $P[A|B] = P[A]/P[B]$,
(b) $P[A|B] = P[A]/(P[A] + P[B])$.
Can you comment on these suggestions? If you do not agree with Al, can you think of any other alternative definition yourself?

2.77 Prove that $P[A^c|B] + P[A|B] = 1$ *directly*, by using the definition of conditional probability.

2.78 Find $P[A|B]$ when (a) $A \cap B = \emptyset$, (b) $A \cap B = A$, and (c) $A \cap B = B$.

2.79 Consider two events A, B. Suppose that $P[A|B] - P[A] = 1$. What does this tell you about A and B?
(a) $A \subseteq B$.
(b) $B \subseteq A$.
(c) A and B have a nonempty intersection.
(d) None of the above.

2.80 If $P[A|B] > P[A]$, is $P[B|A] > P[B]$? Explain.

2.81 If $P[A|B] < P[A]$, we say that event B carries negative information about event A, and we denote it as $B \to A$. Prove or disprove the following statements:
(a) If $B \to A$, then $A \to B$.
(b) If $B \to A$ and $A \to C$, then $B \to C$.
(c) If $B \to A$ and $C \to A$, then $(B \cup C) \to A$.
(d) If $B \to A$ and $C \to A$, then $(B \cap C) \to A$.

2.82 Consider two events A, B with $P[A] \le P[B]$. For an arbitrary event C, is $P[A|C] \le P[B|C]$? Explain.

2.83 Consider two events A, B with $A \subseteq B$. Is $P[A|C] \leq P[B|C]$ for arbitrary C? Explain.

2.84 Consider two events A, B. Which is larger, $P[A \cap B|A]$ or $P[A \cap B|A \cup B]$?

2.85 Consider three events A, B, C such that $P[A|B \cap C] = p$. (Here $p \neq 0$ and $p \neq 1$.) Show by example that $P[A|B] = p$, $P[A|B] > p$, and $P[A|B] < p$ are all possible.

2.86 Consider $n + 1$ events $\{A_i\}_{i=0}^{n}$ such that

$$P[A_n|A_{n-1} \cap \cdots \cap A_0] = P[A_n|A_{n-1}].$$

Does this imply that $P[A_n|A_{n-1} \cap A_{n-2}] = P[A_n|A_{n-1}]$?

2.87 Consider $n + 1$ events $\{A_i\}_{i=0}^{n}$ such that for all $1 \leq k \leq n$,

$$P[A_k|A_{k-1} \cap \cdots \cap A_0] = P[A_k|A_{k-1}] \triangleq p_k.$$

Find $P[A_n \cap A_{n-1} \cap \cdots \cap A_0]$ in terms of p_k.

Conditional Probabilities; Mathematical Skills

2.88 Let $S = \{0, 1, 2, 3, 4, 5, 6, 7\}$, with $P[\{3\}] = P[\{6\}] = 0.25$ and $P[\{i\}] = 1/12$, $i = 0, 1, 2, 4, 5, 7$. Let

$$A = \{0, 1, 2, 3, 4, 5\}, \quad B = \{0, 1, 2, 3, 7\},$$
$$C = \{1, 2, 3, 5, 7\}, \quad D = \{1, 3, 5, 7\}.$$

(a) Calculate $P[A|B \cup C \cap D]$.
(b) Calculate $P[A \cup B|B \cap C]$.

2.89 Consider events A, B, C such that $P[B \cap C] > 0$. Show that

$$P[A \cap B|C] = P[A|C]P[B|C], \quad \text{if and only if} \quad P[A|B \cap C] = P[A|C].$$

2.90 Let $A_1, A_2, \ldots, A_m$ be arbitrary events associated with the same random experiment E. Use induction to prove the *chain rule*:

$$\begin{aligned} P[\cap_{n=1}^{m} A_n] = \; & P[A_1] \cdot P[A_2|A_1] \cdot P[A_3|A_1 \cap A_2] \cdot \\ & P[A_4|A_1 \cap A_2 \cap A_3] \cdots P[A_m|A_1 \cap A_2 \cap \cdots \cap A_{m-1}]. \end{aligned}$$

$$(2.16)$$

2.91 Consider the sample space $S = (0, \infty)$. For any given $a, b > 0$, let $A = [a, a + b]$, $B = [a, \infty)$ and $C = (0, b]$. The probability assignment $P[\cdot]$ satisfies the property $P[A|B] = P[C]$. Show that

$$P[C] = 1 - e^{-cb},$$

where $c > 0$ is a constant.

2.92 We have two coins, one fair and the other loaded. We pick one of the coins at random, and we flip it n times. The coin shows heads n times. Plot the probability we picked the fair coin, as a function of p_l. (*Hint:* Use Bayes' rule.)

2.93 Consider the sample space $S = \{0, 1, 2, \ldots\}$. For each nonnegative integer n, let $A_n \triangleq \{0, 1, 2, \ldots, n\}$. Consider the function

$$f(A_n) = c \cdot \sum_{i \in A_n} \frac{1}{i!}.$$

(a) Find the constant c that will make $f(A_n)$ a valid probability assignment.
(b) Let $D_n = \{n\}$. Find $P[D_n], n = 0, 1, 2, \ldots$, in terms of $f(A_n)$.

2.94 Consider an event A and a family of disjoint events $\{B_i\}$ with the property that $P[A|B_i] = p$, $\forall i$.
(a) Show that $P[A| \cup_i B_i] = p$.
(b) Show that $P[\cup_i B_i | A] = \sum_i P[B_i | A]$.
(c) If the sets B_i form a partition of the sample space, show that for all C,

$$P[A|C] = \sum_i P[B_i|C] P[A|B_i \cap C].$$

Conditional Probabilities; Apply Concepts

2.95 Consider a test of circuit components. Let

$$A \triangleq \{\text{circuit defective}\}, \quad B \triangleq \{\text{test declares defect}\}.$$

Suppose that

$$P[B|A] = p, \quad P[B^c|A^c] = q, \quad P[A] = r, \quad P[B] = s.$$

We know that the test is not perfect, so there are errors of two kinds: (1) a circuit defect is not detected and (2) a correct circuit is declared defective.

Suppose that the manufacturing process and the test itself can be designed such that each of the unknown parameters p, q, r, s can be very close to 0 or to 1.
(a) Discuss which value of each parameter is preferred.
(b) Discuss the meaning of the conditional probabilities $P[B^c|A]$ and $P[B|A^c]$.
(c) Discuss the meaning of the conditional probabilities $P[A^c|B]$ and $P[A|B^c]$.
(d) For given p, q, r, s values, evaluate the probabilities in parts (b) and (c).
(e) What is the sample space S for this experiment?

2.96 Consider the communication channel shown in Figure 2.3.
(a) Find the probability of the events

$$A = \{\text{output symbol} = 0\}, \quad B = \{\text{output symbol} = 1\}.$$

(b) Let $C = \{\text{error in the channel}\}$. Find $P[C|B], P[C|A]$.

2.97 Specify a few conditional probabilities for the *ftp* example. Discuss how they can be used to calculate probabilities of more complex events.

Independence; Clarify Concepts

2.98 How many equations do you need to check in order to establish independence of n events?

2.99 Show that events A and B are independent if and only if $P[A|B] = P[A|B^c]$.

2.100 Consider an event A with $P[A] = 1$.
(a) Argue that A need not be the same as the sample space S. If possible, show this by example.
(b) Consider another arbitrary event B. Are A, B independent?

2.101 Consider two events A, B; we want to calculate $P[A \cup B]$. Which of the following statements is (are) true?
(a) $P[A], P[B]$ suffice to calculate $P[A \cup B]$ when A, B are independent.
(b) $P[A], P[B]$ suffice to calculate $P[A \cup B]$ when A, B are disjoint.
(c) $P[A], P[B]$ suffice to calculate $P[A \cup B]$ in all cases.

2.102 Consider three independent events A, B, C. Their probabilities are known. Can you find the following probabilities, or do you need more information?
(a) $P[A|B \cup C]$
(b) $P[A|B \cap C]$
(c) $P[A|B \cap C^c]$
(d) $P[A \cap B|C^c]$

Independence; Apply Concepts

2.103 Identify independent events in the random experiment in Problem 2.38 when the die is fair. Identify independent events when the die is loaded.

2.104 Let $S = [0, 1] \times [0, 1]$. Assume that for any event A, $P[A]$ is equal to the area of A. Can you specify A, B, C, three subsets of S, that are pairwise independent but for which $P[A \cap B \cap C] \neq P[A]P[B]P[C]$?

2.105 Let $S = [0, 1] \times [0, 1]$. Assume that for any event A, $P[A]$ is equal to the area of A. Can you find two independent events A, B that do *not* have a rectangular form?

2.106 Let $S = \{0, 1, 2\} \times \{0, 1, 2\}$. Assume that all outcomes are equiprobable. Can you find two independent events? Three independent events?

2.107 Can you specify a sample space S with an infinite number of pairwise independent events?

2.108 Can you specify a random experiment E with only independent events?

2.109 Can you specify a random experiment E with only dependent events?

2.110 Can you specify two independent events in the NFL or NBA random experiments?

2.111 Can you specify two independent events in an example from your everyday life?

Independence and Modeling

2.112 Consider the Token Ring local area network in Figure 2.6. Workstation A sends and receives files from the file server FS. Assume that links in the network fail independently of each other. Let p_i be the probability that link i fails.
(a) Calculate the probability a file will be received correctly in station A.
(b) For which station is this probability *maximum*?
(c) For which station is this probability *minimum*?

2.113 The dual Token Ring is an architecture that improves reliability of the ring. It operates as follows: When a link fails, the two computers attached to it wrap the packets around, as Figure 2.7 suggests, creating an alternative route for packet communication. Evaluate the improvement in the probabilities you calculated in Problem 2.112.

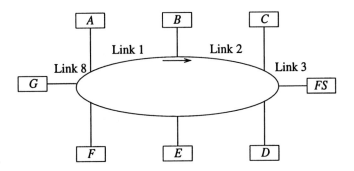

Figure 2.6 Token ring network.

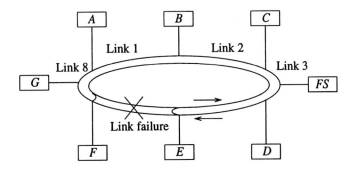

Figure 2.7 Dual token ring network.

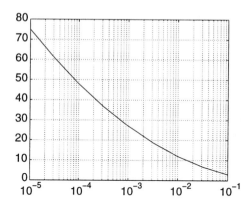

Figure 2.8   Duplicate printer prices.

2.114 To enhance reliability of complex systems, manufacturers *duplicate* certain critical components, having multiple components operate in parallel; if one component fails, the others can continue operating without interruptions. Consider, for example, the CPU that controls a certain function of the space shuttle, say, rocket firings. Suppose that this CPU has failed in tests with probability p. Assume that CPUs fail independently.
(a) How many CPUs must be installed if the probability of failure we can tolerate is at most 10^{-6}?
(b) How many CPUs must be installed if we demand a probability of failure equal to 0?
(c) Discuss how the CPUs must operate and how failure of one could be handled.

2.115 Companies that rely on rapid data processing and timely delivery of information to customers (e.g., banks, credit card companies, stock exchange bureaus) duplicate their printers so that, in case of failure, a standby printer can resume the task of the failed one with minimal lost time. Suppose that the price of a printer, $f(p)$, as a function of p, its probability of failure, is given in Figure 2.8.
(a) Find the cost of a system with one printer only when $p = 10^{-4}$.
(b) What is the cost of a system with one standby printer as a function of p?
c) It takes 1 day to order and receive a new set of printers. If for every day the system is down the company loses x dollars, does it pay to buy a few more printers? If yes, how many?

2.116 Consider an airplane or a nuclear reactor. As a designer, would you make two safety systems fail independently? Explain.

2.117 Consider the binary communication channel in Figure 2.3. Would you make the outputs independent of the inputs? Why? How do you interpret "independent" here?

2.118 The notation $S = \{0, 1\}^n$ is shorthand for the set of all n-dimensional vectors with components 0 or 1. (And S could be the sample space for the random experiment of transmitting n-bit-long packets over a communication channel.) Assume that all outcomes are equiprobable. Specify n independent events on this sample space.

2.119 In the *ftp* example, how are you going to check whether the events "delay less than 3 s" and "throughput larger than 10 kbytes/s" are independent? Explain.

2.120 Consider the events $A = \{$a receiver of a team has more than 100 catches per season$\}$, $B = \{$a receiver of a team has more than 15 touchdowns per season$\}$, and $C = \{$a running back of a team has more than 1800 yd per season$\}$. Are they independent?

Assigning Probabilities to Events

2.121 Calculate some conditional probabilities on the *netstat* or *ftp* experiments, such as $P[$delay less than x seconds$|$file size bigger than y kilobytes$]$, by actually measuring them.

2.122 Do an *ftp* to an Internet site. Transfer the same file repeatedly. Let the random experiment be as follows: The throughput of the file transfer is measured (in kbytes per second). Define Q for this random experiment. Assign probabilities to the events of Q by *measuring* them as relative frequencies. Comment on the difficulties and accuracy of this approach.

2.123 Now assign probabilities to the events of Q by *hypothesizing* about them. Comment on the difficulties and accuracy of this new approach.

2.124 Suppose that the file size changes. Would it be easier to follow the approach of measuring or hypothesizing to calculate probabilities now? Comment.

2.125 Discuss how one can *experimentally* obtain the value of the bit error probability ϵ in the communication channel in Figure 2.3.

2.126 Discuss how probabilities could be assigned in some examples in Section 1.9.3.

2.127 Consider the NBA random experiment you selected in Problem 2.14. Discuss how you would evaluate the probabilities of some events of interest.

2.128 Consider the engineering-related random experiment you selected in Problem 2.9 or 2.10. Discuss how you would evaluate the probabilities of some events of interest.

2.129 On Unix, the command "*ls* − *lR*" gives you a list of file sizes and file attributes (such as read, write, and execute); use it in your workstation to measure probabilities of these events:
(a) File is executable.
(b) File is readable and less than 10 kbytes.
(c) File is both readable and writeable.
(d) File is readable but not executable.
(e) Are there any independent events?

2.130 On Unix, the command "*compress*" or "*compact*" can be used to save disk space by compressing files.
(a) Evaluate experimentally the probability of the event

   A = {compressed file size is at most two-thirds of original file size}.

(b) Evaluate experimentally whether executable files are more compressible than readable ones.
(c) Determine how the compressing efficiency of the algorithm varies as a function of the file size.

Modeling

2.131 Provide a probabilistic model [the triplet (S, Q, P)] for the following experiment: Customers arrive at an automated teller machine at a bank branch.
(a) What are the possible outcomes of the experiment that are elements of S?
(b) What is the collection of subsets A of S that form the collection Q?
c) Discuss how you could assign probabilities to such events. Comment on interesting, difficult, vague, weak, and strong aspects of the model.

2.132 Provide a probabilistic model for an everyday (nontechnical) example. Comment as in Problem 2.131.

2.133 Can you design an experiment to determine whether the communication link or the CPU and disk are the bottleneck in an *ftp* transfer? Vary the file size and observe the transfer delays.

2.134 Would you model airplanes approaching an airport as a random experiment? Think as an air traffic controller. (*Hint:* The issue here is mass phenomena versus safety per single approach.)

2.135 Daily levels of Red Cross blood supplies in a city or state fluctuate randomly. Define a sample space. Develop a model for the daily supply and demand. Devise control policies for keeping supplies above a certain threshold, and describe them in your sample space setup.

2.136 Consider a password selection rule that allows up to eight characters per password. Define the *protection capability* of a given rule as the probability that a computer hacker who chooses a password at random will match the password selected by a legitimate user. Suppose that only lower- and uppercase letters can be selected for a password. Since the rule for selecting a password is known to the public, assume that the hacker knows it as well.
(a) Evaluate the protection capability as a function of the length of the password.
(b) Evaluate the protection capability of an eight-character-long password when numbers are allowed as well.
(c) Suppose that the system has N users; a protection capability of 10^{-20} is desired. How long should the password be?
(d) Suppose that a class of users never use more than two numbers in a password. Evaluate the protection capability for those users.
(e) Repeat parts (a) to (d) when the hacker does not know the password selection rule.
(f) Suppose a class of users uses only nicknames as passwords. Evaluate the protection capability for those users when the hacker knows their habit.
(g) Some password selection rules dictate that one of the characters be a number. Evaluate the reduction in protection capability of such a rule compared to a rule that poses no such restriction.

2.137 ATM, credit, or telephone cards use only numbers for their passwords. Typically, four numbers are used for such passwords. Suppose that a hacker has a program that tries a four-digit combination in x s. How many seconds will it take her or him to break a password?

Projects

You may obtain data for the following projects by yourself, or you may find data in the book Web site.

2.138 Using a suitable compression command (e.g., "compress" on Unix workstations), compress a number N of files in your system. Measure the resulting savings in storage space as follows. Let A_k denote the event "the savings was between k and $k+1$ kilobytes." Interpret and plot the following probabilities:
(a) $P[A_k]$, $k = 0, 1, 2, \ldots$
(b) $P[\cup_{n=0}^{k} A_n]$, $k = 0, 1, 2, \ldots$
(c) $P[\cup_{n=0}^{k} A_n | \cup_{n=0}^{l} A_n]$, $k = 0, 1, 2, \ldots$, $l = 0, 1, 2, \ldots$
(d) What is the probability model for this experiment?

(e) Repeat parts (a) to (c) for $2N$ measurements. Do you observe any differences? Why?

2.139 Measure the number of packets coming through a computer interface (e.g., using the *netstat* 1 command on Unix workstations). Define the event A_k as "the number of packets is equal to k." Repeat parts (a) to (e) of Problem 2.138.

2.140 Measure the delays incurred in file transfers (e.g., using the *ftp* protocol). Define the event A_k as "the delay is between k and $k+1$ tens of a millisecond." Repeat parts (a) to (e) of Problem 2.138.

2.141 Consider the Dow Jones closing index values. Define the event A_k as "the value of the index is between k and $k+1$ dollars." Repeat parts (a) to (e) of Problem 2.138.

2.142 Obtain data for the duration of user log-ins (the Unix command "last -1000" will give you information about the last 1000 user log-ins). Define the event A_k as "the user logged in for k hours." (For simplicity, round fractional values to the nearest integer.) Repeat parts (a) to (e) of Problem 2.138.

2.143 Obtain the data for the MM1 simulator from the Web site. Define the event A_k as "the waiting time is between k and $k+1$ tens of ms." Repeat parts (a) to (e) of Problem 2.138.

3

CONCEPT OF A RANDOM VARIABLE

Practice is the best of all instructors.

Publilius Syrus

In this chapter we introduce the concepts of a *random variable* and its *probability density function*. These two concepts will solve all the problems we mentioned in Section 2.10. In a nutshell, here is the whole story in this chapter. The random variable is a mapping from the sample space S onto the real line R; it allows us to calculate probabilities of events defined on the real line. This "simple trick" will enable us to use integration to calculate such probabilities. The function we will integrate is the probability density function; this function is the probabilistic model of a random experiment, essentially replacing the triplet (S, Q, P). The additivity of the integral will resolve the difficulty we had with checking whether a given function P satisfies Axiom 3. At the same time, the use of a probability density function will answer the question of what sets we should select to assign probabilities to. Last but not least, from the practical point of view, a probability density function is easily calculated through a histogram.

In the next section we introduce special subsets of the real line R. In Section 3.2 we define the concept of a random variable and give some examples. In Section 3.3 we introduce the cumulative distribution function, which is the conceptual link between probabilities in the sample space S and probabilities related to the real line R. In Section 3.4 we present the probability density function—the centerpiece of the subject of random variables and a panacea to all the problems we raised in Chapter 2. Sections 3.1–3.4 are tilted toward mathematical rigor; they may feel a "bit heavy" at first reading. Section 3.5 is short, but we want to emphasize that *from now on, probability density functions will be the probabilistic models of a random experiment*. We can then forget about (S, Q, P), especially the nasty Q. In Section 3.6 we present the

most widely used probability density function models; the ideas of this section
are one of the things you should get out of this book. In Section 3.7 we address
a practical issue, namely, how one can approximate a model from given measurements. In Section 3.8 we investigate how the probability density function
of a random variable changes when the random variable itself is transformed.
In Section 3.9 we introduce the concept of moments of a random variable, and
in particular means and variances; they are very useful in practice. In Section
3.10 we discuss failure rates, a fundamental concept for reliability analysis. In
Section 3.11 we introduce a number of transforms of random variables (not
to be confused with transformations of random variables); they are very powerful theoretical tools. In Section 3.12 we introduce a number of inequalities
regarding "tail probabilities"; these inequalities are calculated via reduced information only (means, variances, etc.). Finally, in Section 3.13 we present a
method to generate values for a random variable; this is very useful in computer
simulations of systems with random variables as inputs.

3.1   SETS OF THE FORM $(-\infty, x]$

3.1.1   From Set Functions to Functions on R

We argued in Section 2.10 that dealing with a function on sets is quite difficult;
the classical tools from calculus (such as integrals and derivatives) cannot be
applied to functions like our probability $P[\cdot]$. It should be apparent then that
we need to escape from a set function and move into a more familiar function,
say, one on the real line R. To do that, we need a mapping (i.e., a function) from
S to R. This function should have a property: Sets in Q should be represented
easily by "something" on the real line R.

Of course, an arbitrary set $A \in Q$ will be mapped onto an arbitrary set
on the real line R. What is then an easy representation of sets in R? This
is where sets of the form $(-\infty, x]$ come into play. We argue that these sets
can represent any[1] subset of the real line R, via set operations such as unions,
intersections, complements, and limits. And the good thing about sets of the
form $(-\infty, x]$ is that they can be represented by a *single argument* x. So, if we
assign probabilities to them, all we have to specify is a function $F(\cdot)$ of a single
argument only, *as opposed to a function on arbitrary sets*, like $P[A]$.

3.1.2   Representation of Subsets of R via $(-\infty, x]$

Here are some examples; you may wish to consult Appendix A first.

---

[1] Later we retract that statement as false.

(a) A semiclosed interval.

(b) The set $\{a\}$.

Figure 3.1

Example 3.1. How can we represent an interval like $(a, b]$, where a, b are given real numbers, with sets of the form $(-\infty, x]$? From Figure 3.1(a) we can see that

$$(a, b] = (-\infty, b] - (-\infty, a] = (-\infty, b] \cap (-\infty, a]^c ,$$

and thus all we need is an intersection and a complement.

Example 3.2. How about a single point $\{a\}$? From Figure 3.1(b) we can see that

$$\{a\} = \lim_{n \to \infty} \left(a - \frac{1}{n}, a + \frac{1}{n} \right].$$

Example 3.3. How about the interval $[a, b]$? That should be easy now.

$$[a, b] = (a, b] \cup \{a\} = [(-\infty, b] \cap (-\infty, a]^c] \cup \left(\lim_{n \to \infty} \left(a - \frac{1}{n}, a + \frac{1}{n} \right] \right).$$

Example 3.4. Here is the representation of $[a, b)$ and (a, b):

$$[a, b) = [a, b] - \{b\} = [a, b] \cap \{b\}^c, \tag{3.1}$$

and we have representations for the last two sets already. Similarly,

$$(a, b) = (a, b] - \{b\} = (a, b] \cap \{b\}^c.$$

Example 3.5. Finally, the semi-infinite intervals $(-\infty, b)$, $[a, \infty)$, and (a, ∞) are also easily represented as

$$
\begin{aligned}
(-\infty, b) &= (-\infty, b] \cap \{b\}^c, \\
[a, \infty) &= \{a\} \cup (-\infty, a]^c, \\
(a, \infty) &= (-\infty, a]^c .
\end{aligned}
$$

So, any union, intersection, complement, or limit of intervals of any kind (open, closed, semiopen, etc.) can be represented via sets of the form $(-\infty, x]$. Are there subsets of the real line R that are *not* representable by such operations on sets of the form $(-\infty, x]$? It is a deep mathematical result of a related branch of mathematics, namely, measure theory (see, for example, [13]) that *indeed there are[2] such sets!* We say then that such operations can represent *almost any* subset of R because nonrepresentable sets are not "that many." It is so difficult to describe such sets that we can feel sure we will never see them in engineering experiments such as *netstat*, quantizations, and *ftp*. We leave the worries to the pure mathematician in us.

The subsets of the real line that can be represented via sets of the form $(-\infty, x]$ are called *Borel sets*.[3] The collection of such subsets is denoted by $\mathcal{B}(R)$ and called the Borel σ field or σ algebra on the real line R.

So, in summary, we need a function that will associate events in Q with Borel sets in $\mathcal{B}(R)$. We develop this function next. As you may have anticipated, this function will be our *random variable*.

3.2   CONCEPT OF A RANDOM VARIABLE

Consider a random experiment E with model (S, Q, P).

Definition: A **random variable** X is a mapping from S onto R such that sets in $\mathcal{B}(R)$ are mapped on sets in Q.

Therefore, $X : S \;\rightarrow\; R$ or, equivalently, for every outcome $\zeta \in S$,

$$X : \zeta \;\rightarrow\; X(\zeta).$$

The subset of R that consists of all values $X(\zeta)$ is called the **range** of X and is usually denoted as S_X. Note that, by definition, $X(\zeta) = \pm\infty$ is not allowed.

Remark 1:   The concept of a random variable has been used almost since the inception of the theory of probability; for example, Laplace used it in numerous problems in his studies. However, he did not name it as such. Kolmogorov formally and rigorously introduced this concept in 1933.

Remark 2: Strictly speaking, X is neither random nor a variable; it is simply a mapping. The strange name is simply a reminder of the randomness associated with the elements of the sample space S.

3.2.1   Measurable Functions

Are *all* functions $f : S \rightarrow R$ random variables? No, as the following examples demonstrate.

---

[2] We now retract our previous false statement.
[3] After Emil Borel, 1871–1956, a famous French mathematician.

Example 3.6. Consider any random experiment with a finite sample space, say, $S = \{a_1, \ldots, a_n\}$. Suppose that $Q = \{\emptyset, S\}$. Let

$$f(a_i) = i, \quad \forall a_i \in S.$$

Since $\{a_i\}$, the inverse image of the set $\{i\}$, is *not* an element of Q, f cannot be a random variable.

Example 3.7. Let $S, f(\cdot)$ be as in the previous example. Suppose that

$$Q = \{\emptyset, S, \{a_1\}, \{a_2, \ldots, a_n\}\}.$$

The inverse image of the set $(-\infty, 1]$ is $\{a_1\}$, and it belongs to Q. However, the inverse image of the set $(-\infty, 2]$ is $\{a_1, a_2\}$, and it does not belong to Q. So f cannot be a random variable.

The functions that map subsets of $\mathcal{B}(R)$ onto subsets in Q in general are called **measurable**. When $Q = 2^{|S|}$, obviously all functions $f : S \rightarrow R$ will be random variables. Since we can (and usually do) take $Q = 2^{|S|}$, whenever S is countable, all functions on such a sample space will be random variables. *The subtlety arises only when S is uncountable*, in which case $Q \neq 2^{|S|}$, by necessity. A detailed handling of measurable functions is beyond the scope of this book. Suffice it to say here that sums, products, and limits of measurable functions are also measurable. For a good exposition, see [25] and [26].

Remark: In the early stages of probability theory, back in the seventeenth through nineteenth centuries, most applications of interest were coin flippings, other games of chance, genetics, and, in general, experiments with sample spaces that were countable. There was no need to rigorously define the concept of a random variable since with the powerset of S as the collection of events Q, *all* functions would be random variables by default! With an arbitrary set as Q, only a subset of all possible functions will be random variables. To avoid this technicality, we can *define* Q to be the collection of images of sets in $\mathcal{B}(R)$. Then *all* functions are guaranteed to be random variables! For conceptual convenience, in the sequel we assume that this is the case.

Let's see now how we can define some random variables and what their ranges are, using the examples in Section 1.9.

3.2.2   Examples of Random Variables from Section 1.9

In all the following examples, we define a single random variable on a random experiment; we show how one can choose the sample space S, and then we determine S_X, the range of X.

Example 3.8. *GPA.* Let's start with with a familiar, hopefully painless case, the college student grade point average (GPA). In this example, let's take as the sample space S the set of all students in the United States. The random experiment is then "a student is selected and his or her GPA, call it X, is observed."[4] What is the range of the random variable X? Some schools compute grades with two decimal points, others with three. Some give extra points for an A^+, so a student's grade can *exceed* the "natural" 4.0. To play it safe, let's take as the range of X

$$S_X = \{0.000, 0.001, 0.002, 0.003, \ldots, 4.000, 4.001, 4.002, \ldots, 10.000\}.$$

Notice that we can also choose a continuous set, instead of the discrete, more natural one; for example, we can take

$$S_X = [0, 10].$$

In both choices, there are of course elements $\zeta \in S_X$ with zero probability.

Example 3.9. *Boyfriend/Girlfriend.* Suppose we observe the behavior of a person who is acting as a boyfriend or girlfriend of another person. In this random experiment, I propose to take the set

$$S = \{\text{entire population in the United States}\}$$

as the sample space. (What are other options? Is my choice restrictive?) The random variable of interest is denoted as B, the "behavior" of a boyfriend or girlfriend. How do we define it? Let's classify behavior as {good, bad, tolerable, obnoxious, excellent}. Remember that for B to be a random variable, it must map S onto R, not onto nonnumeric elements like "good." Let's translate, then, the behavior types into some numbers, say, $\{3, 1, 2, 0, 4\}$. The choice is arbitrary; we could have chosen differently. Then B is the mapping shown in Figure 3.2. Of course, the range is $S_B = \{0, 1, 2, 3, 4\}$.

Example 3.10. *File Size.* Consider a given computer, say, the author's machine ececho.ncsu.edu at NCSU (internet address 152.1.59.96). The users of this machine had about 36,000 files as of April 1994. Let's consider the random experiment "the size of a file is measured." Then $S = \{$all files in ececho.ncsu.edu$\}$. The random variable F, when measured in bytes, will take nonnegative integer values (yes, there are files with 0 bytes). Since we do not know the maximum possible file size, we can take

$$S_F = \{0, 1, 2, \ldots, \infty\}.$$

Of course, you may argue that $S_F = \{0, 1, 2, \ldots, 10^{11}\}$, (100 Gbytes), and $S_F = [0, 10^{11}]$ are also good choices.

---

[4]Notice that when Q is selected as mentioned in Section 3.2.1, any function will be a random variable.

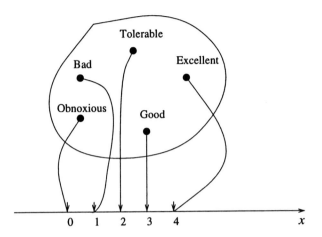

Figure 3.2 Behavior mapping.

Example 3.11. *Number of Packets at a Network Interface.* Consider a workstation connected to the rest of the world via a network interface. Through this interface, you can send and receive information, e.g., transfer files, send e-mail messages, read news groups, and so on. Information is transferred in *packets*, usually less than 10 kilobytes in size. The number of packets a workstation sends or receives, in a unit of time, say, 1 s, is random.

Consider the random experiment in which we count the number N of packets a workstation receives in a 1-s interval. (What is S in this example?) The Unix command *netstat* 1 gives us information about this count. What is the range of N, our random variable? Remember, we count the number of packets in a 1-s interval. What is the maximum number of packets our workstation can receive? If the interface receives packets of the smallest size s (in bits) at the full speed c of the interface (in bits per second), we know that the maximum such number is c/s. Therefore,

$$S_N = \{0, 1, 2, \ldots, \frac{c}{s}\}.$$

I will also suggest two alternatives, $S_N = [0, \infty)$ and $S_N = \{0, 1, 2, \ldots, \infty\}$. Both alternatives will work when the speed of the interface changes, e.g., when the station changes from an Ethernet connection to a Token Ring or to an FDDI (fiber distributed data interface) one. The advantage of the alternate suggestions is that we do not have to change our models!

Example 3.12. *Log-In Duration.* Consider a workstation; the amount of time a user spends from log-in time until log-out is called the *log-in duration*.

The log-in duration varies from user to user (hackers tend to spend a precious percentage of their lives in front of a terminal, computer illiterates log out by accident, etc.).

Let's take the sample space S of the users of a given computer. Then L, the log-in duration random variable, maps S onto the set of nonnegative real numbers:

$$L : S \to [0, \infty).$$

In reality, since the computer clock has a, say, 10-ns resolution, L will take values that are integer multiples of this resolution. Therefore, S_L will be a discrete, proper subset of $[0, \infty)$.

Example 3.13. *Disk Space Usage.* In multiuser computer systems, where a lot of users share the system resources (CPU, disks, network, software tools), *disk space allocation* is a challenging problem. There is a fixed amount of disk space in a computer system (2 gigabytes in the ececho.ncsu.edu machine); the task of the system administrator is to partition this space among a *random* number of users, who require a *random* amount of space. What is a good partitioning strategy that accommodates as big a number of users as possible?

To find such a strategy, one must study disk space usage patterns. Define, therefore, the following experiment. At the end of the day, the total amount of disk space used is observed. Equivalently, the amount of free space by the end of the day can be also observed. (The Unix command *"du -s"* gives you information about disk usage.)

What is S for this random experiment? For the ececho.ncsu.edu machine, I will take S as the set of all ECE students at NCSU. Then the random variable D will map user u onto a nonnegative integer. (The value 0 is allowed since not all ECE students have accounts on ececho.ncsu.edu.) What are the extreme values for D? Every user gets some system default files at user registration time, so the minimum value is *not* 0. In Unix, it will be at least 2048 bytes (your . and .. files) per user, so for D, the minimum value will be a multiple of that. As for the maximum, this will be the system disk capacity. So I will conveniently take

$$S_D = \{0, 1, 2, \ldots, \infty\},$$

as we discussed in Example 3.11.

Example 3.14. *Airline Counter Delays.* Everybody who is not afraid of flying *must* have experienced some delay in front of an airline counter. Even with nobody in front of you when you arrive, it still takes time to check in your baggage, get seat assignments, etc.

Let S be the set of the entire U.S. population. Let's consider the random experiment "the delay D of person ζ is measured at some day xxx." Then the random variable D can take any value from 0 (ζ did not fly that day) to

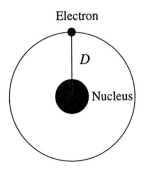

Figure 3.3 Electron orbits.

practically infinity (flight was canceled). So

$$S_D = [0, \infty)$$

is a convenient range.

Example 3.15. *Buffer Delay.* Delays are of interest as performance measures in computer systems and computer networks. The analog of an airline counter here is a *buffer*, i.e., computer memory. Consider, for example, file transfers over the Internet. Let's focus on a given Internet node, say, machine 139.91.1.1 (pythia.csi.forth.gr), which can be an intermediate node through which files may be transferred.

Let D be the random variable that denotes the buffer delay in this experiment. Then S can be taken as the set of all Internet users (humans plus software processes). Quite naturally,

$$S_D = [0, \infty)$$

is a convenient choice, even though this set contains elements that are larger than a human's lifetime!

Example 3.16. *Electron Position.* Consider the system in Figure 3.3. Many properties of the atom are determined by the characteristics of the electron movement around the nucleus. The celebrated Heisenberg principle states that it is not possible to determine *both* the position *and* the momentum of the electron with absolute accuracy. This makes observation of electron movements a random experiment.

Suppose we are interested in the distance D of the electron from the center of the nucleus. Distance D will be a random variable, according to Heisenberg's principle. It is measured in nanometers (nm). The range of D is then

$$S_D = [0, \infty).$$

What do you think is a good choice for the sample space S?

To summarize the main point in all these examples, the range of a random variable may not be chosen in a unique way. Let's proceed now to investigate how we can assign probabilities to events involving the random variable X.

3.3   CUMULATIVE DISTRIBUTION FUNCTION

3.3.1   Definition and Properties

Consider a probability model (S, Q, P) and a random variable X defined on S. For a fixed $x \in R$, define the set $A_x \in Q$:

$$A_x \triangleq \{\zeta \in S : -\infty < X(\zeta) \le x\}.$$

Another convenient notation for A_x, which we use frequently in the sequel, is

$$\{X \le x\} \triangleq \{\zeta \in S : -\infty < X(\zeta) \le x\}. \tag{3.2}$$

Therefore, $P[A_x] = P[\{X \le x\}] \triangleq P[X \le x]$, where we drop the braces for convenience in notation. We will see in Section 3.3.4 that assigning probabilities to the collection of events $\{A_x; x \in R\}$ is all we need to do to solve the initial assignment problem; any other probability of interest can be found via $P[A_x]$.

Definition: The **cumulative distribution function** (**cdf**) of a random variable X, denoted by $F_X(x)$, is defined as

$$F_X(x) \triangleq P[X \le x], \quad \forall x \in R. \tag{3.3}$$

Remark 1: The concept of the cdf was introduced by Kolmogorov in 1933.

Remark 2: We want to stress the fact that, on the left-hand side of Equation 3.3, we have a function of a single argument $x \in R$, while on the right-hand side we have $P[\cdot]$, a function of a set. The hope now is to express an arbitrary set A in Q in terms of sets of the form A_x and to calculate $P[A]$ via algebraic operations on $F_X(x)$.

We did not say yet *how we can calculate* the cdf for a given random experiment E and a given random variable X. We postpone that issue until Section 3.4. Let's investigate next some properties that Equation 3.3 implies.

Lemma 3.1 *(1)* $F_X(x)$ *is nonnegative; that is,*

$$F_X(x) \ge 0, \quad \forall x \in R. \tag{3.4}$$

(2) $F_X(x)$ *is bounded from above by 1; that is,*

$$F_X(x) \le 1, \quad \forall x \in R. \tag{3.5}$$

(3) $F_X(x)$ *is a* nondecreasing *function; i.e., for given* $x, y \in R$, *we have*

$$F_X(x) \le F_X(y), \quad when \quad x < y. \tag{3.6}$$

(4) $F_X(x)$ *is* continuous from the right; *i.e., for any given* $a \in R$ *and any given sequence of real numbers* $\{\epsilon_n\}$ *that converges to 0 from above (that is,* $\lim_{n\to\infty} \epsilon_n = 0$, $\epsilon_n \ge 0$), *we have*

$$F_X(a) = \lim_{n\to\infty} F_X(a + \epsilon_n). \tag{3.7}$$

Proof: Nonnegativity follows directly from Axiom 1. The second property follows from Corollary 2.2. Equality in 3.5 is certainly achieved when $x = \infty$. Property 3.6 follows from Theorem 2.2 since the event $\{X \le x\}$ is a subset of the event $\{X \le y\}$ whenever $x < y$.

Continuity from the right is more difficult to prove rigorously. We provide a sketch of the proof here and refer the interested reader to [68] for a more complete proof.

Consider the event $A_n \triangleq \{a < X \le a + \epsilon_n\}$. We can easily check that

$$P[A_n] = F_X(a + \epsilon_n) - F_X(a).$$

We wish to show that $P[A_n] \to 0$ as $n \to \infty$. The sequence of sets $\{A_n\}$ is a decreasing sequence, with the limit of the sequence being the empty set. From Problem 2.53 (the axiom of continuity), it follows then that $P[A_n] \to 0$, which concludes the proof. $\square$

3.3.2 Examples and Classification of Random Variables

Figures 3.4(a) to 3.6(b) are examples of functions $F(x)$ that possess the properties presented in Lemma 3.1.

Based on their cdf, random variables can be classified as **continuous, discrete,** and **mixed.** A continuous random variable has a cdf that looks like those in Figure 3.4. Mathematically, a random variable is called continuous if and only if its cdf $F_X(x)$ is continuous for all $x \in R$. A discrete random variable has a cdf that looks like those in Figure 3.5; i.e., it has only "flat" regions. Mathematically, a random variable X is called discrete if and only if X maps S on a countable subset of R. The cdf can be written in this case as

$$F_X(x) = \sum_{k \le x} P[X = k],$$

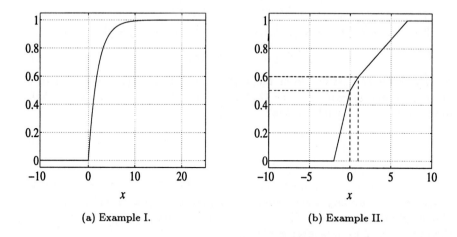

(a) Example I.   (b) Example II.

Figure 3.4   Continuous cdf's.

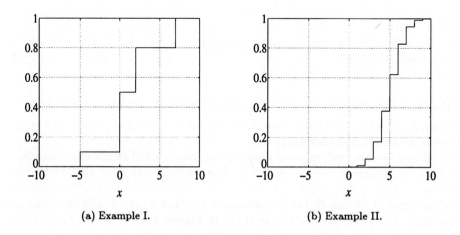

(a) Example I.   (b) Example II.

Figure 3.5   Discrete cdf's.

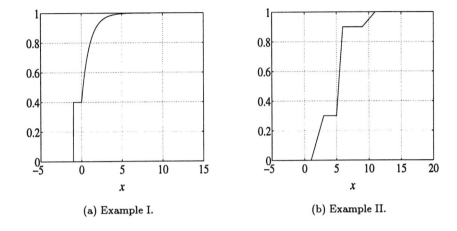

(a) Example I.   (b) Example II.

Figure 3.6   Mixed cdf's.

where k can take discrete values only. Of course, $P[X = k]$ is shorthand notation for the probability $P[\zeta \in S : X(\zeta) = k]$. The function $P[X = k]$ is called the **probability mass function** of X (**pmf** for short). It is equivalent to the cdf, in the sense that we can derive one from the other. As we see later, the pmf is easier to calculate than the cdf, and we take advantage of this fact when we compute it for a realistic experiment. Finally, a mixed random variable has a cdf that looks like those in Figure 3.6. Mathematically, a random variable is of the mixed type if and only if its cdf can be written as a sum of a continuous and a discrete random variable.

3.3.3   An Inverse Problem

Suppose now that we are given an arbitrary function $F(x)$, $x \in R$, which satisfies the four Properties 3.4 to 3.7 of Lemma 3.1. Is there a random experiment E with a model (S, Q, P) and a random variable X with $F(x)$ as its cdf $F_X(x)$? The answer is yes, as might be expected. A proof that utilizes $S = [0, 1]$ as the sample space and P a "uniform" probability on all outcomes of S is given in Section 3.13. We do not pursue this matter further here. The important point is that we can be sure that statements of the form "Suppose that $F(x)$ is the cdf of some random variable" make mathematical sense.

3.3.4   Calculation of Probabilities Through the CDF

Let's see now how the cdf can be used to calculate probabilities of various sets.[5]
Here we extensively use the representations in Section 3.1.

Example 3.17. *Interval $(a, b]$.* Consider the set A described by

$$A \overset{\triangle}{=} \{\zeta \in S : a < X(\zeta) \leq b\} \overset{\triangle}{=} \{a < X \leq b\}. \tag{3.8}$$

Using the notation we introduced in Equation 3.2, we have

$$A = \{X \leq b\} - \{X \leq a\},$$

and thus

$$\{X \leq b\} = A \cup \{X \leq a\}.$$

Then since the sets on the right-hand side are disjoint, from Axiom 3a we get

$$P[a < X \leq b] = F_X(b) - F_X(a). \tag{3.9}$$

Example 3.18.   *Interval $[a, b]$.* Consider now the event $B \overset{\triangle}{=} \{a \leq X \leq b\}$.
Observe that $B = A \cup \{X = a\}$, where A is the set defined in Equation 3.8;
also observe that the two sets on the right-hand side are disjoint. Then, from
Axiom 3a, we get immediately that

$$\begin{aligned} P[a \leq X \leq b] &= P[a < X \leq b] + P[X = a] \\ &= F_X(b) - F_X(a) + P[X = a]. \end{aligned} \tag{3.10}$$

The probability $P[X = a]$ can be readily expressed in terms of the cdf. Using
the result of Example 3.2, we can see that Equation 3.9 reduces to

$$P[X = a] = F_X(a^+) - F_X(a^-).$$

(By convention $a^- < a < a^+$ and a^-, a^+ are arbitrarily close.)

Example 3.19. *Interval $[a, b)$.* Let $C \overset{\triangle}{=} \{a \leq X < b\}$. Easily now

$$C = (A - \{X = b\}) \cup \{X = a\},$$

so from Axiom 3a and Equation 3.9, we get

$$P[a \leq X < b] = F_X(b) - F_X(a) + P[X = a] - P[X = b]. \tag{3.11}$$

---

[5] We still have not said how one can calculate $F_X(x)$.

Example 3.20. *Interval* (a, b). Let $D \triangleq \{a < X < b\}$. We have $D = A - \{X = b\}$, so

$$P[a < X < b] = F_X(b) - F_X(a) - P[X = b]. \qquad (3.12)$$

Example 3.21. *Interval* (a, ∞). Let $E \triangleq \{a < X\}$. Since $E = \{X \le a\}^c$, from Theorem 2.1 we get

$$P[a < X] = 1 - F_X(a). \qquad (3.13)$$

Probabilities of this form, for random variables that take only positive values, are called *tail probabilities*. We see some interesting bounds for tail probabilities in Section 3.12.

Let's group Equations 3.9 through 3.13 in a lemma.

Lemma 3.2

$$
\begin{aligned}
P[a < X \le b] &= F_X(b) - F_X(a), \\
P[a \le X \le b] &= F_X(b) - F_X(a) + P[X = a], \\
P[a \le X < b] &= F_X(b) - F_X(a) + P[X = a] - P[X = b], \\
P[a < X < b] &= F_X(b) - F_X(a) - P[X = b], \\
P[a < X] &= 1 - F_X(a).
\end{aligned}
$$

Of course, whenever it so happens that $P[X = b] = 0$ or $P[X = a] = 0$, the expressions in Lemma 3.2 can be appropriately simplified.

To summarize, we developed a methodology to calculate the probability of an arbitrary event $A' \in Q$, at least in principle: Using a random variable X, we can map A' on some subset A in R. Then we can use Lemma 3.2 and the three axioms to calculate $P[A]$. *The only thing we need* is the assignment for events $P[X \le x]$ or, in other words, the cdf $F_X(x)$.

3.3.5 Conditional CDF

Consider a model (S, Q, P), an arbitrary set $A \in Q$, and a random variable X defined on S. Since the function $F_X(x)$ is the probability of the event $B \triangleq \{X \le x\}$, conditional probabilities of the form $P[B|A]$ make sense.

Definition: The **conditional cumulative distribution function** of a random variable X, given an event A such that $P[A] > 0$, is defined as

$$F_X(x|A) \triangleq \frac{P[\{X \le x\} \cap A]}{P[A]}. \qquad (3.14)$$

Notice that the conditioning event A may or may not be related to the random variable X. The usefulness of the concept of conditional cdf's becomes apparent in Chapter 5.

3.3.6   State of Affairs So Far

Let's summarize what we have developed so far and what questions are still unanswered.

1. We have been able to reduce the probability assignment problem from working with Q to specifying a function $F_X(x)$ of a single argument only. This is a big reduction in complexity, we must admit.

2. We have seen how $F_X(x)$ can be used to calculate probabilities of intervals and their unions, complements etc. Along with Axiom 3 and the theorems and corollaries in Chapter 2, this would suffice to calculate a lot of probabilities.

3. We still do not know how to calculate the cdf itself.

4. Last but not least, even if we calculate one, how do we check that it satisfies Axiom 3?

We consider point 3 in Section 3.7; we answer question 4 in Section 3.4.5.

3.4   PROBABILITY DENSITY FUNCTION

3.4.1   Definition and Properties

Consider again a probability model (S, Q, P) and a random variable X defined on S, with cdf $F_X(x)$.

Definition: The **probability density function (pdf)** of a random variable X, denoted by $f_X(x)$, is defined as the derivative of $F_X(x)$; that is,

$$f_X(x) \triangleq \frac{dF_X(x)}{dx}, \quad \forall x \in R. \tag{3.15}$$

Remark 1: Kolmogorov formally introduced the notion of a pdf in [45]. The notion had appeared as early as the second half of the eighteenth century; Lagrange used it in his treatment of *continuous* errors, in the sense of Equation 3.20, in Section 3.4.7, even though he did not name it as such. Laplace also used it in his studies of error minimization.

Remark 2: The definition makes sense when $F_X(x)$ is a continuous function since the derivative is then well defined for all $x \in R$. When the cdf has a countable number of discontinuities, the definition still makes sense if one considers $\delta(x)$ (Dirac[6] or delta) functions at the points of discontinuity.

---

[6] After the English physicist Paul Dirac, 1902–1984, who introduced them.

The pdf possesses a number of properties that are a direct consequence of Equation 3.15:

Lemma 3.3 *(1)* $f_X(x)$ *is a nonnegative function; that is,*

$$f_X(x) \geq 0, \quad \forall x \in R. \tag{3.16}$$

(2) $f_X(x)$ *integrates to 1; that is,*

$$\int_{-\infty}^{\infty} f_X(x)\,dx = 1. \tag{3.17}$$

Proof: Nonnegativity follows directly from the fact that $F_X(x)$ is a nondecreasing function of x. The second property follows from Axiom 2 since

$$\int_{-\infty}^{y} f_X(x)\,dx = F_X(y)$$

and as $y \to \infty$ the probability on the right-hand side approaches 1. □

3.4.2 Calculation of Probabilities Through the PDF

Properties 3.9–3.13 have direct analogs in terms of pdf's; the proof of the following lemma is a direct consequence of Equation 3.15, so it is omitted. Recall that for any real number a, $a^- < a < a^+$, with a^-, a^+ arbitrarily close to a.

Lemma 3.4

$$P[a < X \leq b] = \int_{a+}^{b+} f_X(x)\,dx,$$

$$P[a \leq X \leq b] = \int_{a-}^{b+} f_X(x)\,dx,$$

$$P[a \leq X < b] = \int_{a-}^{b-} f_X(x)\,dx,$$

$$P[a < X < b] = \int_{a+}^{b-} f_X(x)\,dx,$$

$$P[a < X] = \int_{a+}^{\infty} f_X(x)\,dx.$$

Remark: Be careful in interpreting the limits of integration in the above expressions since, in general, delta functions may be involved. For example, in the first integral, a delta function at b should be included in the probability calculation, but a delta function at a should not.

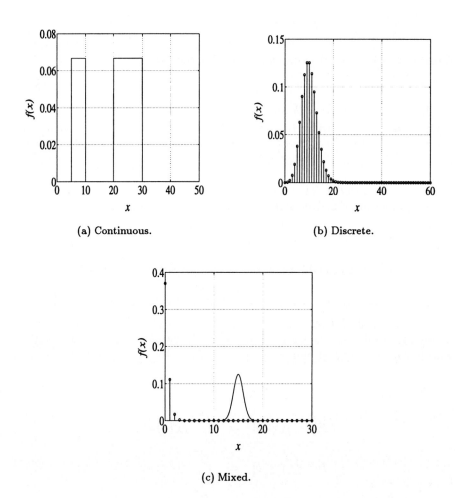

(a) Continuous.

(b) Discrete.

(c) Mixed.

Figure 3.7   Potential pdf's.

3.4.3   Examples of PDF

Some examples of functions that possess such properties are given in Figure 3.7; note that $f_X(x)$ may exceed 1, unlike the cdf, which is always bounded from above by 1. The circles in these plots represent delta functions.

3.4.4 The Inverse Problem

Suppose now that we are given a function $f(x)$ with the two properties of Lemma 3.3. Is there a random variable X with $f(x)$ as its pdf $f_X(x)$? The answer is yes, since from the pdf $f(x)$ we can uniquely compute a cdf.

3.4.5 How Does the PDF Satisfy Axiom 3?

The pdf of a random variable X satisfies Axiom 3 **automatically** *since by its definition, the integral is an additive function.*[7]

Remark: Additivity is easy to see when the sets A, B in Equation 2.3 are disjoint *intervals of nonzero length.* Then easily

$$\int_{A \cup B} f_X(x)\, dx = \int_A f_X(x)\, dx + \int_B f_X(x)\, dx. \tag{3.18}$$

Now any combination of (countable) unions of disjoint intervals $\{A_n\}$ can be represented as a single interval A, to which Equation 3.18 would apply directly.[8]

3.4.6 Conditional PDF

Consider a model (S, Q, P), an arbitrary set $A \in Q$, and a random variable X defined on S.

Definition: The **conditional probability density function** of a random variable X, given an event A such that $P[A] > 0$, is defined as

$$f_X(x|A) \triangleq \frac{dF_X(x|A)}{dx}. \tag{3.19}$$

The usefulness of the concept of conditional pdf's becomes apparent in Chapter 5.

3.4.7 How to Calculate the PDF for a Given Experiment

Now that we have settled all the mathematical worries, it is time to take care of the engineering ones. Let's see how we can apply all these concepts in engineering scenarios.

Who gives us the pdf in practice? From its definition as a derivative, we may consider the following interpretation (for small dx):

$$f_X(x) \quad \approx \quad \frac{P[X \le x + dx] - P[X \le x]}{dx},$$

---

[7] Is it kind of disappointing to have such an easy answer, after so much agonizing? How does this answer apply to the *cdf?*

[8] Mathematicians' alert! What happens if A or B in $\mathcal{B}(R)$ *cannot* be represented through intervals of nonzero length? In this case, we must interpret the integral in Equation 3.18 in the Lebesgue sense; this is beyond the scope of this book, and we will not pursue it further here.

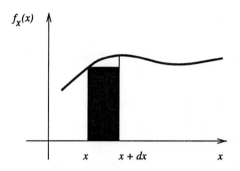

Figure 3.8   Probability density function as a histogram.

$$f_X(x)\,dx \quad \approx \quad P[x < X \le x + dx]. \tag{3.20}$$

As Figure 3.8 denotes, the value of the pdf at a given point x specifies "how much probability mass" exists around that point, which implies that we can calculate $f_X(x)$ as a **histogram**. Histograms are "easily" calculated in most practical applications, as we will see shortly.   Now we present histograms from examples in Section 1.9. We present a general methodology on how to calculate them in Section 3.7.

3.4.8   Examples of Histogram Calculation, Real Experiments

Example 3.22. *Dow Jones Index.* Let X be the random variable that represents the closing value of the Dow Jones index in Section 1.9. We have already seen typical values for X. The histogram for those 1589 values is shown in Figure 3.9(a). A simple summation will give us the cdf shown in Figure 3.9(b) for the same random variable.

Example 3.23. *Netstat Experiment.* The histograms for the various random variables associated with this random experiment were calculated with almost 30,000 measurements.   Figure 3.10(a) shows the histogram for the random variable that measures the number of input packets. Figure 3.10(b) shows the histogram for the random variable that measures the log of the number of output packets. Figure 3.10(c) shows the histogram for the random variable that measures the number of errors in the input packets. Observe that errors are very infrequent; most of the time we get 0 errors as a measurement. In that experiment, we had one interval (only) with 4 errors, and a few with 1, 2, and 3, which explains why the x axis in the histogram extends to 4, even though nothing appears to be plotted. Figure 3.10(d) shows the histogram for the random variable that measures the number of collisions. The peculiar shape of

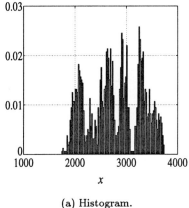

(a) Histogram.

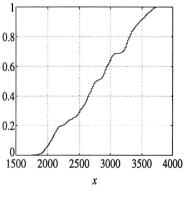

(b) Calculated cdf.

Figure 3.9 Dow Jones closing value.

this histogram can be explained as in the case of the input packet errors. Here, the maximum number of collisions observed was around 60.

Example 3.24. *File Transfer.* The histograms for the various random variables associated with this random experiment were calculated with approximately 400 measurements. Figure 3.11(a) shows the histogram for the delay random variable. Figure 3.11(b) shows the histogram for the throughput random variable. Figure 3.11(c) shows the histogram for the sizes of the files transferred.

Example 3.25. *File Sizes in Machine ececho.ncsu.edu.* This experiment was performed in April 1994, and 36,000 files were registered at that time. Figure 3.12(a) shows the histogram for the file sizes. In that histogram (and the remaining ones in this section), we present the values in the y axis not as probabilities, but as *counts*, as is common in practice. The histogram looks so strange because file sizes are as small as 0 bytes and as large as 7 million bytes. Therefore, a very large range must be plotted. Of course, in such cases, one must resort to logarithmic transformations, which reduce the range of the x axis. Figure 3.12(b) shows a more intelligible plot, in which logarithms are base 10. We will return to this point in Section 3.8. (To avoid taking the log of 0, 1 byte was added to all file sizes.)

Example 3.26. *Netstat from Paris.* To investigate how different machines are loaded, we performed the *netstat* experiment in Paris in December 1994, while this book was being written. We obtained about 20,000 measurements,

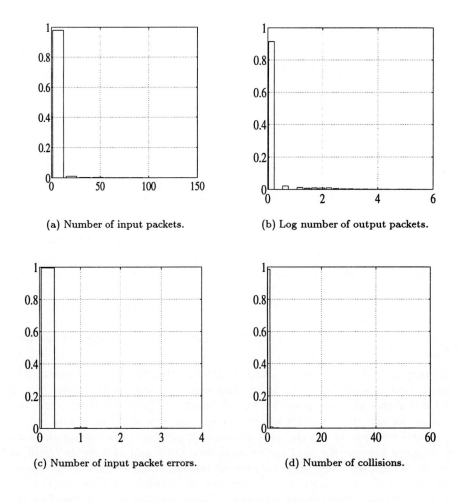

(a) Number of input packets.

(b) Log number of output packets.

(c) Number of input packet errors.

(d) Number of collisions.

Figure 3.10   *Netstat* experiment.

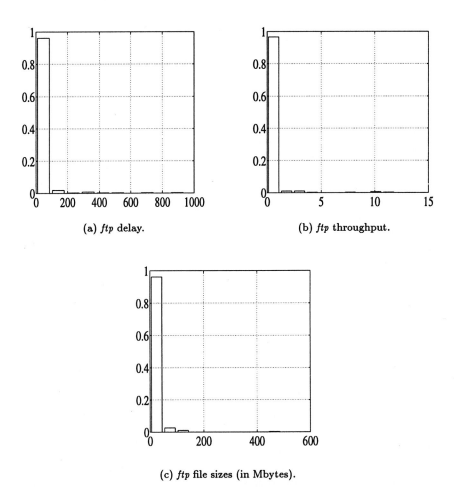

(a) *ftp* delay.

(b) *ftp* throughput.

(c) *ftp* file sizes (in Mbytes).

Figure 3.11   FTP experiment.

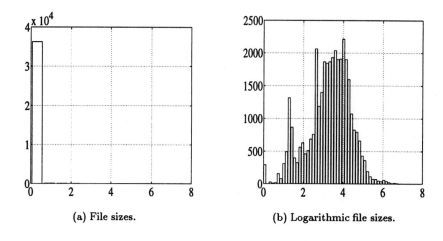

(a) File sizes.

(b) Logarithmic file sizes.

Figure 3.12   File size experiment.

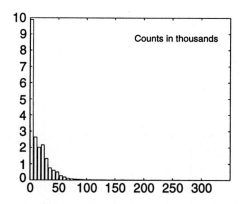

Figure 3.13   Histogram of the Paris *netstat* experiment.

on machine daphne.ibp.fr (Internet address 132.227.61.5). Figure 3.13 shows the histogram for the number of input packets.

Example 3.27. *Quantized Sounds.* On certain workstations, MATLAB provides some sound capabilities. Among the sound files we experimented with, on SUN workstations, the most interesting ones (from the graphics point of view) were the sound an egg makes when it is dropped on the floor and Handel's

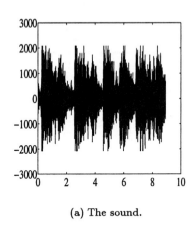

(a) The sound.

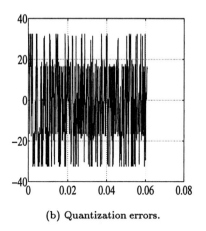

(b) Quantization errors.

Figure 3.14 Handel's *Messiah*.

Messiah. The first 10 s of Handel's *Messiah* is shown in Figure 3.14(a). (Can you tell how many hallelujahs there are?)

When a sound signal (a continuously valued quantity) is stored in computer memory, it must be converted into a discrete valued quantity; as we say, it must be quantized. (We discuss quantizers in greater detail in Section 3.8.) Quantization introduces errors. Figure 3.14(b) displays a few of the errors when a six-level quantizer is used. The histograms of these errors are shown in Figure 3.15(a) for Handel's *Messiah* and Figure 3.15(b) for the egg drop sound (where a two-level quantizer was used).

Example 3.28. *Coin Flip.* I actually had the patience to flip a coin 200 times and record the outcomes. The random variable that represents this experiment takes the value 0 for the outcome "tails" and 1 for the outcome "heads." The histogram is shown in Figure 3.16. We explain why the estimated probability of heads is so close to the expected 0.5 value in Section 6.6.

Example 3.29. *Buffer Size.* This experiment was simulated. The contents of the buffer in Example 1.28 were measured, as Figure 3.17(a) shows. The histograms were calculated for 30,000 and 50,000 measurements. No major differences were noticed, so we present in Figure 3.17(b) the results for 30,000 measurements only.

Example 3.30. *Uniform Random Variable.* This experiment simulated a *uniform* random variable, with parameters 0 and 1, which we discuss in Section 3.6. The theoretical shape of the histogram should be a flat line over the

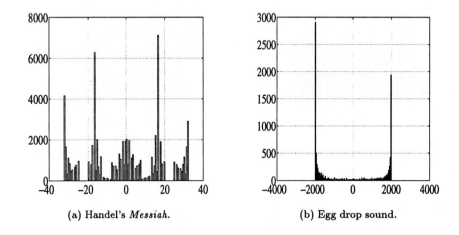

(a) Handel's *Messiah*.   (b) Egg drop sound.

Figure 3.15   Quantization error histograms.

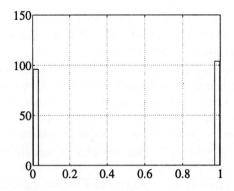

Figure 3.16   Histogram of a coin flip.

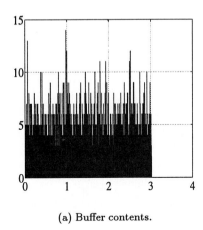
(a) Buffer contents.

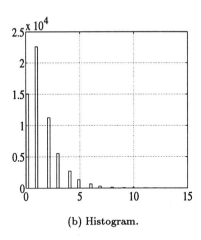
(b) Histogram.

Figure 3.17   Buffer size experiment.

interval $(0, 1)$. Figure 3.18 shows our results for the case of 1000 and 100,000 measurements. You can take this as an indication of how good the random number generator of the machine is (the simulation was performed on machine daphne.ibp.fr in Paris).

Example 3.31. *A Mixed Random Variable.* Finally, we simulated a *mixed* random variable, for the fun of it. Figure 3.19 presents the results for 50,000 measurements. After reading Section 3.6, can you tell what pdf these measurements came from? This question is a special case of the problem of *fitting densities to experimental data,* which we discuss in detail in Section 3.7.

3.4.9   State of Affairs So Far

Let's recap the theoretical developments we have achieved so far. With the concept of the random variable X and its pdf $f_X(x)$,

1. We are able to assign probabilities in a manner that satisfies Axiom 3.

2. We need to assign probabilities to only infinitesimal intervals $(x, x + dx]$.

The implication of point 1 is that we are not going to worry about rigor issues any more (enough is enough). The implication of point 2 is that calculating probabilities is made easy with pdf's: We always deal with assigning probabilities to infinitesimal intervals of the real line R. Of course, in practice, we approximate the density with a histogram; we assign probabilities to "small" in-

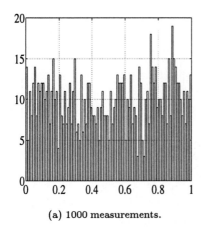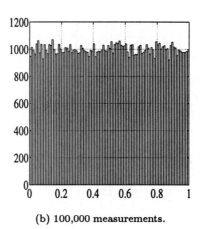

(a) 1000 measurements.   (b) 100,000 measurements.

Figure 3.18   Uniform random variable experiment.

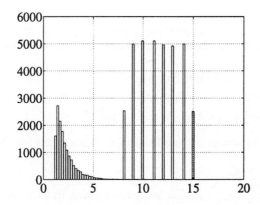

Figure 3.19   Histogram of a mixed random variable.

tervals and calculate such probabilities as relative frequencies (but more about that in Section 3.7).

3.5   PROBABILISTIC MODEL REVISITED

The discussions in Sections 3.2–3.4 were tilted toward rigor. Hopefully, the examples in Section 3.2.1 convinced you that unless we spend some effort on rigor, we are in danger of reaching silly conclusions. From now on, with the theoretical issues settled and put behind us, we will focus more on the application issues. It is a bit sad to leave all this mathematical apparatus behind us, especially after all the trouble we went through. But we must be "lean and mean" in order to play probability ball, i.e., confront the application troubles in front of us.

Consider a random experiment E with initial probability model (S, Q, P). Suppose we define a random variable X on S, with cdf $F_X(x)$ and pdf $f_X(x)$. Essentially, the random variable transforms (S, Q, P) into $[R, \mathcal{B}(R), F_X(x)]$ or, equivalently, into $[R, \mathcal{B}(R), f_X(x)]$. Notice that *regardless of the random experiment and the random variable X, the sample space R and the collection $\mathcal{B}(R)$ are the same for all experiments.* We can then keep them *implicitly* in mind and take the liberty of naming $f_X(x)$ [or $F_X(x)$] as the probabilistic model of a random experiment! This will suit us fine in experiments in which the sample space is nonunique or very difficult to define. Therefore, from now on, the function $f_X(x)$ or $F_X(x)$ *will be the probabilistic model of an experiment!* Let's look at some of the most frequently used models.

3.6   USEFUL RANDOM VARIABLE MODELS

From the applications' point of view, this section is the most important one in this chapter since it provides some of the most widely used models in practice. Make sure you understand how to use them, and apply at least one model in an actual project.

3.6.1   Why Mathematical Models?

The models in this section can be used to *hypothesize* about the pdf of a random experiment. In the design phase of an engineering system (e.g., a quantizer, a computer, a network node), such hypotheses are widely used for comparing alternative designs. (When reading Chapter 11, make sure you understand how such models have been used.) In such an early phase, system prototypes may not be available, so actual experimentation is not possible. Hypothesis is truly the only choice. The models may be also used when some experimental pdf (i.e., a histogram,) must be *interpolated*. For example, storing the entire histogram of an actual experiment is storage-consuming. An interpolation with a pdf model (an abstract function) can considerably reduce these storage requirements. Of course, the price we might pay is loss of accuracy.

We previously classified random variables as discrete, continuous, and mixed. We present the models according to the same classification. All random variable models presented are in terms of the pdf or pmf only. The corresponding cdf's are not known in closed form, in some cases; we ask you to identify cdf's in the problems. Almost all pdf's are expressed in terms of one or more parameters; these parameters can be adjusted to fit experimental data or hypotheses, a property that makes the models more versatile.

3.6.2   Discrete Random Variables

Since discrete random variables take a countable number of values $\{a_k\}$, their pdf is a sum of delta functions:

$$f_X(x) = \sum_{a_k} P[X = a_k]\delta(x - a_k).$$

The alternative presentation, through the pmf $P[X = a_k]$, is more conventional, and we follow it in this chapter.

1. *Bernoulli.*[9] The Bernoulli random variable X, with parameter p, assumes only two values, 0 and 1. Therefore,

$$S_X = \{0, 1\}.$$

Its pmf is given in terms of the parameter p as follows:

$$P[X = 0] = 1 - p, \quad P[X = 1] = p.$$

Using the mapping definition of a random variable, X can be viewed as Figure 3.20 suggests; i.e., all points in a set A, of probability p, are mapped onto the value 1, while all points of A^c are mapped onto 0. So the Bernoulli random variable is nothing else but the indicator function of an event of probability p (see Equation 2.14).

The most famous experiment modeled by a Bernoulli random variable is a coin flip, in which the probability of heads is equal to p. In engineering, this random variable has been applied to model arrivals in discrete time. Consider

---

[9] After James Bernoulli, 1654–1705, a Swiss mathematician and a prominent contributor to the early stages of probability theory.

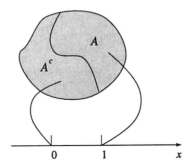

Figure 3.20   The Bernoulli random variable mapping.

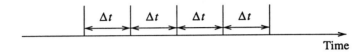

Figure 3.21   A slotted time system.

a "slotted time" system, as Figure 3.21 shows. Suppose that arrivals can occur during a time slot. If the slot size Δt is "small," then either 0 or 1 arrival can happen in any given slot. If the probability of one arrival in a slot is p, then the random variable which represents arrivals during a slot is Bernoulli.

2.   *Binomial.* The binomial random variable X with parameters n, p has a sample space

$$S_X = \{0, 1, 2, \ldots, n\}.$$

Here $p \in [0, 1]$ is a probability, and n is a positive integer. The pmf is

$$P[X = k] = \binom{n}{k} p^k (1 - p)^{n-k}, \quad k = 0, 1, \ldots, n, \tag{3.21}$$

where $\binom{n}{k} = n!/[k!(n-k)!]$ is the so-called binomial coefficient. Figure 3.22(a) shows Equation 3.21 for $n = 10, p = 0.3$. As we discuss in greater detail in Section 4.11, the binomial random variable arises naturally as a sum of n (independent)[10] Bernoulli random variables $\{X_k\}_{k=1}^n$, all with the same parameter p:

$$X = \sum_{k=1}^{n} X_k. \tag{3.22}$$

---

[10] We have not yet defined independence of random variables; see Section 4.7.

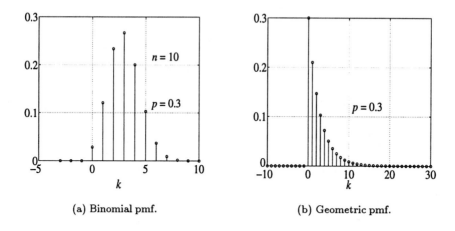

(a) Binomial pmf.   (b) Geometric pmf.

Figure 3.22   The Binomial and Geometric pmf.

In the system in Figure 3.21, a binomial random variable X represents the total number of arrivals during n slots. This random variable was known to Bernoulli; Pascal used the special case $p = 0.5$.

3. *Geometric.* The geometric random variable X, with parameter p, has a sample space

$$S_X = \{0, 1, 2, \ldots\},$$

i.e., all nonnegative integers. The pmf is given by

$$P[X = k] = p(1 - p)^k, \quad k = 0, 1, \ldots \tag{3.23}$$

Here $p \in [0, 1]$ is a probability. Figure 3.22(b) shows Equation 3.23 for $p = 0.3$.

4. *Negative binomial (or Pascal).* The negative binomial or Pascal[11] random variable X, with parameters r, p, has a sample space

$$S_X = \{r, r + 1, r + 2, \ldots\},$$

where r is a positive integer and p is a probability. The pmf is

$$P[X = k] = \binom{k-1}{r-1} p^r (1 - p)^{k-r}, \quad k = r, r + 1, \ldots \tag{3.24}$$

---

[11] After Blaise Pascal, 1623–1662, a French mathematician and the first scientist to report results on probability theory. This pmf was actually derived by Pierre Remond de Montmort, 1678–1719, another French mathematician, in 1713, as a solution to the problem of points (see Section 1.9).

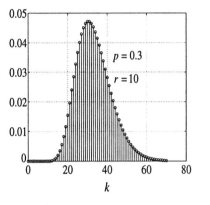
(a) Negative binomial pmf.

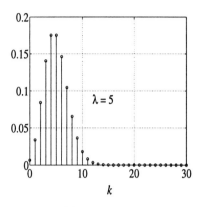
(b) Poisson pmf.

Figure 3.23 The Negative binomial and Poisson pmf.

Figure 3.23(a) shows Equation 3.24 for $r = 10, p = 0.3$.

Consider a sequence of (independent) Bernoulli random variables $\{X_k\}$. The negative binomial random variable represents the number of Bernoulli random variables we have to observe until $X_k = 1$ for the rth time. In the system in Figure 3.21, a negative binomial random variable X represents the number of slots that we wait until we see the rth arrival.

5. *Poisson.* The Poisson[12] random variable X, with parameter λ, has a sample space

$$S_X = \{0, 1, 2, \ldots\}.$$

Its pmf is given by

$$P[X = k] = \frac{\lambda^k}{k!} e^{-\lambda}, \quad k = 0, 1, 2, \ldots \tag{3.25}$$

where $\lambda > 0$ is a positive real number. Figure 3.23(b) shows Equation 3.25 for $\lambda = 5$. This random variable is the most widely used model for arrivals in queuing systems; we study it in great detail in Chapter 8, in connection with the Poisson random process.

6. *Discrete Uniform.* The discrete uniform random variable X, with parameters a, b, has a sample space

$$S_X = \{a, a+1, a+2, \ldots, b-1, b\}.$$

---

[12] Named after Simeon Denis Poisson, 1781–1840, a French mathematician.

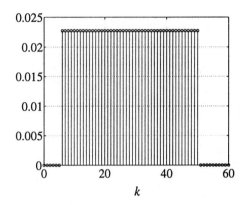

Figure 3.24   The discrete uniform pmf.

Its pmf is given by

$$P[X = k] = \frac{1}{b - a + 1}, \quad k = a, a + 1, \ldots, b. \tag{3.26}$$

Here a, b are integers with $a \leq b$. Figure 3.24 shows Equation 3.26 for $a = 6, b = 50$. This random variable models systems in which a finite number of outcomes are equiprobable; the most famous such system is the throw of a fair die, for which $a = 1, b = 6$. This model is frequently used in the absence of any information about the outcomes of a random experiment, in which case considering them all equiprobable would be a reasonable choice.

3.6.3   Continuous Random Variables

1.   *Uniform.* The uniform random variable X, with parameters a, b, has a sample space

$$S_X = [a, b].$$

Here a, b are real numbers with $a < b$. The pdf is given by

$$f_X(x) = \frac{1}{b - a}, \quad x \in [a, b]. \tag{3.27}$$

Figure 3.25(a) shows Equation 3.27 for $a = -2, b = 4$. This model plays the same role as its discrete counterpart for the case of sample spaces that are continuous.

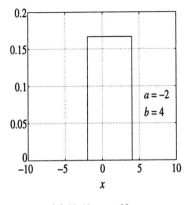

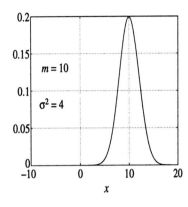

(a) Uniform pdf.   (b) Gaussian pdf.

Figure 3.25 The uniform and Gaussian pdf's.

2. *Gaussian (normal).* The Gaussian[13] random variable X, with parameters m, σ^2, has a sample space

$$S_X = (-\infty, \infty).$$

Here m is any real number, and σ^2 is any positive real number. The pdf is given by

$$f_X(x) = \frac{1}{\sqrt{2\pi}\sigma} e^{-\frac{(x-m)^2}{2\sigma^2}}, \quad x \in (-\infty, \infty). \tag{3.28}$$

Figure 3.25(b) shows Equation 3.28 for $m = 10, \sigma^2 = 4$. For a note about how this model was derived, see Section 5.6. This is the most widely used model for noise description in communication theory; it is also among the most widely used models in statistics and perhaps *the single most important model of a random variable*. This importance can be appreciated by the fact that the Gaussian pdf is the only one *in the world* to earn a position in a banknote (a German one—see the cover of this book). We have ample opportunity to deal with the model throughout the book.

Calculation (or, rather, approximation) of probabilities associated with the Gaussian pdf becomes a lot easier through the so-called Q function, which is defined as follows:

$$Q(x) \stackrel{\triangle}{=} \frac{1}{\sqrt{2\pi}} \int_x^{\infty} e^{-0.5s^2} ds.$$

---

[13] After Carl Gauss, 1777–1855, a German mathematician. The form of the density was first introduced by A. de Moivre, in 1733. Gauss derived it for the case of random errors in observations in 1809. The term *normal* was coined by H. Poincaré.

In other words, $Q(x)$ is a "tail probability." Properties of this function are discussed in the problems.

3. *Exponential.* The exponential random variable X, with parameter λ, has a sample space

$$S_X = [0, \infty).$$

Here λ is a real positive number. The pdf is given by

$$f_X(x) = \lambda e^{-\lambda x}, \quad x \in [0, \infty). \tag{3.29}$$

Figure 3.26(a) shows Equation 3.29 for $\lambda = 2$. This is a very widely used model in performance evaluation studies of computer and network systems (e.g., queuing theory); in that context, it is primarily used to model time durations (e.g., file transmission times). Its main power comes from the *memoryless property* (see Problem 3.118). As we see in Chapter 8, it is closely related to the Poisson process model.

4. *Cauchy.* The Cauchy[14] random variable X, with parameters a, b, has a sample space

$$S_X = (-\infty, \infty).$$

Here a, b are real numbers with $b > 0$. The pdf is given by

$$f_X(x) = \frac{1}{\pi} \frac{b}{(x-a)^2 + b^2}, \quad x \in (-\infty, \infty). \tag{3.30}$$

Figure 3.26(b) shows Equation 3.30 for $a = 1, b = 2$. This random variable is related to the Gaussian one; it is used in communication theory (see Problem 4.92).

5. *Laplace.* The Laplace[15] random variable X, with parameter a, has a sample space

$$S_X = (-\infty, \infty).$$

Here a is a real positive number. The pdf is given by

$$f_X(x) = \frac{a}{2} e^{-a|x|}, \quad x \in (-\infty, \infty). \tag{3.31}$$

Figure 3.27(a) shows Equation 3.31 for $a = 1$. Laplace introduced this random variable in the 1770s, in his effort to model errors in observations, in particular, astronomical observations. According to Laplace, this function captures two

---

[14] After Auguste Louis Cauchy, 1789–1857, a French mathematician.

[15] Named after Marquis Pierre Simon de Laplace, 1749–1827, a notable French mathematician and for many the greatest contributor to the field of probability in its early phase. Among others, the concept of generating functions and the "classical" definition of probability as a relative frequency are due to Laplace.

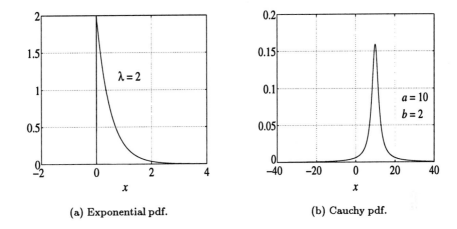

(a) Exponential pdf.   (b) Cauchy pdf.

Figure 3.26   The exponential and Cauchy pdf's.

essential properties of errors: (1) errors are symmetric around 0, and (2) small error values are more likely than large ones when the measuring equipment is "good." (See Section 5.6 for more details.)

6.   *Gamma.* The gamma random variable X, with parameters a, λ, has a sample space
$$S_X = [0, \infty).$$
The parameters a, λ are real positive numbers. The pdf is given by
$$f_X(x) = \frac{\lambda(\lambda x)^{a-1} e^{-\lambda x}}{\Gamma(a)}, \quad x \in [0, \infty), \qquad (3.32)$$
where
$$\Gamma(x) \overset{\Delta}{=} \int_0^\infty s^{x-1} e^{-s}\, ds,$$
is the so-called gamma function. Figure 3.27(b) shows Equation 3.32 for $a = 3.5, \lambda = 2$. The gamma random variable is widely used in statistics and queuing theory. It includes a number of other models as special cases (e.g., Erlang, χ^2).

7.   *Erlang.* The Erlang[16] random variable X, with parameters λ, n, is a special case of the gamma random variable when the parameter $a = n$, a positive

---

[16]Named after A. K. Erlang, 1878–1929, a Danish electrical engineer and a major contributor to early queuing theory.

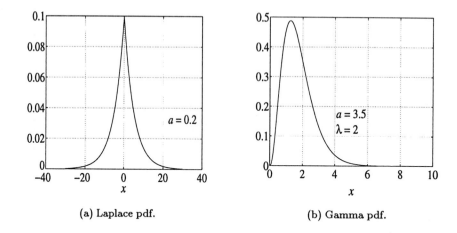

(a) Laplace pdf.   (b) Gamma pdf.

Figure 3.27   The Laplace and gamma pdf's.

integer. It therefore has a sample space

$$S_X = [0, \infty).$$

The pdf is given by

$$f_X(x) = \frac{\lambda(\lambda x)^{n-1}e^{-\lambda x}}{(n-1)!}, \quad x \in [0, \infty). \tag{3.33}$$

Figure 3.28(a) shows Equation 3.33 for $a = 10, \lambda = 2$. The Erlang random variable model is related to the exponential and Poisson models, as we see in Chapter 8.

8.   *Weibull*. The Weibull[17] random variable X, with parameters a, λ, has a sample space

$$S_X = [0, \infty).$$

Here a, λ are positive real numbers. The pdf is given by

$$f_X(x) = a\lambda^a x^{a-1}e^{-(\lambda x)^a}, \quad x \in [0, \infty). \tag{3.34}$$

Figure 3.28(b) shows Equation 3.34 for $a = 2, \lambda = 3$. The Weibull model is widely used in reliability studies because of its versatility in modeling failure rates (see Section 3.10).

---

[17]Named after E. H. W. Weibull, a Swedish physicist and contributor to the theory of reliability.

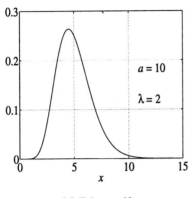

(a) Erlang pdf.

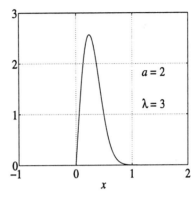

(b) Weibull pdf.

Figure 3.28 The Erlang and Weibull pdf's.

9. χ^2. The χ^2 (chi-squared) random variable X, with parameter a, has a sample space[18]

$$S_X = [0, \infty).$$

Here a is a positive real number. The pdf is given by

$$f_X(x) = \frac{1}{2^{0.5a}\Gamma(0.5a)}x^{0.5a-1}e^{-0.5x}, \quad x \in [0, \infty), \qquad (3.35)$$

where $\Gamma(x)$ is the gamma function. Figure 3.29(a) shows Equation 3.35 for $a = 5.5$. This model is very widely used in statistics. We encounter it in Section 3.7 when we discuss fitting histograms.

10. F. The F random variable X, with parameters (degrees of freedom) a, b, has a sample space[19]

$$S_X = [0, \infty).$$

Here a, b are real numbers. The pdf is given by

$$f_X(x) = \frac{\Gamma(\frac{a+b}{2})(\frac{a}{b})^{\frac{a}{2}}x^{\frac{a-2}{2}}}{\Gamma(\frac{a}{2})\Gamma(\frac{b}{2})(1 + \frac{a}{b}x)^{\frac{a+b}{2}}}, \quad x \in [0, \infty). \qquad (3.36)$$

Figure 3.29(b) shows Equation 3.36 for $a = 2, b = 1$. This model arises very frequently in statistics.

---

[18] The model was introduced by Karl Pearson, 1857–1936, a prominent English statistician.

[19] The model was proposed by G. W. Snedecor, a U.S. mathematician, in the early twentieth century. He named it F in honor of Sir Ronald Fisher, a prominent statistician.

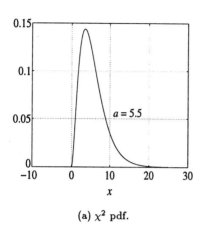

(a) χ^2 pdf.

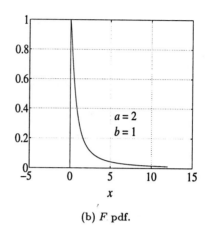

(b) F pdf.

Figure 3.29   The χ^2 and F pdf's.

11. *Student's t.* The Student's t^{20} random variable X, with parameter (or degrees of freedom) a, has a sample space

$$S_X = (-\infty, \infty).$$

Here a is a real number with $a > 0$. The pdf is given by

$$f_X(x) = \frac{\Gamma(\frac{a+1}{2})}{\sqrt{a\pi}\Gamma(\frac{a}{2})}\left(1 + \frac{x^2}{a}\right)^{-\frac{a+1}{2}}, \quad x \in (-\infty, \infty). \tag{3.37}$$

Figure 3.30(a) shows Equation 3.37 for $a = 2$. This model also arises very frequently in statistics. The Gaussian, χ^2, F, and Student's t models are the four most widely used models in statistics.

12. *Rice.* The Rice random variable X, with parameters m, σ^2, has a sample space

$$S_X = [0, \infty).$$

Here m, σ^2 are real numbers with $\sigma^2 > 0$. The pdf is given by

$$f_X(x) = \frac{x}{\sigma^2}e^{-\frac{1}{2}\frac{m^2+x^2}{\sigma^2}}I_0(\frac{mx}{\sigma^2}), \quad x \in [0, \infty). \tag{3.38}$$

---

[20] After W. S. Gossett, 1876–1937, an English mathematician who used the pseudonym "Student" for his papers, because his employer did not permit publication of research results.

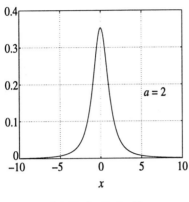

(a) Student's *t* pdf.

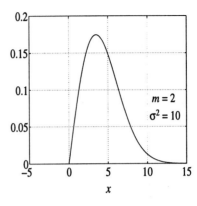

(b) Rice pdf.

Figure 3.30 The Student's *t* and Rice pdf's.

In Equation 3.38, $I_0(\cdot)$ is the zero-order, modified Bessel function of the first kind. Figure 3.30(b) shows Equation 3.38 for $m = 2, \sigma^2 = 10$. The Rice model was developed in the 1940s to study noise in communication channels (see Section 4.8.2).

13. *Rayleigh.* The Rayleigh[21] random variable X with parameter a has a sample space

$$S_X = [0, \infty).$$

Here a is a positive real number. The pdf is given by

$$f_X(x) = \frac{x}{a^2} e^{-\frac{x^2}{2a^2}}, \quad x \in [0, \infty). \tag{3.39}$$

Figure 3.31(a) shows Equation 3.39 for $a = 7$. This random variable is related to the Gaussian model; it was developed in the late nineteenth century in order to study statistical mechanics. It has also found application in communication theory (see Section 4.8.2).

14. *Pareto.* The Pareto random variable X, with parameters a, b, has a sample space

$$S_X = [b, \infty).$$

---

[21] Named after J. W. Rayleigh, 3d baron, 1842–1919, an English physicist.

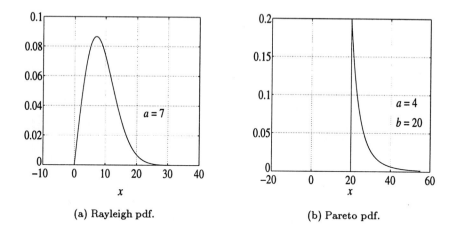

(a) Rayleigh pdf.   (b) Pareto pdf.

Figure 3.31   The Rayleigh and Pareto pdf's.

Here a, b are positive real numbers. The pdf is given by

$$f_X(x) = \frac{a}{b}\left(\frac{b}{x}\right)^{a+1} , \quad x \in [b, \infty). \tag{3.40}$$

Figure 3.31(b) shows Equation 3.40 for $a = 4, b = 20$. This random variable has found application in economics and reliability.

3.6.4   Mixture PDF

Consider a random experiment E with model $(S, \mathcal{Q}, P)$. Consider two random variables X_1 and X_2, defined on S, with pdf's $f_{X_1}(x)$ and $f_{X_2}(x)$, respectively. Observe that when $\alpha + \beta = 1$, for positive numbers α, β, the function

$$f(x) \overset{\triangle}{=} \alpha f_{X_1}(x) + \beta f_{X_2}(x) \tag{3.41}$$

is also a density function since it satisfies Equations 3.16 and 3.17. When the densities $f_{X_1}(x)$ and $f_{X_2}(x)$ are of different types, the density $f(x)$ given by Equation 3.41 is called a **mixture pdf**. Observe that Equation 3.41 does *not* imply that $X = \alpha X_1 + \beta X_2$!

The idea in Equation 3.41 can be generalized: Let $\{\alpha_i\}_{i=1}^n$ be a sequence of nonnegative real numbers such that $\sum_{i=1}^n \alpha_i = 1$. Consider a collection of pdf's $\{f_{X_i}(x),\ i = 1, 2, \ldots, n\}$, not all of them of the same type. Then

$$f(x) \triangleq \sum_{i=1}^{n} \alpha_i f_{X_i}(x)$$

is a mixture pdf.

Example 3.32. As an example, the histogram of the logarithm of the random variable X in the file size experiment [see Figure 3.12(b) in Section 3.4.8] can be fitted with a mixture pdf that contains a continuous and a discrete part. The discrete part could be placed at various locations; for example, there are three "spikes" (one at $x = 0$ that corresponds to about 300 measurements, one at about $x = 1.1$ that corresponds to about 1300 measurements, and one at about $x = 2.7$ that corresponds to about 2000 measurements). These spikes could be modeled with a discrete random variable that takes on the values $x = 0, 1.1, 2.7$. The rest of the histogram could be modeled with a continuous random variable.

3.7   HISTOGRAMS

Now let's look closer on how one can calculate pdf's (i.e., approximate them) from actual measurements of a random variable. Here is the framework: A random experiment E is repeated n times. The outcomes $\zeta_1, \ldots, \zeta_n$ are observed, and the values

$$X_1 \triangleq X(\zeta_1), \ldots, X_n \triangleq X(\zeta_n)$$

are measured. The problem we have then is the following:

Problem:[22] Suppose we have $n < \infty$ measurements of a random variable X, call them $\{X_i\}_{i=1}^{n}$. How do we find the pdf $f_X(x)$ of the random variable X?

As Equation 3.20 suggests, we can approximate the value of the pdf at any given point x with the probability that the random variable X lies in the interval $(x, x + dx]$. To connect the pdf with the measurements $\{X_i\}$, we can evaluate $P[x < X \leq x + dx]$ as a *relative frequency*, as discussed in Section 2.8. To do that, let $X_{\max}$ and $X_{\min}$ represent the largest and smallest values in the measurement set, i.e.,

$$X_{\max} \triangleq \max_{1 \leq i \leq n} X_i, \quad X_{\min} \triangleq \min_{1 \leq i \leq n} X_i.$$

Consider a partition of the range $[X_{\min}, X_{\max}]$ into K *bins* (subintervals) of equal length,[23] as Figure 3.32 suggests.

---

[22] It appears that a very similar problem was known to John Bernoulli, a relative of James Bernoulli, toward the end of the eighteenth century. His solution, however, was not rigorous.

[23] We may consider variable lengths as well. We choose equal lengths here for simplicity of presentation.

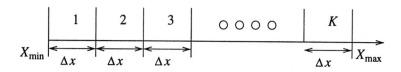

Figure 3.32 K bins of length Δx.

Given a desired number of bins K, we can calculate Δx as follows:

$$\Delta x = \frac{X_{\max} - X_{\min}}{K} \tag{3.42}$$

Inversely, given a desired length for Δx, we can calculate the necessary number of bins K as follows:

$$K = \frac{X_{\max} - X_{\min}}{\Delta x}.$$

With Δx fixed, let $n(x)$ denote the number of measurements such that

$$x < X_i \le x + \Delta x, \quad i = 1, \dots, n.$$

Then $n(x)/n$ is the desired relative frequency, and thus our approximation is

$$f_X(x) \approx \frac{n(x)}{n}, \quad x \in R. \tag{3.43}$$

Equation 3.43 is the basis for our algorithm for calculating the pdf of a random variable X in practice, from a set of measurements of X. A few questions arise here:

1. What is a good value for K or Δx?

2. What is a good value for n?

3. Is it true that the more measurements, the better? In other words, if we have two sets of measurements with n_1 and n_2 measurements in which $n_1 > n_2$, can we say that the histogram based on n_1 measurements is "closer" to the real pdf?

4. Given two sets of measurements that do *not* produce the same histograms, how do we judge what is a good fit?

5. How do we really *design* an experiment to measure the values of a random variable?

These are very important questions, so it is no surprise that a whole branch of the theory, called *statistics*, is devoted to their answers. We attempt to give some quick answers to questions 1 and 2 in this section. For a more in-depth analysis, see [22]. The answer to question 3 in general is no; the material in Section 6.6 is relevant and explains why this is so. For question 5, look at [46]. In general, for a good design, our measurements must be "independent." We introduce the concept of independence in Section 4.7, so we do not pursue this question here further, except to say the following. Consider an experiment that takes a lot of labor (like the *ftp* one). Suppose the person assigned to perform the measurements gets lazy one day and, instead of actually performing the experiment, simply copies a set of previous measurements. There is certainly a form of dependence now in the measurements, and intuitively this cannot be good. (If we stretch it to the limit, one can take only a single measurement and repeat it over and over!) If you think this is a far-fetched example, consider the scenario in which, in the *ftp* transfers, you always take measurements when the network is idle. This will give you the wrong idea about the quality of your transfers.

3.7.1 χ^2 Tests

How do we choose an appropriate value for K? *Inspection* is the first choice. We can try with a few bins first and gradually increase their number until no "noticeable" refinement is achieved. There are of course ways to formalize this intuitive idea [22]. A very popular test, which we can use to select K and at the same time to decide whether a suggested, theoretical pdf actually fits the histogram "suf-ciently" well, is the so-called χ^2 test. This test was developed in 1900 by K. Pearson.

This test works as follows. We select first a "level of significance" l, usually 1 percent or 5 percent. (Since it is unlikely that a theoretical pdf will fit the data exactly, the level of significance specifies how willing we are to accept errors. So 5 percent is stricter than 1 percent.) From the chosen level of significance and the χ^2 pdf (see the model in Section 3.6) we can choose a "threshold" t_l such that

$$P[E \geq t_l] = l, \tag{3.44}$$

where E is a random variable whose meaning will become apparent in a moment, from the description of the test. Figure 3.33 shows the probability in Equation 3.44 as the shaded area under the pdf curve (for $t_l = 20$, in this example). From this figure, we can see that as l decreases, t_l increases. We simply present the test here; for a proof that it actually works, see [22].

Let $\{X_i, i = 1, \ldots, n\}$ be the set of measurements we have from a random experiment. Let X be a random variable with pdf $f_X(x)$ and range S_X. We wish to test the hypothesis that the measurements actually came from a ran-

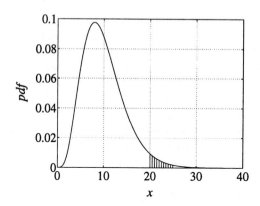

Figure 3.33   The probability $P[E \geq t_l] = l$.

dom variable with a pdf $f_X(x)$; that is, we want to fit $f_X(x)$ to the relative frequencies calculated according to Equation 3.43.

χ^2 **test:**

1. Choose a "level of significance" l.

2. Choose K, the number of bins.

3. Partition the range of the measurements into K bins $\{B_i\}$, where $B_1 = [X_{\min}, X_{\min} + \Delta x)$, $B_2 = [X_{\min} + \Delta x, X_{\min} + 2\Delta x)$, etc.

4. Calculate the theoretical probability

$$p_i \triangleq \int_{B_i} f_X(x)\, dx,$$

for each bin B_i, $i = 1, 2, \ldots, K$.

5. Calculate n_i, the number of measurements that fall in bin B_i.

6. Calculate $m_i = np_i, i = 1, 2, \ldots, K$; this is the number of measurements that "should" belong[24] to the ith bin (that should remind you of the binomial random variable).

7. Calculate

$$E \triangleq \sum_{i=1}^{K} \frac{(n_i - m_i)^2}{m_i}. \tag{3.45}$$

---

[24] We have not defined what this term means; to do that, we need the concept of average values, which is developed later in Section 3.9.

Table 3.1   Threshold values t_l for the χ^2 test.

K	$l = 1\%$	$l = 5\%$	K	$l = 1\%$	$l = 5\%$
1	6.6349	3.8415	18	34.8053	28.8693
2	9.2103	5.9915	20	37.5662	31.4104
3	11.3449	7.8147	25	44.3141	37.6525
4	13.2767	9.4877	30	50.8922	43.7730
5	15.0863	11.0705	35	56.0609	49.8018
6	16.8119	12.5916	40	63.6907	55.7585
7	18.4753	14.0671	50	76.1539	67.5048
8	20.0902	15.5073	60	88.3794	79.0819
9	21.6660	16.9190	70	100.4252	90.5312
10	23.2092	18.3070	80	112.3288	101.8795
12	26.2170	21.0261	90	124.1163	113.1453
14	29.1412	23.6848	100	135.8067	124.3421
16	31.9999	26.2962	1000	1106.9690	1074.6794

8. In the ideal case of a perfect fit, $E = 0$ since all numerators in Equation 3.45 are zero. So if the fit is good, E will have a value smaller than the threshold t_l, which we can read from Table 3.1 (see [22] for a more complete listing).

9. Therefore, our decision is the following:

    If $E < t_l$, we will accept the fit of the suggested pdf.

    If $E \geq t_l$, we will reject the fit of the suggested pdf.

Here is an intuitive explanation of this test, which is best understood after the material in Chapter 4 has been covered. Now $n_i - m_i$ is a random variable that represents the error in approximating the histogram with a theoretical pdf. It is "reasonable" to assume that this error is a Gaussian random variable. From Problem 3.176 we know that the square of a Gaussian random variable is a χ^2 distributed random variable. Sums of χ^2 random variables, and therefore the random variable E in Equation 3.45, are also χ^2 distributed, from Problem 4.180. Then the decision in step 9 of the test makes sense.

Let's see now how we can apply the χ^2 test to an experiment presented earlier. For the example that follows, we have chosen a level of significance $l = 5$ percent and number of bins $K = 50$.

Example 3.33. *Netstat Experiment.* Consider the histogram of the random variable X that represents the number of input packets measured in 1-s inter-

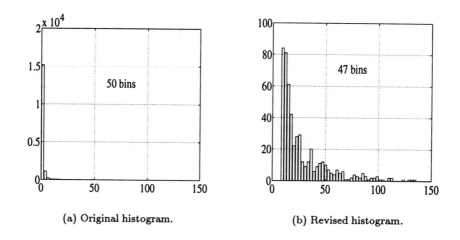

(a) Original histogram.   (b) Revised histogram.

Figure 3.34   *Netstat* experiment.

vals, shown in Figure 3.34(a). And 17,039 measurements were used to generate
it. Suppose that we want to check whether a binomial random variable would
fit this set of data.

The issue then is how to determine the two parameters n, p of this random
variable. From the histogram, observe that we have some measurements at
$X \approx 25$. (The maximum value of X was equal to 136, but the resolution of the
figure is not that small.) We choose therefore $n = 25$. For the value of p, we
chose arbitrarily $p = 0.1$. We have $K = 50$, and since $l = 5$ percent, from Table
3.1, the threshold value $t_l = 67.50481$. Based on Equation 3.45 and the given
data for this experiment, we have calculated $E = \infty$, and the test failed since
$E > t_l$. The test failed as well for $p = 0.2, 0.3, \ldots, 0.95$, giving again $E = \infty$.
(The reason for having $E = \infty$ was a division by zero; can you explain why?)

We tried a Poisson random variable with various parameters, but the test
kept failing. We then decided to model this data set with a mixed random
variable, the discrete part of which models the first three "spikes" of the given
histogram. With those parts removed, the new histogram is as shown in Figure 3.34(b). Can you fit now a continuous density (e.g., gamma) to this new
histogram?

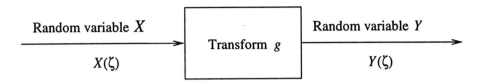

Figure 3.35 Abstract model for a transform.

3.8 TRANSFORMATIONS OF A RANDOM VARIABLE

3.8.1 What Transformations Will We Consider?

An engineer's life is full of transformations: from blue jeans during college years to a nice dress or suit on interview day; from the free spirit of high school years to the mature style of college life; etc. This is not the kind of transformations we consider here; instead, we analyze the case where the outcomes of a random experiment are transformed into something different. We show this idea in an abstract form, i.e., in a black box notation, as in Figure 3.35.

In general, we call this black box transformation *data processing*. Special cases are called *filtering, quantizing, clipping, predicting*, etc. Transforms are classified as *memoryless* and *with memory*. For a memoryless transform, $Y(\zeta) = g(X(\zeta))$; that is, the value of the output depends on only one value of the input. Two examples of memoryless transforms are quantizing and clipping. We consider only memoryless transforms in this chapter. Examples of transforms with memory are in Chapters 7 and 9 (e.g., ARMA process).

Suppose that the input is a random variable X. Is the output of a transformation $g(\cdot)$ a random variable as well? Yes, if and only if g is a measurable function (see our discussion in the beginning of Chapter 3). We consider only such functions here.

3.8.2 PDF and CDF of the Transformed Random Variable

What is the pdf or cdf of the output variable Y, given the pdf or cdf of the input X and the processing function $g(\cdot)$? In principle, the answer is easy. The basic idea here is to relate the event $A = \{Y \leq y\}$ to an equivalent event that involves the given random variable X. This event is loosely denoted as $B = \{X \leq g^{-1}(y)\}$ in the expression below. [Observe that, in general, $g(\cdot)$ may not be invertible.] The two events are related through their images in the original sample space S. Figure 3.36 depicts this idea a bit more clearly.

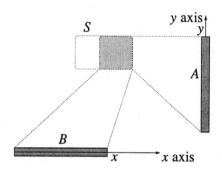

Figure 3.36   Equivalent events.

Then we can immediately write

$$F_Y(y) \stackrel{\Delta}{=} P[Y \le y] = P[g(X) \le y] = P[X \le g^{-1}(y)], \qquad (3.46)$$

and the latter probability can be (at least in principle) evaluated from the cdf of X. Let's look at this idea more rigorously.

Theorem 3.1 *Consider a random variable X, with pdf $f_X(x)$ and a given transformation $Y = g(X)$. Denote the real roots of the equation $y - g(x) = 0$, (when solved for x) by $x_l, l = 1, 2, \ldots, n$. Then the pdf of the random variable $Y = g(X)$ is given by*

$$f_Y(y) = \sum_{l=1}^{n} f_X(x_l) \cdot \frac{1}{|dg(x_l)/dx|} \qquad (3.47)$$

If, for a given y, the equation $y - g(x) = 0$ has no real roots, then

$$f_Y(y) = 0.$$

Proof: Consider a given, fixed value of $y \in R$. To simplify presentation, suppose that the equation $y - g(x) = 0$ has four roots (see Figure 3.37). Let x_l denote the lth root of the equation $y - g(x) = 0$ when we solve it for x (with y given). Consider the event $\{y < Y \le y + dy\}$ and its equivalent events involving X, as shown in Figure 3.37. Since the events $\{x_l < X \le x_l + dx_l\}, l = 1, 2, 3, 4$, are disjoint, we must have

$$\{y < Y \le y + dy\} = \cup_{l=1}^{2} \{x_l < X \le x_l + dx_l\} \cup \left(\cup_{l=3}^{4} \{x_l + dx_l < X \le x_l\} \right),$$

and thus

$$f_Y(y)\,|dy| \approx \sum_{l=1}^{n} f_X(x_l)\,|dx_l|.$$

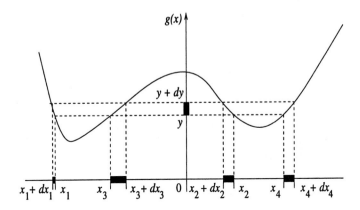

Figure 3.37 The events $\{y < Y \leq y + dy\}$ and $\cup_{l=1}^{4}\{x_l < X \leq x_l + dx_l\}$.

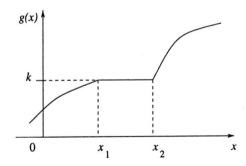

Figure 3.38 Uncountably infinite roots in $y = g(x)$.

Equation 3.47 follows since $y = g(x_l)$. In this argument, we implicitly assumed that equation $y - g(x) = 0$ has $n > 0$ roots. If there are no roots, then it must be that $f_Y(y) = 0$ since the random variable X is real-valued. Moreover, we assumed that n is finite. If n is countably infinite, the same argument holds. □

Remark 1: If n is uncountably infinite, the function $g(x)$ must be "flat" over some region(s) of x (see Figure 3.38). Then we can write

$$P[Y = k] = P[x_1 \leq X \leq x_2],$$

and if the latter probability is not zero, we can easily see that the random variable Y must have a discrete part.

Remark 2: In Equation 3.47 it is not possible to have $dg/dx = 0$. Why?

3.8.3 Some Important Transformations

Let's now apply Theorem 3.1 to some special cases of transformations.

1. Linear (affine) transformation

Let $Y = aX + b$, where a and b are real numbers. To avoid trivialities, assume that $a \neq 0$ and X is continuous.

Theorem 3.2 *Let $Y = aX + b$. Then*

$$f_Y(y) = \frac{1}{|a|} f_X \left(\frac{y - b}{a} \right).$$

Moreover,

$$F_Y(y) = \begin{cases} F_X(\frac{y-b}{a}), & a > 0, \\ 1 - F_X(\frac{y-b}{|a|}), & a < 0. \end{cases}$$

Proof: The function $g(x)$ in Theorem 3.1 is given by $g(x) = ax + b$. Therefore the equation $y - g(x) = 0$ has only one root, namely, $x_1 = \frac{y-b}{a}$. Moreover,

$$\left| \frac{dg(x_1)}{dx} \right| = |a|,$$

and thus Equation 3.47 specializes to

$$f_Y(y) = \frac{1}{|a|} f_X \left(\frac{y - b}{a} \right).$$

A linear (affine) function is invertible. From Equation 3.46, observe that when $a > 0$, $\{Y \leq y\}$ is equivalent to $\{X \leq \frac{y-b}{a}\}$. Similarly, when $a < 0$, $\{Y \leq y\}$ is equivalent to $\{X \geq \frac{y-b}{|a|}\}$.   □

2. Cosine transformation of uniform random variables

This transformation arises often in communication systems, as we see in Chapters 7 and 9. In this case, $Y = \cos(X)$, so $g(x) = \cos(x)$. We plot the $g(x)$ function over the region $[0, 2\pi)$ in Figure 3.39(a).

Suppose that X is a uniform random variable with range $[0, 2\pi)$. From Figure 3.39(a) we see that, for $|y| < 1$, the sum in Equation 3.47 contains two terms only, so $n = 2$ and $x_1 = \cos^{-1}(y)$, $x_2 = 2\pi - \cos^{-1}(y)$. Since $f_X(x) = 1$, $x \in [0, 2\pi)$, and $\frac{dg}{dx} = -\sin(x)$, Equation 3.47 gives us

$$f_Y(y) = \frac{1}{2\pi\sqrt{1 - y^2}} + \frac{1}{2\pi\sqrt{1 - y^2}} = \frac{1}{\pi\sqrt{1 - y^2}}, \quad -1 < y < 1.$$

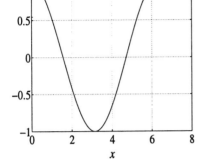

(a) The cosine function.

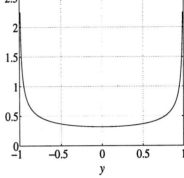

(b) The pdf.

Figure 3.39 The $g(x) = \cos(x)$, $x \in [0, 2\pi)$, transformation.

The pdf of Y is shown in Figure 3.39(b). Observe that the pdf approaches ∞ when y approaches ± 1.

3. Clipping transformation

This transformation arises often in signal processing applications. Consider a random variable X with a range $S_X = (-\infty, \infty)$. In this case the transformation is as shown in Figure 3.40. In algebraic form,

$$g(x) = \begin{cases} x, & \text{if } |x| \leq a, \\ -a, & \text{if } x < -a, \\ a, & \text{if } x > a. \end{cases}$$

From Figure 3.40 we can see that when $|y| < a$, there is only one root, and therefore the sum in Equation 3.47 contains only one term. When $y = a$, there are an infinite number of roots since every $x \in [a, \infty)$ is mapped onto the point a. Similarly, there are an infinite number of roots when $y = -a$. Therefore, Y will be a *mixed* type random variable. To summarize,

$$f_Y(y) = \begin{cases} f_X(y), & \text{if } |y| < a, \\ \delta(y - a) \cdot P[X \geq a], & \text{if } y = a, \\ \delta(y + a) \cdot P[X \leq a], & \text{if } y = -a. \end{cases}$$

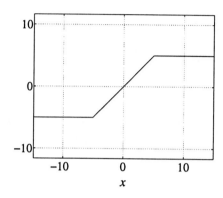

Figure 3.40   The clipping transformation, $a = 5$.

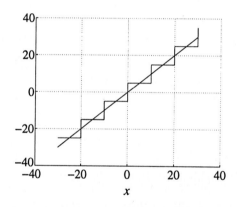

Figure 3.41   A quantizer with parameters $d = 10, a = d/2$.

4. Quantizer transformation

A quantizer is a device that converts its input (continuous or discrete-valued) to discrete-valued outputs. When sound, video, or voltage signals are stored in computers, they are quantized first since they must be represented with finite-precision numbers. The generic form of a quantizer with parameters d, a is shown in Figure 3.41.

The mathematical expression for the quantizer transformation with parameters d, a is given by

$$g(x) = \begin{cases} a, & x \in [0, d), \\ d + a, & x \in [d, 2d), \\ \vdots & \vdots \\ kd + a, & x \in [kd, (k+1)d), \\ -d + a, & x \in [-d, 0), \\ -2d + a, & x \in [-2d, -d), \\ \vdots & \vdots \\ -(k+1)d + a, & x \in [-(k+1)d, -kd). \end{cases} \qquad (3.48)$$

Consider a (continuous or discrete) random variable X with a range $S_X = (-\infty, \infty)$. Since $g(x)$ takes discrete values only, the pdf $f_Y(y)$ will contain only $\delta(\cdot)$ functions. The weight of the impulse at point $y = kd + a$ can be easily calculated from Equation 3.48 as

$$P[Y = kd + a] = P[kd \leq X < (k+1)d], \quad k = 0, 1, 2, \ldots$$

Similarly, the weight of the impulse at point $y = -(k+1)d + a$ can be easily found as well:

$$P[Y = -(k+1)d + a] = P[-(k+1)d \leq X < -kd], \quad k = 0, 1, 2, \ldots$$

5. *Logarithmic transformation*

In many experiments, the scale of the resulting values and relative frequencies covers several orders of magnitude. The file size experiment for machine ece-cho.ncsu.edu in Section 3.2 is such an example since file sizes (the values of X) range from 0 to about 7 million bytes, while relative frequencies range from 0.1 to 10^{-5}. If we plot such data on a linear scale, we do not see good results, as Figure 3.12(a) depicts. *Logarithmic plots* are called for here.

So let $g(x) = \log(x)$ be the transformation of interest; let X range over $(0, \infty)$. In general, since the logarithm is a monotonic function (see Figure 3.42), we have only one root in Equation 3.47, namely, $x_1 = e^y$. Moreover, $\frac{dg}{dx} = \frac{1}{x}$, and thus

$$f_Y(y) = e^y f_X(e^y), \quad y \in (-\infty, \infty).$$

3.9   MOMENTS OF RANDOM VARIABLES

3.9.1   Need for Partial Information About the PDF

The pdf $f_X(x)$ of a random variable X is the *complete* source of information about a random variable, in the sense that probabilities of *any* event involving

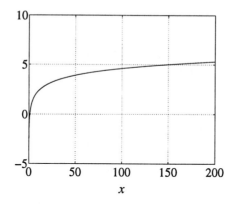

Figure 3.42   The logarithmic transformation.

the random variable can be determined. Sometimes, however, we may *not* be interested in all those events, or we may need or want a lot less information about a random experiment, for two reasons: We do not care, or we simply cannot provide (or digest) so much information.

As an example, consider the random experiment "measure the calories of a Coca Cola[25] can." Because the amount of sugar that goes in each individual can cannot be exactly the same (why?), the calorie content of a can is a random variable indeed. The bottling factory *can* determine a pdf, like that in Figure 3.43, and print it on the can label. This is *complete* information about the experiment, but as a product manufacturer, you would rather use the precious can space for better purposes, such as revenue generating advertising, and as a cola consumer, you do not really want that much information, do you? You would rather have a single number.

Come to think of it, we *do* get single numbers in all the food labels: cholesterol content, fat, carbohydrates, vitamins, etc. (In diet cola cans, we get something like "1 calorie per serving.") We *do* get a single number for car mileage, such as "this Chevrolet Geo gets 55 miles per gallon." The *consumer price index* and the *inflation index* in a given year are also single numbers. How trustworthy could those numbers be, given that all these are random experiments? Of course, the same question could be valid in more technical situations, such as "the throughput of this protocol is 2.5 Mbits/s" or "the bit error rate of this link is 10^{-6}."

If you think about it, what we do in all the above examples is give some *reduced* or *partial* information about the pdf. The question is *what* to provide

---

[25] Pepsi would also do fine.

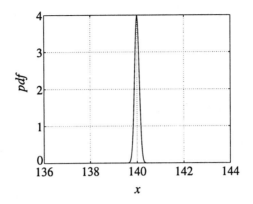

Figure 3.43 Hypothetical pdf of calorie contents.

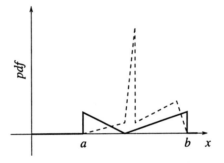

Figure 3.44 Two hypothetical pdf's with the same extreme values a, b.

as partial information. We could use *extreme* values of the pdf, such as "the maximum throughput of the protocol is 5 Mbits/s" or "the smallest bit error rate is 10^{-12}." But these extreme values could be quite misleading, as Figure 3.44 shows; sometimes such values are trivial (for example, 0 or ∞).

Useful partial information is indeed provided by the *moments* of the pdf, which we introduce next. It turns out that moments play a more fundamental role than that implied by the above examples, as we see in Chapters 5 and 7. The notion of first moment (average) has been widely used since the very early stages of the theory. It appears that Chebyshev was the first to put higher-order moments in serious use; he used them in his study of the central limit theorem.

3.9.2   Average and Variance of a PDF

Definition: The **average** of a random variable X, denoted by EX, is defined as

$$EX \triangleq \int_{-\infty}^{\infty} x f_X(x)\, dx. \tag{3.49}$$

EX is also known as the *expected value, expectation, first moment,* or *mean* of X.

Remark 1: We see from Equation 3.49 that two random variables X_1 and X_2 with the same pdf $f_X(x)$ have the same average EX. Therefore, the average is actually associated with the pdf, not the random variable.

Remark 2: The concept of the average was known to Huygens,[26] who mentioned in his treatise "De Ratiociniis in Ludo Aleae," in 1658, that "... if a player has p chances of gaining a and q chances of gaining b, his *expectation* is $\frac{pa+qb}{p+q}$" [62]. Laplace used its more general form and coined the term *mean* (French *moyen*) in his seminal book *Théorie Analytique des Probabilités,* published in 1812.

Is the mean well defined for *any* pdf $f_X(x)$? From basic calculus, we know that Equation 3.49 is well defined only when

$$\int_{-\infty}^{\infty} |x f_X(x)|\, dx = \int_{-\infty}^{\infty} |x| f_X(x)\, dx < \infty, \tag{3.50}$$

i.e., only when the integral in Equation 3.50 converges absolutely. For example, the Cauchy random variable does not have a mean (see Problem 3.213). The Pareto random variable may not have a mean (see Problem 3.215). All other models in Section 3.6 have well-defined means.

Definition: The **variance** of a random variable X (with a finite mean), denoted by $\mathrm{var}(X)$, is defined as

$$\mathrm{var}(X) \triangleq \int_{-\infty}^{\infty} (x - EX)^2 f_X(x)\, dx. \tag{3.51}$$

The square root of the variance of X is called the **standard deviation** of X and is usually denoted as σ_X. The variance is alternately denoted as σ_X^2.

Is the variance integral well defined for *any* pdf $f_X(x)$? Since the integrand in Equation 3.51 is always positive, the answer is yes, even though it may be equal to $+\infty$. Of course, when we approximate pdf's from experimental data, there is no question about means and variances being well defined since $f_X(x)$ will vanish outside a finite interval.

Let's now examine some properties of the mean and variance.

---

[26]Christian Huygens, 1629–1695, a Dutch mathematician and physicist and the first author of a book dedicated to probability theory.

Lemma 3.5   *1.* Shifting a random variable by a constant $c \in R$.

$$E(X + c) \;=\; EX + c, \tag{3.52}$$
$$\mathrm{var}(X + c) \;=\; \mathrm{var}(X). \tag{3.53}$$

2. Scaling a random variable by a constant $c \in R$.

$$E(c \cdot X) \;=\; c \cdot EX, \tag{3.54}$$
$$\mathrm{var}(c \cdot X) \;=\; c^2 \cdot \mathrm{var}(X). \tag{3.55}$$

3. The relationship between mean and variance.

$$\mathrm{var}(X) = \int_{-\infty}^{\infty} x^2 f_X(x)\, dx - (EX)^2. \tag{3.56}$$

Proof: We prove all properties by direct application of the definition of the mean and variance. Observe that, from Theorem 3.2, the random variable $Y = X + c$ has pdf $f_Y(y) = f_X(y - c)$. Therefore, with the obvious change of variables,

$$
\begin{aligned}
EY \;&\triangleq\; \int_{-\infty}^{\infty} y f_Y(y)\, dy = \int_{-\infty}^{\infty} y f_X(y - c)\, dy \\
&=\; \int_{-\infty}^{\infty} (x + c) f_X(x)\, dx = \int_{-\infty}^{\infty} x f_X(x)\, dx + \int_{-\infty}^{\infty} c f_X(x)\, dx \\
&=\; EX + c \int_{-\infty}^{\infty} f_X(x)\, dx = EX + c.
\end{aligned}
$$

Similarly, we can show that

$$
\begin{aligned}
\mathrm{var}(X + c) \;&\triangleq\; \int_{-\infty}^{\infty} [x + c - (EX + c)]^2 f_X(x)\, dx \\
&=\; \int_{-\infty}^{\infty} (x - EX)^2 f_X(x)\, dx = \mathrm{var}(X).
\end{aligned}
$$

Assume now that $c \neq 0$; from Theorem 3.2, the random variable $Y = cX$ has pdf

$$f_Y(y) = \frac{1}{|c|} f_X\!\left(\frac{y}{c}\right).$$

Therefore, omitting a few simple steps, we have

$$E(c \cdot X) \;=\; \int_{-\infty}^{\infty} y f_Y(y)\, dy = \int_{-\infty}^{\infty} cx f_X(x)\, dx = cEX.$$

$$
\begin{aligned}
\mathrm{var}(c \cdot X) \;&=\; \int_{-\infty}^{\infty} (cx - cEX)^2 f_X(x)\, dx = c^2 \int_{-\infty}^{\infty} (x - EX)^2 f_X(x)\, dx \\
&=\; c^2 \cdot \mathrm{var}(X).
\end{aligned}
$$

Finally,

$$
\mathrm{var}(X) \triangleq \int_{-\infty}^{\infty} (x - EX)^2 f_X(x)\, dx = \int_{-\infty}^{\infty} [x^2 - 2x EX + (EX)^2] f_X(x)\, dx
$$

$$
= \int_{-\infty}^{\infty} x^2 f_X(x)\, dx - 2EX \cdot EX + (EX)^2
$$

$$
= \int_{-\infty}^{\infty} x^2 f_X(x)\, dx - (EX)^2.
$$

□

3.9.3   Higher Moments

Definition: The **nth moment** of a random variable X, denoted by EX^n, is defined as

$$
EX^n \triangleq \int_{-\infty}^{\infty} x^n f_X(x)\, dx \tag{3.57}
$$

provided that the integral exists. Based on Equation 3.57, we can rewrite Equation 3.56 as

$$
\mathrm{var}(X) = EX^2 - (EX)^2.
$$

This form is useful in theoretical studies.

Definition: The **nth moment around the mean** of a random variable X, denoted by $\overline{EX^n}$, is defined as

$$
\overline{EX^n} \triangleq \int_{-\infty}^{\infty} (x - EX)^n f_X(x)\, dx.
$$

We have already seen the second such moment, the variance. Higher moments play an important role primarily in theoretical studies regarding pdf's. Therefore, we will not have ample opportunities to see them in practice.

3.9.4   Average and Variance of the Main Models

We can calculate the moments of the given models in Section 3.6. Table 3.2 summarizes this information. We leave the calculations as exercises; all of them are straightforward, even though some are tedious algebraic manipulations of integrals.

3.9.5   Means of Transformations

What about the mean of the transformation random variable $Y = g(X)$? Of course, once we find the density $f_Y(y)$, we can apply Definition 3.49; another approach, which avoids calculating $f_Y(y)$ altogether, is given by the following:

Table 3.2 Average and variance of important random variables.

Random Variable	Average	Variance
Bernoulli	p	$p(1-p)$
Binomial	np	$np(1-p)$
Geometric	$\frac{1-p}{p}$	$\frac{1-p}{p^2}$
Negative Binomial	$\frac{r}{p}$	$\frac{r(1-p)}{p^2}$
Poisson	λ	λ
Discrete Uniform	$\frac{b+a}{2}$	$\frac{1}{12}(b-a)(b-a+2)$
Uniform	$\frac{b+a}{2}$	$\frac{(b-a)^2}{12}$
Gaussian	m	σ^2
Exponential	$\frac{1}{\lambda}$	$\frac{1}{\lambda^2}$
Cauchy	Does not exist	Does not exist
Laplace	0	$\frac{2}{a^2}$
Gamma	$\frac{a}{\lambda}$	$\frac{a}{\lambda^2}$
Erlang	$\frac{n}{\lambda}$	$\frac{n}{\lambda^2}$
Weibull	$\frac{1}{\lambda}\Gamma\left(\frac{a+1}{a}\right)$	$\frac{1}{\lambda^2}[\Gamma(\frac{a+2}{a})-\Gamma(\frac{a+1}{a})]^2$
χ^2	a	$2a$
F	$\frac{b}{b-2}$, $b>2$	$\frac{2b^2(a+b-2)}{a(b-2)^2(b-4)}$, $b>4$
Student's t	0	$\frac{a}{a-2}$, $a>2$
Rayleigh	$a\sqrt{\frac{\pi}{2}}$	$(2-\frac{\pi}{2})a^2$
Pareto	$\frac{ba}{a-1}$, $a>1$	$\frac{ab^2}{(a-1)^2(a-2)}$, $a>2$.

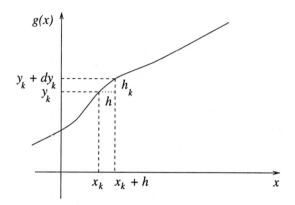

Figure 3.45   The mean of $Y = g(X)$.

Theorem 3.3 *Let $Y = g(X)$ be a function of the random variable X with pdf $f_X(x)$. Then*

$$EY = \int_{-\infty}^{\infty} g(x) f_X(x)\, dx. \tag{3.58}$$

Proof: We sketch only a proof since the main idea is identical to that in Theorem 3.1. To simplify presentation, we further assume that the function $g(x)$ is monotonic increasing. The general case is left as an exercise (see Problem 3.149).

Consider Figure 3.45. From its definition, we can approximate EY as

$$EY \approx \sum_k y_k f_Y(y_k) h_k.$$

Fix a y_k. From Theorem 3.1, since the function $g(x)$ is monotonic, we have

$$f_Y(y_k) = f_X(x_k) \frac{1}{|g'(x_k)|}$$

where $y_k = g(x_k)$. When h is small, we can write approximately $|g'(x_k)| = \frac{h_k}{h}$, and thus

$$f_Y(y_k) h_k = f_X(x_k) h.$$

Equation 3.58 now follows as we take the limit $h_k \to 0$.   □

3.10 RELIABILITY AND FAILURE RATES

The notion of reliability is common in everyday life. Suppose you buy a product such as a car, a computer, or a calculator today (i.e., at time 0). How *reliable* the product is determines the chances (or probability) it will be operational t days or years in the future. On the other hand, suppose you are t years old today. What are your chances of living an *extra* x years? Life insurance companies base their fortunes on accurate calculation of such chances. Mortalities and life insurance were, as a matter of fact, two of the first problems to be studied by probability theory in its infancy. Daniel Bernoulli[27] was the first to publish a work on reliability.

Let's make these ideas more precise. In all these examples, we are interested in a random variable L, which can take only *nonnegative* values. It is customary to call such a random variable a **lifetime** since typically the theory of reliability applies to lifetimes of products, persons, etc.

Let $t > 0$ be any fixed, real number. Let $L : S \to [0, \infty)$ be a random variable defined on some probability space $(S, \mathcal{Q}, P)$.

Definition: The probability $P[L > t]$, denoted by $R_L(t)$, is called the **reliability** associated with the random variable L.

Remark: Reliability $R_L(t)$ is just the tail probability of the random variable L. Obviously, if L represents the lifetime of a car or computer, we want it to be as large as possible. On the other hand, if L represents the delay we encounter with a bank teller, we want it to be as small as possible. Reliability has a natural frequency interpretation since it is a probability. Suppose that L represents the lifetime of a computer chip. Then $R_L(t)$ can be considered as the percentage of chips that will function for more than t time units.

Consider now two systems S_1 and S_2 with lifetimes represented by two random variables L_1 and L_2 with reliabilities $R_{L_1}(t)$ and $R_{L_2}(t)$, respectively. Consider a time instant t_0 for which

$$R_{L_1}(t_0) \geq R_{L_2}(t_0). \tag{3.59}$$

We say then that at time t_0, system S_1 is more reliable than system S_2. If Inequality 3.59 holds true for all t_0 in an interval of time $[a, b]$, then we say that system S_1 is more reliable than system S_2 in the time interval $[a, b]$.

Is it possible for a system to be more reliable than another for all $t \in (0, \infty)$? Yes, but before we jump to any hasty conclusions, recall the frequency interpretation of reliability. This only means that the fractions are ordered as in Inequality 3.59. When we compare *any two* samples of systems S_1 and S_2, they can "break down" in any order.

---

[27] Daniel Bernoulli, 1700–1782, a Swiss mathematician and a relative of James Bernoulli.

The reliability of a random variable L is easily calculated from its cdf:

$$R_L(t) = 1 - F_L(t). \tag{3.60}$$

Differentiating Equation 3.60, we get

$$R'_L(t) = -f_L(t). \tag{3.61}$$

Note that $R'_L(t)$ in Equation 3.61 is *not* a pdf.

In reliability lingo, the expected value of the random variable L is usually called the **average lifetime** or **mean time to failure**. Assuming it is well defined, the average lifetime can be calculated as follows:

$$EL \overset{\triangle}{=} \int_0^\infty t f_L(t)\, dt = -\int_0^\infty t R'_L(t)\, dt$$
$$= -\int_0^\infty t\, dR_L(t) = -tR_L(t)|_0^\infty + \int_0^\infty R_L(t)\, dt. \tag{3.62}$$

To arrive at Equation 3.62, we used integration by parts. Observe now that since $EL < \infty$,

$$tR_L(t) = t\int_t^\infty f_L(s)\, ds < \int_t^\infty s f_L(s)\, ds \to_{t\to\infty} 0$$

and thus $tR_L(t) \to 0$ as $t \to \infty$. Therefore,

$$EL = \int_0^\infty R_L(t)\, dt$$

3.10.1   Residual Lifetimes

Conditional probabilities involving lifetimes are of special interest. For example, say that the average lifetime of a person is about 70 to 75 years. We expect then that a newborn would live that long. However, we do not expect the additional life span of a 50-year-old person to be another 70 to 75 years. Medicine has not advanced this much, and nobody has found the elixir of life.

Definition: The **residual lifetime** L_R associated with a lifetime L is a random variable whose cdf for fixed t is given by the following conditional probability:

$$F_{L_R}(s|t) \overset{\triangle}{=} F_L(s|L > t). \tag{3.63}$$

Using the definition of conditional probabilities and Equation 3.63, we have

$$F_{L_R}(s|t) = \begin{cases} \frac{F_L(s) - F_L(t)}{1 - F_L(t)}, & s \geq t, \\ 0, & s < t. \end{cases} \tag{3.64}$$

Table 3.3 Reliability and failure rate of common models.

Model	Reliability $R_L(t)$	Failure Rate $r_L(t)$
Discrete Uniform	$(b-t)/(b-a+1)$	$1/(b-t)$
Geometric	$(1-p)^{t+1}$	$p/(1-p)$
Uniform	$(b-t)/b-a)$	$1/(b-t)$
Exponential	$e^{-\lambda t}$	λ
Erlang	$e^{-\lambda t}\sum_{k=0}^{n-1}\frac{(\lambda t)^k}{k!}$	$\lambda(\lambda t)^{n-1}/\left[(n-1)!\sum_{k=0}^{n-1}\frac{(\lambda t)^k}{k!}\right]$
Weibull	$e^{-(\lambda t)^a}$	$a\lambda(\lambda t)^{a-1}$
Pareto	$(b/t)^a$	a/t

[If $F_L(t) = 1$, we define $F_{L_R}(s|t) = 1$.] The pdf of L_R can be easily found by differentiating Equation 3.64 with respect to s:

$$f_{L_R}(s|t) = \frac{f_L(s)}{1 - F_L(t)} = \frac{f_L(s)}{R_L(t)}, \quad s \geq t. \tag{3.65}$$

Of course, $f_{L_R}(s|t) = 0$ when $s < t$. Note that Equation 3.65 defines a *family* of pdf's, indexed by t. When $s = t$ in Equation 3.65, the pdf has a special name.

Definition: The **failure rate function**, denoted as $r_L(t)$, is defined as

$$r_L(t) \overset{\triangle}{=} f_{L_R}(t|t).$$

From Equation 3.61 we immediately get

$$r_L(t) = \frac{-R'_L(t)}{R_L(t)}. \tag{3.66}$$

Based on Equation 3.66, lifetime random variable models are usually classified with respect to whether $r_L(t)$ is an increasing, decreasing, or constant function. We talk about IFR (increasing failure rate), DFR (decreasing failure rate), and CFR (constant failure rate) distributions.

Example 3.34. Table 3.3 contains the reliability and failure rates for some of the models in Section 3.6. The calculations are fairly straightforward and left as an exercise.

3.10.2 System Reliability

When a complex system, consisting of a number of individual subsystems or components is designed, the overall system lifetime or reliability will of course

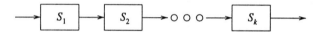

Figure 3.46   Series interconnection of components.

depend on those of the individual components.[28] Now we examine the reliability of two generic systems.

1. Systems with components in series

Consider a *series system*, as shown in Figure 3.46. Each box S_i can be described by a lifetime $L_i(t)$ and a reliability $R_{L_i}(t), i = 1, 2, \ldots, k$. Define L, the lifetime of the series system, as

$$L = \min_{i=1,2,\ldots,k}\{L_i\}. \tag{3.67}$$

The interpretation of Equation 3.67 is that the series system is functioning *if and only if all* its components are functioning (hence the min operation). So, there is no component that is less or more critical in the sense described in the footnote.

To calculate the reliability and failure rate of the series system, we must find the cdf of the minimum of k random variables. We show rigorously how to do that in Section 4.7. Intuitively, when the systems fail "independently," we can provide an argument to calculate $R_L(t)$ as follows. Consider the events

$$
\begin{aligned}
F_i &= \{\text{system } S_i \text{ is functional at time } t\}, \\
F &= \{\text{whole system is functional at time } t\}.
\end{aligned} \tag{3.68}
$$

[This is defined on what (S, Q, P)? That is the problem we handle in Chapter 4.] Then

$$P[F_i] \stackrel{\triangle}{=} R_{L_i}(t), \quad i = 1, 2, \ldots, k, \qquad P[F] \stackrel{\triangle}{=} R_L(t),$$

and from the definition of the series system in Equation 3.67, we get

$$P[F] = P[F_1 \cap F_2 \cap \cdots \cap F_k]. \tag{3.69}$$

If systems fail independently, Equation 3.69 gives us

$$P[F] = P[F_1] \cdot P[F_2] \cdots P[F_k].$$

Thus the desired reliability of the series system is given by

$$R_L(t) = R_{L_1}(t) \cdot R_{L_2}(t) \cdots R_{L_k}(t).$$

---

[28] Of course, some of the components may be less critical than others; their failure may not cause the whole system to fail (e.g., failure of a speedometer does not prevent the car from running). We do not consider such situations here; if you wish, in our definition of a system, all components are critical.

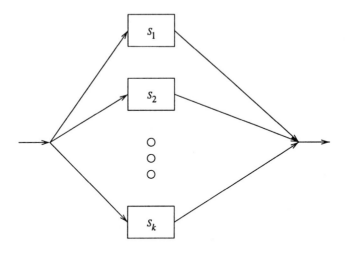

Figure 3.47   Parallel interconnection of components.

We summarize this result as a theorem.

Theorem 3.4 *Consider the series interconnection of k systems with reliability functions $R_{L_i}(t)$, $i = 1, \ldots, k$. Assuming that systems fail independently, the reliability of the overall system is given by*

$$R_L(t) = R_{L_1}(t) \cdot R_{L_2}(t) \cdots R_{L_k}(t).$$

2. Systems with components in parallel

Consider next a system with components in parallel, as shown in Figure 3.47. Again, each box S_i can be described by a lifetime $L_i(t)$ and a reliability $R_{L_i}(t), i = 1, 2, \ldots, k$. Define L, the lifetime of the parallel system, as

$$L = \max_{i=1,2,\ldots,k} \{L_i\}. \tag{3.70}$$

The interpretation of Equation 3.70 is that the parallel system will function *even when only one* of its components is functioning (hence the max operation).

Consider again the events in Equation 3.68. From the definition of a parallel system, we can see that when individual components fail independently,

$$P[F^c] = P[F_1^c \cap F_2^c \cap \cdots \cap F_k^c],$$

and thus

$$P[F^c] = P[F_1^c] \cdot P[F_2^c] \cdots P[F_k^c].$$

So,

$$1 - R_L(t) = [1 - R_{L_1}(t)] \cdot [1 - R_{L_2}(t)] \cdots [1 - R_{L_k}(t)],$$

and finally

$$R_L(t) = 1 - [1 - R_{L_1}(t)] \cdot [1 - R_{L_2}(t)] \cdots [1 - R_{L_k}(t)].$$

Theorem 3.5 *Consider the parallel interconnection of k systems with reliability functions $R_{L_i}(t)$, $i = 1, \ldots, k$. Assuming that the systems fail independently, the reliability of the overall system is given by*

$$R_L(t) = 1 - [1 - R_{L_1}(t)] \cdot [1 - R_{L_2}(t)] \cdots [1 - R_{L_k}(t)].$$

Design of highly reliable systems, (called **fault-tolerant systems**) from unreliable components is a challenging task. From Theorem 3.4 we see that putting components in series decreases the overall system reliability. On the other hand, from Theorem 3.5 we see that putting components in parallel increases the overall system reliability. So, other than using better quality components, *replicating* components is the only way to provide highly reliable systems. Critical airplane control systems and important files in database systems are duplicated or triplicated in order to increase reliability. Independence of the failure events is fundamental here.[29]

3.11   TRANSFORMS OF PROBABILITY DENSITY FUNCTIONS

As you may recall from linear systems or circuit theory courses, transform methods (such as Fourier and Laplace transforms) are clever techniques that let you solve equations that involve derivatives or integrals of functions. Generating functions (such as Z transforms) are indispensable in solving equations with an infinite number of unknowns [41]. Probability density functions may also appear in (an infinite number of) equations that involve differentiation or integration. We see such equations in later chapters; examples are the convolution integral 4.37 in Section 4.8 and the sums of random variables in Section 6.3. Solution of such equations becomes much easier through transforms.[30]

---

[29] The crash of a United Airlines plane in Sioux City, Iowa, in the late 1980s was blamed on a total failure of the three hydraulic systems of the plane. Since all three passed through the same area, their failure events were *dependent*, reducing the reliability of the system, despite the fact that it was triply backed up. Do you store your backup diskettes in a fault-tolerant fashion?

[30] Do not confuse the *transform of a pdf* with the *transformation of a random variable* that we studied in Section 3.8.

In this chapter we develop three transforms; the first one, the **characteristic function,** is defined for an arbitrary random variable, which can take positive and/or negative values, integer or noninteger. The **Laplace transform** is defined for random variables that take only positive real values. The **probability generating function** is defined for discrete random variables that take nonnegative integer values. Each one has its own advantages, as we see later.

3.11.1 Characteristic Function

Definition: The **characteristic function** of the pdf $f_X(x)$ is defined as

$$\Phi_X(\omega) \triangleq E e^{j\omega X}. \tag{3.71}$$

In Equation 3.71, j is the usual imaginary number $j = \sqrt{-1}$, and $\omega \in (-\infty, \infty)$ is a parameter. Since $e^{j\omega X}$ is a (complex) function of the random variable X, from Theorem 3.3 we have[31]

$$\Phi_X(\omega) = \int_{-\infty}^{\infty} f_X(x) e^{j\omega x}\, dx, \quad \omega \in R, \tag{3.72}$$

and from Equation 3.72 we recognize the characteristic function as the Fourier[32] transform of the pdf $f_X(x)$ (except for the sign of the exponent). Notice that $\Phi_X(\omega)$ is well defined, for all pdf's since $|e^{j\omega x}| = 1$ and

$$\left| \int_{-\infty}^{\infty} f_X(x) e^{j\omega x}\, dx \right| \le \int_{-\infty}^{\infty} \left| f_X(x) e^{j\omega x} \right|\, dx = \int_{-\infty}^{\infty} f_X(x)\, dx = 1.$$

Remark 1: Observe from Equation 3.72 that two different random variables X_1 and X_2 with the *same* pdf $f_X(x)$ have the *same* $\Phi_X(\omega)$. Characteristic functions are therefore associated with pdf's rather than random variables. Abusing the terminology a little, we talk about the characteristic function of a random variable as well as of a pdf.

Remark 2: It is well known in Fourier transform theory that there is a one-to-one correspondence between a function and its Fourier transform. Therefore, the *inverse transform formula* allows us to recover the pdf once $\Phi_X(\omega)$ is known. The formula is the following:

$$f_X(x) = \frac{1}{2\pi} \int_{-\infty}^{\infty} \Phi_X(\omega) e^{-j\omega x}\, d\omega, \quad x \in R. \tag{3.73}$$

---

[31] Notice that we have only defined a real function of a random variable. We must therefore interpret Equation 3.71 as a pair of random variables, one for the real and one for the imaginary part.

[32] After Jean Baptiste Joseph Fourier, 1768–1830, a French mathematician.

Equation 3.73 therefore provides *complete* information about a random variable since it identifies a pdf uniquely. The idea of going back and forth between a pdf and its characteristic function has seen some use in the early stages of probability theory, but it was Lévy who formalized the notions in the early twentieth century. In practice, solving Equation 3.73 is rather dif-cult. However, we can recover *partial* information about X, namely, its moments, rather easily, as the following theorem shows.

Theorem 3.6 *Let EX^n denote the nth moment of a random variable X with characteristic function $\Phi_X(\omega)$. Assume that $\Phi_X(\omega)$ is n times differentiable at the point $\omega = 0$. Then*

$$EX^n = \frac{1}{j^n} \frac{d^n \Phi_X(\omega)}{d\omega^n}\bigg|_{\omega=0}$$

Proof: Differentiating both sides of Equation 3.72 n times with respect to ω, we have

$$\frac{d^n \Phi_X(\omega)}{d\omega^n} = \int_{-\infty}^{\infty} f_X(x)(jx)^n e^{j\omega x}\, dx$$

and the desired result follows by letting $\omega = 0$. □

Let's see now two examples where $\Phi_X(\omega)$ can be calculated easily.

Example 3.35. *Exponential pdf.* Consider an exponential random variable X with parameter λ. Let $u(x)$ denote the usual step function. Then the characteristic function of X is given by

$$\Phi_X(\omega) = \int_{-\infty}^{\infty} \lambda e^{-\lambda x} u(x) e^{j\omega x}\, dx = \int_0^{\infty} \lambda e^{-\lambda x} e^{j\omega x}\, dx = \frac{\lambda}{\lambda - j\omega}.$$

Example 3.36. *Gaussian pdf.* Consider a Gaussian random variable with parameters m, σ^2. Its characteristic function is given by

$$\Phi_X(\omega) = \int_{-\infty}^{\infty} \frac{1}{\sqrt{2\pi}\sigma} e^{-\frac{(x-m)^2}{2\sigma^2}} e^{j\omega x}\, dx \qquad (3.74)$$

The integral in Equation 3.74 can be computed easily *after* we complete the square in the exponent (a very handy trick when one deals with the Gaussian pdf; see also Problem 3.115). Omitting a few simple algebraic steps, we have

$$\Phi_X(\omega) = e^{mj\omega - \frac{\sigma^2 \omega^2}{2}} \int_{-\infty}^{\infty} \frac{1}{\sqrt{2\pi}\sigma} e^{-\frac{(x-m-j\omega\sigma^2)^2}{2\sigma^2}}\, dx. \qquad (3.75)$$

Now the integral in Equation 3.75 is the density of another Gaussian random variable, with parameters $m + j\omega\sigma^2$ and σ^2, so it is equal to 1. Therefore,

$$\Phi_X(\omega) = e^{mj\omega - \frac{\sigma^2 \omega^2}{2}}.$$

Table 3.4 (see page 140) summarizes the characteristic functions for some other random variables (as well as Laplace transforms and probability generating functions). Its entries are from standard mathematical handbooks [1].

3.11.2 Laplace Transform

Definition: The **Laplace transform** of the pdf $f_X(x)$ of a positive random variable X is defined as

$$L_X(s) \triangleq Ee^{-sX} = \int_0^\infty f_X(x)e^{-sx}\,dx. \tag{3.76}$$

In Equation 3.76, s is a complex number with a positive real part. This condition on s suf-ces for $L_X(s)$ to be well defined.

An inverse formula also holds true for the Laplace transform, and there is a one-to-one-correspondence between $f_X(x)$ and $L_X(s)$.

$$f_X(x) = \frac{1}{2j\pi} \int_{c-j\infty}^{c+j\infty} L_X(s)e^{sx}\,ds, \quad x \in R.$$

There is also an equivalent version of Theorem 3.6.

Theorem 3.7 *Let EX^n denote the nth moment of a random variable X with Laplace transform $L_X(s)$ that is n times differentiable at the point $s = 0$. Then*

$$EX^n = (-1)^n \left.\frac{d^n L_X(s)}{ds^n}\right|_{s=0} \tag{3.77}$$

Its proof is omitted. Laplace introduced both the transform and this theorem in the early nineteenth century.

Remark: Let's expand $L_X(s)$ near the origin $s = 0$; using Equation 3.77, we get

$$L_X(s) = \sum_{n=0}^\infty \frac{EX^n}{n!}s^n. \tag{3.78}$$

Equation 3.78 shows that the Laplace transform, and hence the pdf of X, can be computed, at least in principle, from knowledge of the moments of the random variable X.

Let's see now two examples where $L_X(s)$ can be calculated easily.

Example 3.37. *Gamma pdf.* Let X be a gamma random variable with parameters λ, a. Its Laplace transform is then given by

$$L_X(s) = \int_0^\infty \frac{\lambda(\lambda x)^{a-1}e^{-\lambda x}}{\Gamma(a)}e^{-sx}\,dx = \frac{\lambda^a}{\Gamma(a)}\int_0^\infty x^{a-1}e^{-(\lambda+s)x}\,dx. \tag{3.79}$$

Now change variables from $(\lambda + s)x$ to y; then, from Equation 3.79, we get

$$L_X(s) = \frac{\lambda^a}{\Gamma(a)} \frac{1}{(\lambda + s)^a} \int_0^\infty y^{a-1} e^{-y}\, dy = \frac{\lambda^a}{(\lambda + s)^a}; \qquad (3.80)$$

here we used the definition of the $\Gamma(\cdot)$ function in the last step. Note that the Laplace transforms for the Erlang, exponential, and χ^2 random variables are easily obtained from Equation 3.80.

Example 3.38. *Bernoulli pmf.* Let X be a Bernoulli random variable with parameter p. Its Laplace transform is then given by

$$L_X(s) = \int_0^\infty [(1 - p)\delta(x) + p\delta(x - 1)]e^{sx}\, dx = (1 - p) + pe^s. \qquad (3.81)$$

Table 3.4 (see page 140) summarizes the Laplace transforms for useful models.

3.11.3   Probability Generating Function

Laplace formally introduced another transform as well. Observe that the transform is defined for nonnegative random variables; for possible extensions to arbitrary random variables, see Problems 3.273 and 3.274. For ease of notation, let's define $p_X(k) \overset{\Delta}{=} P[X = k]$.

Definition: The **probability generating function** of the pmf $P[X = k]$ is defined as

$$G_X(z) \overset{\Delta}{=} Ez^X = \sum_{k=0}^\infty p_X(k)z^k. \qquad (3.82)$$

The parameter z is in general a complex number; note that when $|z| < 1$, the series in Equation 3.82 is guaranteed to converge. Of course, an inverse formula is also available, and the moment theorem analog for the probability generating function is the following:

Theorem 3.8 *Let $P[X = k]$ denote the pmf of a random variable X with probability generating function $G_X(z)$ that is differentiable at the point $z = 0$. Then*

$$P[X = k] = \frac{1}{k!} \left. \frac{d^k G_X(z)}{dz^k} \right|_{z=0}, \quad k = 0, 1, 2, \dots$$

James Bernoulli used this theorem in his proof of the Weak Law of Large Numbers (see Section 6.5); Laplace used the theorem in many of his problems. Its proof is omitted. Note that instead of generating the moments, as we did with $\Phi_X(\omega)$ and $L_X(s)$, we recover the entire pmf. If we differentiate $G_X(z)$

at $z = 1$, we recover moments of the pmf, as the following equations show:

$$\frac{dG_X(z)}{dz}\bigg|_{z=1} = \sum_{k=0}^{\infty} p_X(k)kz^{k-1}\bigg|_{z=1} = \sum_{k=0}^{\infty} kp_X(k) = EX.$$

$$\frac{d^2G_X(z)}{dz^2}\bigg|_{z=1} = \sum_{k=0}^{\infty} p_X(k)k(k-1)z^{k-2}\bigg|_{z=1} = \sum_{k=0}^{\infty} k(k-1)p_X(k)$$

$$= E[X(X-1)] = EX^2 - EX.$$

To summarize:

Theorem 3.9 *Consider an integer-valued random variable X with probability generating function $G_X(z)$. Then*

$$EX = G'_X(1),$$
$$\text{var}(X) = G''_X(1) + G'_X(1) - [G'_X(1)]^2.$$

Let's see two examples where calculating $G_X(z)$ is not that terrible.

Example 3.39. *Poisson pmf.* Let X be a Poisson random variable with parameter λ. Its probability generating function is given then by

$$G_X(z) = \sum_{k=0}^{\infty} \frac{\lambda^k}{k!}e^{-\lambda}z^k = e^{-\lambda}\sum_{k=0}^{\infty} \frac{(\lambda z)^k}{k!} = e^{-\lambda}e^{\lambda z} = e^{\lambda(z-1)}.$$

Example 3.40. *Binomial pmf.* Let X be a binomial random variable with parameters n, p. Its probability generating function is given then by

$$G_X(z) = \sum_{k=0}^{n} \binom{n}{k}p^k(1-p)^{n-k}z^k = \sum_{k=0}^{n} \binom{n}{k}(pz)^k(1-p)^{n-k} = (1-p+pz)^n,$$

where the last equality follows directly from the binomial theorem.

3.12   TAIL INEQUALITIES

Probabilities of the form $P[X \geq a]$ and $P[|X| \geq a]$ are known as tail probabilities, as Figure 3.48 suggests. Consider the case in which X represents delay in a file transfer experiment. If such probabilities are large, then we incur unacceptably large delays quite frequently. It is of both theoretical and practical interest to get an upper bound on such probabilities so that an estimate can

Table 3.4   Transforms of some useful models.

X	$\Phi_X(\omega)$	$L_X(s)$	$G_X(z)$
Bernoulli	$1 - p + pe^{j\omega}$	$1 - p + pe^s$	$1 - p + pz$
Binomial	$(1 - p + pe^{j\omega})^n$	$(1 - p + pe^s)^n$	$(1 - p + pz)^n$
Geometric	$\frac{p}{1-(1-p)e^{j\omega}}$	$\frac{p}{1-(1-p)e^s}$	$\frac{p}{1-(1-p)z}$
Negative Binomial	$\left[\frac{pe^{j\omega}}{1-(1-p)e^{j\omega}}\right]^r$	$\left[\frac{pe^s}{1-(1-p)e^s}\right]^r$	$\left[\frac{p}{1-(1-p)z}\right]^r$
Poisson	$e^{\lambda(e^{j\omega}-1)}$	$e^{\lambda(e^s-1)}$	$e^{\lambda(z-1)}$
Uniform	$\frac{e^{j\omega b}-e^{j\omega a}}{j\omega(b-a)}$	$\frac{e^{sb}-e^{sa}}{s(b-a)}$	–
Gaussian	$e^{jm\omega-\frac{\sigma^2\omega^2}{2}}$	–	–
Exponential	$\frac{\lambda}{\lambda-j\omega}$	$\frac{\lambda}{\lambda+s}$	–
Laplace	$\frac{a^2}{a^2+\omega^2}$	–	–
Gamma	$\frac{\lambda^a}{(\lambda-j\omega)^a}$	$\frac{\lambda^a}{(\lambda+s)^a}$	–

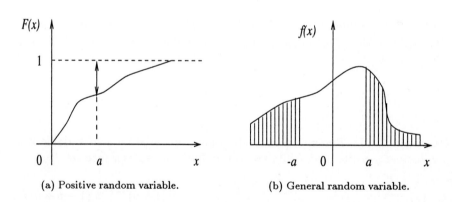

(a) Positive random variable.　　　　(b) General random variable.

Figure 3.48   Tail probabilities associated with a random variable X.

be obtained without the need to actually evaluate them. The following three bounds (coincidentally all discovered by Russian mathematicians) provide us with various estimates. They are based on partial information about the pdf, except for the Chernoff bound, which requires knowledge of the Laplace transform. In general, the Chernoff bound is tighter.

3.12.1 Markov Inequality

Consider[33] a random variable X with range $[0, \infty)$. We have

$$EX = \int_0^\infty x f_X(x)\, dx = \int_0^a x f_X(x)\, dx + \int_a^\infty x f_X(x)\, dx \geq$$

$$\int_a^\infty x f_X(x)\, dx \geq a \int_a^\infty f_X(x)\, dx = aP[X \geq a].$$

Therefore, we have proved the following useful inequality:

Lemma 3.6 *Markov inequality. Consider a nonnegative random variable X with finite mean EX. Then, for any $a > 0$, we have*

$$P[X \geq a] \leq \frac{EX}{a}. \tag{3.83}$$

Example 3.41. Let's apply Inequality 3.83 to some of the models in Section 3.6, for which the exact value of the left-hand side can be computed exactly. For an exponential random variable X, we have $EX = 1/\lambda$ and thus

$$P[X \geq a] = \int_a^\infty \lambda e^{-\lambda x}\, dx = e^{-\lambda a} \leq \frac{1}{\lambda a}.$$

Figure 3.49 shows a plot of the function $1 - xe^{-x}$ versus x. From this graph we can see that the bound is, in general, pretty loose.

Example 3.42. For a binomial random variable X with parameters n, p, we have $EX = np$. Let $a = 5, n = 10$, and $p = 0.3$. Then from Inequality 3.83, we get

$$P[X \geq 5] = \sum_{k=5}^{10} \binom{10}{k} 0.3^k 0.7^{10-k} = 0.15 \leq \frac{3}{5}.$$

3.12.2 Chebyshev Inequality

This bound[34] applies to any random variable, positive or not, and deals with tail probabilities of the form $P[|X - EX| \geq a]$. It involves knowledge of the mean and variance of the random variable X.

We can derive this bound directly from the definition of the variance[35] of X (which is assumed finite), as follows. First observe from Figure 3.50 that in the shaded set $a^2 \leq (x - EX)^2$. Based on that,

---

[33] After A. A. Markov, 1856–1922, the originator of *Markov Chains*, an invaluable modeling tool for dynamic systems. This inequality is another of his contributions.

[34] After Pafnuti Chebyshev (or Tshebycheff or Tchebycheff), 1821–1894, another Russian mathematician, also known for his polynomials. He developed this bound in 1866.

[35] When Chebyshev gave his original proof, he did not use the concept of the variance of a random variable; his proof occupies several pages.

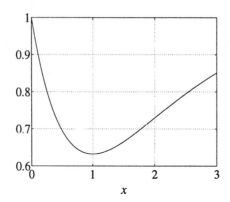

Figure 3.49   The function $1 - xe^{-x}$.

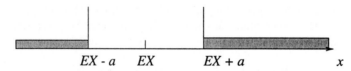

$$EX - a \quad EX \qquad EX + a \qquad\qquad x$$

Figure 3.50   The set $\{|X - EX| \geq a\}$.

$$\text{var}(X) \;=\; \int_{-\infty}^{\infty} (x - EX)^2 f_X(x)\,dx \;\geq\; \int_{EX+a}^{\infty} (x - EX)^2 f_X(x)\,dx$$

$$+ \int_{-\infty}^{EX-a} (x - EX)^2 f_X(x)\,dx \;\geq\; a^2 P[(X - EX)^2 \geq a^2].$$

Therefore, we have the following inequality:

Lemma 3.7 *Chebyshev inequality. Consider a random variable X with finite mean EX and finite variance $\text{var}(X)$. Then, for any $a > 0$, we have*

$$P[|X - EX| \geq a] \leq \frac{\text{var}(X)}{a^2}. \qquad\qquad (3.84)$$

Sometimes the following form is also convenient:

$$P[|X - EX| < a] \geq 1 - \frac{\text{var}(X)}{a^2}.$$

Finally, we can express a in units of the variance, to get a feeling for the spread of the pdf. Let $\sigma^2 = \text{var}(X)$; let $a = n\sigma$ in Inequality 3.84. Then

$$P[|X - EX| \geq n\sigma] \leq \frac{1}{n^2}.$$

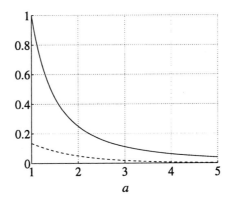

Figure 3.51 Chebyshev bound, exponential random variable, $\lambda = 1$.

Example 3.43. Let's apply Inequality 3.84 to the same models we used in Markov's inequality. For an exponential random variable X,

$$EX = \frac{1}{\lambda}, \quad \text{var}(X) = \frac{1}{\lambda^2};$$

therefore, the Chebyshev bound becomes

$$P[|X - \frac{1}{\lambda}| \geq a] \leq \frac{1}{\lambda^2 a^2}.$$

Let $h \triangleq \max\{0, \frac{1}{\lambda} - a\}$. Then the exact probability can be calculated as follows:

$$
\begin{aligned}
P[|X - EX| \geq a] &= \int_0^h \lambda e^{-\lambda x}\, dx + \int_{\frac{1}{\lambda}+a}^\infty \lambda e^{-\lambda x}\, dx \\
&= 1 - e^{-\lambda h} + e^{-(\lambda a + 1)}.
\end{aligned}
$$

Figure 3.51 shows the Chebyshev bound and the exact probability (dashed line) for various values of a and $\lambda = 1$.

Example 3.44. For a binomial random variable X we have $EX = np$ and $\text{var}(X) = np(1-p)$. Assuming again that $a = 5$, $n = 10$, and $p = 0.3$, we have

$$P[|X - 3| \geq 5] = \sum_{k=8}^{10} \binom{10}{k} 0.3^k 0.7^{10-k} = 0.0016 \leq \frac{10 \cdot 0.3 \cdot 0.7}{5^2} = 0.0840.$$

For $a = 7$,

$$P[|X - 3| \geq 7] = 0.3^{10} = 0.0000059049 \leq \frac{10 \cdot 0.3 \cdot 0.7}{7^2} = 0.0428.$$

3.12.3   Chernoff Bound

This bound also applies to any random variable, positive or not. It can be derived as follows. Observe that for any $s > 0$, we have

$$e^{s(x-a)} \geq u(x - a), \tag{3.85}$$

where $u(\cdot)$ is the usual step function. Then since

$$P[X \geq a] = \int_a^\infty f_X(x)\,dx = \int_{-\infty}^\infty f_X(x)u(x - a)\,dx,$$

we obtain from Inequality 3.85 that for all $s > 0$,

$$P[X \geq a] \leq \int_{-\infty}^\infty f_X(x)e^{s(x-a)}\,dx \triangleq e^{-as}\phi(s). \tag{3.86}$$

Observe that the right-hand side of Inequality 3.86 is a function of the parameter s, so we can minimize it with respect to s. The tightest such bound is known as the *Chernoff bound*.

Lemma 3.8 *Consider a random variable X. Then, for any a,*

$$P[X \geq a] \leq \min_{s>0} \{e^{-as}\phi(s)\}. \tag{3.87}$$

In Problem 3.298 we explore a similar lemma for the probability $P[X \leq a]$.

Example 3.45. Consider a Gaussian random variable X, with $EX = 0$ and $\text{var}(X) = 1$. We can easily check that $\phi(s) = e^{\frac{s^2}{2}}$, and from Inequality 3.86 we have

$$P[X \geq a] \leq e^{-as}e^{\frac{s^2}{2}} = e^{-s(a-\frac{s}{2})}.$$

Differentiating with respect to s, we see that the minimum is achieved when $s = a$, and thus

$$P[X \geq a] \leq e^{-\frac{a^2}{2}}.$$

Example 3.46. Consider now a Poisson random variable X with parameter λ. From Table 3.4 we have $\phi(s) = e^{\lambda(e^s - 1)}$, and therefore

$$P[X \geq a] \leq e^{-as}e^{\lambda(e^s - 1)}. \tag{3.88}$$

Differentiating with respect to s, we see that the minimum occurs for $s = \log(\frac{a}{\lambda})$; after a few simple steps, we get

$$P[X \geq a] \leq \frac{\lambda^a e^{(a-\lambda)}}{a^a}.$$

For example, for $a = 3$ and $\lambda = 1$

$$P[X \geq 3] \leq \frac{e^2}{27} = 0.27366874,$$

while the exact value is 0.018988156876. For $a = 2$ and $\lambda = 1$, we get $P[X \geq 2] \leq 0.67957046$, while the exact value is 0.26424112.

3.13 GENERATION OF VALUES OF A RANDOM VARIABLE

3.13.1 Need for Simulating the Values of a Random Variable

When we have an actual system, we can of course perform experiments in order to measure a pdf. We saw this with the *netstat*, file size, and *ftp* experiments. In the *design phase* of a system, however, there is no system available, and therefore we cannot resort to measurements for generating a pdf. Moreover, it is sometimes simply too expensive or time-consuming to take actual measurements. If you have tried the *netstat* or *ftp* experiments already, you understand pretty well how time-consuming this is! In all these cases, what is called for is modeling. *Computer simulation modeling* in particular is handy.

Once we have a computer simulation model of the actual system, we need to generate values for the random quantities that are part of the system input. As an example, consider simulation of a communication network node, depicted in Figure 3.52. Packets of information arrive at this node, from various users. Packets have random sizes; we must somehow generate actual values for packet sizes. The packets arrive at random instants; we must also generate those instants.

What we need, in general, is a facility or algorithm to generate values of a random variable X, given its pdf $f_X(x)$ or cdf $F_X(x)$. We call such values **variates of X** (or **deviates**[36] **of X**). Of course, the brute force way to generate variates of a given cdf is to devise an algorithm for each given cdf. This approach can work well if one is interested in only a few cdf's, but not otherwise.

A viable approach is suggested by our transformation ideas in Section 3.8. Suppose we have available an algorithm to generate variates V_n, $n = 1, 2, \ldots,$ of a random variable V with a specific cdf $F_V(x)$. Then we can use an appropriate transformation $g(\cdot)$ in order to generate a variate of X, call it $X_n = g(V_n)$.

---

[36]But consider *normal deviates*, for Gaussian random variables; I had a really hard time trying to justify this term to a French professor of statistics.

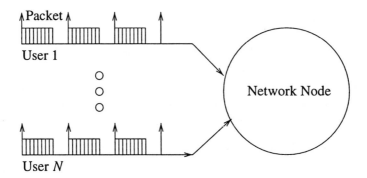

Figure 3.52   A model for a network node.

What is a good choice for $F_V(x)$? Theoretically, it does not matter what V is. In practice, of course, we have to consider how fast variates of V can be generated, since in simulations it is not uncommon to generate billions of variates.

Example 3.47. To simulate the stream of bits being transmitted on a communication line with speed 155 Mbits/s, for 30 min, we need to generate $155 \times 10^6 \times 30 \times 60 = 2.79 \times 10^{11}$ Bernoulli (0–1 valued) numbers!

Historically, a long time before computers or calculators, variates of the *uniform* random variable with parameters 0 and 1 were tabulated and then used to generate variates of other random variables *by hand*; older books on probability theory included such laboriously produced tables (e.g., [57], [56]). As it turns out, generating uniform variates on computers is fast, so the uniform random variable is still used today. All computer operating systems come equipped with a *uniform random number generator*. This is usually a built-in algorithm that uses modulo operations [46]. A typical such algorithm looks like

$$V_i = aV_{i-1} \bmod (2^{32}) \tag{3.89}$$

for computers with 32–bit CPUs. Typical values for the parameter a are $a = 7^5, a = 663,608,941$.

Our problem is then reduced to the following: How do we generate variates of a random variable X, with a given cdf $F_X(x)$, from variates of the uniform random variable V? In the next subsections we describe two of the most frequently used methods. Method 1 uses exactly one uniform variate in order to generate a variate of X. Method 2 uses at least two uniform variates and a "rejection" idea. A third, common method uses again at least two uniform variates and a transformation to generate the desired variate. For example, since it is known that the sum of n Bernoulli random variables follows a binomial distribution, one way of generating a binomial variate is to first generate n Bernoulli variates and then sum them. We do not describe this method further since it can be thought of as an extension of the first two.

3.13.2   An Approach That Uses the CDF (The Transformation Method)

The following theorem is the basis of an algorithm.

Theorem 3.10 *Let V be a random variable with a uniform cdf $F_V(x) = x$, $x \in [0, 1]$. Then the transformation $X = F_X^{-1}(V)$ gives variates X with cdf $F_X(x)$.*

Proof: Fix an $a \in R$. Then

$$P[X \le a] = P[F_X^{-1}(V) \le a] = P[V \le F_X(a)] = F_V(F_X(a)) = F_X(a).$$

The second equality follows from the monotonicity of the cdf. The fourth one follows from the fact that $0 \le F_X(a) \le 1$ and $F_V(v) = v$, for all $v \in [0, 1]$.   □

Based on Theorem 3.10, we can develop the following algorithm for generating variates of a random variable X (known as the *transformation method*):

1. Invert the given cdf to find $F_X^{-1}(x)$.

2. Generate a uniform variate V in $[0, 1]$.

3. Generate a variate via the transformation $X = F_X^{-1}(V)$.

Let's now look at the necessary transformations for some of the basic models in Section 3.6. Note that the method can be used only when the cdf is known in closed form since an inversion must take place in step 1 of the algorithm.

Example 3.48. *Bernoulli Random Variable.* Let p denote the parameter of this random variable. From Figure 3.53(a), we can easily see that the inverse of the cdf is

$$X = F_X^{-1}(V) = u(V - 1 + p),$$

where, as usual, $u(\cdot)$ is the step function.

Example 3.49. *Binomial Random Variable.* Let n, p be the parameters of this random variable. In this case the inverse transform is shown in Figure 3.53(b). We simply split the interval $[0, 1]$ into $n + 1$ subintervals, with the length of the ith such subinterval equal to $p_i \triangleq P[X = i], i = 0, 1, \ldots, n$. For example, if V belongs to the shaded area in Figure 3.53(b), then $X = F_X^{-1}(V) = n$.

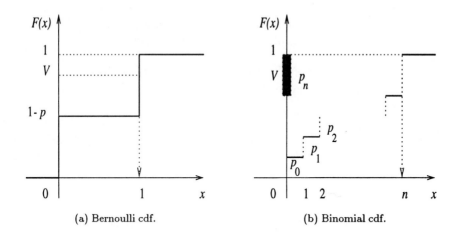

(a) Bernoulli cdf.　　　　　(b) Binomial cdf.

Figure 3.53   Inverse transformations.

Another way to generate a binomial random variable value is to make use of Equation 3.22. We simply generate n Bernoulli random variables and then sum them. This method is frequently used in practice since it does not require calculation of the probabilities p_i.

Example 3.50. *Exponential Random Variable.* Let $V = 1 - e^{-\lambda X}$. We can easily invert this transformation algebraically; solving for X, we get

$$X = -\frac{1}{\lambda} \log(1 - V),$$

and since the random variables V and $1 - V$ are both uniformly distributed in $[0, 1]$, we may use the transformation

$$X = -\frac{1}{\lambda} \log(V),$$

which saves us one subtraction.

Example 3.51. *Laplace Random Variable.* Consider a Laplace random variable with cdf

$$F_X(x) = \begin{cases} 0.5e^x, & x < 0, \\ 1 - 0.5e^{-x}, & x \geq 0. \end{cases}$$

We can invert this formula to yield the transformation

$$X = F_X^{-1}(V) = \begin{cases} \log(2V), & V < 0.5, \\ -\log(2(1 - V)), & V \geq 0.5. \end{cases}$$

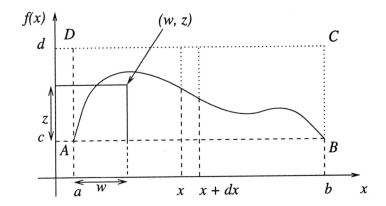

Figure 3.54 The rejection method.

Example 3.52. *Cauchy Random Variable.* Consider a Cauchy random variable with cdf

$$F_X(x) = \frac{1}{2} + \frac{1}{\pi} \tan^{-1}(x).$$

(What are the parameters of this random variable?) The desired transformation is

$$X = F_X^{-1}(V) = \tan(\pi(V - 0.5)).$$

3.13.3 An Approach That Uses the PDF (The Rejection Method)

The algorithm in the previous section assumed that the cdf is known. The method we describe below makes use of only the pdf, and thus it can be used when cdf's are not explicitly known (e.g., Gaussian and gamma random variables).

The rejection method is best explained through Figure 3.54. Consider a pdf $f_X(x)$, where $x \in [a, b]$ and $f_X(x) \in [c, d]$; that is, both x and $f_X(x)$ are bounded from above and from below. Then the rectangle $ABCD$ bounds the pdf as shown in Figure 3.54.

Suppose we perform the following "experiment": We throw darts that always land in the rectangle $ABCD$. When the dart lies "below" the $f_X(x)$ curve, we record the value X of the x coordinate of the landing point. When the dart lies "above" the curve, we repeat the experiment. This process will generate variates of a random variable X; we now claim[37] that the pdf of this random variable is equal to the desired $f_X(x)$.

The landing of the dart, at the point (w, z), is equivalent to selecting the coordinates W and Z as uniform random variables in the intervals $[a, b]$ and $[c, d]$, respectively. Consider a small interval $(x, x + dx]$, as shown in Figure 3.54. Let's define

$$E_1 \triangleq \{x < X \leq x + dx\}, \quad E_2 \triangleq \{Z \leq f_X(W)\}.$$

And E_1 is the event that the variate is in the range $(x, x + dx]$, while E_2 is the event that the algorithm will not reject the selected (W, Z) point. Consider now the probability that the algorithm *will indeed* generate a variate in the range $(x, x + dx]$. This can be expressed as the conditional probability $P[E_1|E_2]$, and thus

$$P[E_1|E_2] = \frac{P[E_1 \cap E_2]}{P[E_2]}.$$

Now $P[E_2]$ can be expressed as the ratio of the area under the pdf curve to the area of the bounding rectangle, and thus

$$P[E_2] = \frac{1}{(d - c)(b - a)}.$$

On the other hand,

$$P[E_1 \cap E_2] = \frac{f_X(x)\, dx}{(d - c)(b - a)},$$

and thus we finally get that $P[E_1|E_2] = f_X(x)\, dx$. From the frequency interpretation of the density, we can immediately see that the random variable X that the algorithm produces *does* follow the desired pdf.

In the above argument, we made the restricting assumption that the pdf is bounded in the interval $[c, d]$ and that the range of the random variable X is also bounded in some interval $[a, b]$. For a good discussion about relaxing these assumptions and improving the speed of the rejection method, see [59].

3.14   SUMMARY OF MAIN POINTS

- A **random variable** is a (measurable) function from S on R. It is a convenient way to avoid dealing with the subtleties of Q.

- Random variables are classified as **continuous, discrete**, and **mixed**.

---

[37] A rigorous proof that this algorithm works would require the notion of independence of two random variables and the notion of a function of two random variables; we introduce both notions in Chapter 4.

- The **cumulative distribution function** of X provides the conceptual link between probabilities in the original sample space S and probabilities in the real line R.

- The **probability density function** is the central concept of this chapter. On the theoretical side, it solves the questions of how we satisfy Axiom 3 and to which events we assign probabilities. On the practical side, it enables us to approximate probabilities via **histograms**.

- The most common random variable **models** are those described in Section 3.6 (Poisson, exponential, Gaussian, etc.).

- If $Y = g(X)$ is a given **transformation** of a random variable X, the pdf of Y is given by Theorem 3.1.

- **Means** and **variances** summarize information about the pdf of a random variable.

- **Transforms** of pdf's are powerful theoretical tools for dealing with pdf's.

- The **rejection** and **transformation** methods and a uniform random number generator suffice to generate **variates** of a random variable with a desired cdf or pdf.

3.15   CHECKLIST OF IMPORTANT TOOLS

- The pdf models in Section 3.6

- Histograms to approximate pdf's

- χ^2 tests to check how good models fit experimental data

- Transforms of pdf's

- Transformation Theorem 3.1

- The rejection method to generate variates of a random variable

- The transformation method to generate variates of a random variable

3.16   PROBLEMS

Sets $(-\infty, x]$

3.1 Express the set in Equation 3.1 in terms of sets of the form $(-\infty, x]$ *directly.*

3.2 Figure 3.1(b) is a graphical proof that $\{a\} = \lim_{n \to \infty} \left(a - \frac{1}{n}, a + \frac{1}{n}\right]$. Prove this result formally by using the definition of equal sets in Appendix A.

3.3 *What If?* The choice of sets $(-\infty, x]$ is not the only choice to jump away from Q. Consider, for example, the alternatives: (a) $(-\infty, x)$, (b) (x, ∞), (c) $[x, \infty)$, (d) $(0, x)$, and (e) $(0, x]$. Show that all alternatives are equivalent to our choice of $(-\infty, x]$, in the sense that they generate exactly the same subsets of R.

3.4 *Tricky.* Assigning probabilities to elements in Q can be nasty when S is "complex." Consider $S = [0, 1]$ and $Q = 2^{|S|}$. For any $s \in S$ and any $A \in Q$, let $f_A(s)$ be the indicator function of A [that is, $f_A(s) = 1$ when $s \in A$ and 0 otherwise]. It is tempting and so natural to define $P[A]$ as

$$P[A] = \int_S f_A(s)\, ds.$$

The Riemann integral will work well for sets A that are intervals; the troubles arise when we try subsets of S that are more complex than intervals.
(a) Plot the function $f_A(s)$ for $A = [a, b]$.
(b) Try to plot $f_A(s)$ for $A = \{$irrational numbers in $[0, 1]\}$.
(c) Try to plot $f_A(s)$ for $A = \{$rational numbers in $[0, 1]\}$.
(d) Is the integral well defined over the sets in parts (b) and (c)?

Concept of a Random Variable; Clarify Concepts

3.5 Let $S = \{a, b\}$ and $Q = 2^{|S|}$, and take $P[\cdot]$ arbitrarily. Consider the function $f : S \to R$ defined by

$$f(\zeta) = \begin{cases} 1, & \zeta = a, \\ 1, & \zeta = b. \end{cases}$$

(a) Is this function a random variable?
(b) How many random variables can you define on this sample space?
(c) Let $f(b) = \infty$ instead. Is this function a random variable?

3.6 Consider a discrete, finite sample space S. Take $Q = 2^{|S|}$; define a few random variables of your own.

3.7 Consider a sample space S with $n < \infty$ elements. How many random variables can you specify on this sample space?

3.8 Consider the sample space $S = \{1, 2, \ldots, n\}$. The set of events of interest is given by $Q = \{\emptyset, S, \{1, 2\}, \{3, \ldots, n\}\}$. Construct a nonmeasurable function on S.

3.9 Let $S = R$, the real line. Let

$$Q = \{A \subseteq R : A \text{ is countable or } A^c \text{ is countable}\}.$$

(a) Show that the function $f(\zeta) = \zeta$ is not a random variable.
(b) Show that the function

$$f(\zeta) = \begin{cases} 1, & \zeta \text{ is irrational,} \\ 0, & \zeta \text{ is rational,} \end{cases}$$

is a random variable.

3.10 Specify a triplet (S, Q, P) with S discrete and Q *not* the powerset of S. Define a few simple functions on S. Are they random variables?

Concept of a Random Variable; Apply Concepts

3.11 Consider the lock example in Section 1.9. Let $S = \{$all possible lock combinations$\}$. Let $Q = 2^{|S|}$. Consider a function $f : S \to R$ that describes choosing a combination. In other words, assuming combination 123 was chosen,

$$f(\zeta) = \begin{cases} 1, & \zeta = 123, \\ 0, & \zeta \neq 123. \end{cases}$$

Is f a random variable?

3.12 Consider a thief who tries to break your lock combination. How would you describe mathematically the following stealing strategies?
(a) Choose all even combinations serially, starting with 000.
(b) Choose all odd combinations serially, starting with 001.
(c) Choose 100–199 and 900–999; then 200–299 and 800–899; then 300–399 and 700–799; and so on.
(d) Choose all odd combinations serially, starting with 999.
(e) Choose a combination at random.

3.13 Define a random experiment that involves all the Internet-connected computers. Define a random variable on this experiment. State your sample space S.

3.14 Define a few random variables in some of the examples in Section 1.9.3.

3.15 Consider the GPA example in Section 1.9. Take $S = \{$all college students in the United States$\}$ and $Q = 2^{|S|}$. Let $G_i(\zeta)$ denote the GPA of student ζ during her or his ith semester. Assume it takes eight semesters to graduate and the threshold for a passing GPA is 2.000.
(a) Consider the function $f : S \to R$, defined as

$$f(\zeta) = \begin{cases} 1, & \text{if } G_i(\zeta) \geq 2.000, \\ 0, & \text{if } G_i(\zeta) < 2.000, \end{cases}$$

for any fixed value of i. Is this function a random variable? Can you describe this function in words?
(b) Consider now

$$h(\zeta) = \begin{cases} +2, & \text{if } G_i(\zeta) \geq 2.000, \\ -2, & \text{if } G_i(\zeta) < 2.000. \end{cases}$$

Is this function a random variable? Can you describe this function in words?
(c) Define convenient random variables to represent the total amount spent on tuition, in the entire United States, during a given semester as a random variable.

3.16 Consider the Dow Jones example in Section 1.9. We do not really have an exact expression of S for this random experiment; so let S be a vague sample space. Let C, O be random variables representing the closing and opening values of the Dow Jones index, respectively. Suppose that investors *always* buy their stock at opening time and sell it at closing time.
(a) Consider the function $f : S \to R$ defined as $f(\zeta) = C(\zeta) - O(\zeta)$. What does this random variable represent?
(b) Consider another function $h : S \to R$, defined as

$$h(\zeta) = \begin{cases} +1, & \text{if } C(\zeta) - O(\zeta) > 0, \\ 0, & \text{if } C(\zeta) - O(\zeta) = 0, \\ -1, & \text{if } C(\zeta) - O(\zeta) < 0. \end{cases}$$

What does this random variable represent?

3.17 *Fun!* In this problem we assume that you know the game of backgammon. Define three random variables related to that game. If you rate yourself as a backgammon *master*, define five.

3.18 If you are an NFL fan, define five random variables related to that game.

3.19 If you are an NBA fan, define five random variables related to that game.

3.20 All you ACC fans, define five random variables related to a game (basketball or football) in that conference. One of the variables must be related to

NCAA basketball championships. Fans of other leagues, define three random variables.

3.21 Define a few random variables for NL and AL.

The CDF

3.22 *What If?* Suppose you use the following sets instead of $(-\infty, x]$ to generate subsets of R: (a) $(-\infty, x)$, (b) (x, ∞), (c) $[x, \infty)$. Choose and match how the definition of the cdf would change:
(1) $P[X < x]$, (2) $P[X \geq x]$, (3) $P[X > x]$, (4) $P[X \leq x]$, (5) none of the above.

3.23 *What If?* Our friend Al Steinein suggests that we change a property of the cdf, just for fun. Are any of his suggestions catastrophic? Explain.
(a) $F_X(\infty) = 0.5$.
(b) $F_X(\infty) = -0.5$.
(c) $F_X(\infty) = 100$.
(d) $F_X(\infty) = \infty$.

3.24 Suppose you flip a coin four times, and each time you observe the face of the coin.
(a) Define the sample space S and a random variable X that counts the number of heads.
(b) Determine the range S_X.
(c) Calculate the cdf $F_X(x)$ for a loaded coin with $P[\text{heads}] = 0.8$.
(d) Can you see a pattern and generalize for n flips?

3.25 Calculate the cdf for the random variable that describes the number of heads in n tosses of a fair coin.

CDF Construction

3.26 Let $S = [0, 1] \times [0, 1]$. Let $P[A]$ be equal to the area of the subset $A \subset S$. Define the random variable X as follows:

$$X(\zeta_1, \zeta_2) = \begin{cases} \zeta_1, & \zeta_1 \geq \zeta_2, \\ \zeta_2, & \zeta_1 < \zeta_2. \end{cases}$$

(a) Plot this function.
(b) Find the minimum and maximum values that X can take.
(c) Plot the cdf $F_X(x)$ for all $x \in R$.

3.27 Repeat Problem 3.26 for the random variable $X(\zeta_1, \zeta_2) = |\zeta_1 - \zeta_2|$.

3.28 Consider the sample space $S = \{(a, b) \in R^2 : a^2 + b^2 \leq 1\}$. In words, S is the unit circle with center at the origin. For any subset $A \in Q$, we define its probability as the normalized area of A, or

$$P[A] = \frac{\text{area of } A}{\pi}.$$

Consider the random variable X defined by $X(a, b) = \sqrt{a^2 + b^2}$; that is, X is the distance of the point (a, b) from the origin.
(a) Determine the range S_X.
(b) Show graphically the subset of S that corresponds to the event $\{X \leq x\}$.
(c) Calculate and plot the cdf of X.

CDF; Mathematical Skills

3.29 Consider an event A with $P[A] > 0$ and a random variable X defined on some sample space S (the event does not have to be associated with X). Let $F_X(x|A)$ denote the conditional cdf of X. Show that Bayes' rule takes the form

$$F_X(x|A) = \frac{P[A|X \leq x]F_X(x)}{P[A]}, \quad \forall x \in R.$$

3.30 Consider two events A, B and a random variable X defined on a countable sample space S. Let x_0 be a fixed real constant. Suppose that for all $x \leq x_0$, we have $P[X = x] \neq 0$, and $P[A|X = x] = P[B|X = x]$. Show that

$$P[A|X \leq x_0] = P[B|X \leq x_0].$$

3.31 The pth percentile of a random variable X is denoted by x_p and defined via $F_X(x_p) = p$, where $p \in [0, 1]$.
(a) Interpret the meaning of the pth percentile.
(b) Does x_p exist for any given cdf?

3.32 Is it possible to find values for the constants a, b, c, d, e that make the function $F(x) = a + b \cdot x^c$, $x \in (d, e)$, a proper cdf? Explain.

3.33 Consider a random variable X with the property that

$$a \leq X(\zeta) \leq b, \quad \forall \zeta \in S.$$

Which statement(s) is (are) true?
(a) $F_X(x) = 0, \quad x < a,$
(b) $F_X(x) = 0, \quad x > b,$

(c) $F_X(x) = 1, \quad x > a,$
(d) $F_X(x) = 1, \quad x > b.$

3.34 Let $S = R$ and $A = [0, 1] \cup [3, 6] \cup [10, 20]$, with $P[A] = 0.3$. Consider a random variable X on S with the property that

$$X(\zeta) \begin{cases} = 0, & \zeta \in A, \\ \neq 0, & \zeta \in A^c. \end{cases}$$

Sketch a valid cdf for X for all $x \in R$. Which statement(s) is (are) true?
(a) $F_X(0) = 0.3.$
(b) $F_X(20) = 0.3.$
(c) $F_X(20) \geq 0.3.$
(d) $F_X(20) - F_X(0) = 0.3.$
(e) $F_X(6) < 0.3.$

3.35 *Tricky!* Let $S = [0, 1]$; define

$$X(\zeta) = \begin{cases} 0, & \zeta \text{ rational,} \\ 1, & \zeta \text{ irrational.} \end{cases}$$

Is X a random variable? If yes, can you sketch a valid cdf for X for all $x \in R$?

3.36 Consider the function

$$F(x) = \begin{cases} 1, & x \geq 2, \\ \frac{1}{4} + \frac{c}{2}(x+2)^2, & 0 \leq x < 2, \\ 0, & \text{otherwise.} \end{cases}$$

(a) Plot this function for various values of the constant c.
(b) Determine the value(s) of c that makes (make) $F(x)$ the cdf of a random variable X.
(c) Find the probabilities $P[X > 0.5], P[0.2 < X < 1.3]$.
(d) Now change the definition of $F(x)$ to

$$F(x) = \begin{cases} 1, & x > 2, \\ \frac{1}{4} + \frac{c}{2}(x+2)^2, & 0 \leq x \leq 2, \\ 0, & \text{otherwise.} \end{cases}$$

Repeat part (b).

3.37 Consider the cdf in Figure 3.55. What type of a random variable is X?

3.38 Consider the function

$$F(x) = \begin{cases} 0, & x \leq -1, \\ c[1 - \sin(\pi x)], & -1 < x \leq 1, \\ 1, & 1 < x. \end{cases}$$

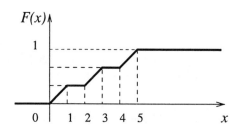

Figure 3.55   The cdf $F(x)$.

Determine, if possible, the constant c that makes $F(x)$ the cdf of a random variable X.

3.39 Consider the function

$$F(x) = \begin{cases} 0, & x \le 0, \\ c \, \sin(\pi x), & 0 < x \le 0.5, \\ 1, & 0.5 < x. \end{cases}$$

(a) Plot this function for various c values.
(b) Determine the constant c that makes $F(x)$ the cdf of a random variable X.
(c) What type of a random variable is X?
(d) Find the probabilities $P[X > 0.4]$, $P[0.2 < X < 1.5]$.

CDF; Conceptual Skills

3.40 Consider the set $S = \{$all college students in the United States$\}$. Consider the random variable X that describes the income of student ζ. Specify another random variable Y that is always greater than X, that is, $X(\zeta) < Y(\zeta), \quad \forall \zeta \in S$. Describe Y in words.

3.41 Consider a model (S, Q, P) with two random variables X and Y defined on S that satisfy

$$F_X(x) \le F_Y(x), \quad \forall x \in R. \tag{1}$$

We then call the random variable X *stochastically larger* than the random variable Y.
(a) Show that the condition

$$X(\zeta) \ge Y(\zeta), \quad \forall \zeta \in S, \tag{2}$$

implies that X is stochastically larger than Y.
(b) Show by counterexample that Equation (1) does not imply Equation (2).

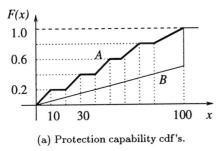
(a) Protection capability cdf's.

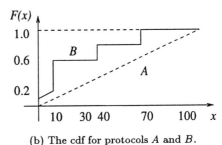
(b) The cdf for protocols A and B.

Figure 3.56

3.42 Let $S = \{0, 1, 2, \ldots, 10\}$. Can you specify one discrete, one mixed, and one continuous random variable on this sample space? Explain.

3.43 *Tricky!* Let $S = [-10, 20]$.
(a) Specify a continuous random variable X on this sample space.
(b) Specify a discrete random variable Y.
(c) Specify a mixed random variable Z.

3.44 Can you define a continuous random variable on the *ftp* experiment? Can you ever get a continuous random variable from experimental data?

3.45 Suppose a computer scientist-mathematician proposes two password generation algorithms A and B. Their protection capability is evaluated via the cdf of a random variable X that represents the time until a password is broken. The candidate cdf's are shown in Figure 3.56. Which algorithm would you choose for your system? Why?

3.46 You are hired as a consultant to evaluate the delay performance of two file transfer protocols A and B. The experimental cdf's for the transfer delay are shown in Figure 3.56(b). Discuss the following:
(a) Which protocol gives "smoother" delays?
(b) Which protocol is more likely to give smaller delays?
(c) Which protocol is more likely to give a delay of 30 time units exactly?
(d) Which protocol is more likely to give a delay less than 15 time units?

PDF

3.47 *What If?* Our friend Al Steinein does it for the pdf's, too; this time he suggests that for all $x \in R$:

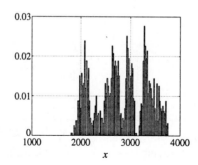

Figure 3.57   Experimental pdf for high value of Dow Jones.

(a) $f_X(x) \geq -1$ or (b) $f_X(x) \geq 2$ or (c) $0 \leq f_X(x) \leq 10$.
Any luck for our friend Al this time, or do you think there are problems with his suggestions?

3.48 *What If?* Suppose you choose the sets (a) $(-\infty, x)$, (b) (x, ∞), or (c) $[x, \infty)$ instead of $(-\infty, x]$ in order to define the cdf. Would you change the definition of the pdf? Why?

3.49 *Do It!* Consider the experimental pdf shown in Figure 3.57 (from the Dow Jones high value H; the unit on the x axis is cents). Calculate (either approximately, from this graph, or better yet, by generating a similar pdf from the data in the Web site) the following:
(a) $P[H > \$2000]$, (b) $P[H < \$3000]$, (c) $P[\$1900 < H < \$2500]$.

PDF; Mathematical Skills

3.50 Consider a random variable X with a symmetric pdf, that is, $f_X(x) = f_X(-x)$. Show that its percentiles satisfy $x_{1-p} = -x_p$. (See Problem 3.31.)

3.51 Consider a random variable X with pdf $f_X(x)$ given by

$$f_X(x) = \begin{cases} \frac{3}{4}(1 - x^2), & -1 \leq x \leq 1, \\ 0, & \text{otherwise.} \end{cases}$$

Calculate $P[X > 0], P[X < 0.5]$, and $P[|X| > 0.75]$.

3.52 Repeat Problem 3.51 when

$$f_X(x) = \begin{cases} \frac{3}{16}(1 - x^2), & -1 \leq x \leq 1, \\ 0.25\delta(x - 10) + 0.5\delta(x + 100), & \text{otherwise.} \end{cases}$$

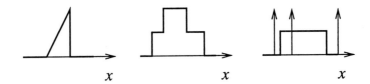

Figure 3.58   The pdf's in Problem 3.55.

3.53 Is it possible to find values for the constants a, b, c, d in order to make the function $f(x) = a\cos(x) + b$, $x \in (c, d)$, a proper pdf? Explain.

3.54 Consider the function

$$f(x) = \begin{cases} 0, & |x| \geq 1, \\ c[b + \sin(\pi x)], & -1 < x < 1. \end{cases}$$

(a) Plot this function for some values b, c.
(b) Can you determine the constants b, c that make it a proper pdf?

3.55 Determine the type of the three random variables whose pdf's are shown in Figure 3.58. The arrows specify delta functions.

3.56 In each of the following cases, sketch (if possible) a valid pdf $f_X(x)$ for a random variable X that satisfies the following properties. Determine the type of the random variable.
(a) $f_X(x)$ is increasing in $[0, 1]$ and is 0 everywhere else.
(b) $f_X(x)$ is decreasing in $(0, \infty)$ and is 0 everywhere else.
(c) The derivative of $f_X(x)$ is positive in $(0, \infty)$.
(d) The derivative of $f_X(x)$ is negative in $(0, \infty)$.
(e) The second derivative of $f_X(x)$ is negative in $(0, 1)$ and positive in $(2, 3)$.

3.57 (a) Can you find a constant c such that

$$f(x) = \frac{c}{|x|}, \quad x \in (-\infty, \infty),$$

is a valid pdf? Explain.
(b) Can you find a constant c such that

$$f(x) = \frac{c}{x^n}, \quad x \in (1, \infty),$$

is a valid pdf? Here n is a given positive integer. Explain.
(c) Can you find constants a, b such that

$$f(x) = \frac{1}{|x|}, \quad x \in [a, b],$$

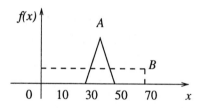

(a) Protection capability pdf's.

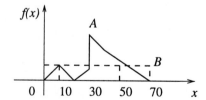

(b) The pdf's for protocols A and B.

Figure 3.59

is a valid pdf? Explain.

3.58 Consider the function

$$f(x) = \begin{cases} cx^2(x-1), & 0 \le x \le 1, \\ 0, & \text{otherwise.} \end{cases}$$

(a) Plot this function.
(b) Determine the constant c that makes $f(x)$ the pdf of a random variable X.
(c) Determine the corresponding cdf.
(d) Find $P[X > 0.3]$, $P[0.1 \le X \le 1.5]$.

3.59 Consider the function

$$f(x) = \begin{cases} cx^2, & |x| \le 1, \\ 0, & \text{otherwise.} \end{cases}$$

(a) Plot this function.
(b) Determine the constant c that makes $f(x)$ the pdf of a random variable X.
(c) Determine the cdf.
(d) Determine and plot $P[X > a]$ for $a \in (0, \infty)$.

PDF; Conceptual Skills

3.60 Consider Problem 3.45 again. The protection capability of the proposed algorithms is now evaluated via the pdf's for a random variable X that represents the time until a password is broken. These pdf's are shown in Figure 3.59(a). (Note that they are *not* the pdf's associated with the cdf's in Problem 3.45.) Which algorithm would you choose for your system? Why?

3.61 Consider Problem 3.46 again. The experimental pdf's for the transfer delay are shown in Figure 3.59(b). (They are *not* the pdf's associated with the cdf's in Problem 3.46.) Repeat parts (a) to (d) of Problem 3.46.

Models; *General*

3.62 For all discrete models in Section 3.6, verify that

$$P[X = k] \geq 0, \quad \sum_k P[X = k] = 1.$$

3.63 Verify that

$$f_X(x) \geq 0, \quad \int_{-\infty}^{\infty} f_X(x)\, dx = 1,$$

for as many continuous models in Section 3.6 as possible.

Bernoulli Model

3.64 Consider a random variable X with $S_X = \{0, 2\}$. Is X a Bernoulli random variable?

3.65 Can you model the flip of a loaded coin with a Bernoulli random variable? Why or why not?

3.66 Define a Bernoulli random variable on the experiment "draw a card from a 52-card deck."

3.67 Define two Bernoulli random variables on the *ftp* experiment.

3.68 Consider the *netstat* experiment. Errors are very infrequent, so we may suppose that within a 1-s interval we never have more than one error. (If we do get more than one, we can discard the measurement as an experimental error.) We can then use a Bernoulli random variable to model the number of errors occurring within 1 s. How can you determine the parameter p of this random variable from your experimental data?

3.69 Let $S = \{a_1, \ldots, a_n\}$ be a sample space with a finite number of elements n. Suppose that $P[a_i] = p_i$ is known for all i.
(a) Can you define two Bernoulli random variables on S with the same parameter p?
(b) How about different p's?
(c) Suppose that $p_i = \frac{1}{n}$. How many different Bernoulli random variables can you define on S that have $p = \frac{1}{n}$? That have $p = \frac{2}{n}$? Explain.

3.70 Let $S = \{a, b, c, d\}$. Let $P[a] = P[b] = P[c] = P[d] = 0.25$. Is the number of Bernoulli random variables you can define on S *larger* than 4^4? Explain.

3.71 Can you construct an *infinite* number of Bernoulli random variables $\{X_n\}_{n=1}^{\infty}$, each with parameter p_n, where $0 < p_n < 1$ is a given sequence, when the sample space S is:
(a) Discrete, (b) continuous, (c) the union of a discrete and a continuous set?

Binomial Model

3.72 Define, if possible, a binomial random variable with parameters $n = 2$ and $p = \frac{1}{52}$ on the card experiment.

3.73 Define, if possible, a binomial random variable on the *ftp* experiment.

3.74 Define, if possible, a binomial random variable on the *netstat* experiment.

3.75 Let S be a sample space with a finite number of elements k.
(a) Can you define two binomial random variables on S with the same n, p?
(b) How about different p's?
(c) How many binomial random variables can you define on S that have $n = k$ and $p = 0.1$? Explain.

3.76 Derive a recursive expression for calculating $P[X = k]$ for the binomial random variable.

3.77 Let $B(k, n, p) = P[X \leq k]$ denote the cdf of a binomial random variable X with parameters n, p. Show that

$$B(k, n, p) = 1 - B(n - k - 1, n, 1 - p).$$

3.78 Plot the binomial pmf for various n, p values. Observe how it approaches the Poisson distribution with parameter $\lambda = 1$ as n increases and p decreases such that $np = 1$.

3.79 Let X be a binomial random variable with parameters n, p (your choice). Plot (for various values of n, p):
(a) The cdf of X,
(b) $P[X > k]$, for various k,
(c) $P[a \leq X \leq b]$, for various a, b,
(d) $P[X = k | X = l]$ and $P[X = k | X \leq l]$ for various k, l.

Geometric Model

3.80 Define (if possible) a geometric random variable on the card deck experiment.

3.81 Define (if possible) a geometric random variable on the *ftp* experiment.

3.82 Define (if possible) a geometric random variable on the *netstat* experiment.

3.83 *Tricky!* Let S be a sample space with a finite number of elements n.
(a) Can you define two geometric random variables on S with the same p?
(b) How about different p's?

3.84 Let X be a geometric random variable with parameter $p = 0.7$. Find and plot:
(a) The cdf of X,
(b) $P[X > k]$,
(c) $P[a \le X \le b]$ for various a, b,
(d) $P[X = k | X \le l]$ for various k, l, p values.

3.85 Let X be a geometric random variable with parameter p. Show that

$$P[X > a + b | X > a] = P[X \ge b]$$

for all nonnegative integers a, b.

Negative Binomial Model

3.86 Define (if possible) a negative binomial random variable on the card deck experiment.

3.87 Define (if possible) a negative binomial random variable on the *ftp* experiment.

3.88 Define (if possible) a negative binomial random variable on the *netstat* experiment.

3.89 *Tricky!* Let S be a sample space with a finite number of equiprobable elements n.
(a) Can you define two negative binomial random variables on S with the same parameters r, p?
(b) With different r, p?

3.90 Let X be a negative binomial random variable with parameters r, p (your choice).
(a) Plot the cdf of X.
(b) Plot $P[X > k]$.
(c) Plot $P[a \le X \le b]$ for various a, b.
(d) Plot $P[X = k | X \ge l]$ for various k, l.

Poisson Model

3.91 Plot the Poisson pmf for various λ. Observe how the maximum value of the pmf behaves as a function of λ.

3.92 Let X be a Poisson random variable with parameter λ (your choice). Evaluate numerically and plot:
(a) The cdf of X,
(b) $P[X > k]$,
(c) $P[a \leq X \leq b]$ for various a, b.
(d) $P[X = k | X \geq l]$ for various k, l.

3.93 Let X be a Poisson random variable with parameter $\lambda = 1, 5, 10$. Calculate numerically and plot $h(k) \overset{\triangle}{=} P[X = k | X < 2k]$.

3.94 Consider a Poisson random variable X with parameter λ. Prove that for all $k = 0, 1, \ldots,$ $p_{k+1} = \frac{\lambda}{k+1} p_k$, where $p_k = P[X = k]$.

3.95 Let X be a Poisson random variable with parameter λ. Find the maximum value of $P[X = k]$ when $\lambda > 1$ and when $\lambda < 1$.

3.96 Let X be a Poisson random variable with parameter λ. Prove that

$$P[X \leq n] = \frac{1}{n!} \int_{\lambda}^{\infty} e^{-x} x^n \, dx.$$

3.97 Let X be a Poisson random variable with parameter λ. Prove that for all $n \geq 0$,

$$\frac{\lambda^n e^{-\lambda}}{n!} \leq P[X \geq n] \leq \frac{\lambda^n}{n!}.$$

3.98 Calculate $h(k) \overset{\triangle}{=} P[X = k | X > k - 1]$, where X is a Poisson random variable, with parameter λ; plot it as a function of k. What is $\lim_{k \to \infty} h(k)$?

3.99 Let's denote $P[X = k] = p(k, \lambda)$ when X is a Poisson random variable with parameter λ and $P[X = k] = b(k, n, p)$ when X is a binomial random variable with parameters n, p, where $\lambda = np$. Verify that for all $k \leq n$,

$$\frac{\lambda^k}{k!} \left(1 - \frac{k}{n}\right)^k \left(1 - \frac{\lambda}{n}\right)^{n-k} \leq b(k, n, p) \leq \frac{\lambda^k}{k!} \left(1 - \frac{\lambda}{n}\right)^{n-k}.$$

3.100 Calculate the percentiles x_p for various values of p for the Poisson random variable with parameter $\lambda = 0.1, 1, 10$. (See Problem 3.31.)

Uniform Model

3.101 For the uniform random variable X, with parameters $a = -1, b = 1$, calculate the following probabilities:
(a) $P[X \geq 0]$, (b) $P[X < -0.3]$, (c) $P[|X| \leq 0.5]$, (d) $P[|X| > -0.01]$.

3.102 Consider a uniform random variable X with parameters $a = 0, b = 1$.
(a) Calculate and plot $P[c < X < d]$ as a function of $c, d \in (-\infty, \infty)$.
(b) Calculate and plot $P[|X - c| > d]$ as a function of $c, d \in (-\infty, \infty)$.
(c) Calculate and plot $P[X > c | X < d]$ as a function of $c, d \in (-\infty, \infty)$.

3.103 Consider a uniform random variable X with parameters $a = 0, b = 1$. Calculate its pth percentile. Repeat for arbitrary a, b. (See Problem 3.31.)

Gaussian Model

3.104 Consider a Gaussian random variable X with parameters $m = -1, \sigma^2 = 2$. Calculate the following probabilities:
(a) $P[X \geq 0]$, (b) $P[X < -0.3]$, (c) $P[|X| \leq 0.5]$, (d) $P[|X| > 10]$.

3.105 Consider a Gaussian random variable with $m = 0, \sigma^2 = 1$. Show that for any $x > 0$

$$P[|X| \leq x] = 2 \int_{-\infty}^{x} \frac{1}{\sqrt{2\pi}} e^{-\frac{t^2}{2}} \, dt - 1.$$

3.106 Consider a Gaussian random variable X with parameters $m = 0, \sigma^2 = 1$. Calculate and plot the Q function.

3.107 Show that the Q function satisfies the property $Q(x) = 1 - Q(-x)$.

3.108 Show that for all $x > 0$,

$$\frac{x}{1 + x^2} e^{-\frac{x^2}{2}} < \int_{x}^{\infty} e^{-\frac{y^2}{2}} \, dy < \frac{1}{x} e^{-\frac{x^2}{2}}.$$

3.109 Based on the bounds in Problem 3.108, show that

$$Q(x) \approx \frac{1}{x\sqrt{2\pi}} e^{-\frac{x^2}{2}}.$$

Plot $Q(x)$ and this approximation.

3.110 Show that for all $x > 0$,

$$\left(\frac{1}{x} - \frac{1}{x^3} \right) e^{-\frac{x^2}{2}} < \int_{x}^{\infty} e^{-\frac{y^2}{2}} \, dy$$

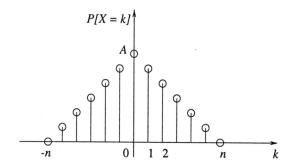

Figure 3.60   A triangular pmf.

3.111 Show that for all constants c,

$$\frac{Q(x + \frac{c}{x})}{Q(x)} \to_{x \to \infty} e^{-c}.$$

3.112 Calculate the pth percentile (see Problem 3.31) of a Gaussian random variable by approximating its Q function.

3.113 Consider a Gaussian random variable X with parameters m, σ^2. Evaluate $P[|X - m| < k\sigma]$ for $k = 1, 2, \ldots, 10$, through the Q function.

3.114 Calculate the conditional pdf and cdf of a Gaussian random variable X with parameters m, σ^2 when the conditioning event is $A = \{X < 2m\}$; when $A = \{|X - m| > 2\sigma\}$. Plot for various values of m, σ^2.

3.115 Show that

$$\int_{-\infty}^{\infty} e^{-ax^2 \pm bx}\, dx = \sqrt{\frac{\pi}{a}} e^{\frac{b^2}{4a}}.$$

Miscellaneous Models

3.116 The triangular pmf in Figure 3.60 was proposed by Thomas Simpson in the eighteenth century as a model of errors in observations. Given the value of the integer n, determine the proper value of A. Discuss the advantages and/or disadvantages of this model.

3.117 Consider an exponential random variable X with parameter λ. Calculate and plot, as a function of $c, d \in (0, \infty)$:
(a) $P[c < X < d]$, (b) $P[X > c | X < d]$, (c) $P[X > 2c | c < X < d]$.

3.118 Consider an exponential random variable X with parameter λ. Show that for all $c, d > 0,$

$$P[X > c + d|X > d] = P[X > c].$$

This is known as the *memoryless property*.

3.119 Can the memoryless property hold for a pdf that vanishes in $[b, \infty)$? For a pdf that is nonzero for negative x?

3.120 Consider an exponential random variable X with parameter λ. Find its pth percentile (see Problem 3.31).

3.121 Observe how the Cauchy pdf changes as its parameters vary.

3.122 Consider a Cauchy random variable X with parameters $a = 0, b = 1$.
(a) Calculate and plot $P[c < X < d]$ as a function of $c, d \in (-\infty, \infty)$.
(b) Calculate and plot $P[|X - c| > d]$ as a function of $c, d \in (-\infty, \infty)$.
(c) Calculate and plot $P[X > c|X < d]$ as a function of $c, d \in (-\infty, \infty)$.

3.123 Consider a Laplace random variable X with parameter $a = 1, 10, 100$.
(a) Calculate and plot $P[c < X < d]$ as a function of $c, d \in (-\infty, \infty)$.
(b) Calculate and plot $P[|X - c| > d]$ as a function of $c, d \in (-\infty, \infty)$.
(c) Calculate and plot $P[X > c|X < d]$ as a function of $c, d \in (-\infty, \infty)$.

3.124 Observe how the gamma pdf changes as its parameters vary.

3.125 Consider a gamma random variable X with parameters $a = 1, \lambda = 1, 5.5, 10$.
(a) Calculate and plot $P[c < X < d]$ as a function of $c, d \in (-\infty, \infty)$.
(b) Calculate and plot $P[|X - c| > d]$ as a function of $c, d \in (-\infty, \infty)$.
(c) Calculate and plot $P[X > c|X < d]$ as a function of $c, d \in (-\infty, \infty)$.

3.126 Plot the χ^2 pdf for various values of its parameter.

3.127 Consider a χ^2 random variable X with parameter $a = 1, 10, 100$.
(a) Calculate and plot $P[c < X < d]$ as a function of $c, d \in (-\infty, \infty)$.
(b) Calculate and plot $P[|X - c| > d]$ as a function of $c, d \in (-\infty, \infty)$.
(c) Calculate and plot $P[X > c|X < d]$ as a function of $c, d \in (-\infty, \infty)$.

3.128 Consider a random variable Y that is χ^2-distributed with parameter a an even number and a random variable X that is Poisson with parameter λ. Show that

$$F_Y(2\lambda) = 1 - F_X\left(\frac{a}{2} - 1\right).$$

3.129 Plot the F pdf and cdf for various values of its parameters.

3.130 Consider a random variable X that is F-distributed with parameters a, b. Show that

$$\lim_{a,b\to\infty} F_X(x) = \frac{1}{\sqrt{2\pi}} \int_{-\infty}^{x} e^{-\frac{t^2}{2}} \, dt.$$

3.131 Compute the cdf of the Rayleigh random variable.

Models; Apply Concepts

3.132 You have an instrument that measures voltages across a resistor. You know that the instrument is *not* calibrated; when there is no current across the resistor, the instrument shows a reading of 1 V. Suppose that you take a very large number of measurements and record the measurement errors. These errors are of course random; the following characteristics can be summarized from your measurements:

- Positive and negative errors are likely to occur.

- The extreme observed error values are -10 and 20.

- Most of the error values fall into the interval $[10, 11]$.

Based on the information given *and only that information:*
(a) Identify three pdf's that are *not* suitable fits for your data. Explain why they are not suitable fits.
(b) Consider the functions shown in Figure 3.61. Identify the one(s) that would fit your data well.

3.133 Suppose that an electron moves around a nucleus in a circle of radius R, where $r_1 < R < r_2$, for some constants $0 < r_1 < r_2$. The radius R of the orbit is a random variable that has a given pdf $f_R(r)$. Which two discrete and three continuous random variables from the models in Section 3.6 cannot be good models? Explain.

3.134 You want to model interarrival times of packets at a network node. You know that two successive packets will *never* arrive within 2 ms. Which of the pdf's shown in Figure 3.62 is (are) not appropriate models? (An arrow represents a delta function.) Explain.

Histograms

3.135 *Clarify Concepts.* Explain how you would calculate the histogram of a random variable, using variable length bins instead of the bins in Equation 3.42.

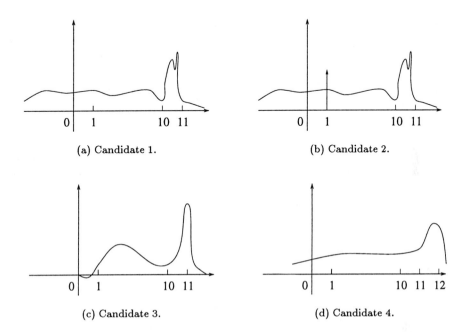

Figure 3.61 Candidate pdf's for the error data.

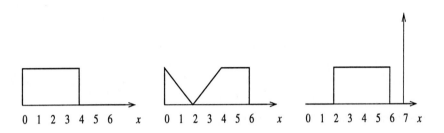

Figure 3.62 Candidate arrival models.

3.136 Replot a histogram (of, say, the Dow Jones data), using two different bin sizes. Comment on why and how the histograms differ.

3.137 *Do It!* Produce histograms for a given experiment (say, *netstat*) and for various numbers of measurements n. Decide experimentally when it's a good time to stop.

3.138 *Do It!* Calculate histograms for bit error rates, using a channel simulator or actual measurements (say, on a LAN in your environment).

3.139 *Do It!* Produce a histogram of Handel's *Messiah* (or any other sound file you have access to).

3.140 *Fun!?* In directory *CHISQUARE*, in the book Web site, are the files *dataset1.dat, dataset2.dat, dataset3.dat*. They contain some simulated random data.
(a) Choose a data set; create a histogram of this data set.
(b) Fit a model to it, using the χ^2 test. You have to choose the confidence level and degrees of freedom.

3.141 In the same directory is file *filesize.dat*. It contains the file sizes of all files in machine eceyv.ncsu.edu as of April 1994.
(a) Create a histogram of this data set.
(b) Fit a model to it, using the χ^2 test. You may want to take a log transformation before you fit a model. You are allowed to lower the confidence level, to facilitate your efforts.

3.142 Choose any random variable from the Dow Jones data; repeat Problem 3.141.

3.143 Consider the data set you chose in Problem 3.142. Fit a mix of Gaussian random variables, as best as you can.

Transformations of Random Variables; Simple Skills

3.144 Classify the three transformations in Figure 3.63(a) as linear or nonlinear, memoryless or not. State whether the transformations result in mixed random variables. You may assume that $f_X(x)$ is continuous and nonzero everywhere.

3.145 Prove Theorem 3.3 when the function $g(x)$ is not monotonic.

3.146 Let $g(x) = 5x + 2$. Find the cdf and pdf of $Y = g(X)$ when
(a) X is exponential, with $\lambda = 2$.
(b) X is uniform, with $a = 1, b = 10$.
(c) X is Gaussian, with $m = 10, \sigma^2 = 3$.

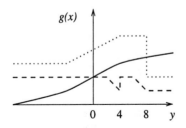

(a) Various transformations.

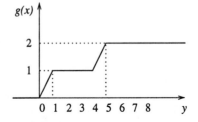

(b) The transformation in Problem 3.150.

Figure 3.63

(d) X is Bernoulli, with $p = 0.3$.
(e) X is Pascal, with $p = 0.3, r = 10$.

3.147 Let N be a Poisson random variable, and let k be a positive integer. Is $Y = N + k$ also a Poisson random variable? How about $Z = Nk$? Find the pmf of Z; find the pmf of Y.

3.148 Let $g(x) = x^2$.
(a) Apply the transformation $Y = g(X)$ to three discrete random variables in Section 3.6. Plot the pmf of Y.
(b) Apply the transformation $Y = g(X)$ to some of the continuous random variables in Section 3.6. Plot the pdf of Y.

3.149 Repeat Problem 3.148 for $g(x) = x^n$, n odd.

3.150 Consider the transformation described in Figure 3.63(b). Let X be a Poisson random variable. Find and plot the pdf of the random variable $Y = g(X)$.

3.151 Repeat Problem 3.150 for X an exponential random variable; a Gaussian random variable.

Transformations of Random Variables; Advanced Skills

3.152 Let X be an exponential random variable with parameter $\lambda = 1$. Find a transformation $Y = g(X)$ such that Y is a mixed random variable with the following properties:
(a) Its continuous part is nonzero only in the range $[-10, -5]$.
(b) Its discrete part consists of three terms only, namely, $\{5, 10, 15\}$, all equiprobable, with $P[Y = 5] = a$, where a has a suitable value.

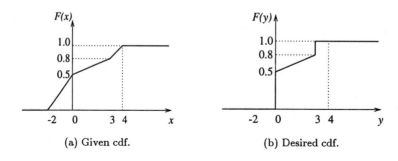

(a) Given cdf.   (b) Desired cdf.

Figure 3.64

3.153 Suppose you are given a random variable X with range $S_X = R$ and the cdf shown in Figure 3.64(a). What transformation $Y = g(X)$ will make the cdf of Y as shown in Figure 3.64(b)?

Specific Transformations

3.154 Let c be a positive constant. Consider the transformation

$$Y = \max\{0, X - c\}.$$

(a) Show this transformation graphically for various c values.
(b) Find the pmf of Y when X is the discrete uniform random variable with parameters $a = -c, b = 2c$.

3.155 Let c be a positive integer constant. Consider the transformation $Y = R(X, c)$, where the function $R(X, c)$ calculates the remainder of the division X/c.
(a) Show this transformation graphically for $c = 5$.
(b) Find the pmf of Y when X is a Poisson random variable.
(c) Find the pmf of Y when X is a binomial random variable.

3.156 Consider a "full-wave rectifier," described by the transformation $Y = |X|$. Find $f_Y(y), F_Y(y)$.

3.157 Disk space is allocated to files in the form of *blocks* of size d bytes; thus a file of size $d + 1$ bytes will actually be allocated $2d$ bytes. Let X represent the file size in bytes. The actual disk space a file would take is then given by the transformation:

$$Y = d \cdot \text{ceil}\left(\frac{X}{d}\right),$$

where ceil is a function that rounds a positive real number to the next integer. [More specifically, $\text{ceil}(x) = \min_{n=0,1,2,...}(n : x \le n)$.]

(a) Find the pmf of Y when the pdf of X is approximated by an exponential pdf.

(b) Find the pmf of Y when the file size follows the experimental pdf of the file size example in Section 1.9.

3.158 In Problem 3.157, $Y - X$ is the allocation "overhead." Determine the pdf of the random variable $Y - X$ experimentally.

3.159 Generate N values of an exponential random variable with parameter λ (see Section 3.13); quantize those values and determine experimentally the pmf of the quantizer output. Use various values for a, d and set $\lambda = k/a$, where $k = 1, 10, 100$. Interpret your results.

3.160 Repeat Problem 3.159 for a Poisson random variable.

3.161 Consider the clipping transformation

$$g(x) = \begin{cases} x, & |x| \le a, \\ a, & |x| > a. \end{cases}$$

(a) Find the cdf and pdf of $Y = g(X)$.

(b) What type of random variable is Y when X is binomial? Exponential?

3.162 Generate N values of an exponential random variable with parameter λ (see Section 3.13); clip those values and determine experimentally the pdf of the clipper output. Use various values for a, λ. Interpret your results.

3.163 Repeat Problem 3.162 for a Gaussian random variable.

3.164 The a-centered clipper is described by the following transformation:

$$g(x) = \begin{cases} x - a, & x > a, \\ 0, & |x| \le a, \\ x + a, & x < -a. \end{cases}$$

(a) Plot this function for various values of a.

(b) Find the cdf and pdf of $Y = g(X)$.

(c) What type of random variable is Y when X is continuous?

3.165 Choose a (nonuniform) random variable X; let $Y = \cos(X)$. Find the pdf or pmf of Y, theoretically or experimentally. Compare your results to Figure 3.39(b).

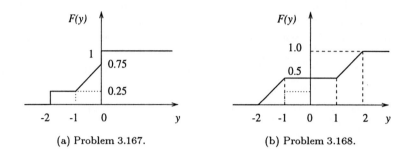

(a) Problem 3.167.   (b) Problem 3.168.

Figure 3.65   The desired cdf.

Transformations of Random Variable Models

3.166 Suppose X is a uniform random variable with parameters $a = 0, b = 1$. What is the pdf of $X - 0.5$? Of $1 - X$?

3.167 Let X be a uniform random variable with parameters $a = -1, b = 1$. Find a transformation $Y = g(X)$ that will make $F_Y(y)$ as shown in Figure 3.65(a).

3.168 Let X be a uniform random variable with parameters $a = -1, b = 1$. Find a transformation $Y = g(X)$ that will make $F_Y(y)$ as shown in Figure 3.65(b).

3.169 Suppose X is a uniform random variable with $a = -1, b = 1$. What transformation $g(x)$ makes the random variable Y exponential?

3.170 Consider a uniform random variable X with parameters a, b. Find the pdf of $Y = \log(X)$; find the pdf of $Z = \log(\log(X))$ theoretically or experimentally.

3.171 Let $Y = \tan(X)$, where X is a uniform random variable in $(-\pi, \pi)$. Show that Y is Cauchy-distributed.

3.172 Consider a Cauchy random variable X and the transformation $Y = \arctan(X)$. Find the pdf of Y.

3.173 Let X be a Cauchy random variable. Find the pdf of $Y = X^n$, for n a positive integer.

3.174 Let X be a Cauchy random variable. Find the pdf of $Y = 1/X$.

3.175 Let X be a uniform random variable with parameters $a = 0, b = 6$. Consider the random variable $Y = g(X)$, where $g(x)$ is shown in Figure 3.66.

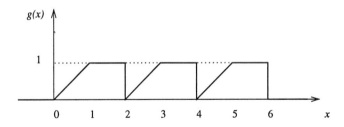

Figure 3.66 The function $g(x)$.

(a) Is Y a continuous, discrete, or mixed random variable?

(b) Find $P[Y > 0.75]$.

3.176 Let X be a Gaussian random variable with parameters $m = 0, \sigma^2 = 1$. Show that the random variable $Y = X^2$ is χ^2-distributed.

Transformations of Random Variables; Miscellaneous Problems

3.177 Show that the random variable $Z = F_X(X)$ is always uniformly distributed in $[0, 1]$, regardless of the random variable X.

3.178 Consider any experimental data set (such as the *netstat*, Dow Jones, and *ftp* measurements in the Web site, or your own). Find a transformation that will make the histogram associated with this data set as close to the histogram of a uniform random variable with parameters 0, 1 as possible. (*Hint:* Use the result of Problem 3.177.)

3.179 Consider a random variable X with a pdf symmetric around $x = 0$; let $Y = X^2$. Show that $F_Y(y) = 2F_X(\sqrt{y}) - 1$.

Moments of Random Variables; Simple Skills

3.180 Use the triangle inequality to show that $|EX| \le E|X|$.

3.181 (a) Show that the condition $EX < \infty$ implies that $\lim_{x \to \pm\infty} x f_X(x) = 0$.

(b) Show that if $E|X| < \infty$, then $|x| P[|X| \ge x] \to 0$ as $x \to \infty$.

3.182 Can you construct a random variable X which takes only a few values (say, no more than 5) and has the following properties?

(a) A given mean m.

(b) A given variance σ^2.

(c) A given mean m and variance σ^2.

3.183 Consider a discrete-valued random variable X with

$$P[X = 0] = c, \quad P[X = n\pi] = \frac{3}{n^2\pi^2}, \quad n = 1, 2, \ldots$$

Find c and EX.

3.184 Consider a sample space S (continuous or discrete, your choice). Construct an infinite number of random variables with a given pdf (your choice again) with $EX_n \to 0$, $EX_n^2 \to \infty$.

3.185 If $EX^2 < \infty$, is it true that always $EX < \infty$? That $EX^3 < \infty$?

3.186 Show that if $E|X|^a$ exists, $E|X|^b$ exists as well for all positive $b < a$.

3.187 Show that if $F_X(x) \geq F_Y(x), \forall x \in R$, then $EX \leq EY$. Show that the inverse is not true in general.

3.188 Show that if $E|X| < \infty$, then for every $\epsilon > 0$, there is a discrete random variable X_ϵ, with a finite range S_{X_ϵ}, such that $E|X - X_\epsilon| < \epsilon$.

Moments of Random Variables; Advanced Skills

3.189 Show that for all $\epsilon > 0$,

$$E\left(\frac{|X|}{1 + |X|}\right) - \frac{\epsilon}{1 + \epsilon} \leq P[|X| \geq \epsilon] \leq \frac{1 + \epsilon}{\epsilon} E\left(\frac{|X|}{1 + |X|}\right).$$

3.190 Consider an integer, nonnegative random variable X. Show that

$$EX = \sum_{k=0}^{\infty} P[X > k].$$

3.191 Consider an integer, nonnegative random variable X. Show that

$$EX^2 = \sum_{k=0}^{\infty} (2k + 1)P[X > k].$$

3.192 Consider a nonnegative random variable X. Show that

$$EX = \int_0^{\infty} [1 - F_X(x)]\, dx.$$

3.193 Let $\{a_n, \; n = 0, 1, 2, \ldots\}$ be a strictly increasing sequence of nonnegative numbers with $a_n \to \infty$. Let $a(x)$ be a strictly monotonic increasing, continuous function on $[0, \infty)$ such that $a(n) = a_n$. For any nonnegative random variable X, show that

$$\sum_{n=1}^{\infty} P[X \geq a_n] \leq Ea^{-1}(X) \leq \sum_{n=0}^{\infty} P[X > a_n].$$

3.194 *Total Probability Theorem for Expected Values.* Consider a sample space S and a random variable defined on S. Consider a partition $\{B_k, \ k = 1, \ldots, n\}$ of S. Define

$$E(X|B_k) \triangleq \int_{-\infty}^{\infty} x f_X(x|B_k) \, dx, \quad k = 1, \ldots, n.$$

Show that

$$EX = E(X|B_1)P[B_1] + E(X|B_2)P[B_2] + \cdots + E(X|B_n)P[B_n].$$

3.195 A tourist wants to visit n cities and so chooses one at random. After visiting a city, the tourist selects the next one among the rest (regardless of whether a city has been visited already) with equal probability. Find the average number of trips required until all n cities are visited.

Moments of Specific Models

3.196 *Bernoulli.* Evaluate EX^n, $n > 1$, for the Bernoulli model.

3.197 Verify that for the Bernoulli model with parameter p and for all integers $n \geq 1$,
$$E(X - EX)^n = p(1 - p)[(1 - p)^{n-1} - (-p)^{n-1}].$$

3.198 *Binomial.* Let X be a binomial random variable with parameters n, p. Show that
$$E(X - np)^4 = npq(3npq - 6pq + 1).$$

3.199 Evaluate EX^n, $n > 1$, for the binomial random variable.

3.200 *Tough!* Let X be a binomial random variable with parameters n, p. Show that
$$E|X - np| = 2np \binom{n-1}{m} p^m q^{n-m}$$

where m represents the integer part of np.

3.201 Consider a binomial random variable X with parameters n and p. Show that for all positive integers $r \leq n$,

$$\sum_{j=0}^{n} j(j-1)\cdots(j-r+1)P[X=j] = n(n-1)\cdots(n-r+1)p^r.$$

(*Hint:* Evaluate Ez^X and differentiate r times.)

3.202 For any given integer $n > 0$, let X_n be a binomial random variable with parameters n, p. Show that for every given $\epsilon > 0$,

$$\sum_{n=1}^{\infty} P\left[|\frac{X_n}{n} - p| \geq \epsilon\right] < \infty.$$

(*Hint:* Use the result of Problem 3.290 to bound this probability.)

3.203 *Poisson.* Verify that for the Poisson model with parameter λ

$$EX^3 = \lambda^3 + 3\lambda^2 + \lambda, \quad E(X-\lambda)^6 = 15\lambda^3 + 25\lambda^2 + \lambda.$$

3.204 Evaluate numerically and plot EX^n for the Poisson model.

3.205 Find the nth moment around the mean for the Poisson model.

3.206 *Geometric.* Evaluate EX^3 and $EX(X-1)(X-2)$ for the geometric model with parameter p.

3.207 *Gaussian.* Evaluate EX^3 for the Gaussian model. Can you eneralize for EX^{2n+1}, where n is a nonnegative integer?

3.208 Evaluate $E|X|$ for the Gaussian model with $m = 0, \sigma = 1$.

3.209 Consider a Gaussian random variable X with zero mean and unit variance. Show that for any positive integer n,

$$EX^{2n} = 1 \cdot 3 \cdots (2n-1).$$

3.210 *Cauchy.* Show that the mean of a Cauchy random variable X does not exist.

3.211 *Exponential.* Evaluate EX^n for the exponential random variable.

3.212 *Erlang.* Evaluate EX^m for the Erlang pdf as a function of its parameters.

3.213 *Student's t.* For a Student's t random variable X with parameter $a > 2$, show that

$$EX^{2n+1} = 0, \quad EX^2 = \frac{a}{a-2}.$$

3.214 *Pareto.* Consider the Pareto random variable with parameters a, b.
(a) Show that EX^n exists if and only if $n < a$.
(b) Find EX and EX^2.

3.215 *Miscellaneous.* Consider the Log-normal random variable with pdf

$$f_X(x) = \begin{cases} \frac{1}{\sqrt{2\pi}\sigma x} e^{-\frac{1}{2}(\frac{\log(x)-m}{\sigma})^2}, & x > 0, \\ 0, & \text{otherwise.} \end{cases}$$

Show that
$$EX = e^{m+\frac{\sigma^2}{2}}, \qquad \text{var}(X) = e^{2m+\sigma^2}(e^{\sigma^2} - 1).$$

3.216 *Do It!* Generate N variates of a random variable with a given pdf (see Section 3.13). Calculate var(X) experimentally, and compare its value to the theoretical one.

Medians of Random Variables

3.217 The *median* m_X of a random variable X is defined as the solution of the equation $0.5 = F_X(x)$.
(a) Does a solution exist for any continuous random variable? If yes, is it unique?
(b) How about a discrete random variable? A mixed one?
(c) For any real number a, show that

$$E|X - a| = E|X - m_X| + |(m_X - a)(1 - 2P[X < a])|$$

3.218 Consider a random variable X with median m_X. Suppose that for some constant c, $P[|X| \geq c] < a \leq 0.5$. Show that $|m_X| \leq c$.

3.219 Consider a random variable X with the property that X and $-X$ have the same cdf. Show that $m_X = 0$.

3.220 Let m_X denote the median of the random variable X. Find the median of the random variables $X + c$ and cX, where $c \neq 0$ is a real constant.

Moments of Transformed Random Variables

3.221 Let $Y = X\cos(\omega t) + c$, where c, ω, t are constants. Find EY and var(Y) in terms of EX and var(X).

3.222 Consider Problem 3.154. Find the mean and variance of the random variable $Y = g(X)$ when X is exponential.

3.223 Consider Problem 3.155. Find the mean and variance of the random variable $Y = g(X)$ when X is uniform with parameters $0, 2c$.

3.224 Find the mean and variance of the clipping transformation in Problem 3.161 when X is uniformly distributed with parameters b, c.

3.225 Find the mean and variance of the a-centered clipper in Problem 3.164 when X is uniformly distributed.

3.226 Consider an exponential random variable X with parameter λ. Evaluate and plot the second and third moments of the random variable $Y = \log(X)$ as a function of λ.

3.227 Show that for any nonnegative random variable X and constant $c > 0$,

$$E\min(X, c) = \int_0^c [1 - F_X(x)]\, dx, \quad E\max(X, c) = c + \int_c^\infty [1 - F_X(x)]\, dx.$$

Moments and Inequalities

3.228 *Jensen's Inequality.* Consider an arbitrary random variable X, with $E|X| < \infty$. To avoid unnecessary complexities, assume that X is discrete. Prove that for any convex function $g(x)$,

$$Eg(X) \geq g(EX).$$

Then use this inequality to prove that, for any real $s \geq 1$,

$$E|X|^s \geq (E|X|)^s$$

and, for any real numbers $0 < s < r < \infty$,

$$(E|X|^s)^{1/s} \leq (E|X|^r)^{1/r}.$$

3.229 Suppose the random variables X, Y have pdf's $f_X(x)$ and $f_Y(x)$, respectively, that are nonzero for all $x \in R$. Consider the transformations $V = \log(f_X(X))$ and $W = \log(f_Y(X))$. Show that $EV \geq EW$. (*Hint:* Use Jensen's inequality.)

3.230 Consider a random variable X with median m_X and variance σ^2. Show that

$$|EX - m_X| \leq \frac{\sigma}{\sqrt{2}}.$$

(*Hint:* Use Jensen's inequality.)

3.231 Consider a random variable X. For any given real numbers $a < b$, define truncated versions of X as the random variables Y, V, W, such that

$$Y = \begin{cases} X, & X \in [a, b], \\ 0, & \text{otherwise,} \end{cases} \quad ; \quad V = \begin{cases} X, & X > a, \\ 0, & \text{otherwise,} \end{cases} \quad ; \quad W = \begin{cases} X, & X < a, \\ 0, & \text{otherwise.} \end{cases}$$

Calculate $\text{var}(Y), \text{var}(V), \text{var}(W)$ and compare, if possible, to $\text{var}(X)$.

3.232 Consider a nonnegative random variable X with $E|X|^p < \infty$ for all $p > 0$. Define $f(p) \triangleq \log(EX^p)$. Show that $f(p)$ is a convex function.

3.233 Consider a discrete random variable X with range $\{0, 1, 2, \ldots\}$; suppose that its pmf is a nonincreasing function. Show that for all $k > 0$,

$$P[X = k] \le \frac{2EX}{k(k+1)}, \qquad P[X = k] \le \frac{6EX^2}{(2k+1)(k+1)k}$$

Experimental Calculation of Moments

3.234 Let X be a random variable you get from the *netstat* experiment. Find EX from your experimental data, as follows. First, obtain a pmf $f(k)$ from the histogram. Then calculate EX as $\sum_k k f(k)$. Now calculate

$$\bar{X} = \frac{1}{n} \sum_{i=1}^{n} X_i,$$

where $\{X_i, i = 1, 2, \ldots, n\}$ represent your measurements. Do the two values EX and $\bar{X}$ agree? Can you explain why?

3.235 Repeat Problem 3.234 for the variance of X. First calculate $\text{var}(X)$ through the pmf $f(k)$ and then calculate $\bar{V}$ as

$$\bar{V} = \frac{1}{n} \sum_{i=1}^{n} (X_i - EX)^2,$$

where EX is calculated in Problem 3.234. Are $\text{var}(X)$ and $\bar{V}$ the same? Why or why not?

3.236 Calculate the mean and variance of the closing, high, and opening values of the Dow Jones index. Calculate the mean volume of purchased shares.

3.237 Calculate the mean *ftp* transfer throughput, using your own measurements or the data provided in the Web site.

3.238 Calculate the mean error rate from the *netstat* measurements.

Reliability

3.239 Carefully define the range of the parameter t for the reliability and failure rate functions in Table 3.3.

3.240 Find the failure rate and reliability of the Erlang random variable. Plot them for all values of $t \in R$.

3.241 Find and plot the failure rate and reliability of the Rayleigh random variable.

3.242 Find and plot the failure rate and reliability of the Weibull random variable.

3.243 Evaluate numerically and plot the failure rate and reliability of the Poisson random variable as a function of λ.

3.244 Evaluate numerically and plot the failure rate of the binomial random variable as a function of n, p.

3.245 Evaluate numerically and plot the failure rate of the gamma random variable as a function of a, λ.

3.246 Evaluate and plot the reliability function of the Pareto pdf as a function of its parameters.

3.247 Consider a Weibull random variable X. Try to identify a transformation $Y = g(X)$ such that the failure rate of Y is greater than that of X. Plot the two failure rates.

3.248 Generate N variates of a random variable with a given pdf and known failure rate (see Section 3.13). Calculate the failure rate experimentally, and compare its value to the theoretical one.

3.249 Two systems S_1 and S_2 have respective reliability functions

$$R_1(t) = \begin{cases} 1 - \frac{t}{10}, & 0 \le t \le 10, \\ 0, & \text{otherwise,} \end{cases} \quad ; \quad R_2(t) = \begin{cases} 1 - 10t, & 0 \le t \le 0.1, \\ 0, & \text{otherwise.} \end{cases}$$

Which system is more reliable?

3.250 Consider a series interconnection of n subsystems. Each subsystem has reliability given by

$$R(t) = \begin{cases} 1 - t^2, & 0 \le t \le 1, \\ 0, & \text{otherwise.} \end{cases}$$

Find the mean time to failure.

Table 3.5 Life expectancies of U.S. population. (*Source:* National Center for Health Statistics, 1996.)

Age	Male	Female	All People in U.S.
0	72.3	79.1	75.4
10	63.2	69.8	66.6
20	53.7	60.1	56.8
30	44.5	50.4	47.5
40	35.5	40.9	38.3
50	26.8	31.6	29.3
60	18.9	23.1	21.1
70	12.4	15.5	14.2
71	11.8	14.8	13.5
72	11.2	14.1	12.9
73	10.7	13.6	12.3

3.251 The power supply of a system has a constant failure rate of 1 per month. We want to have a mean time to failure of 5 months. How many supplies should we have in standby mode?

3.252 *Do It!* Table 3.5 gives statistical data on life expectancies of people in the United States, based on their age.
(a) Fit a model (from those given in Table 3.3 or your own) that can best capture the given data.
(b) Based on your choice, predict the life expectancies for ages not given in the table, especially the ones in the 80–100 range.
(c) Fit a model using ages up to 70 only; then predict the life expectancies for ages 71, 72, and 73. Compare to the actual data.

Characteristic Functions; Simple Skills

3.253 Evaluate analytically the characteristic function of as many models as possible in Section 3.6.

3.254 Let $Z = aX + c$, where a, c are real-valued constants. Find the characteristic function of Z.

3.255 Find the pdf of a random variable X with characteristic function

$$\Phi_X(\omega) = \frac{e^{j\omega}}{c - e^{j\omega}}.$$

3.256 Consider two random variables X_1, X_2 with respective pdf's $f_{X_1}(x)$ and $f_{X_2}(x)$. Let $f_X(x) = af_{X_1}(x)+(1-(a)f_{X_2}(x)$. Find the characteristic function of X.

3.257 The random variable X_n is discrete uniform with parameters 1, n. Find its characteristic function; evaluate the limit as $n \to \infty$.

3.258 Suppose that $\Phi_X(\omega)$ is a characteristic function. The functions Re and Im return the real and imaginary parts of the argument, respectively. Find which of the following functions can also be characteristic functions.
(a) $|\Phi_X(\omega)|^2$, (b) $|\Phi_X(\omega)|^4$, (c) $\text{Re}(\Phi_X(\omega))$, (d) $\text{Im}(\Phi_X(\omega))$, (e) $[\text{Re}(\Phi_X(\omega))]^2$.

3.259 Can you determine whether the following functions are characteristic functions?

$$\frac{1}{1+\omega^4}, \quad \frac{10}{1+\omega^2}, \quad e^{j\,\cos(\omega)}, \quad e^{j\,\sin(\omega)}, \quad e^{j\,\cos^2(\omega)}, \quad e^{j\,\sin^2(\omega)}, \quad \frac{1}{1+\cos(\omega^4)}.$$

3.260 Find $\Phi_X(\omega)$ for the χ^2 random variable.

Characteristic Functions; Advanced Skills

3.261 Consider a random variable X with characteristic function $\Phi_X(\omega)$. Suppose that $|\Phi_X(\omega_0)| = 1$ for some $\omega_0 \neq 0$. Show that X is a discrete type of random variable.

3.262 Show that a characteristic function $\Phi_X(\omega)$ is real-valued if and only if the cdf of the random variable X is a symmetric function.

3.263 Show that if $\phi(\omega)$ is a characteristic function, then $\phi^2(\omega)$ is also a characteristic function.

3.264 Describe how you could (in principle) find the density of the random variable $Y = g(X)$ by using transform methods. Then use your ideas to find the pdf of $Y = X^2$ when X is Gaussian.

3.265 Consider an integer-valued random variable X. Show that

$$P[X = n] = \frac{1}{2\pi} \int_{-\pi}^{\pi} e^{-j\omega n} \Phi_X(\omega)\, d\omega.$$

Laplace Transforms; Simple Skills

3.266 Show that the condition "real part of $s > 0$" is sufficient for the Laplace transform to be well defined.

3.267 Can the following function be the Laplace transform of a pdf? If yes, which one?

$$L_X(s) = \frac{a}{s+a} \cdot \frac{b}{s+b}.$$

Probability Generating Functions; Simple Skills

3.268 Show that the condition $|z| < 1$ is sufficient for the probability generating function to be well defined.

3.269 Evaluate the probability generating function of as many models as possible in Section 3.6.

3.270 Find EX and var(X) for some of the models in Section 3.6, using Theorem 3.9.

3.271 Consider a nonnegative, integer-valued random variable X.
(a) Calculate EX^2, using the generating function $G_X(z)$ of X.
(b) Show that, in general,

$$E[X(X-1)(X-2)\cdots(X-n+1)] = \frac{d^n G_X(z)}{dz^n}\Big|_{z=1}.$$

c) Show that

$$G_X(z) = \sum_{i=0}^{\infty} \frac{1}{i!} E[X(X-1)(X-2)\cdots(X-i+1)](z-1)^i.$$

(d) Show that

$$P[X = k] = \sum_{n=k}^{\infty} (-1)^{n-k} \frac{1}{n!(n-k)!} E[X(X-1)(X-2)\cdots(X-n+1)].$$

(e) Consider a geometric random variable X with parameter p. Show that

$$E[X(X-1)(X-2)\cdots(X-n+1)] = n! \left(\frac{1-p}{p}\right)^n.$$

3.272 Find the pmf for the random variable X with $G_X(z) = z(1-z^3)^{-1}$. Can you generalize for $G_X(z) = z^n(1-z^3)^{-1}$?

3.273 Can you extend Theorem 3.9 for the case of a random variable X that can take negative values as well?

3.274 Prove Theorem 3.8. Can you extend it for a discrete random variable X that takes negative values as well?

Probability Generating Functions; Advanced Skills

3.275 Consider a discrete, positive random variable X. Consider the *shifted* version of X, defined for every outcome $\zeta \in S$ and a fixed integer $k > 0$ by

$$Y(\zeta) = \begin{cases} X(\zeta) - 1, & X(\zeta) \geq k, \\ 0, & X(\zeta) < k. \end{cases}$$

(a) Show that when $k = 1$,

$$G_Y(z) = P[X = 0] + \frac{G_X(z) - P[X = 0]}{z}.$$

(b) Generalize this result for the case of arbitrary k.

3.276 Consider a random variable X that takes values in $\{0, 1, 2, \ldots\}$. Let $\tilde{G}(z) \triangleq \sum_{k=0}^{\infty} P[X > k]z^k$. Show that for $|z| < 1$,

$$\tilde{G}(z) = \frac{1 - G_X(z)}{1 - z}.$$

Moreover,

$$EX = \tilde{G}(1), \quad \text{var}(X) = 2\frac{d\tilde{G}}{dz}|_{z=1} + \tilde{G}(1) - \tilde{G}^2(1).$$

3.277 Consider a random variable X with known probability generating function. Define

$$\tilde{G}_1(z) \triangleq \sum_{k=0}^{\infty} P[X \leq k]z^k, \quad \tilde{G}_2(z) \triangleq \sum_{k=0}^{\infty} P[X = ak]z^k,$$

where a is a fixed positive integer. Find $\tilde{G}_1(z), \tilde{G}_2(z)$ in terms of $G_X(z)$.

3.278 *Fun!* Florida is a crowded place during spring break; three friends, from Los Angeles, New York, and Raleigh got rooms in three different hotels in Fort

Lauderdale. They want to meet at a given hotel lobby and go surfing, at a time that we can conveniently call time $n = 0$. One of the friends, an expert in probability theory *and a spring break veteran*, proposes the following simple and fun algorithm for the rendezvous. Every hour and on the hour (i.e., at times $n = 1, 2, 3, \ldots$), they should show up at a hotel lobby. To make it more fun, the friend suggests that if one is missing, each should choose which hotel to go to next *totally randomly*.

(a) Let T be the time until all three meet at a lobby. Find the probability generating function of T.

(b) On the average, how many hours will the three friends waste if they use this rendezvous algorithm?

(c) If you have not yet guessed, the friend's name was Al Steinein. What would you suggest to improve on Al's algorithm?

(d) If only *two* instead of three friends are involved, how much time will they waste on average before the fun begins?

Transforms; *Nice Plots*

3.279 Plot the characteristic function of the Gaussian random variable as a function of ω. You may plot the phase and magnitude separately; or if you have a nice graphical package, you may plot it as a function over the complex plane.

3.280 Repeat Problem 3.279 for the Laplace transform of the gamma random variable as a function of s.

3.281 Repeat Problem 3.279 for the probability generating function of the binomial random variable as a function of z.

Tail Inequalities; *Clarify Concepts*

3.282 Compare the Markov and Chebyshev bounds with the exact probabilities for the following models: uniform, exponential, Bernoulli, Poisson, and discrete uniform.

3.283 Find, if possible, a discrete random variable X that satisfies the Markov bound with equality for a fixed, given a. Can you find a discrete random variable X that satisfies the Markov bound with equality for *all a*?

3.284 Repeat Problem 3.283 for the Chebyshev bound.

3.285 Repeat Problem 3.283 for the Chernoff bound.

3.286 Is it possible to find a continuous random variable X that satisfies the Chebyshev bound with equality for all a? Explain.

3.287 Plot the Chernoff bound in Inequality 3.88 for the Poisson random variable.

3.288 Investigate the tightness of the Chernoff bound for a gamma random variable.

Tail Inequalities; Mathematical Skills

3.289 Let $g(x)$ be a nonnegative function of x with the property that $x \geq a$ implies that $g(x) \geq b$ for some $b > 0$. Show that for any random variable X,

$$P[X \geq a] \leq \frac{Eg(X)}{b}.$$

3.290 Prove the following generalization of Chebyshev's inequality:

$$P[|X - c| \geq a] \leq \frac{E[|X - c|^n]}{a^n}.$$

Here c is an arbitrary real number, $a > 0$, and n is a positive integer.

3.291 Consider a positive random variable X with mean a. Show that

$$P[X \geq \sqrt{a}] \leq \sqrt{a}.$$

3.292 Prove the following version of the Chernoff bound:

$$P[X \leq a] \leq \min_{s<0} \{e^{-as} \phi(s)\}.$$

3.293 Do you think that by using moments in the set $\{EX, EX^2, EX^3, EX^4\}$ you can come up with better bounds for $P[X \geq a]$? Explain.

Generating Variates of Random Variables

3.294 Consider a communication line with capacity 1.5 Mbits/s. We transfer packets of variable size, with an average of 40 kbytes, over this line via some file transfer protocol. We wish to simulate the operation of this line, say, for h hours. How many random numbers do we need to generate?

3.295 Show that the transformation needed to generate a binomial random variable X from a uniform variate U is given by

$$X = \sum_{k=0}^{n} u\left(U - \sum_{i=0}^{k} p_i\right);$$

here n is the parameter of the binomial random variable and $p_i = P[X = i]$.

3.296 Suggest transformation algorithms for some of the random variable models in Section 3.6 that we did not discuss already in Section 3.13.

3.297 Suggest an algorithm to generate values for mixed random variables.

3.298 Implement the random number generator given by

$$U_i = 7^5 U_{i-1} \mathrm{mod}(2^{31} - 1).$$

Plot histograms of 100, 1000, and 100,000 numbers and observe how close to the uniform distribution they get.

3.299 Implement a random number generator of *your* choice, based on an equation similar to the one given in Problem 3.298. Plot histograms of the generated values, and check how close to a uniform distribution they get.

3.300 Write your own generator algorithms for given pdf's. Generate a few million variates. Compare the runtimes of your algorithm to the built-in ones for your system.

3.301 Generate n variates of a random variable X with a given distribution, using both the transformation and the rejection methods. Compare the speeds of the two methods. Compare the speed of the faster of the two to the speed of the generators provided in the Web site.

General Problems and Projects

3.302 An antismoking agency proposes the following "game" to entice smokers to quit smoking: Each smoker is provided with two boxes of matches, each containing initially $n = 20$ matches. The smokers may randomly choose a box, in order to light up a cigarette, without knowing the number of leftover matches. They *must quit* when they choose an empty box for the next light-up, in which case they are allowed to smoke a last cigarette and are given a prize equal to $1000 times the number of remaining matches in the other box.
(a) Find the average number of cigarettes a smoker will smoke before quitting.
(b) Determine the average cost of this game to the agency. (A variation of this problem, not on smoking, was solved by Stefan Banach, the Polish mathematician who invented Banach spaces.)

3.303 Take noisy measurements in a laboratory with (a) resistors and oscilloscopes or (b) transistors or (c) electrical motors. Discuss how you would process your measurements to "remove" the noise (more on this in Chapter 9).

3.304 Consider a manufactured product (e.g., a chip or a car). The lifetime of the product is a random variable Y. The lifetime can be "improved" if quality control and inspection equipment are used, at a cost of $X per product unit. With quality control, the lifetime of the product is another random variable Y'. The replacement of a failed product costs $R per product unit. The manufacturer wants to offer a guarantee of Z years on the product.

(a) Discuss under what conditions it is beneficial for the manufacturer to install quality control and inspection equipment.

(b) Suppose that this equipment is installed. The financial department insists that the guarantee costs be kept below 1 percent of the sale price S. Suggest how the manufacturer could choose the "optimal" value for Z.

3.305 Find about the *Nielsen ratings* that advertising companies use to assess how many people watch a given TV show. Discuss the advantages and disadvantages of this rating system.

3.306 Consider the Red Cross blood supplies for a given city. Certainly the blood supply and demand per day are random quantities.

(a) Define random variables for the blood supply and demand; discuss what models you could assume for pdf's.

(b) Suggest "good" policies for keeping blood supplies above a certain threshold. Describe your policies in terms of random variables.

3.307 Consider the letter frequencies in file *letfreq* in the Web site. Suppose we want to send large messages over a slow channel (such as a 19.2-Kbits/s modem). If we use the standard ASCII representation of characters that uses 8 bits per character, sending 1 million characters over this line will take approximately 800 s. We are interested in *encoding* the letters so that faster communication can take place.

One possibility that capitalizes on the randomness of the letters is the following: Suppose we use a *variable* length to encode the letters, with some letters using only 1 bit, others using 2 bits and so on. Can you finish this idea? In other words, can you specify an encoding algorithm that results in a faster communication? Can you quantify the speed up of your algorithm? For simplicity in your calculations, consider lowercase letters only.

3.308 Perform N file transfers from an archive machine on the Internet (e.g., ftp.nic.com to your account). Log in with user-id *anonymous* and password *your e-mail address*.

(a) Tabulate the N file size, delay, and throughput numbers.

(b) Derive a relative frequency histogram from the tabulated data.

(c) Fit a continuous pdf model from the ones in Section 3.6.

(d) Fit a discrete pdf model from the ones in Section 3.6.

(e) Fit a mixed pdf model from the ones in Section 3.6.

(f) Discuss any reasons why you should choose a continuous or a discrete pdf model to fit these data.

3.309 Let D be the delay of a file transfer in Problem 3.308 (measured in units of seconds). Compute the probabilities of the following two events: (a) $A = \{D \leq 3\}$; (b) $B = \{1 \leq D \leq 10\}$.

3.310 Perform N transfers of the same file from a local machine to a remote one. Derive histograms of the transfer delays as a function of the file size. If you are patient enough, observe the behavior of the histograms as N increases. Anything interesting?

3.311 Write a small program that opens a file and writes a fixed number of lines X in it. Measure the execution time of this program. Now make X a random variable that follows your favorite distribution from Section 3.6. Can you design an experiment that will determine the disk arm movement overhead from the observed program execution delays?

3.312 Consider Problem 3.12 again. Suppose that the thief will always try one of the mentioned attacks (e.g., the serial trial of all combinations from 000 to 999) and nothing else. The thief chooses an attack with equal probability. Let X be the number of trials before the thief breaks the combination. What combination would you choose in order to "maximize" X? How do you interpret the word *maximize*?

3.313 Can you evaluate a probability generating function from experimental data (say, a histogram) without having knowledge of the theoretical pmf? Explain. Simulate N values of a Bernoulli and a Poisson random variable; calculate the corresponding probability generating function from your data and explain any differences.

4

A VECTOR RANDOM VARIABLE

In the theory of Chapter 3, we always associated a random experiment with a *single* random variable. In many random experiments, however, we may naturally have to deal with *multiple* random variables. For example, in the *netstat* experiment, we actually monitor 20 random variables; we show them in the next section. In the communication channel experiment of Section 1.9, two random variables might be noise and demodulation errors. We will see more such examples in the next section.

On the other hand, in many random experiments, we may observe the value of a single random variable, but *repeatedly*, in consecutive trials of the experiment. For example, in the *netstat* experiment, we observe packet arrivals in repeated, consecutive 1-s intervals. Notice that before we actually perform the experiment, we may not know how many random variables we must deal with. In such cases, it is customary to talk about an infinite number of random variables. To summarize, we may have to deal with a finite or infinite collection of random variables. We postpone discussion of the latter case until Chapters 6 and 7. We deal with the former case in this chapter. We call a collection of N random variables an N-dimensional **random vector**, or a **vector random variable** with N elements.

Why bother about N-dimensional random vectors? Can't we simply "apply the theory of Chapter 3 N times," one for each random variable? As we will shortly see, no!

In Section 4.1 we describe some experiments with multiple random variables, in order to motivate the need for the theory of this chapter. In Section 4.2 we define some basic subsets of the plane R^2, which will help us in developing the theory. In Sections 4.3 and 4.4 we introduce the **joint cumulative distribution function** and **joint probability density function**, which, as their one-dimensional counterparts, play a key role in answering the theoretical questions about Axiom 3 and dealing with Q. In Section 4.5 we emphasize the fact that the joint cumulative distribution function or joint probability density function will be the probabilistic model of a random experiment with multiple random variables; again we can "forget" about the mathematical intricacies of Q and Axiom 3. In Section 4.6 we introduce a variety of **conditional probability functions**. They are quite useful in both theory (estimates) and practice (they are easier to calculate sometimes since conditioning removes uncertainty somehow). In Section 4.7 we establish criteria that guarantee **independence** of random variables. In the next section we develop a number of theorems regarding transformations of random vectors. In Section 4.9 we introduce the concepts of **covariance** and **correlation** of random variables; these concepts are heavily used in later chapters (estimation and analysis of random signals). In Section 4.10 we provide some **useful models**; in Section 4.11 we discuss the case of more than two random variables briefly. Finally, in Section 4.12 we describe an algorithm that generates variates of (dependent) vector random variables.

4.1   EXPERIMENTS WITH MORE THAN ONE MEASUREMENT

Example 4.1. *Netstat Experiment.* Consider the *netstat* experiment in Section 1.9. The output of this Unix command gives us actual measurements of 10 random variables. The command *netstat* 1 will give us repeated measurements for as long as we wish, so, in principle, we can get an infinite number of measurements. Moreover, we get a summary of each of the 10 random variables every 20 s, so this represents another 10 random variables, if you wish. Here is a sample of the *netstat* 1 command output; the summary is given in the third line.

input		(1e0)	output		input		(Total)		output
packets	errs	packets	errs	colls	packets	errs	packets	errs	colls
5550140	119	6970795	449	54593	5720943	119	7141598	449	54593
4	0	3	0	0	4	0	3	0	0
1	0	1	0	0	1	0	1	0	0
14	0	8	0	0	14	0	8	0	0
1	0	1	0	0	1	0	1	0	0
1	0	1	0	0	1	0	1	0	0
2	0	2	0	0	2	0	2	0	0
1	0	1	0	0	1	0	1	0	0
1	0	1	0	0	1	0	1	0	0
1	0	1	0	0	1	0	1	0	0
1	0	1	0	0	1	0	1	0	0

input (1e0)		output			input (Total)		output		
packets	errs	packets	errs	colls	packets	errs	packets	errs	colls
3	0	2	0	0	3	0	2	0	0
3	0	1	0	0	3	0	1	0	0
1	0	1	0	0	1	0	1	0	0
1	0	1	0	0	1	0	1	0	0
2	0	2	0	0	2	0	2	0	0
2	0	2	0	0	2	0	2	0	0
1	0	1	0	0	1	0	1	0	0
2	0	2	0	0	2	0	2	0	0
1	0	1	0	0	1	0	1	0	0
2	0	2	0	0	2	0	2	0	0
5550191	119	6970841	449	54593	5720994	119	7141644	449	54593
24	0	17	0	0	24	0	17	0	0
7	0	4	0	0	7	0	4	0	0
20	0	17	0	0	20	0	17	0	0
2	0	2	0	0	2	0	2	0	0
12	0	14	0	0	12	0	14	0	0
11	0	11	0	0	11	0	11	0	0
8	0	3	0	0	8	0	3	0	0
28	0	25	0	0	28	0	25	0	0
2	0	1	0	0	2	0	1	0	0

Example 4.2. *The ftp Experiment.* Consider next the *ftp* experiment. A typical output of this command gives us values for three random variables, namely, delay (in seconds), throughput (in Kbytes per second), and file size (in kilobytes). The following is a sample of the *ftp* command output.[1]

```
2192 bytes received in 0.11 seconds (20 Kbytes/s)
26573 bytes received in 0.46 seconds (56 Kbytes/s)
22 bytes received in 0.048 seconds (0.45 Kbytes/s)
852 bytes received in 0.1 seconds (8.3 Kbytes/s)
2332 bytes received in 0.14 seconds (16 Kbytes/s)
32075 bytes received in 0.63 seconds (50 Kbytes/s)
5258 bytes received in 0.23 seconds (22 Kbytes/s)
104215 bytes received in 1.5 seconds (69 Kbytes/s)
250770 bytes received in 3 seconds (81 Kbytes/s)
```

Example 4.3. *Teacher Evaluations.* In your teacher evaluation forms, you provide the school administrators with measurements of a few tens of random

---

[1] Transfer performed between Université de Paris 6 and Université de Paris 5, May 28, 1994.

variables. At North Carolina State University, our forms have 99 entries! Your SAT and GRE answer forms contain a few tens of random variables as well.

Example 4.4.   *National Elections.*   In national elections, you provide the country with "measurements" about random variables such as "the President," "the Vice-President," and "the congressperson representing state x." Notice that the first two random variables are very dependent, in the sense that both must come from the same running ticket.

Example 4.5.   *Transformed Random Variables.* Suppose that X and Y are two uniformly distributed random variables on $[0, 1]$. Then $X + Y$ and $\sqrt{X^2 + Y^2}$ are two new random variables.   We study such transformations of random variables in Section 4.8.

Remark 1: It is intuitively clear from the above examples that the elements of the random vector have some form of interrelationship: We might expect that high income (call it random variable X) implies better GPA (call it random variable Y); that large file size implies large delay. In a loose language, let $A = \{\text{high income}\}$, $B = \{\text{good GPA}\}$.[2] Both A and $B \in \mathcal{Q}$; A is defined in terms of the random variable X only, while B is defined in terms of the random variable Y only. To find $P[A]$ or $P[B]$, we can apply the theory in Chapter 3. But what is $P[A \cap B]$? How do we evaluate it? Neither $F_X(x)$ nor $F_Y(y)$ can give us an answer! What is $P[A|B]$? What we need is a *new* cdf, which assigns probabilities to *joint* events like these.

4.2   THE SETS $(-\infty, x] \times (-\infty, y]$

In Chapter 3 we saw that the set $(-\infty, x]$ played a very fundamental role in the transition phase from $\mathcal{Q}$ to the easily manageable $\mathcal{B}(R)$. The same problem (namely, the difficulty of assigning probabilities to arbitrary sets) is of course present with N-dimensional random variables. As in Chapter 3, we wish to map S on R^N now and then assign probabilities to a smaller, more structured collection of subsets of R^N. We will concentrate on the case $N = 2$ until Section 4.11, to simplify our presentation.

This smaller collection, which can generate (almost) all other subsets of R^2 (definitely the Borel ones), is now the collection of sets of the form $(-\infty, x] \times (-\infty, y]$. A typical set of this form is shown in Figure 4.1(a). Notice that this set is the Cartesian product of two subsets of R (see Appendix A).

Let's see how these sets can generate some other subsets of R^2, via unions, intersections, complements, and limits.

---

[2] More accurately, we should define $A \triangleq \{\zeta \in S : \text{student } \zeta\text{'s family income is high}\}$ and $B \triangleq \{\zeta \in S : \text{student } \zeta \text{ has a good GPA}\}$, for whatever *high* and *good* mean to you.

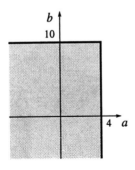
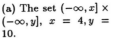
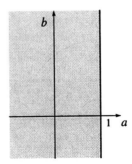

(a) The set $(-\infty, x] \times (-\infty, y]$, $x = 4, y = 10$.

(b) A vertical semi-plane at $x = 1$.

Figure 4.1   Two subsets of R^2.

4.2.1   Representation of Various Subsets of R^2 via $(-\infty, x] \times (-\infty, y]$

You may wish to consult Appendix A before you study this section. In the remainder of this section, we use the symbol $A_{x,y}$ to denote the set $(-\infty, x] \times (-\infty, y]$.

Example 4.6. *Horizontal and Vertical Semiplanes.* The *horizontal semiplane* $\mathrm{HS}_y \triangleq \{(a, b) \in R^2 : b \le y\}$ can be represented as

$$\mathrm{HS}_y = \lim_{n \to \infty} A_{n,y} = \lim_{n \to \infty} (-\infty, n] \times (-\infty, y].$$

Similarly, the *vertical semiplane* $\mathrm{VS}_x \triangleq \{(a, b) \in R^2 : a \le x\}$, shown in Figure 4.1(b), can be represented as

$$\mathrm{VS}_x = \lim_{n \to \infty} A_{x,n} = \lim_{n \to \infty} (-\infty, x] \times (-\infty, n].$$

Example 4.7. *Semi-infinite Strips.* The *vertical semi-infinite strip*

$$\mathrm{VSS}_{x_1,x_2,y} \triangleq \{(a, b) \in R^2 : x_2 < a \le x_1, b \le y\},$$

shown in Figure 4.2(a), can be represented as

$$\mathrm{VSS}_{x_1,x_2,y} = A_{x_1,y} - A_{x_2,y} = A_{x_1,y} \cap A_{x_2,y}^c.$$

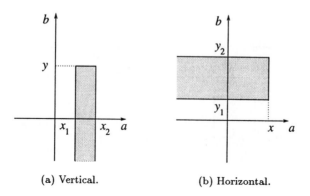

(a) Vertical.   (b) Horizontal.

Figure 4.2   Semi-infinite strips.

Notice that the line at $x = x_2$ does not belong to the set $\text{VSS}_{x_1,x_2,y}$. The *horizontal semi-infinite strip* $\text{HSS}_{x,y_1,y_2} \triangleq \{(a,b) \in R^2 : a \le x, y_1 < b \le y_2\}$, shown in Figure 4.2(b), can be similarly represented as

$$\text{HSS}_{x,y_1,y_2} = A_{x,y_2} - A_{x,y_1} = A_{x,y_2} \cap A^c_{x,y_1}.$$

Example 4.8. *Infinite Strips.* The *vertical infinite strip* $\text{VIS}_{x_1,x_2} \triangleq \{(a,b) \in R^2 : x_2 < a \le x_1\}$, can be represented as

$$
\begin{aligned}
\text{VIS}_{x_1,x_2} &= \lim_{n\to\infty} \text{VSS}_{x_1,x_2,n} \\
&= \lim_{n\to\infty} \left((-\infty, x_1] \times (-\infty, n] \cap ((-\infty, x_2] \times (-\infty, n])^c \right).
\end{aligned}
$$

The *horizontal infinite strip* $\text{HIS}_{y_1,y_2} \triangleq \{(a,b) \in R^2 : y_1 < b \le y_2\}$, can be represented as

$$
\begin{aligned}
\text{HIS}_{y_1,y_2} &= \lim_{n\to\infty} \text{HSS}_{n,y_1,y_2} \\
&= \lim_{n\to\infty} \left((-\infty, n] \times (-\infty, y_1] \cap ((-\infty, n] \times (-\infty, y_2])^c \right).
\end{aligned}
$$

Example 4.9. *Semi-infinite and Infinite Lines.* The *vertical semi-infinite line* $\text{VSL}_{x,y} \triangleq \{(a,b) \in R^2 : a = x, b \le y\}$ can be represented as

$$\text{VSL}_{x,y} = \lim_{n\to\infty} \text{VSS}_{x-\frac{1}{n},x+\frac{1}{n},y}.$$

The *horizontal semi-infinite line* $\text{HSL}_{x,y} \triangleq \{(a,b) \in R^2 : a \le x, b = y\}$ can be represented as

$$\text{HSL}_{x,y} = \lim_{n\to\infty} \text{HSS}_{x,y-\frac{1}{n},y+\frac{1}{n}}.$$

The *vertical infinite line* $\text{VIL}_x \triangleq \{(a, b) \in R^2 : a = x\}$ can be represented as

$$\text{VIL}_x = \lim_{n \to \infty} \text{VSL}_{x,n}.$$

The *horizontal infinite line* $\text{HIL}_y \triangleq \{(a, b) \in R^2 : b = y\}$ can be represented as

$$\text{HIL}_y = \lim_{n \to \infty} \text{HSL}_{n,y}.$$

Example 4.10. *Semi-infinite Lines Pointing Up.* The lines in the previous example were all extending to $-\infty$. Here is the expression for a vertical line $\text{VSLU}_{x,y}$ that starts at the point (x, y) and extends to $+\infty$:

$$\text{VSLU}_{x,y} = \text{VIL}_x - \text{VSL}_{x,y} = \text{VIL}_x \cap \text{VSL}_{x,y}^c.$$

The point (x, y) does not belong to the set $\text{VSLU}_{x,y}$.

Example 4.11. *Line Segments.* The *vertical line segment* $\text{VLS}_{x,y_1,y_2} \triangleq \{(a, b) \in R^2 : a = x, \; y_1 < b \le y_2\}$ can be represented as

$$\text{VLS}_{x,y_1,y_2} = \text{VSL}_{x,y_2} - \text{VSL}_{x,y_1} = \text{VSL}_{x,y_2} \cap \text{VSL}_{x,y_1}^c.$$

The *horizontal line segment* $\text{HLS}_{x_1,x_2,y} \triangleq \{(a, b) \in R^2 : x_2 < a \le x_1, \; b = y\}$ can be represented as

$$\text{HLS}_{x_1,x_2,y} = \text{HSL}_{x_1,y} - \text{HSL}_{x_2,y} = \text{HSL}_{x_1,y} \cap \text{HSL}_{x_2,y}^c.$$

Example 4.12. *Points.* The set $P_{x,y} \triangleq \{(a, b) \in R^2 : a = x, \; b = y\}$ contains the single point (x, y) and can be represented as

$$P_{x,y} = \lim_{n \to \infty} \text{VLS}_{x, y - \frac{1}{n}, y + \frac{1}{n}}.$$

Example 4.13. *Rectangles.* The rectangle $R_{x_1,x_2,y_1,y_2} \triangleq \{(a, b) \in R^2 : x_1 < a \le x_2, \; y_1 < b \le y_2\}$, shown in Figure 4.3, can be represented as

$$
\begin{aligned}
&R_{x_1,x_2,y_1,y_2} \\
&\quad = \left((-\infty, x_2] \times (-\infty, y_2] - (-\infty, x_2] \times (-\infty, y_1]\right) - (-\infty, x_1] \times (-\infty, y_2]
\end{aligned}
$$

Notice that the boundary lines AB, AD do not belong to the set R_{x_1,x_2,y_1,y_2}.

Example 4.14. *Circles.* Consider the unit-radius circle with origin at the point (x, y), shown in Figure 4.4(a). Let's divide the radius OA into $n \ge 2$ segments of equal length $\frac{1}{n}$. Call the segments $OS_1, S_1S_2, S_2S_3, \ldots, S_{n-1}A$. Let $I_{k,n}$ be the kth rectangle, obtained by projecting S_k on the circle. [In Figure

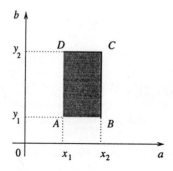

Figure 4.3   Rectangles.

4.4(a) we show $I_{1,n}$ and $I_{2,n}$.] From the previous expression for rectangles, we have

$$I_{k,n} = R_{x+\frac{k-1}{n},x+\frac{k}{n},y-\sqrt{1-(\frac{k}{n})^2},y+\sqrt{1-(\frac{k}{n})^2}}, \quad k = 1,\ldots,n.$$

Then the semicircle $BOCAB$ is obtained as

$$\lim_{n \to \infty} \cup_{k=1}^n I_{k,n}. \tag{4.1}$$

The other semicircle is obtained similarly; the full circle is of course the union of the two semicircles.

Example 4.15. *Triangles.* Consider the triangle AOB, shown in Figure 4.4(b). For an arbitrary integer $n \geq 1$, inscribe n rectangles H_k, $k = 1,\ldots,n$. (The width and height of the triangles are, of course, functions of n.) Then

$$H = \lim_{n \to \infty} \cup_{k=1}^n H_k$$

is the required expression for the triangle.

The point in all the above examples is that all these sets can be expressed as unions, intersections, complements, and/or limits of the sets $(-\infty, x] \times (-\infty, y]$. The collection of all subsets of R^2 generated by $(-\infty, x] \times (-\infty, y]$ (via unions, intersections, complements, and limits) is called the **Borel σ field** on R^2 and is denoted as $\mathcal{B}(R^2)$. With this field in mind, we can now formally define a two-dimensional random variable.

4.2.2   Two-Dimensional Random Variables

Consider a random experiment E with probabilistic model $(S, \mathcal{Q}, P)$. Consider two random variables X and Y defined on it, as shown in Figure 4.5, i.e.,

$$X \; : \; S \to R, \quad Y \; : \; S \to R.$$

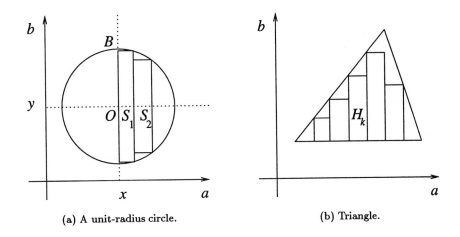

(a) A unit-radius circle.   (b) Triangle.

Figure 4.4   Nonrectangular sets.

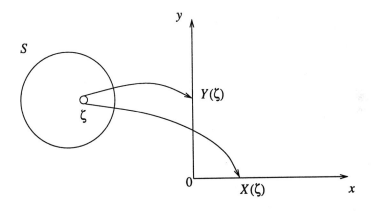

Figure 4.5   Two random variables on S.

For notational simplicity, we use boldface $\mathbf{X} \triangleq (X, Y)$ to represent the vector of the two random variables.[3]

Definition: **A two-dimensional random variable** is a function $\mathbf{X}$ from S on R^2

$$\mathbf{X} \ : \ S \to R^2 \tag{4.2}$$

---

[3] The first known instance where two random variables appear in the study of a problem is in a work on *joint* lives of siblings, in 1693.

such that sets in $\mathcal{B}(R^2)$ are mapped onto $\mathcal{Q}$.

The *range* S_{XY} contains all possible values that X, Y can take. It is important to observe here that both elements of the random vector are defined *on the same sample space.* If we have two random variables defined on two different spaces, we do not consider that as a two-dimensional random variable in the sense of Equation 4.2. If we take, for instance, X, the number-of-packets random variable from the *netstat* example, and Y, the GPA random variable from the student example, (X, Y) is *not* a two-dimensional random variable since its elements do not refer to the same random experiment. See our discussion in Subsection 4.7.3 for more information on how we can deal with unrelated experiments.

4.3   JOINT CUMULATIVE DISTRIBUTION FUNCTION

4.3.1   Definition and Properties

We need to establish some notation first. To simplify things a little, the event (subset of R^2) $(-\infty, x] \times (-\infty, y]$ will be denoted also as $\{X \leq x, \ Y \leq y\}$. Again, when denoting probabilities, we simplify $P[\{A\}]$ to $P[A]$, for any event A.

Definition: The **joint cumulative distribution function (jcdf)** of the two random variables X and Y, denoted by $F_{XY}(x, y)$, is defined as

$$F_{XY}(x, y) \triangleq P[X \leq x, \ Y \leq y], \quad \forall (x, y) \in R^2. \tag{4.3}$$

Remark 1: The right-hand side of Equation 4.3 is shorthand notation for the probability

$$P[\zeta \in S : \{X(\zeta) \leq x\} \cap \{Y(\zeta) \leq y\}]. \tag{4.4}$$

Note that the comma in the expression $X \leq x, \ Y \leq y$ has the same meaning as intersection.

Remark 2: *Conceptually, it is simple to calculate* $F_{XY}(x, y)$ *for a given* (x, y) *as a probability in the original sample space.* Indeed, observe from 4.4 that

$$P[X \leq x, \ Y \leq y] \triangleq P[\zeta \in S : \{X(\zeta) \leq x \cap Y(\zeta) \leq y\}],$$

and since the last event is an element in $\mathcal{Q}$, its probability can be determined (at least in principle) from the function P.

Table 4.1 A sample of *netstat* measurements.

Row	X	Y	Row	X	Y
1	11	15	12	17	20
2	11	13	13	13	18
3	5	9	14	18	22
4	9	11	15	27	32
5	14	19	16	21	26
6	18	21	17	19	23
7	11	17	18	29	35
8	22	26	19	24	26
9	16	20	20	89	37
10	16	21	21	58	23
11	10	15	22	23	7

Example 4.16. *Netstat Experiment.* Consider the data from the *netstat* experiment, shown in Table 4.1, where X represents the number of packets and Y represents collisions. For argument's sake, assume that all pairs are equiprobable. Let X denote the number of input packets and Y denote the number of collisions. Then the probability $P[X \leq 15, Y \leq 11]$ can be calculated from the tabular data as the sum of probabilities for pairs in rows 3 and 4 (that is, $1/22 + 1/22 = 1/11$).

In practice, who gives us the function $F_{XY}(x, y)$? We will come back to this question later, in Section 4.4.4. For now, let's investigate some of the properties of this function [compare them to Properties 3.4–3.7 of the single random variable cdf $F_X(x)$].

Lemma 4.1 *(1) $F_{XY}(x, y)$ is nondecreasing in both arguments, that is,*

$$F_{XY}(x_1, y_1) \leq F_{XY}(x_2, y_2), \quad \text{for } x_1 \leq x_2 \text{ and } y_1 \leq y_2. \qquad (4.5)$$

(2) Boundary values:

$$F_{XY}(-\infty, y) = F_{XY}(x, -\infty) = 0, \quad \forall x, y \in R. \qquad (4.6)$$

(3) $F_{XY}(x, y)$ is bounded from above by 1, that is,

$$F_{XY}(x, y) \leq 1, \quad \forall x, y \in R. \qquad (4.7)$$

(4) $F_{XY}(x, y)$ is continuous from the right, in both x and y, that is,

$$\lim_{x \to a^+} F_{XY}(x, y) = F_{XY}(a^+, y), \quad \forall y \in R. \qquad (4.8)$$

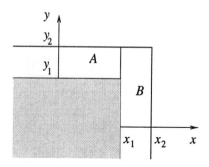

Figure 4.6   The sets in Property 1.

$$\lim_{y \to b+} F_{XY}(x, y) = F_{XY}(x, b^+), \quad \forall x \in R. \tag{4.9}$$

Proof: For fixed x_1, y_1, x_2, y_2, let $A \triangleq \text{HSS}_{x_1,y_1,y_2}$, $B \triangleq \text{VSS}_{x_1,x_2,y_2}$, as shown in Figure 4.6, $C \triangleq (-\infty, x_1] \times (-\infty, y_1]$, and $D \triangleq (-\infty, x_2] \times (-\infty, y_2]$. Since A, B, C are disjoint sets, we have $D = C \cup A \cup B$, and thus

$$P[D] = P[C] + P[A] + P[B] \geq P[C].$$

This proves Property 1. For Property 2, observe that for any fixed x or y, both sets $(-\infty, -\infty] \times (-\infty, y]$ and $(-\infty, x] \times (-\infty, -\infty]$ are equivalent to the empty set, and thus their probability must be equal to 0. Property 3 follows from Axiom 2 and Property 1. Equality is of course achieved for $x = \infty, y = \infty$. Continuity is much more difficult to derive in a rigorous fashion; the proof is similar to that in Lemma 3.1, so we omit it. □

If we are given a function $F(x, y)$, $x, y \in R^2$, which satisfies Properties 4.5–4.9, *are there* two random variables X and Y with F as their jcdf? Yes, and the proof is similar to that given for a single random variable in Section 3.3.3, so we omit it.

4.3.2   Marginal CDF

Once we are given the jcdf, we can easily recover the cdf's of the single random variables X and Y as follows. Observe that for any $x \in R$, we have

$$\{X \leq x\} = \{X \leq x, \ Y \leq \infty\}.$$

Therefore,

$$P[X \leq x] \triangleq F_X(x) = F_{XY}(x, \infty). \tag{4.10}$$

In order to distinguish $F_X(x)$ from the jcdf, we call the cdf of the single random variable the **marginal cumulative distribution function**. Similarly, for any $y \in R$, we have

$$P[Y \le y] \triangleq F_Y(y) = F_{XY}(\infty, y). \tag{4.11}$$

We can summarize these two facts as a lemma.

Lemma 4.2 *Consider two random variables X, Y with jcdf $F_{XY}(x, y)$. Then the marginal cdf's of the random variables X and Y are given by*

$$F_X(x) = F_{XY}(x, \infty), \quad \forall x \in R, \quad F_Y(y) = F_{XY}(\infty, y), \quad \forall y \in R.$$

4.3.3   Calculation of Probabilities of Various Events

Let's see now how the jcdf can be used to calculate probabilities of various events. The events of course will be subsets of R^2. Here we make extensive use of the representations in Section 4.2.

1. Rectangular sets

We will consider a variety of rectangles, depending on whether boundary lines or corner points are included in the event. Since the proofs of the lemmas in this subsection are quite similar, we leave most as exercises.

Rectangle $\{a < X \le b, \ c < Y \le d\}$. This rectangle is shown in Figure 4.7(a). Notice that the boundary lines AB, AD do not belong to this rectangle since Y is strictly greater than c and X is strictly greater than a.

Lemma 4.3

$$P[a < X \le b, \ c < Y \le d] = F_{XY}(b, d) - F_{XY}(b, c) - F_{XY}(a, d) + F_{XY}(a, c). \tag{4.12}$$

Proof: Observe that the event $\{X \le b, \ Y \le d\}$ can be written as

$$\begin{aligned}
\{X \le b, \ Y \le d\} &= \{a < X \le b, \ c < Y \le d\} \cup \{X \le a, \ Y \le d\} \\
&\quad \cup \{X \le b, \ Y \le c\} - \{X \le a, \ Y \le c\}.
\end{aligned}$$

(Sketch these sets in a two-dimensional plane.) The result follows easily.   □

Rectangle $\{a \le X \le b, \ c < Y \le d\}$. This rectangle is shown in Figure 4.7(b). Notice that the boundary line AD is not part of the event.

Lemma 4.4

$$\begin{aligned}
P[a \le X \le b, \ c < Y \le d] &= F_{XY}(b, d) - F_{XY}(b, c) - F_{XY}(a, d) \\
&\quad + F_{XY}(a, c) + P[X = a, \ c < Y \le d].
\end{aligned} \tag{4.13}$$

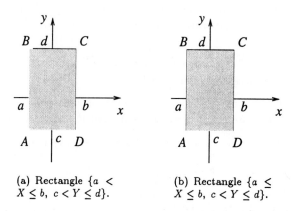

(a) Rectangle $\{a <$
$X \le b,\ c < Y \le d\}$.

(b) Rectangle $\{a \le$
$X \le b,\ c < Y \le d\}$.

Figure 4.7   Two rectangular subsets of R^2.

Rectangle $\{a < X < b,\ c < Y \le d\}$. This rectangle is similar to that in Figure 4.7(b) except that boundary lines AB, AD, and CD are not part of the event.

Lemma 4.5

$$P[a < X < b,\ c < Y \le d] \;=\; F_{XY}(b,d) - F_{XY}(b,c) - F_{XY}(a,d)$$
$$+\, F_{XY}(a,c) - P[X = b,\ c < Y \le d].$$
$$(4.14)$$

Of course, the probability $P[X = b,\ c < Y \le d]$ can be expressed in terms of the jcdf only, in a manner similar to that of Example 3.18.

Rectangle $\{a < X < b,\ c < Y < d\}$. This rectangle is again similar to that in Figure 4.7(b), except that no boundary line is part of the event.

Lemma 4.6

$$P[a < X < b,\ c < Y < d] \;=\; F_{XY}(b,d) - F_{XY}(b,c) - F_{XY}(a,d)$$
$$+\, F_{XY}(a,c) - P[X = b,\ c < Y \le d]$$
$$-\, P[a < X < b,\ Y = d].$$
$$(4.15)$$

Semi-infinite strip $\{a < X \le b,\ Y \le c\}$. This strip is similar to that in Figure 4.2(a). Observe that no points in the boundary line $x = a$ belong to the strip.

Lemma 4.7

$$P[a < X \le b,\ Y \le c] = F_{XY}(b,c) - F_{XY}(a,c).$$
$$(4.16)$$

Line segment $\{X = a, \ b < Y \le c\}$. Notice that the point (a, b) is not part of the segment.

Lemma 4.8

$$P[X = a, \ b < Y \le c] = \lim_{n \to \infty} F_{XY}\left(a + \frac{1}{n}, c\right) - \lim_{n \to \infty} F_{XY}\left(a + \frac{1}{n}, b\right)$$
$$- \lim_{n \to \infty} F_{XY}\left(a - \frac{1}{n}, c\right) + \lim_{n \to \infty} F_{XY}\left(a - \frac{1}{n}, b\right).$$

$$(4.17)$$

2. Nonrectangular sets

Probabilities for all the other subsets in Section 4.2.1 can be derived as limits, in a similar manner to the probability of a line segment. We show here how to derive an expression for a circular subset, and we leave the rest as exercises.

Circle $\{X^2 + Y^2 \le 1\}$. This circle is shown in Figure 4.4(a), where now $x = y = 0$; that is, the origin is at the point $(0, 0)$. Let $I'_{k,n}$ denote the generic rectangle used to represent the left semicircle in Figure 4.4(a). Since the sets in Equation 4.1 are disjoint, from Axiom 3a and Lemma 4.4 we get immediately the following:

Lemma 4.9

$$P[X^2 + Y^2 \le 1] = \lim_{n \to \infty} \sum_{k=1}^{n} P[I_{k,n}] + \lim_{n \to \infty} \sum_{k=1}^{n} P[I'_{k,n}].$$

4.3.4 Examples of Valid JCDFs

Let's see some examples of two-dimensional functions now that are valid jcdf's (i.e., they satisfy Properties 4.5 through 4.9).

Example 4.17. Figures 4.8(a) through 4.10(b) present the jcdf's for exponential, Poisson, and uniform random variables (see Section 4.10). The graphs are necessarily three-dimensional, so you must use your imagination a little. We show two different views of each graph, in order to get a feel for what they actually look like.

In Figure 4.8 we show the function

$$F(x, y) = \begin{cases} (1 - e^{-x})(1 - e^{-2y}), & x \ge 0, \ y \ge 0, \\ 0, & \text{otherwise.} \end{cases}$$

In Figure 4.9 we show the function

$$F(i, j) = \begin{cases} e^{-2}e^{-5} \sum_{k=0}^{i} \sum_{l=0}^{j} \frac{2^k}{k!} \frac{5^l}{l!}, & i, j = 0, 1, 2, \ldots, \\ 0, & \text{otherwise.} \end{cases}$$

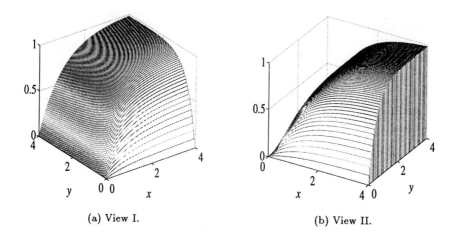

(a) View I.   (b) View II.

Figure 4.8   Two exponential random variables.

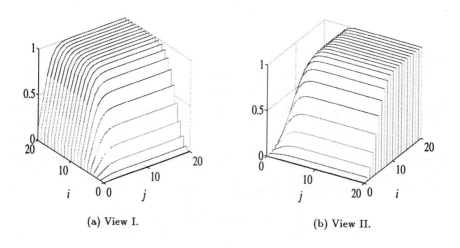

(a) View I.   (b) View II.

Figure 4.9   Two Poisson random variables.

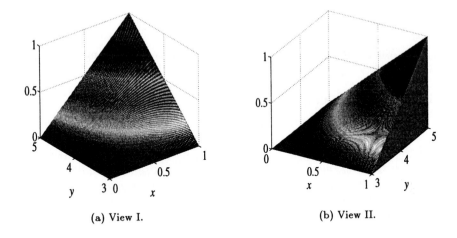

(a) View I.   (b) View II.

Figure 4.10   Two uniform random variables.

(This figure is a bit misleading, since i and j are discrete-valued.)

In Figure 4.10 we show the function

$$F(x, y) = \begin{cases} x \cdot \frac{y-3}{2}, & 0 \leq x \leq 1,\ 5 \geq y \geq 3, \\ 0, & \text{otherwise.} \end{cases}$$

4.3.5   Classification of Two-Dimensional Random Variables

In analogy to Chapter 3, we can classify the random vectors of this chapter into three categories:

Jointly continuous random vectors. These are random vectors with jcdf's that contain no jumps. Figure 4.8(a) shows an example of such a vector.

Jointly discrete random vectors. These are random variables that take only discrete values, typically integers. Figure 4.9(a) shows such an example. The joint probability mass function, which we introduce next, is typically used instead of the jcdf to describe such random variables.

Consider two random variables X and Y that take discrete values $\{x_i\}$ and $\{y_j\}$, respectively.

Definition: The **joint probability mass function (jpmf)** of two random variables X and Y is defined as

$$p_{XY}(i, j) \triangleq P[X = x_i,\ Y = y_j], \quad i, j = 0, 1, \ldots.$$

Mixed random vectors. These are vectors in which one component is discrete and the other continuous. We show such a random variable in Figure 4.16(a).

We describe jointly continuous and jointly discrete random vectors with more than two elements in Section 4.10.

4.3.6   State of Affairs So Far

Let's summarize the theoretical advances we have achieved so far: We once again avoided the problems of dealing with functions on arbitrary sets. Instead, we managed to reduce this problem to specifying a function of two arguments only. Still, however, we have not shown that such a function satisfies Axiom 3! We have exactly the same situation as with a single random variable. The joint probability density function, which we introduce next, is, as expected, the answer to that question. Again, after we settle all problems pertaining to rigor, we can tackle the implementation issues (Section 4.4.4).

4.4   JOINT PROBABILITY DENSITY FUNCTION

4.4.1   Definition and Properties

Definition: The **joint probability density function (jpdf)** of two random variables X and Y, denoted as $f_{XY}(x, y)$, is defined as

$$f_{XY}(x, y) \triangleq \frac{\partial^2 F_{XY}(x, y)}{\partial x \, \partial y}, \quad \forall (x, y) \in R^2. \tag{4.18}$$

Remark: The definition makes sense when the two random variables X and Y are jointly continuous. When one of them has a discrete component, however, $\delta(\cdot)$ functions must be considered in the derivatives.

Definition 4.18 implies some properties for the jpdf. Their proof is similar to that for the single random variable, so it is left as an exercise.

Lemma 4.10 *(1) The jpdf is nonnegative, that is,*

$$f_{XY}(x, y) \geq 0, \quad \forall x, y \in R. \tag{4.19}$$

(2) The jpdf integrates to 1, that is,

$$\int_{-\infty}^{\infty} \int_{-\infty}^{\infty} f_{XY}(x, y) \, dx \, dy = 1. \tag{4.20}$$

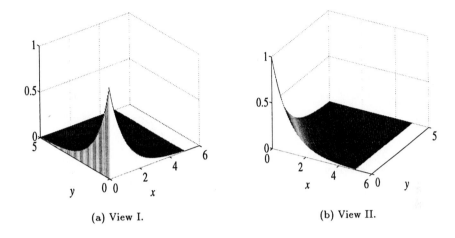

(a) View I. (b) View II.

Figure 4.11 Two exponential random variables.

Again, given a function $f(x, y)$, $(x, y) \in R^2$, which satisfies Properties 4.19 and 4.20, *are there two random variables* X *and* Y with $f(x, y)$ as their jpdf? That answer should be an easy yes now.

Why is Axiom 3 satisfied by the jpdf? For the same reason the pdf of a single random variable satisfies it! We do not discuss the issue further here; consult [25] for more details.

Here are some two-dimensional graphs for the jpdf of a variety of random variables (continuous, discrete, and mixed).

Example 4.18. In Figure 4.11 we plot the function

$$f(x, y) = \begin{cases} e^{-x}e^{-y}, & x \geq 0, \ y \geq 0, \\ 0, & \text{otherwise.} \end{cases}$$

Example 4.19. In Figure 4.12 we plot the function

$$f(k, l) = \begin{cases} \frac{2^k}{k!} \frac{5^l}{l!} e^{-2} e^{-5}, & k, l = 0, 1, 2, \ldots, \\ 0, & \text{otherwise.} \end{cases}$$

Example 4.20. In Figure 4.13 we plot the function

$$f(x, y) = \begin{cases} 0.5, & 0 \leq x \leq 1, \ 1 \leq y \leq 3, \\ 0, & \text{otherwise.} \end{cases}$$

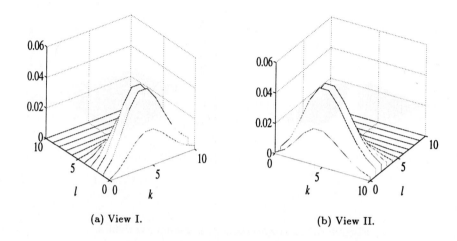

(a) View I.   (b) View II.

Figure 4.12   Two Poisson random variables.

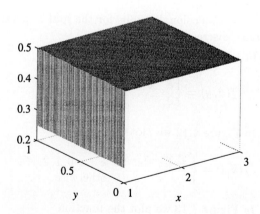

Figure 4.13   Two uniform random variables.

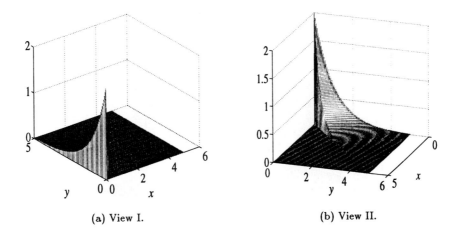

(a) View I. (b) View II.

Figure 4.14 Two random variables on a triangular region.

Example 4.21. In Figure 4.14 we plot the function

$$f(x,y) = 2e^{-x}e^{-y}u(y-x), \quad x,y \in R, \tag{4.21}$$

where $u(x)$ is the step function. Notice that the function $f(x,y)$ is identical to 0 whenever $y > x$ (that is, it is only defined on a triangular region of R^2).

Example 4.22. In Figure 4.15 we plot the function

$$f(x,y) = \frac{1}{9\pi}, \quad 16 \le x^2 + y^2 \le 25.$$

Example 4.23. In Figure 4.16 we plot the mixed function

$$f(x,k) = e^{-x}\frac{2^k}{k!}e^{-2}, \quad x \ge 0, \; k = 0,1,\ldots.$$

4.4.2 Marginal PDF

We can easily derive the marginal pdf's from the jpdf. Differentiating Equation 4.10 with respect to x (and using Definition 4.18), we can easily prove the following:

Lemma 4.11 *The marginal pdf's of the random variables X, Y are given by*

$$f_X(x) = \int_{-\infty}^{\infty} f_{XY}(x,y)\, dy, \quad f_Y(y) = \int_{-\infty}^{\infty} f_{XY}(x,y)\, dx.$$

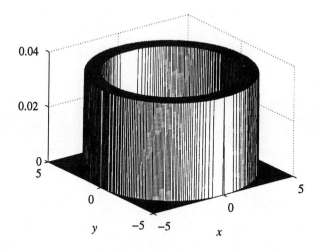

Figure 4.15   Two random variables defined on a ring.

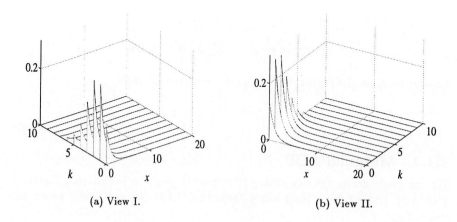

(a) View I.   (b) View II.

Figure 4.16   Mixed (exponential and Poisson) random variables.

4.4.3 Calculation of Probabilities Through the JPDF

In general, the probability of A, a subset of R^2, is given as the integral of the jpdf over A, or

$$P[A] = \int\int_A f_{XY}(x,y)\,dx\,dy.$$

Formulas 4.12–4.17 can be easily restated through the jpdf. Care should be exercised, though, with the limits of integration when discrete or mixed random variables are involved. You may have to integrate a $\delta(\cdot)$ function that sits right at the end on the integration interval! In analogy to Lemma 3.4, we have the following:

Lemma 4.12

$$P[a < X \le b,\ c < Y \le d] = \int_{a+}^{b+}\int_{c+}^{d+} f_{XY}(x,y)\,dx\,dy,$$

$$P[a \le X \le b,\ c < Y \le d] = \int_{a-}^{b+}\int_{c+}^{d+} f_{XY}(x,y)\,dx\,dy,$$

$$P[a < X < b,\ c < Y \le d] = \int_{a+}^{b-}\int_{c+}^{d+} f_{XY}(x,y)\,dx\,dy,$$

$$P[a < X < b,\ c < Y < d] = \int_{a+}^{b-}\int_{c+}^{d-} f_{XY}(x,y)\,dx\,dy,$$

$$P[a < X \le b,\ c \le Y] = \int_{a+}^{b+}\int_{c-}^{\infty} f_{XY}(x,y)\,dx\,dy,$$

$$P[X = a,\ b < Y \le c] = \int_{a-}^{a+}\int_{b+}^{c+} f_{XY}(x,y)\,dx\,dy.$$

4.4.4 Two-Dimensional Histograms as an Approximation of JPDF

Now that our mathematical arsenal is well stocked and our mathematical worries are (hopefully) settled, it is time to play probability ball once again. Finally, the interesting question: How can we determine the jpdf for a given random experiment? The **histogram**, two-dimensional now, is the answer.

Let's take a closer look at the following interpretation of the jpdf:

$$f_{XY}(x,y)\,dx\,dy \approx P[x < X \le x + dx,\ y < Y \le y + dy]. \tag{4.22}$$

Figure 4.17(a) explains the idea behind Equation 4.22.

What Equation 4.22 suggests then is the following: If we want to approximate the value of the jpdf at a given point (x,y), we must find the probability

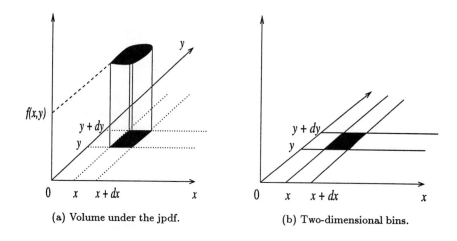

(a) Volume under the jpdf.   (b) Two-dimensional bins.

Figure 4.17   Two-dimensional histograms.

of the rectangle $\{x < X \le x + \Delta x, \; y < Y \le y + \Delta y\}$. The probability can of course be approximated in practice as a relative frequency; the measurements now must be sorted in two-dimensional bins, as shown in Figure 4.17(b).

The questions of Section 3.7 arise here as well. Fitting models to experimental histograms by inspection is now a lot more difficult (and of course *impossible* for more than two random variables).

4.4.5   Experimental Histograms

Based on the above ideas, we have evaluated experimental histograms for some of the experiments in Section 1.9. With the MATLAB code provided, it is very simple to generate your own histograms, for your own defined experiments as well. Try it!

The histograms in Figures 4.18 through 4.20(b) come from the relevant examples in Section 1.9. The histogram in Figure 4.18 appears to be "concentrated" around the (0,0) point since most measurements indicate a very low number of errors. The histograms in Figure 4.19 appear to be concentrated along the diagonals since the closing and opening values of the Dow Jones index do not differ by much. For better resolution, we took the logarithm of delay and throughput values in the histograms in Figure 4.20.

The histograms in Figures 4.21 through 4.24(b) have been created by generating 5000 samples for each of the random variables in question. In all cases, we generated uniform variates in the interval $[0, 1]$. We then transformed the

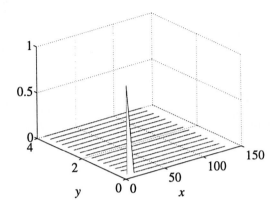

Figure 4.18 *Netstat* histogram, number of packets (X) and error (Y) random variables.

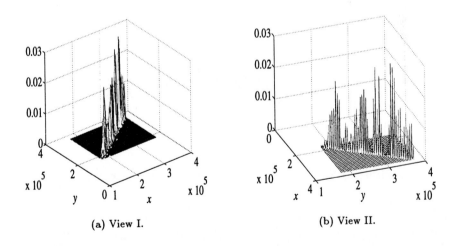

(a) View I.

(b) View II.

Figure 4.19 Dow Jones closing (X) and opening (Y) random variables.

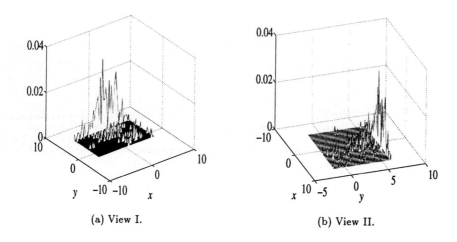

(a) View I.   (b) View II.

Figure 4.20   The *ftp* delay (X) and throughput (Y) random variables.

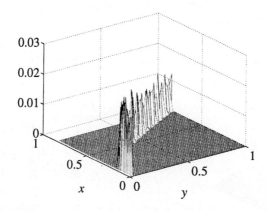

Figure 4.21   The random variables X and $Y = X\sin(X)$.

variates as the figure suggests. In some of the experiments, it is apparent that the sample size was not large enough. (Can you tell which ones?) In all of them, we used 50 bins for sorting the measurements.

In Figure 4.21, we plot the histogram of the uniform random variable X and the random variable $Y = X\sin(X)$.

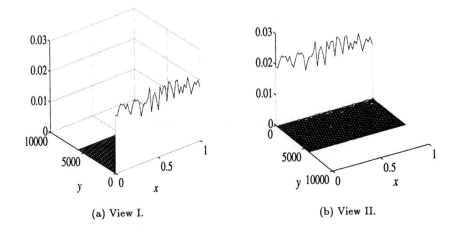

(a) View I. (b) View II.

Figure 4.22 The random variables $X = \bar{X}$ and $Y = \bar{X}/\bar{Y}$.

For the histograms in Figure 4.22, we created two uniform[4] random variables $\bar{X}, \bar{Y}$, each with a range $[0, 1]$. We then plot the histogram of the random variables $X = \bar{X}$ and $Y = \bar{X}/\bar{Y}$. Most of the values of Y are around 1, hence the shape of this histogram.

The histogram in Figure 4.23(a) plots the random variables $X = \bar{X}$ and $Y = \bar{X} \cdot \bar{Y}$. The shape of this histogram can be explained from the fact that $X > Y$. The histogram in Figure 4.23(b) depicts the random variables $X = \bar{X}$ and $Y = \bar{X}^2 + \bar{Y}^2$. Figure 4.24 depicts the histogram of the two random variables $X = \bar{X} \cdot \bar{Y}$ and $Y = \bar{X}/\bar{Y}$.

4.5   PROBABILISTIC MODEL REVISITED

Once again, we will leave the apparatus of $(S, \mathcal{Q}, P)$ behind us and focus on $f_{XY}(x, y)$ or $F_{XY}(x, y)$ as our probabilistic model for any random experiment with two random variables defined on it.

If we have to fit experimental data pertaining to two random variables we must now fit two-dimensional functions. This is easier said than done since inspectionlike methods will not work as easily as they did with the single random variable. We must resort to purely algebraic methods such as the χ^2 tests. (See Problem 4.61.) The abstract models become more and more important; this im-

---

[4] As we will discuss in Section 4.7, the two random variables were independent.

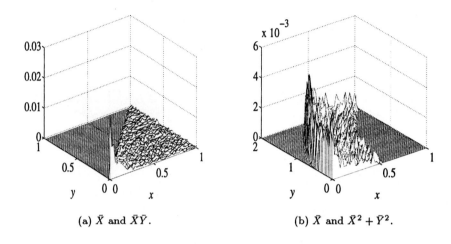

(a) $\bar{X}$ and $\bar{X}\bar{Y}$.   (b) $\bar{X}$ and $\bar{X}^2 + \bar{Y}^2$.

Figure 4.23   Two more pairs.

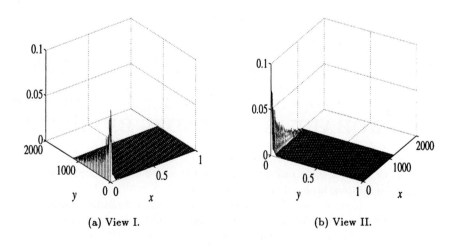

(a) View I.   (b) View II.

Figure 4.24   The random variables $\bar{X}\bar{Y}$ and $\bar{X}/\bar{Y}$.

portance is even more profound in the general case of more than two random variables. The difficulty of dealing with multidimensional data explains also why the IID sequences of Chapter 6 and the second-order statistics of Chapter 7 are so popular. They essentially reduce the problem to the one-dimensional case, in which a single pdf suffices to describe the experiment.

4.6 CONDITIONAL PROBABILITIES AND DENSITIES

4.6.1 Conditional Probabilities

In Chapter 2 we introduced the notion of $P[A|B]$, the conditional probability of event A given event B, where both $A, B \in \mathcal{Q}$. Consider the case where event A is specified in terms of the random variable X only, while event B is specified in terms of the random variable Y only. To emphasize this point, we use the notation $X \in A$ and $Y \in B$.

In principle, we can calculate conditional probabilities of this type *easily* once we have $F_{XY}(x, y)$ or $f_{XY}(x, y)$:

$$P[X \in A|Y \in B] = \frac{P[X \in A, \ Y \in B]}{P[Y \in B]}. \tag{4.23}$$

The numerator in Equation 4.23 can be computed now from $F_{XY}(x, y)$. The denominator can be computed from the marginal cdf of Y.

Remark: If A and/or B contains *both* random variables, we cannot evaluate the conditional probability (yet). Consider, for example, the event $A \triangleq \{\zeta \in S : X(\zeta) + Y(\zeta) \geq 0\}$, shown in Figure 4.25. Now A cannot be brought into a form to which Equation 4.23 applies. We develop the theory for such cases soon, in Section 4.8. Of course, such events can still be evaluated, at least in principle, through the original sample space S and probability assignment P.

4.6.2 Conditional PDF and CDF

We can tailor Equation 4.23 for various forms of A and B and for the cases when X and/or Y are discrete. The notation becomes a bit cumbersome, so' be patient. The new definitions will be handy in later chapters.

1. $A = \{X \leq x\}$, $B = \{Y = y\}$, Y *discrete*

To avoid trivialities, we assume that $P[Y = y] \neq 0$.

Definition: The **conditional cumulative distribution function of the random variable** X **given the event** $\{Y = y\}$, denoted by $F_X(x|y)$, is

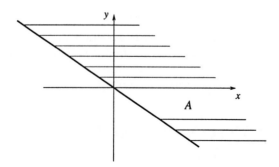

Figure 4.25   The event $A = \{\zeta \in S : X(\zeta) + Y(\zeta) \geq 0\}$.

defined as

$$F_X(x|y) \triangleq \frac{P[X \leq x,\ Y = y]}{P[Y = y]}.$$

When the derivative exists, we may define the following:

Definition: The **conditional probability density function of the random variable** X **given the event** $\{Y = y\}$, denoted by $f_X(x|y)$, is defined as

$$f_X(x|y) \triangleq \frac{dF_X(x|y)}{dx}.$$

2. $A = \{X \leq x\}$, $B = \{Y = y\}$, Y *continuous*

Suppose that X and Y are jointly continuous, and that the jpdf exists. Consider the (well-defined) conditional cdf $F_X(x|y < Y \leq y + h)$ for some small $h > 0$. From the definition of this function, we get

$$
\begin{aligned}
F_X(x|y < Y \leq y + h) &= \frac{P[X \leq x,\ y < Y \leq y + h]}{P[y < Y \leq y + h]} \\[2mm]
&= \frac{\int_{-\infty}^{x} \int_{y}^{y+h} f_{XY}(s,t)\ ds\,dt}{\int_{y}^{y+h} f_Y(t)\,dt} \\[2mm]
&\approx \frac{h \int_{-\infty}^{x} f_{XY}(s,y)\ ds}{h f_Y(y)}.
\end{aligned}
$$

As $h \to 0$, we get the following:

Definition: The **conditional cumulative distribution function of the random variable** X **given the event** $\{Y = y\}$, denoted by $F_X(x|y)$, is defined as

$$F_X(x|y) = \frac{\int_{-\infty}^{x} f_{XY}(s,y)\ ds}{f_Y(y)}. \tag{4.24}$$

Differentiating Equation 4.24 with respect to x, we get the following:

Definition: The **conditional probability density function of the random variable X given the event $\{Y = y\}$**, denoted by $f_X(x|y)$, is defined as

$$f_X(x|y) \triangleq \frac{dF_X(x|y)}{dx} = \frac{f_{XY}(x, y)}{f_Y(y)}. \tag{4.25}$$

Of course, by symmetry we have:

Definition: The **conditional probability density function of the random variable Y given the event $\{X = x\}$**, denoted by $f_Y(y|x)$, is defined as

$$f_Y(y|x) \triangleq \frac{dF_Y(y|x)}{dy} = \frac{f_{XY}(x, y)}{f_X(x)}. \tag{4.26}$$

4.6.3 Examples

Let's see now various conditional cdf's and pdf's, approximated through histograms, for the examples in the previous section.

Example 4.24. *Netstat Experiment.* Consider the *netstat* example; we chose the number of packets and the number of collisions as the two random variables of interest (X and Y, respectively). Figure 4.26(a) shows the conditional pdf of X given Y. Figure 4.26(b) shows the conditional pdf of Y given X.

Example 4.25. *Dow Jones Index.* Consider the Dow Jones example. Figure 4.27(a) shows the experimental conditional pdf of the closing random variable, given the opening random variable. Figure 4.27(b) shows the experimental conditional pdf of the opening random variable, given the closing random variable.

Example 4.26. *File Sizes.* In the file size example, the two random variables are the logarithm of the file size (X) and the x (executable) attribute of the file (Y). Figure 4.28 shows the conditional cdf of the random variable X given Y.

Example 4.27. *FTP.* In the *ftp* example, let X denote the logarithm of the file size and Y the delay in milliseconds. Figure 4.29 shows the two conditional experimental densities.

Example 4.28. In this example, the random variables are

$$X, \quad Y = \sin(10X) + 2.$$

Figure 4.30(a) shows the conditional pdf of Y given X. Figure 4.30(b) shows the conditional pdf of X given Y.

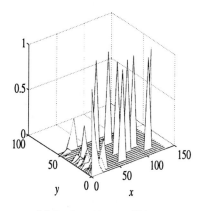 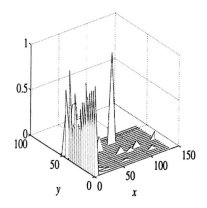

(a) Packets given collisions.   (b) Collisions given packets.

Figure 4.26   *Netstat* experiment.

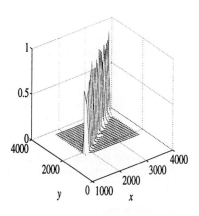 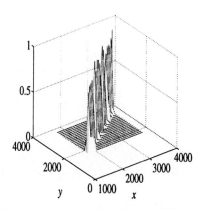

(a) Closing given opening.   (b) Opening given closing.

Figure 4.27   Dow Jones experiment.

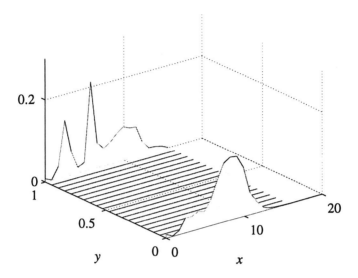

Figure 4.28 File sizes.

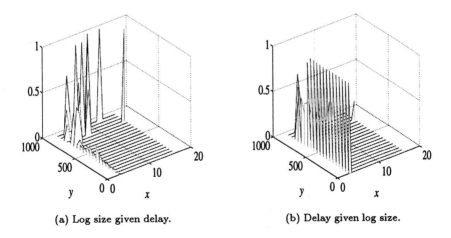

(a) Log size given delay.

(b) Delay given log size.

Figure 4.29 The *ftp* experiment.

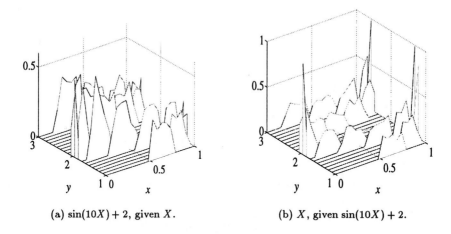

(a) $\sin(10X)+2$, given X.   (b) X, given $\sin(10X)+2$.

Figure 4.30   Conditional pdf's for X and $\sin(10X)+2$.

Example 4.29.   Let $S = [0,1]^2$, with a uniform probability assignment on subsets of S. In Figure 4.31 we plot the experimental conditional densities for two random variables that we simulated as follows:

$$X(\zeta,\theta) = \zeta^2, \quad Y(\zeta,\theta) = \theta^2,$$

where (ζ,θ) is an element of the sample space.

Example 4.30.   In this example, the random variables are generated as follows: X is the product of two independent, uniform random variables, while Y is their ratio. Figure 4.32 shows the two conditional densities.

Example 4.31.   In this example, the random variables are generated as follows: X is a uniform random variable, while Y is the ratio of X and another uniform random variable independent of X. Figure 4.33 shows the two conditional densities.

4.6.4   Total Probability Theorem Revisited

The total probability theorem of Section 2.6 can be restated in terms of the various conditional pdf's and cdf's we have introduced. We state two expressions (one for discrete and one for continuous random variables) without proof.

Theorem 4.1 *Consider two random variables X, Y with Y discrete. Then*

$$P[X \le x] = \sum_{y=-\infty}^{\infty} F_X(x|y)P[Y = y].$$

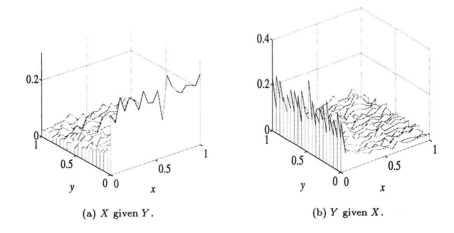

(a) X given Y. (b) Y given X.

Figure 4.31 The conditional pdf's in Example 4.29.

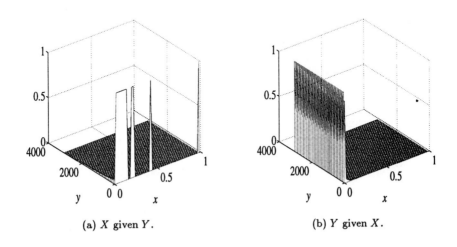

(a) X given Y. (b) Y given X.

Figure 4.32 The conditional pdf's in Example 4.30.

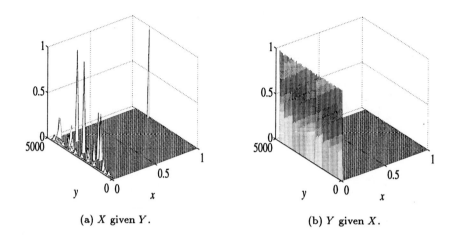

(a) X given Y.   (b) Y given X.

Figure 4.33   The conditional pdf's in Example 4.31.

Theorem 4.2 *Consider two random variables X, Y with Y continuous. Then*

$$P[X \leq x] = \int_{-\infty}^{\infty} F_X(x|y) f_Y(y) \, dy.$$

4.7   INDEPENDENCE

4.7.1   Definition

The notion of independent events in Section 2.7 generalizes to that of independent random variables, similar to the notion of conditional probabilities generalizing to conditional pdf's.

Consider a random experiment E with model $(S, \mathcal{Q}, P)$. Let $A \in \mathcal{Q}$ $(B \in \mathcal{Q})$ be an event specified in terms of (or, as we also say, generated by) the random variable X only[5] (Y only). If A and B are independent, we must have

$$P[A \cap B] = P[A] \cdot P[B]. \tag{4.27}$$

Suppose that we can find an event C (D) generated by the random variable X (Y) only, such that

$$P[C \cap D] \neq P[C] \cdot P[D].$$

---

[5]That is, $\{X \leq a\}$, $\{b < X < c\}$, or in general $\{\zeta \in S : X(\zeta) \in C\}$, where C is any Borel subset of R.

It is intuitively clear, in this case, that the two random variables X and Y are dependent on each other. On the other hand, if Equation 4.27 holds true for *all* events A and B (generated by X only and Y only), it is again intuitively obvious that X and Y are independent. This discussion leads to the following:

Definition: Let A be *any* event generated by the random variable X only, and let B be *any* event generated by the random variable Y only. The two random variables X and Y are called **independent**[6] if and only if

$$P[A \cap B] = P[A] \cdot P[B] \tag{4.28}$$

for all such events A, B.

4.7.2   Tests for Independence

How do we determine whether two random variables are independent? Applying the definition seems too time-consuming (or even impossible) when the random variables generate many events. Since the jpdf or jcdf contains all the information about the two random variables, we must find another way to test independence. Indeed, we have the following:

Theorem 4.3 *The random variables X and Y are independent if and only if*

$$F_{XY}(x,y) = F_X(x) \cdot F_Y(y), \quad \forall (x,y) \in R^2. \tag{4.29}$$

Sketch of a proof: If X and Y are independent, Equation 4.29 follows directly from Definition 4.27 with sets $A \triangleq \{X \leq x\}$, $B \triangleq \{Y \leq y\}$. The reverse would require us to show that Equation 4.27 holds true for any sets $A, B \in Q$. A rigorous proof of this part is beyond the scope of this book. The interested reader may consult [8]. We outline below how Equation 4.29 can be used to show independence of some specific sets in Q. Consider, for example, the rectangle in Figure 4.3. Let $C \triangleq P[a < X \leq b, \ c < Y \leq d]$. We have

$$
\begin{aligned}
C &= F_{XY}(b,d) - F_{XY}(b,c) - F_{XY}(a,d) + F_{XY}(a,c) \\
&= F_X(b)F_Y(d) - F_X(b)F_Y(c) - F_X(a)F_Y(d) + F_X(a)F_Y(c) \\
&= [F_X(b) - F_X(a)][F_Y(d) - F_Y(c)] \\
&= P[a < X \leq b]P[c < Y \leq d].
\end{aligned}
$$

$\square$

---

[6] Also **statistically independent** or **stochastically independent**.

Theorem 4.4 *The random variables X and Y are independent if and only if*

$$f_{XY}(x,y) = f_X(x) \cdot f_Y(y), \quad \forall (x,y) \in R^2. \tag{4.30}$$

Proof: Equation 4.30 follows immediately from Equation 4.29 by simple differentiation. $\qquad \square$

Remark 1: The above theorems can be restated in an obvious way through the pmf's when X and Y are discrete. We omit this restatement.

Remark 2: It is quite important to note that the expressions in both Equations 4.29 and 4.30 *must* hold true for *all* (x,y) in R^2. Consider, for instance, the function in Example 4.21. Even though $f_{XY}(x,y) = f_X(x) \cdot f_Y(y)$ when $x < y$, the random variables X and Y are not independent since when $x > y$, we have

$$f_{XY}(x,y) = 0 \neq f_X(x) \cdot f_Y(y).$$

Let's see some examples next.

Example 4.32. *Two Independent Random Variables.* Consider the function

$$F_{XY}(x,y) = \begin{cases} 1, & 0 \le x, y, \\ 0, & \text{otherwise.} \end{cases}$$

(Plot this function!) We can easily see that X, Y are discrete degenerate random variables with $P[X = 0,\ Y = 0] = 1$. Since $P[X = 0] = P[Y = 0] = 1$, the random variables X, Y are independent.

Example 4.33. *Two Dependent Random Variables.* Consider the function

$$f_{XY}(x,y) = \begin{cases} g(x,y) > 0, & 4 \ge x^2 + y^2 \ge 3, \\ 0, & \text{otherwise.} \end{cases}$$

The exact form of the function g does not matter here—only its property that it is nonzero whenever $4 \ge x^2 + y^2 \ge 3$. The random variables X and Y must be dependent, since Equation 4.30 fails for, say, $x = 0, y = 0$. Indeed, we can easily check that

$$f_X(0) = \int_{-\infty}^{\infty} f_{XY}(0,y) \, dy > 0, \quad f_Y(0) = \int_{-\infty}^{\infty} f_{XY}(x,0) \, dx > 0.$$

Example 4.34. *Two Dependent Random Variables.* Consider the function

$$P[X = k,\ Y = l] = c \frac{1}{k!} \frac{1}{l!}, \quad 0 \le l + k \le 10,$$

where $c > 0$ is a normalizing constant. Again, even though $P[X = k,\ Y = l]$ seems to factor out as $\sqrt{c}\frac{1}{k!} \times \sqrt{c}\frac{1}{l!}$, X and Y are dependent, since Equation 1.30 (more accurately, its counterpart for discrete random variables) fails for, say, $k = 10, l = 10$.

4.7.3 Statistical Independence of "Unrelated" Random Variables

Consider the *netstat* and GPA experiments. They are performed separately or, one might say, "independently" of each other. ("Independently" is in quotes because we do not have a single experiment here.) Let X_1 denote the random variable that counts packets and X_2 denote the random variable that monitors the GPA of a student. It is clear that the two random variables are not related at all. Are they independent of each other in the statistical sense of Definition 4.28? Intuitively, the answer must be yes! But let's see why.

Here is a more rigorous statement of the whole problem. Let E_1 (or E_2) denote the *netstat* (or GPA) experiment. Let (S_1, Q_1, P_1) (or simply F_{X_1}) be the probability model associated with the random variable X_1. Let (S_2, Q_2, P_2) (or simply F_{X_2}) be the probability model associated with the random variable X_2. For independence, we *must* have a single random experiment on which both random variables must be defined. So, consider a "new" experiment E with outcomes that are *pairs* of the form

$$(\zeta_1, \zeta_2), \quad \text{with } \zeta_1 \in S_1, \ \zeta_2 \in S_2.$$

We see that the sample space for E is the Cartesian product of the two sample spaces S_1 and S_2:

$$S = S_1 \times S_2 \tag{4.31}$$

(see Appendix A for details on Cartesian products). We can define Q, the new collection of events of interest, as the set

$$Q = \{A \times B : A \in Q_1, \ B \in Q_2\}. \tag{4.32}$$

(In Problem A.41, we ask you to formally investigate whether $Q = Q_1 \times Q_2$.)

And what is the probability assignment function? Well, since we only have $F_{X_1}(x)$ and $F_{X_2}(y)$, we take

$$F_{X_1 X_2}(x, y) = F_{X_1}(x) \cdot F_{X_2}(y), \quad \forall (x, y) \in R^2. \tag{4.33}$$

Now, on this "new" random experiment E with probability model (S, Q, P) (or simply $F_{X_1 X_2}$), given by Equations 4.31, 4.32, and 4.33, X_1 and X_2 are *formally and trivially* independent!

Remark: We saw above how easy it is to construct independent random variables on product spaces. Are any of the random variables in the examples in Section 4.6.3 independent? It should be easy to answer this question based on the graphs in that section.

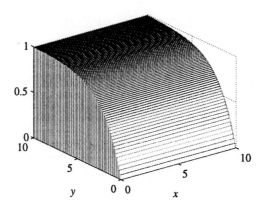

Figure 4.34   Two independent random variables.

4.7.4   More Tests for (In)Dependence

An immediate consequence of Theorem 4.3 and the definition of conditional probabilities is that when the random variables X, Y are independent, we have $F_X(x|y) = F_X(x)$, $\forall x, y \in R$. We can then state the following:

Corollary 4.1 *The random variables X and Y are independent if and only if*

$$F_X(x|y_1) = F_X(x|y_2),$$

for all $x, y_1, y_2 \in R$.

Similarly, from Theorem 4.4 we get the following:

Corollary 4.2 *The random variables X and Y are independent if and only if*

$$f_X(x|y_1) = f_X(x|y_2),$$

for all $x, y_1, y_2 \in R$.

Corollaries 4.1 and 4.2 provide a quick test for dependence. If the graph of $F_X(x|y)$ or $f_X(x|y)$ is given, we can determine dependence by inspection.

Example 4.35. The random variables in Figure 4.30(a) are dependent since, for instance,

$$f_X(1.5|0) \neq f_X(1.5|1).$$

The random variables in Figure 4.34, where we plot $F_X(x|y)$, are independent.

4.7.5 Why Bother About Independence?

Anticipating the material in Chapter 5 we mention here that *dependence between two random variables is desirable* when we want to extract "information" about one random variable (X) from given knowledge (e.g., measurements) about the other variable (Y). You may ask Why not measure X directly? The answer is that, in a lot of situations, it may simply be impossible to do so! In all communication channels, for example, the value of the input bits (considered as a random variable X) is *never available* at the receiver because of noise or channel effects on the electric signals that represent the bits. In this case, we sincerely hope that the output random variable Y, which is directly measurable at the receiver, *will be very dependent on X!* As a matter of fact, we hope for the "strongest" possible dependence, namely, $Y = X$!

Finally, anticipating the material in Chapter 6, we mention here that *independence is desirable* when we have to deal with repeated trials of experiments, with one trial being unrelated to the rest.

4.7.6 Independence of Functions of Random Variables

The transformations of Section 3.8 raise the following question: Suppose we have two independent random variables X and Y; we transform each one of them to, say, $V = g(X)$ and $W = h(Y)$. Are V, W independent? The answer is given by the following theorem.

Theorem 4.5 *Let X, Y be two independent random variables. Let $V = g(X)$ and $W = h(Y)$. Then V, W are also independent random variables.*

Proof: Consider an arbitrary event $\{V \in A, W \in B\}$, where A, B are subsets of R. This event is equivalent to an event that involves the random variables X and Y only, which we denote as $\{X \in C, Y \in D\}$ (see Figure 4.35). Therefore,

$$
\begin{aligned}
P[V \in A, W \in B] &= P[X \in C, Y \in D] = P[X \in C] \cdot P[Y \in D] \\
&= P[V \in A] \cdot P[W \in B],
\end{aligned}
$$

since the events $\{V \in A\}$ and $\{X \in C\}$ are equivalent (and so are the events $\{W \in B\}$ and $\{Y \in D\}$). □

The opposite question is also of interest: Given two *dependent* random variables X and Y, can we transform them to two independent ones? We address this question in a more general setting next.

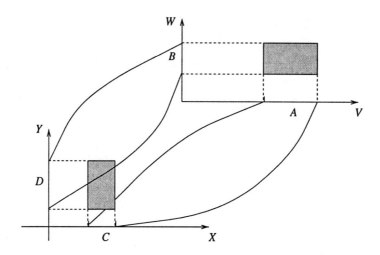

Figure 4.35   Independent random variables.

4.8   TRANSFORMATIONS OF A RANDOM VECTOR

4.8.1   Various Possibilities

Outcomes of random variables are processed in various ways, as we saw in Section 3.8. Since now we have more than one random variable to deal with, we can specify a variety of processing scenarios. We focus again on the case of $N = 2$ random variables, for simplicity. We can classify the various possibilities for processing as

- One function of two random variables, say, $Z = g(X,Y)$

- Two functions of two random variables, say, $Z_i = g_i(X,Y), i = 1,2$

- More than two functions of two random variables, say, $Z_i = g_i(X,Y), i = 1,2,\ldots,K$

In all cases, the functions can be linear or nonlinear, and the random variables X and Y can be independent or not. We treat the linear case first because it is simpler conceptually and quite important in practice (i.e., in signal processing, as we see in Chapter 9).

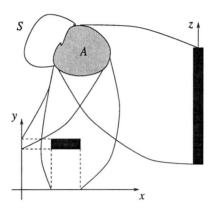

Figure 4.36 The basic idea.

4.8.2 A Single Function of Two Random Variables $Z = g(X, Y)$

In this subsection we determine the cdf or pdf of the transformed random variable Z in terms of the pdf of the given random variables X, Y and the function g. The general approach is a simple extension of the approach used for the single-variable case in Section 3.8. Since we have a two-dimensional space to deal with here, our notation will be a bit more cumbersome; however, as Figure 4.36 suggests, the principle remains the same.

The event of interest in evaluating the distribution of the new random variable Z is still $\{Z \leq z\}$. We can evaluate $P[Z \leq z]$ by referring to the original sample space, "bypassing" in effect the random variables X, Y (see Figure 4.36). We now see this idea applied in a few special cases.

1. A linear function $Z = aX + bY$

Consider first the simple case of a linear function of two random variables $Z = aX + bY$, where $a, b \neq 0$, in order to avoid trivialities. This function is shown in Figure 4.37 for the special case $a = 1, b = 2$.

Theorem 4.6 *Let* $Z = aX + bY$. *Then*

$$F_Z(z) = \int_{-\infty}^{\infty} \int_{-\infty}^{(z-ax)/b} f_{XY}(x, y) \, dy \, dx \tag{4.34}$$

and

$$f_Z(z) = \int_{-\infty}^{\infty} f_{XY}\left(x, \frac{z - ax}{b}\right) dx. \tag{4.35}$$

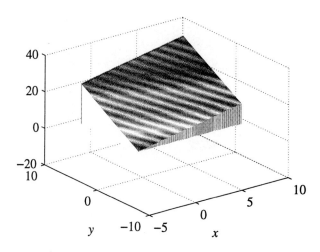

Figure 4.37   The function $z = x + 2y$ over the domain $[-1, 10]^2$.

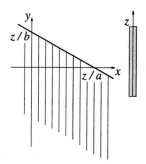

Figure 4.38   Equivalent events.

Proof: The event $\{Z \leq z\}$ is equivalent to the event $\{aX + bY \leq z\}$, shown as the shaded area below the line $z = ax + by$ in Figure 4.38. Integration over this area can be done in the following way: We fix a value of x, and then we let y range from $-\infty$ to the point $(z - ax)/b$, as the figure suggests. But these are exactly the limits in the integral of Equation 4.34. We can easily then get Equation 4.35 by a simple differentiation (assuming that the integration and differentiation operations can be interchanged).   □

Note that we make no assumption regarding the form of the joint density of the random variables X and Y. The case $a = 1, b = 1$, and X, Y independent is very common and important in practice. We restate Theorem 4.6 separately for this case, since we make repeated use of it in later chapters, e.g., in Chapter 6.

Theorem 4.7 *Let $Z = X + Y$, where X and Y are independent random variables. Then*

$$F_Z(z) = \int_{-\infty}^{\infty} \int_{-\infty}^{z-x} f_X(x) f_Y(y) \, dy \, dx \qquad (4.36)$$

and

$$f_Z(z) = \int_{-\infty}^{\infty} f_X(x) f_Y(z - x) \, dx. \qquad (4.37)$$

Proof: Equation 4.36 is a direct consequence of Equation 4.34 since by independence the joint pdf is the product of the marginal pdf's. Equation 4.37 follows again by a simple differentiation of Equation 4.36. □

Equation 4.37 is a *convolution integral,* like the ones we encounter in linear systems courses. *Efficient calculation* of this integral is important in applications since there is a lot of work involved in carrying out the integration numerically. Laplace, fast Fourier transforms, and Z transforms are methods that can be used for calculating convolution integrals efficiently. We presented Laplace and Z transforms in Section 3.11. For fast Fourier transforms consult [53].

Let's see some examples next.

Example 4.36. *Sum of Two Dependent Random Variables.* Consider two random variables X and Y with jpdf

$$f_{XY}(x, y) = \frac{1}{2\pi 0.75^2} e^{-\frac{1}{2(1-0.75)} \frac{x^2 - xy + y^2}{2}}.$$

(Verify that this is a valid jpdf and that the random variables are dependent!) The pdf of the sum $Z = 3X + 2Y$ is given by Equation 4.35 as

$$
\begin{aligned}
f_Z(z) &= \int_{-\infty}^{\infty} f_{XY}\left(x, \frac{z - 3x}{2}\right) dx \\
&= \frac{1}{2\pi 0.75^2} \int_{-\infty}^{\infty} e^{-\frac{1}{2(1-0.75)} \frac{1}{2}\{x^2 - x(z-3x)/2 + [(z-3x)/2]^2\}} \, dx.
\end{aligned}
$$

After tedious but elementary operations, we get the desired pdf

$$f_Z(z) = \frac{1}{\sqrt{2\pi}} e^{-z^2/2},$$

which is the density of a Gaussian random variable. As we see in Section 4.10, each of the two random variables X, Y is Gaussian. This example, then, says that sums of Gaussian random variables (even dependent ones) are also Gaussian, a quite important fact! (If you defined Z as $aX + bY$, you would still get a Gaussian pdf; we prove that more formally later, in Theorem 4.17.)

Example 4.37. *Sum of Independent Poisson Random Variables.* Let $Z = X + Y$, where X and Y are independent Poisson random variables, with parameters λ_1 and λ_2, respectively. Their joint pmf is given by (see Equation 4.58)

$$P[X = k, \, Y = l] = \frac{\lambda_1^k \, \lambda_2^l}{k! \, l!} e^{-\lambda_1} e^{-\lambda_2}.$$

Let $n \geq 0$ be a given integer. The discrete analog of Equation 4.37 is now

$$
\begin{aligned}
P[Z = n] &= \sum_{k=0}^{n} \frac{\lambda_1^k}{k!} e^{-\lambda_1} \frac{\lambda_2^{n-k}}{(n-k)!} e^{-\lambda_2} \\
&= e^{-(\lambda_1 + \lambda_2)} \sum_{k=0}^{n} \frac{1}{k!} \frac{1}{(n-k)!} \lambda_1^k \lambda_2^{n-k} \\
&= \frac{e^{-(\lambda_1 + \lambda_2)}}{n!} \sum_{k=0}^{n} \binom{n}{k} \lambda_1^k \lambda_2^{n-k} = \frac{(\lambda_1 + \lambda_2)^n}{n!} e^{-(\lambda_1 + \lambda_2)},
\end{aligned}
$$

which we recognize as the pmf of another Poisson random variable, with parameter $\lambda_1 + \lambda_2$. This property explains (in part) the popularity of the Poisson model; for example, in modeling packet arrivals at a computer node, multiplexed packet streams will still be Poisson if individual arrival streams are modeled as Poisson. Thus analysis (and results) that holds true for the individual streams holds true for the multiplexed one as well.

Example 4.38. *Sum of Independent Exponential Random Variables.* Let $Z = X + Y$, where X and Y are independent exponential random variables with common parameter λ. Their joint pdf is given by

$$f_{XY}(x, y) = \lambda^2 e^{-\lambda x} e^{-\lambda y} u(x) u(y).$$

Then

$$
\begin{aligned}
f_Z(z) &= \int_{-\infty}^{\infty} f_{XY}(x, z - x) \, dx = \int_{-\infty}^{\infty} \lambda^2 e^{-\lambda x} e^{-\lambda(z-x)} u(x) u(z - x) \, dx \\
&= \int_{0}^{z} \lambda^2 e^{-\lambda x} e^{-\lambda(z-x)} \, dx = \lambda^2 e^{-\lambda z} \int_{0}^{z} dx.
\end{aligned}
$$

Therefore,

$$f_Z(z) = \lambda^2 e^{-\lambda z} z, \quad z > 0.$$

From the models in Section 3.6, we recognize this as the pdf of an **Erlang** random variable, with parameters $n = 2, \lambda$.

Example 4.39. *Sum of Independent Gamma Random Variables.* Now let $Z = X + Y$, where X and Y are independent gamma random variables with parameters a, λ and b, λ. Their joint pdf is given by

$$f_{XY}(x, y) = \frac{\lambda(\lambda x)^{a-1}e^{-\lambda x}}{\Gamma(a)} \frac{\lambda(\lambda y)^{b-1}e^{-\lambda y}}{\Gamma(b)}, \quad x, y \geq 0.$$

Then

$$\begin{aligned}
f_Z(z) &= \int_{-\infty}^{\infty} f_{XY}(x, z - x)\, dx \\
&= \int_0^z \frac{\lambda(\lambda x)^{a-1}e^{-\lambda x}}{\Gamma(a)} \frac{\lambda[\lambda(z - x)]^{b-1}e^{-\lambda(z-x)}}{\Gamma(b)}\, dx \\
&= \frac{\lambda e^{-\lambda z}}{\Gamma(a)\Gamma(b)} \int_0^z (\lambda x)^{a-1}\lambda[\lambda(z - x)]^{b-1}\, dx.
\end{aligned}$$

After some tedious calculations we get

$$f_Z(z) = \frac{\lambda(\lambda z)^{a+b-1}e^{-\lambda z}}{\Gamma(a + b)},$$

which we recognize as a gamma pdf with parameters λ and $a+b$. This property makes the gamma random variable a popular model in many situations, for reasons similar to those explained for the Poisson model.

Example 4.40. *Sum of Independent Binomial Random Variables.* Let $Z = X + Y$, where X and Y are independent binomial random variables with the same parameters n, p. Since the random variables are discrete, we evaluate the pmf of Z. In Figure 4.39 we show the event $\{Z = k\}$ for $k = 3$ and $k = n + 2$. We see that we must distinguish two cases, namely, $k \leq n$ and $k > n$. In the first case, the index of summation runs from 0 to k, while in the second it runs from $k - n$ to n. Combining the two cases in one formula, we can write

$$\begin{aligned}
P[Z = k] &= \sum_{l=\max(0,k-n)}^{\min(n,k)} \binom{n}{l}p^l(1 - p)^{n-l}\binom{n}{k - l}p^{k-l}(1 - p)^{n-(k-l)} \\
&= \binom{2n}{k}p^k(1 - p)^{2n-k}, \quad k = 0, 1, 2, \ldots, 2n,
\end{aligned}$$

where we use a property of the combinatorial factors that we ask you to prove in Problem 4.197. It is easy now to recognize the last equality as the pmf of another binomial random variable with parameters $2n, p$.

Example 4.41. Let's consider now an experimental pdf from Section 3.4.8. Take the Dow Jones case, for instance. Suppose we want to transform the

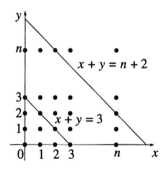

Figure 4.39   The equivalent event $\{Z = k\}$.

closing and opening random variables, call them X and Y, to $Z = X - Y$, which could represent the gain of an investment strategy. How do we evaluate the pdf of Z?

The experimental way is to pore through the data again and construct a histogram for the new random variable Z since for all outcomes ζ,

$$Z(\zeta) = X(\zeta) - Y(\zeta).$$

The indirect way is to use a fitted two-dimensional density to the jpdf of X, Y and then apply Theorem 4.6.

2. A nonlinear function $Z = g(X, Y)$

From the proof of Theorem 4.6 in the linear case, it should be apparent that we can evaluate the cdf of Z quite easily, at least in principle.[7] The basic idea behind Figure 4.36 still holds true in the nonlinear case since events that involve random variables can be evaluated over the original sample space. In the following theorem, let $g(\cdot, \cdot)$ be an arbitrary, real-valued function of two real arguments (well, not entirely arbitrary since it must map Borel subsets of R^2 on Q).

Theorem 4.8 *Let* $Z = g(X, Y)$. *For a given* $z \in R$, *let* $R_z \triangleq \{(x, y) \in R^2 : g(x, y) \leq z\}$. *Then*

$$F_Z(z) = \int\int_{R_z} f_{XY}(x, y)\, dx\, dy. \tag{4.38}$$

---

[7] Again we are talking as mathematicians. As you will see from the examples in this section, it takes a very good understanding of integration over R^2 to carry out this simple principle.

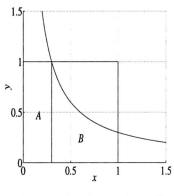

(a) $z = x/y$, $(x,y) \in (0, 1.5)^2$.

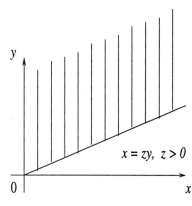

(b) R_z for $g(x,y) = x/y$.

Figure 4.40 The function and region in Example 4.42.

Remark: The integral extends over all $(x, y) \in R_z$. In general, we cannot give a formula for the pdf of Z; we cannot differentiate Equation 4.38 since the dependence on z is *implicit*, through the set R_z. Every case must be treated separately!

Let's see now some examples in detail.

Example 4.42. Let $Z = X/Y$, where X, Y are independent and exponentially distributed with parameters λ_1 and λ_2, respectively. The function $z = x/y$ is shown in Figure 4.40(a). (When $x = y = 0$, we set $z = 0$.) The region R_z in Equation 4.38 is given by

$$R_z \triangleq \{(x, y) \in [0, \infty)^2 : \frac{x}{y} \leq z\}$$

and is shown in Figure 4.40(b). Therefore, the appropriate limits in the integral of Equation 4.38 are

$$F_Z(z) = \int_0^\infty \int_0^{yz} \lambda_1 e^{-\lambda_1 x} \lambda_2 e^{-\lambda_2 y} \, dx \, dy, \quad z \geq 0.$$

After some simple calculation, we get

$$F_Z(z) = 1 - \frac{\lambda_2}{\lambda_2 + \lambda_1 z}, \quad z \geq 0.$$

This cdf is shown in Figure 4.41 for the special case $\lambda_1 = 1$, $\lambda_2 = 2$.

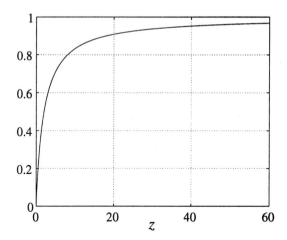

Figure 4.41   The cdf of Z in Example 4.42.

Example 4.43. Let now $Z = XY$, with X, Y independent and uniformly distributed in $[0, 1]$. The experimental joint pdf of the random variables Z, X was studied in Section 4.4.5 [see Figure 4.23(a)]. Let's evaluate $F_Z(z)$ theoretically.

The function $z = x \cdot y$ is shown in Figure 4.42(a). The region R_z in Equation 4.38 is given by

$$R_z \triangleq \{(x, y) \in [0, 1]^2 : \ x \cdot y \le z\} \tag{4.39}$$

and it is shown in Figure 4.42(b) for $z = 0.3$.

Therefore, we can split the integral into two pieces, as shown in Figure 4.42(b). Then easily

$$F_Z(z) = z + \int_z^1 \frac{z}{x} \, dx = z - z \log(z), \ \ 0 \le z \le 1.$$

This cdf is shown in Figure 4.43.

Example 4.44. Let again $Z = XY$, but now consider two dependent random variables X, Y with a jpdf given by

$$f_{XY}(x, y) = 2e^{-x}e^{-y}u(x)u(x - y), \ \ x, y \in R,$$

where $u(\cdot)$ is the usual step function. The region R_z in Equation 4.38 is of course still given by Equation 4.39. From Figure 4.44(a), the integral can be written as

$$F_Z(z) = \int_0^{\sqrt{z}} \int_0^x 2e^{-x}e^{-y} \, dy \, dx + \int_{\sqrt{z}}^\infty \int_0^{\frac{z}{x}} 2e^{-x}e^{-y} \, dy \, dx, \ \ z \ge 0.$$

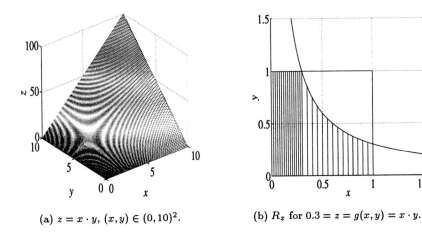

(a) $z = x \cdot y$, $(x, y) \in (0, 10)^2$.

(b) R_z for $0.3 = z = g(x, y) = x \cdot y$.

Figure 4.42 The function and region in Example 4.43.

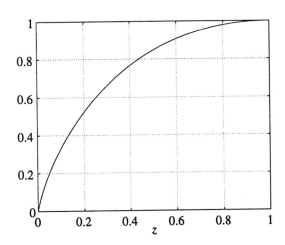

Figure 4.43 The cdf in Example 4.43.

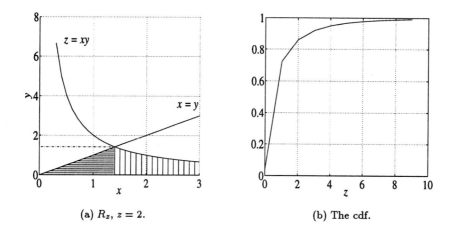

(a) R_z, $z = 2$.　　　　　　　　(b) The cdf.

Figure 4.44   The region and cdf in Example 4.44.

After some calculations, we get

$$F_Z(z) = 1 + e^{-2\sqrt{z}} - 2 \int_{\sqrt{z}}^{\infty} e^{-(x+\frac{z}{x})}\, dx.$$

This cdf (calculated numerically) is shown in Figure 4.44(b).

Example 4.45.   Consider now $Z = \sqrt{X^2 + Y^2}$, with X, Y independent and uniformly distributed in $[0, 1]$. An experimental pdf related to this example was studied in Section 4.4.5, Figure 4.23(b).

The function $z = \sqrt{x^2 + y^2}$ is shown in Figure 4.45(a). The region R_z in Equation 4.38 is now given by

$$R_z \triangleq \{(x, y) \in [0, 1]^2 : \ x^2 + y^2 \leq z^2\},$$

and it is shown in Figure 4.45(b). For $0 \leq z \leq 1$ we easily get

$$F_Z(z) = \frac{\pi}{4}z^2.$$

When $z > \sqrt{2}$, from Figure 4.45(b) we can easily see that $F_Z(z) = 1$. When $1 < z \leq \sqrt{2}$, from the same figure and after some algebra, we have

$$F_Z(z) = \sqrt{z^2 - 1} + \frac{z^2}{2}\left(\frac{\pi}{2} - 2\tan^{-1}\sqrt{z^2 - 1}\right).$$

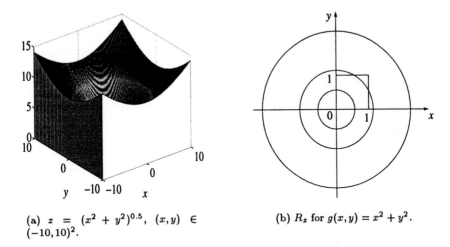

(a) $z = (x^2 + y^2)^{0.5}$, $(x,y) \in (-10, 10)^2$.

(b) R_z for $g(x,y) = x^2 + y^2$.

Figure 4.45 The function and region in Example 4.45.

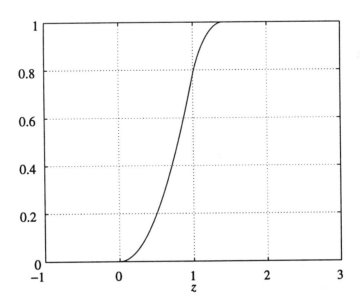

Figure 4.46 The cdf in Example 4.45.

The cdf is shown in Figure 4.46.

Example 4.46. Consider again the random variable $Z = \sqrt{X^2 + Y^2}$, but now X and Y are two independent Gaussian random variables, with unit variance and means m and 0, respectively. Therefore,

$$F_Z(z) = \frac{1}{2\pi} \int \int_{R_z} e^{-\frac{(x-m)^2}{2}} e^{-\frac{y^2}{2}} \, dx \, dy, \quad z \ge 0. \tag{4.40}$$

It is more convenient to work with polar coordinates, so let's apply the change of variables $x = r\cos\theta, y = r\sin\theta$ to Equation 4.40. We then have

$$F_Z(z) = \frac{1}{2\pi} e^{-\frac{1}{2}m^2} \int_0^z e^{-\frac{1}{2}r^2} \left(\int_0^{2\pi} e^{rm\cos\theta} \, d\theta \right) r \, dr, \quad z > 0. \tag{4.41}$$

We recognize the function inside parentheses as $2\pi I_0(rm)$, the zero-order modified Bessel function of the first kind. Therefore, we can rewrite Equation 4.41 as

$$F_Z(z) = e^{-\frac{1}{2}m^2} \int_0^z r e^{-\frac{1}{2}r^2} I_0(rm) \, dr, \quad z > 0. \tag{4.42}$$

From the distributions in Section 3.6, we recognize now the function in Equation 4.42 as the cdf of a Rice random variable. We show it in Figure 4.47.

Example 4.47. Let Z, X, Y be as in the previous example, but now take $m = 0$. We have

$$F_Z(z) = \frac{1}{2\pi} \int \int_{R_z} e^{-\frac{x^2}{2}} e^{-\frac{y^2}{2}} \, dx \, dy. \tag{4.43}$$

With the same Cartesian-to-polar coordinate change, as in the previous example, we can rewrite Equation 4.43 as

$$F_Z(z) = \frac{1}{2\pi} \int_0^z r e^{-\frac{r^2}{2}} \, dr \int_0^{2\pi} d\theta = 1 - e^{-\frac{1}{2}z^2}.$$

Differentiating the above equation, we can recognize Z as a Rayleigh random variable with parameter 1.

4.8.3   Two Functions of Two Random Variables

Let's move on now to address the more difficult problem of two functions of two random variables. The basic principle of evaluating probabilities with equivalent events in the original space will be helpful here as well.

1. Two linear functions $Z_1 = aX + bY$, $Z_2 = cX + dY$

Consider the transformation

$$Z_1 = aX + bY, \quad Z_2 = cX + dY, \tag{4.44}$$

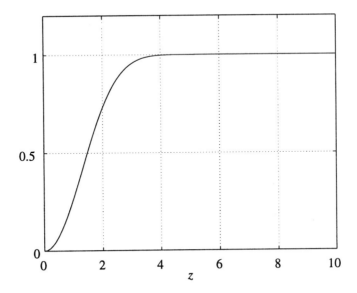

Figure 4.47 The cdf of a Rice random variable.

where a, b, c, d are real constants. It is customary and more compact in notation to rewrite Equation 4.44 using matrix notation. Let, therefore,

$$A \overset{\triangle}{=} \begin{bmatrix} a & b \\ c & d \end{bmatrix}, \quad \mathbf{Z} \overset{\triangle}{=} \begin{bmatrix} Z_1 \\ Z_2 \end{bmatrix}, \quad \mathbf{z} \overset{\triangle}{=} \begin{bmatrix} z_1 \\ z_2 \end{bmatrix}, \quad \mathbf{X} \overset{\triangle}{=} \begin{bmatrix} X \\ Y \end{bmatrix}.$$

Then Equation 4.44 can be rewritten more compactly as

$$\mathbf{Z} = A\mathbf{X}.$$

To avoid trivial cases, we assume that matrix A is invertible. The case of a singular A is considered in Problem 4.63.

Theorem 4.9 *Let* $\mathbf{Z} = A\mathbf{X}$. *Let* $[xy]^T = A^{-1}\mathbf{z}$. *Then*

$$f_{Z_1 Z_2}(z_1, z_2) = \frac{f_{XY}(x, y)}{|A|}, \quad z_1, z_2 \in R. \tag{4.45}$$

Remark: A note on notation is called for here. Even though the right-hand side of Equation 4.45 appears to be a function of x, y, it is not! Keep in mind

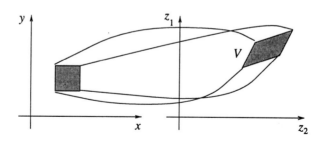

Figure 4.48   The equivalent parallelogram.

that x, y are expressed as $A^{-1}\mathbf{z}$. Moreover, $|A|$ denotes the absolute value of the determinant.

Proof: Instead of evaluating the probability of the basic event $\{Z \leq z\}$, as we have done so far, we evaluate the jpdf directly now, in order to see that the general approach is very flexible.

The event of interest now will involve the new random variable $\mathbf{Z}$ in an infinitesimal "rectangle" $\{z_i < Z_i \leq z_i + dz_i\}$; the equivalent event in the X, Y space is in general a parallelogram (check that!). Instead of evaluating probabilities with such infinitesimal rectangles, it is customary to work in the opposite way: We start with a rectangle in the X, Y space and transform it to a parallelogram in the Z plane, as Figure 4.48 shows.

Let's denote the parallelogram by V and its area by dV. Since the events are equivalent, we must have

$$f_{XY}(x, y)\, dx\, dy \approx f_{Z_1 Z_2}(z_1, z_2)\, dV.$$

The area of the parallelogram is given by (see reference [3] for more details)

$$dV = |A|\, dx\, dy,$$

where $|A|$ is the absolute value of the determinant of matrix A. Therefore, Equation 4.45 follows by a simple division (assuming that the determinant is nonzero).   □

Let's see some examples now.

Example 4.48. Let X and Y be jointly Gaussian with parameters

$$m_1 = 0, \quad \sigma_1 = 1, \quad m_2 = 0, \quad \sigma_2 = 1, \quad \rho_{XY} = \rho.$$

(The jpdf of X and Y is given in Equation 4.60, in Section 4.10, where the meaning of the parameters is also explained.) Let

$$Z_1 = X + Y, \quad Z_2 = X - Y.$$

In order to apply Theorem 4.9, we need to evaluate $|A|$ and $A^{-1}z$, where

$$A = \begin{bmatrix} 1 & 1 \\ 1 & -1 \end{bmatrix}.$$

Therefore, $|A| = 2$, and the solution x, y is given by

$$\begin{bmatrix} x \\ y \end{bmatrix} = \begin{bmatrix} 0.5 & 0.5 \\ 0.5 & -0.5 \end{bmatrix} \begin{bmatrix} z_1 \\ z_2 \end{bmatrix} = \begin{bmatrix} z_1 + z_2 \\ z_1 - z_2 \end{bmatrix}.$$

Let

$$h(z_1, z_2) \triangleq \frac{1}{2(1 - \rho^2)} \left[\frac{(z_1 + z_2)^2}{8} - 2\rho \frac{z_1 + z_2}{4} \frac{z_1 - z_2}{4} + \frac{(z_1 - z_2)^2}{8} \right]$$

Then the jpdf in Equation 4.45 is given by

$$f_{Z_1 Z_2}(z_1, z_2) = \frac{1}{2} \frac{e^{-h(z_1, z_2)}}{2\pi \sqrt{(1 - \rho^2)}}.$$

Example 4.49. Let X and Y be independent, jointly Gaussian random variables, with parameters

$$m_1 = m_2 = 0, \quad \sigma_1 = \sigma_2 = 1.$$

Let

$$Z_1 = X \cos(a) + Y \sin(a), \quad Z_2 = X \sin(a) - Y \cos(a),$$

where a is a fixed angle. We have

$$A = \begin{bmatrix} \cos(a) & \sin(a) \\ \sin(a) & -\cos(a) \end{bmatrix}.$$

Therefore, $|A| = |-\cos^2(a) - \sin^2(a)| = 1$, and

$$x = z_1 \cos(a) + z_2 \sin(a), \quad y = z_1 \sin(a) - z_2 \cos(a).$$

From Theorem 4.9

$$f_{Z_1 Z_2}(z_1, z_2) = \frac{1}{2\pi} e^{-\frac{1}{4} \{ [z_1 \cos(a) + z_2 \sin(a)]^2 + [z_1 \sin(a) - z_2 \cos(a)]^2 \}}.$$

Expanding the exponent and simplifying, we get

$$f_{Z_1 Z_2}(z_1, z_2) = \frac{1}{2\pi} e^{-\frac{1}{2}(z_1^2 + z_2^2)},$$

which shows that Z_1, Z_2 are also Gaussian and independent.

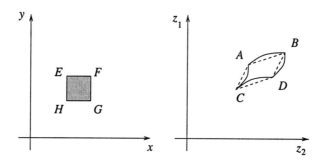

Figure 4.49   Infinitesimal equivalent events.

2. Two nonlinear functions $Z_i = g_i(X, Y), i = 1, 2$

Consider now the more general case of two nonlinear functions

$$Z_1 = g_1(X, Y), \qquad Z_2 = g_2(X, Y). \tag{4.46}$$

Again, in principle, all we have to do is evaluate the equivalent event (in terms of X and Y) for the infinitesimal event shown in Figure 4.49. Since the functions are not linear, the infinitesimal event in the $Z_1 Z_2$ plane will not be a parallelogram in general. (For simplicity, in Figure 4.49 we assume that the rectangle $EFGH$ is equivalent to only one nonrectangular event in the $Z_1 Z_2$ plane. See Example 4.51 for a different case.)

The curves AB, AC, BD, CD in Figure 4.49 can be approximated via straight lines, from the Taylor series expansion of the functions $g_1(x, y)$ and $g_2(x, y)$. Indeed, as a first-order approximation,

$$g_i(x + dx, y) \approx g_i(x, y) + \frac{\partial g_i(x, y)}{\partial x}\, dx, \quad i = 1, 2,$$

$$g_i(x, y + dy) \approx g_i(x, y) + \frac{\partial g_i(x, y)}{\partial y}\, dy, \quad i = 1, 2.$$

Therefore,

$$f_{XY}(x, y)\, dx\, dy = f_{Z_1 Z_2}(z_1, z_2)\, dV, \tag{4.47}$$

where dV is the area of the quadrilateral $ABDC$. From geometry, such an area is approximately given by

$$dV = |A|\, dx\, dy,$$

where A, called the *Jacobian* of the transformation, represents the linear transformation

$$Z_1 = \frac{\partial g_1(x, y)}{\partial x} X + \frac{\partial g_1(x, y)}{\partial y} Y,$$

$$Z_2 = \frac{\partial g_2(x,y)}{\partial x} X + \frac{\partial g_2(x,y)}{\partial y} Y.$$

A word of caution on notation: The derivatives in the above equations are considered constants since they are assumed to be evaluated at the generic point (x, y). Therefore,

$$A = \begin{bmatrix} \frac{\partial g_1(x,y)}{\partial x} & \frac{\partial g_1(x,y)}{\partial y} \\ \frac{\partial g_2(x,y)}{\partial x} & \frac{\partial g_2(x,y)}{\partial y} \end{bmatrix}. \tag{4.48}$$

From Equation 4.47 we get finally

$$f_{Z_1 Z_2}(z_1, z_2) = \frac{f_{XY}(x,y)}{|A|}.$$

We express this as a theorem with a slight change in notation, to emphasize that the right-hand side is a function of (z_1, z_2):

Theorem 4.10 *Let $Z_1 = g_1(X, Y)$, $Z_2 = g_2(X, Y)$; let A be given as in Equation 4.48. Then*

$$f_{Z_1 Z_2}(z_1, z_2) = \frac{f_{XY}(h_1(z_1, z_2), h_2(z_1, z_2))}{|A|}, \tag{4.49}$$

where $h_1(z_1, z_2)$ and $h_2(z_1, z_2)$ denote the solution of System 4.46 in terms of x, y. If we have multiple solutions, the right-hand side in Equation 4.49 must be interpreted as a sum.

Notice the similarity between this theorem and Theorem 3.1. Let's see some examples next.

Example 4.50. Let X and Y be independent, jointly Gaussian random variables, with parameters

$$m_1 = m_2 = 0, \quad \sigma_1 = \sigma_2 = 1.$$

Consider the polar coordinate transformation

$$Z_1 = \sqrt{X^2 + Y^2}, \quad Z_2 = \tan^{-1} \frac{X}{Y}. \tag{4.50}$$

We can easily check that the System 4.50 has a single solution, and therefore

$$x = h_1(z_1, z_2) = z_1 \cos(z_2), \quad y = h_2(z_1, z_2) = z_1 \sin(z_2).$$

From Equation 4.48, the determinant of the Jacobian of this system is given by

$$|A| = \frac{1}{z_1}.$$

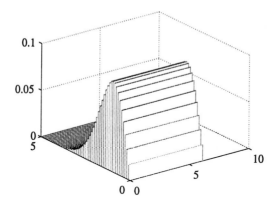

Figure 4.50   The joint pdf of the polar coordinates.

Therefore, for $z_1 > 0$ and $z_2 \in [0, 2\pi)$, we have from Theorem 4.10

$$f_{Z_1 Z_2}(z_1, z_2) = \frac{z_1}{2\pi} e^{-\frac{1}{2}[z_1^2 \cos^2(z_2) + z_1^2 \sin^2(z_2)]} = \frac{z_1}{2\pi} e^{-\frac{1}{2} z_1^2}.$$

From the last equation, we recognize that Z_1 and Z_2 are independent, with Z_1 following a Rayleigh distribution and Z_2 a uniform one. This joint pdf is shown in Figure 4.50.

Example 4.51.   Let X and Y be as in the above example; this time, let

$$Z_1 = \sqrt{X^2 + Y^2}, \qquad Z_2 = \frac{X}{Y}.$$

Now this system has two solutions in x, y, namely,

$$x = \frac{z_1 z_2}{\sqrt{1 + z_2^2}}, \quad y = \frac{z_1}{\sqrt{1 + z_2^2}},$$

$$x = -\frac{z_1 z_2}{\sqrt{1 + z_2^2}}, \quad y = -\frac{z_1}{\sqrt{1 + z_2^2}}.$$

The determinant of the Jacobian for both cases is

$$|A| = \frac{1 + z_2^2}{z_1},$$

and thus

$$f_{Z_1 Z_2}(z_1, z_2) = \frac{z_1}{1 + z_2^2} \frac{1}{\pi} e^{-\frac{1}{2} z_1^2} = \frac{1}{\pi} \frac{1}{1 + z_2^2} \cdot z_1 e^{-\frac{1}{2} z_1^2}.$$

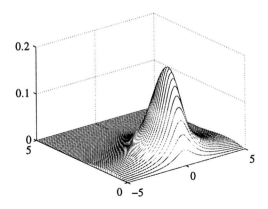

Figure 4.51 The joint pdf in Example 4.51.

Therefore Z_1 is a Rayleigh random variable, while Z_2 is Cauchy. Moreover, the two random variables are independent. Their joint pdf is shown in Figure 4.51.

Example 4.52. Let X and Y be two independent, jointly uniform random variables with range $[0, 1]$. Let

$$Z_1 = \max\{X, Y\}, \quad Z_2 = \min\{X, Y\}.$$

We will evaluate the joint cdf of Z_1, Z_2. We cannot use our theorem here since the max and min functions are not differentiable everywhere. We have to resort to the basics. From Figure 4.52(a), we can see that $\{Z_1 \leq z_1, \ Z_2 \leq z_2\}$ is a nontrivial event only when $0 < z_1, \ z_2 \leq 1$. Let $z_1 \geq z_2$. From the same figure, we recognize this event as the L-shaped region covered with squares and thus

$$P[Z_1 \leq z_1, \ Z_2 \leq z_2] = 2z_1 z_2 - z_2^2 = z_2(2z_1 - z_2).$$

Let next $z_1 < z_2$. Then we can easily see that

$$\{Z_1 \leq z_1, \ Z_2 \leq z_2\} = \{Z_1 \leq z_1\},$$

and thus

$$P[Z_1 \leq z_1, \ Z_2 \leq z_2] = P[Z_1 \leq z_1] = P[X \leq z_1]P[Y \leq z_1] = z_1^2.$$

This cdf is plotted in Figure 4.52(b).

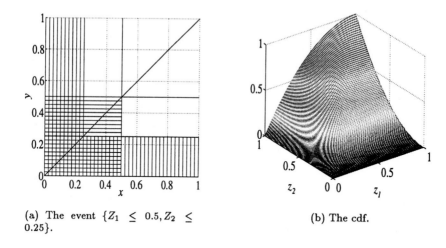

(a) The event $\{Z_1 \leq 0.5, Z_2 \leq 0.25\}$.

(b) The cdf.

Figure 4.52   The equivalent event and cdf of Z_1, Z_2.

4.9   EXPECTATION, COVARIANCE, AND CORRELATION COEFFICIENT

4.9.1   Expectation of a Function $g(X, Y)$

Consider a random experiment E with model (S, Q, P) and two random variables X and Y defined on it. Let $Z = g(X, Y)$ be an arbitrary function of two random variables X and Y that is itself a random variable. In this section, the generic problem of interest is to find the expected value of Z. Of course, one way is to calculate $f_Z(z)$ via Equation 4.38, and then easily

$$EZ = \int_{-\infty}^{\infty} z f_Z(z)\, dz,$$

provided of course that the integral exists (for example, Z is not, say, a Cauchy random variable). However, similar to the result in Section 3.9, we can show (but do not do so here; see [7] for a proof) that

$$EZ = \int_{-\infty}^{\infty} \int_{-\infty}^{\infty} g(x, y) f_{XY}(x, y)\, dx\, dy; \tag{4.51}$$

i.e., the desired expected value can be found directly from the jpdf of X and Y, without the need to evaluate $f_Z(z)$. In principle, this is good, but one has to consider how easy it is to evaluate EZ through Equation 4.51. Let's see some examples next.

Example 4.53. *Expected Value of a Sum $Z = X + Y$.* From Equation 4.51 we have

$$
\begin{aligned}
E(X + Y) &= \int_{-\infty}^{\infty} \int_{-\infty}^{\infty} (x + y) f_{XY}(x, y) \, dx \, dy \\
&= \int_{-\infty}^{\infty} \int_{-\infty}^{\infty} x f_{XY}(x, y) \, dx \, dy + \int_{-\infty}^{\infty} \int_{-\infty}^{\infty} y f_{XY}(x, y) \, dx \, dy \\
&= \int_{-\infty}^{\infty} x f_X(x) \, dx + \int_{-\infty}^{\infty} y f_Y(y) \, dy = EX + EY.
\end{aligned}
\tag{4.52}
$$

Equation 4.52 is used very frequently in practice. Note that it holds true regardless of whether the two variables are independent. In the case where the two random variables are independent, we can calculate the variance of the sum as follows:

$$
\begin{aligned}
\text{var}(X + Y) &= E[(X + Y) - EX - EY]^2 \\
&= E(X - EX)^2 + E(Y - EY)^2 + 2E(X - EX)(Y - EY) \\
&= \text{var}(X) + \text{var}(Y) + 2E(X - EX)E(Y - EY) \\
&= \text{var}(X) + \text{var}(Y),
\end{aligned}
\tag{4.53}
$$

since $E(X - EX) = 0$.

Example 4.54. *Expected Value of Functions of Independent Random Variables $Z = h_1(X) \cdot h_2(Y)$.* Suppose that X and Y are two independent random variables; let $h_1(X)$ and $h_2(Y)$ be two transformations of those random variables. Then

$$
\begin{aligned}
E[h_1(X) \cdot h_2(Y)] &= \int_{-\infty}^{\infty} \int_{-\infty}^{\infty} h_1(x) h_2(y) f_X(x) f_Y(y) \, dx \, dy \\
&= \int_{-\infty}^{\infty} h_1(x) f_X(x) \, dx \cdot \int_{-\infty}^{\infty} h_2(y) f_Y(y) \, dy \\
&= E h_1(X) \cdot E h_2(Y)
\end{aligned}
$$

Example 4.55. *A Useful Property of Conditional Expectations.* Consider two random variables X and Y, not necessarily independent, and two functions $h_1(X)$ and $h_2(Y)$. Then a formula that is quite useful in many situations (and that's reminiscent of the total probability theorem) is the following:

$$
E[h_1(X) h_2(Y)] = E\{E[h_1(X) h_2(Y)|Y]\} = E\{h_2(Y) E[h_1(X)|Y]\}.
\tag{4.54}
$$

4.9.2   Joint Moments of X and Y

Let i, j be fixed positive integers, and let $Z \triangleq X^i Y^j$.

Definition: The *ij*th **joint moment** of the random variables X and Y is

$$E(X^i Y^j) = \int_{-\infty}^{\infty} \int_{-\infty}^{\infty} x^i y^j f_{XY}(x, y)\, dx\, dy,$$

provided of course that the integral exists. Let now $Z = (X - EX)^i (Y - EY)^j$.

Definition: The *ij*th **joint central moment** of the random variables X and Y is

$$E[(X - EX)^i (Y - EY)^j]. \tag{4.55}$$

A certain central moment is very useful, as we see in Chapter 5. We give it a special name.

Definition: The **covariance** of the random variables X and Y, denoted as $\text{cov}(X, Y)$, is defined as

$$\text{cov}(X, Y) \triangleq E[(X - EX)(Y - EY)]. \tag{4.56}$$

Definition: The **correlation coefficient** ρ_{XY} of the random variables X and Y is

$$\rho_{XY} = \frac{\text{cov}(X, Y)}{\sigma_X \sigma_Y}. \tag{4.57}$$

It is easy to check that for any random variables X, Y, $|\rho_{XY}| \le 1$ (see Problem 4.119). Of course, in the definition we assume that $\sigma_X, \sigma_Y > 0$. If $\sigma_X \sigma_Y = 0$, we define $\rho_{XY} = 0$.

Remark: The term *correlation coefficient* was formally introduced by Galton[8] in the late nineteenth century, in his seminal studies on linear regression. Laplace used it in his analysis-of-error studies, even though he did not foresee its importance in estimation.

As expected, the moments and central moments are related. The following relationship is very often used, so we state it as a theorem. It can be derived quite easily by expanding the bracketed term in Equation 4.56, so we omit its proof.

Theorem 4.11 *For any arbitrary random variables X and Y,*

$$\text{cov}(X, Y) = E(XY) - EX \cdot EY.$$

---

[8] Sir Francis Galton, 1822–1911, an English statistician.

Whenever $\text{cov}(XY) = 0$, the two random variables are called *uncorrelated*; when $E(XY) = 0$, the two random variables are called *orthogonal*.

Let's see some examples now.

Example 4.56. Consider the random variables X, Y with the following jpdf:

$$f_{XY}(x, y) = 2e^{-x}e^{-y}u(x)u(x - y).$$

We will calculate $\text{cov}(X, Y)$ and ρ_{XY}. We have

$$
\begin{aligned}
EXY &= \int_{-\infty}^{\infty} \int_{-\infty}^{\infty} xy \left(2e^{-x}e^{-y}\right) u(x)u(x - y) \, dy \, dx \\
&= \int_{0}^{\infty} 2xe^{-x} \int_{0}^{x} ye^{-y} \, dy \, dx = \int_{0}^{\infty} 2xe^{-x}(1 - e^{-x}x - e^{-x}) \, dx.
\end{aligned}
$$

After some calculations, we get that $EXY = 1$. Moreover,

$$EX = \int_{-\infty}^{\infty} \int_{-\infty}^{\infty} x \left(2e^{-x}e^{-y}\right) u(x)u(x - y) \, dy \, dx = 1.5,$$

$$EY = \int_{-\infty}^{\infty} \int_{-\infty}^{\infty} y \left(2e^{-x}e^{-y}\right) u(x)u(x - y) \, dy \, dx = 0.5.$$

$$EX^2 = \int_{-\infty}^{\infty} \int_{-\infty}^{\infty} x^2 \left(2e^{-x}e^{-y}\right) u(x)u(x - y) \, dy \, dx = 3.5,$$

$$EY^2 = \int_{-\infty}^{\infty} \int_{-\infty}^{\infty} y^2 \left(2e^{-x}e^{-y}\right) u(x)u(x - y) \, dy \, dx = 0.5,$$

and thus $\sigma_X = \sqrt{EX^2 - (EX)^2} = 1.118$, $\sigma_Y = \sqrt{EY^2 - (EY)^2} = 0.5$. Now,

$$\text{cov}(X, Y) = 1 - 1.5 \cdot 0.5 = 0.25,$$

and

$$\rho_{XY} = \frac{\text{cov}(X, Y)}{\sigma_X \sigma_Y} = \frac{0.25}{1.118 \cdot 0.5} = 0.447.$$

Example 4.57. Consider the *netstat* example. Calculate the correlation between the number of packets and the number of collisions.

Using the MATLAB function corrcoef() and the data in the Web site, we can calculate for this experiment and these two random variables, $\rho_{XY} = 0.4667$.

Example 4.58. Consider the *ftp* experiment. Calculate the correlation coefficient between all pairs of the random variables involved.

The three random variables in this experiment are the file size, delay of transfer, and throughput. Let F, D, and T denote these random variables, respectively. From the given measurements in the Web site (approximately 400), we get

$$\rho_{FD} = 0.6789, \quad \rho_{FT} = 0.0540, \quad \rho_{DT} = -0.0162.$$

Example 4.59. Let X and Y be defined through $X = \cos(\Theta), Y = \sin(\Theta)$, where Θ is a uniform random variable with range $[-\pi, \pi)$. Calculate ρ_{XY}.

We have

$$EX = \int_{-\pi}^{\pi} \cos(\theta) \frac{1}{2\pi} \, d\theta = 0,$$

$$EY = \int_{-\pi}^{\pi} \sin(\theta) \frac{1}{2\pi} \, d\theta = 0,$$

$$EXY = \int_{-\pi}^{\pi} \cos(\theta) \sin(\theta) \frac{1}{2\pi} \, d\theta = 0,$$

$$EY^2 = \int_{-\pi}^{\pi} \sin^2(\theta) \frac{1}{2\pi} \, d\theta = 0.5,$$

$$EX^2 = \int_{-\pi}^{\pi} \cos^2(\theta) \frac{1}{2\pi} \, d\theta = 0.5,$$

and thus $\rho_{XY} = 0$.

4.10   USEFUL JOINT DISTRIBUTIONS

In this section we present in detail some of the most widely used models; as expected, these models are variations of the one-dimensional Gaussian, Poisson, etc., models, with or without dependence between the individual elements of the random vector. We treat the Gaussian and Poisson models in detail since they are very widely used.

4.10.1   Independent Random Variables

When the two random variables are independent, the joint pdf (or pmf) is simply the product of two of the pdf's (or pmf's) in Section 3.6. For brevity, we present only the Poisson and Gaussian models here.

1. The Poisson model

Let X and Y be independent Poisson random variables with parameters λ_1 and λ_2. Then the two-dimensional random variable (X, Y) has a jpmf given by

$$P[X = i, \ Y = j] = \begin{cases} e^{-\lambda_1} e^{-\lambda_2} \frac{\lambda_1^i}{i!} \frac{\lambda_2^j}{j!}, & i, j = 0, 1, 2, \ldots \\ 0, & \text{otherwise.} \end{cases} \tag{4.58}$$

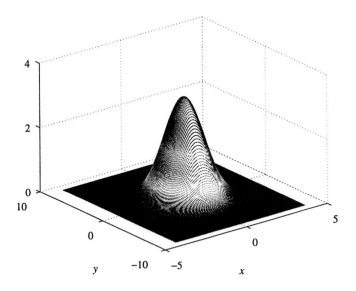

Figure 4.53 Two independent Gaussian random variables.

This jpmf is shown in Figure 4.12 for the special case $\lambda_1 = 2, \lambda_2 = 5$.

2. *The Gaussian model*

Let X and Y be independent Gaussian random variables with parameters m_1, σ_1^2 and m_2, σ_2^2. The two-dimensional random variable (X, Y) has a jpdf given by

$$f_{XY}(x, y) = \frac{1}{\sqrt{2\pi}\sigma_1} e^{-\frac{1}{2\sigma_1^2}(x-m_1)^2} \frac{1}{\sqrt{2\pi}\sigma_2} e^{-\frac{1}{2\sigma_2^2}(y-m_2)^2}, \quad x, y \in R.$$

This jpdf is shown in Figure 4.53.

4.10.2 Dependent Random Variables

1. *The Poisson model*

Let X and Y be two dependent Poisson random variables. The joint pmf of the vector (X, Y) is defined over all nonnegative integers (i, j) and given by

$$P[X = i, Y = j] = e^{-\lambda_X - \lambda_Y - \lambda_{XY}} \sum_{k=0}^{\min(i,j)} \frac{\lambda_{XY}^k \lambda_X^{i-k} \lambda_Y^{j-k}}{k!(i-k)!(j-k)!}, \quad i, j = 0, 1, \ldots.$$

$$(4.59)$$

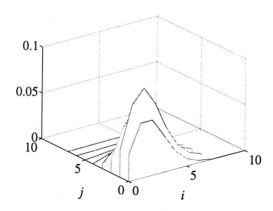

Figure 4.54 The dependent Poisson model.

Here $\lambda_{XY}, \lambda_X, \lambda_Y$ are nonnegative constants. As we ask you to show in Problem 4.75, the parameter of X (or Y) is $\lambda_{XY} + \lambda_X$ (or $\lambda_{XY} + \lambda_Y$). Figure 4.54 shows this model for $\lambda_{XY} = 0.5, \lambda_X = 1, \lambda_Y = 2$.

2. The Gaussian model

Let X and Y be two dependent Gaussian random variables with parameters $m_1, \sigma_1^2, m_2, \sigma_2^2$, and correlation coefficient ρ. The jpdf of the vector (X, Y) is defined over all $(x, y) \in R^2$ and given by

$$f_{XY}(x, y) = \frac{e^{-\frac{1}{2(1-\rho^2)}\left\{\frac{(x-m_1)^2}{2\sigma_1^2} - 2\rho\frac{(x-m_1)}{\sigma_1}\frac{(y-m_2)}{2\sigma_2} + \frac{(y-m_2)^2}{\sigma_2^2}\right\}}}{2\pi\sigma_1\sigma_2\sqrt{(1-\rho^2)}}, \quad x, y \in R. \quad (4.60)$$

Figure 4.55 shows Equation 4.60 for $m_1 = 10, m_2 = 0, \sigma_1 = \sigma_2 = 1$, and $\rho = 0.2$ and 0.95, respectively.

The conditional pdf's associated with this Gaussian model can be easily found from Equation 4.60 via Equations 4.25 and 4.26 in Section 4.6.2:

$$f_X(x|y) = \frac{f_{XY}(x, y)}{f_Y(y)} = \frac{e^{-\frac{1}{2(1-\rho^2)\sigma_1^2}\left[x - \rho\frac{\sigma_1}{\sigma_2}(y-m_2) - m_1\right]^2}}{\sqrt{2\pi\sigma_1^2(1-\rho^2)}}, \quad (4.61)$$

and by symmetry,

$$f_Y(y|x) = \frac{f_{XY}(x, y)}{f_X(x)} = \frac{e^{-\frac{1}{2(1-\rho^2)\sigma_2^2}\left[y - \rho\frac{\sigma_2}{\sigma_1}(x-m_1) - m_2\right]^2}}{\sqrt{2\pi\sigma_2^2(1-\rho^2)}}.$$

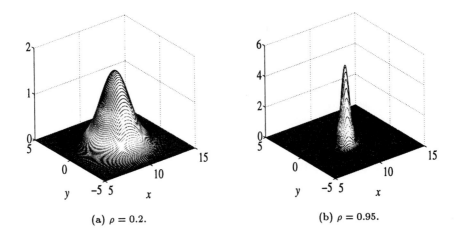

(a) $\rho = 0.2$. (b) $\rho = 0.95$.

Figure 4.55 Two dependent Gaussian random variables.

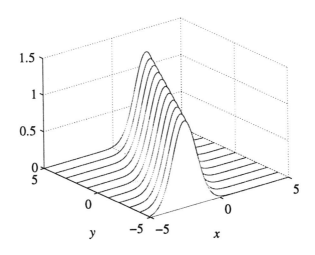

Figure 4.56 Conditional probability of X given Y.

Figure 4.56 shows the pdf in Equation 4.61 for various values of x and $m_1 = m_2 = 0$, $\sigma_1^2 = \sigma_2^2 = 1$, and $\rho = 0.5$. (Can you tell which axis represents x?)

What is the meaning of the parameters $m_1, \sigma_1^2, m_2, \sigma_2^2$, and ρ? As implied by their symbols, m_1 and m_2 must be the means of X and Y, respectively, σ_1^2, σ_2^2 must be the variances, and finally ρ must be the correlation coefficient. We formally show now that this is the case.

For the means and variances, we must first find the marginal pdf's. The easiest way is to complete the square in the exponent of Equation 4.60; we leave this as a tedious but straightforward exercise. Let's calculate the correlation coefficient next since it involves use of the conditional expectation properties we mentioned in Section 4.9.1. Using Equation 4.54, we have

$$\text{cov}(X, Y) = E(X - m_1)(Y - m_2) = E[E(X - m_1)(Y - m_2)|Y]. \quad (4.62)$$

The innermost expectation is calculated easily from the conditional pdf in Equation 4.61 as follows:

$$
\begin{aligned}
E[(X - m_1)(Y - m_2)|Y = y] &= (y - m_2)E[(X - m_1)|Y = y] \\
&= (y - m_2)(E[X|Y = y] - m_1) \\
&= (y - m_2)\left[\rho\frac{\sigma_1}{\sigma_2}(y - m_2)\right]. \quad (4.63)
\end{aligned}
$$

We used Property 4.54 to get the first equality and Property 3.52 to get the second one. We can get the last equality by direct calculation of the integral in $E[X|Y = y]$ or by using the definition of the correlation coefficient. Substituting Equation 4.63 into Equation 4.62, we get

$$\text{cov}(X, Y) = E\left\{(Y - m_2)[\rho\frac{\sigma_1}{\sigma_2}(Y - m_2)]\right\} = \rho\sigma_1\sigma_2,$$

which we recognize as the definition of the correlation coefficient.

4.11   MORE THAN TWO RANDOM VARIABLES

In this section we generalize the definitions and theorems in Sections 4.1–4.10 for the more general case of $n > 2$ random variables. We will need some additional definitions (such as the one in Equation 4.65 since we can condition on more than one random variable now). Since the proofs of the theorems are fairly straightforward extensions of the two-dimensional case, we omit them. Notice that the notation becomes much more cumbersome now; unfortunately, there is no way we can simplify it without introducing ambiguities and confusion.

4.11.1   Definitions

Consider again a probabilistic model $(S, \mathcal{Q}, P)$. Suppose now we define $n > 2$ random variables on it, denoted as $X_1, \ldots, X_n$:

$$X_i \; : \; S \to R, \quad i = 1, \ldots, n.$$

For notational convenience we may sometimes write $\mathbf{X} \overset{\triangle}{=} (X_1, \ldots, X_n)$.

Definition: $\mathbf{X}$ is called an n-**dimensional random vector** or a **random vector with n elements**.

Definition: The jcdf of $\mathbf{X} = (X_1, \ldots, X_n)$ is defined as

$$F_{X_1 X_2 \cdots X_n}(x_1, x_2, \ldots, x_n) \overset{\triangle}{=} P[X_1 \le x_1, X_2 \le x_2, \ldots, X_n \le x_n]. \qquad (4.64)$$

In a more compact notation,

$$F_{\mathbf{X}}(\mathbf{x}) \overset{\triangle}{=} P[\mathbf{X} \le \mathbf{x}],$$

where $\mathbf{x} = (x_1, \ldots, x_n) \in R^n$.

Definition: The jpdf of $\mathbf{X} = (X_1, \ldots, X_n)$ is defined as

$$f_{X_1 X_2 \cdots X_n}(x_1, x_2, \ldots, x_n) \overset{\triangle}{=} \frac{\partial^n F_{X_1 X_2 \cdots X_n}(x_1, x_2, \ldots, x_n)}{\partial x_1 \, \partial x_2 \ldots \partial x_n},$$

provided of course the derivative exists.

For notational convenience, when the n random variables are discrete, we can define the joint probability mass function thus:

Definition: The **joint probability mass function** of $\mathbf{X} = (X_1, \ldots, X_n)$ is

$$P_{X_1 X_2 \cdots X_n}(X_1 = x_1, X_2 = x_2, \ldots, X_n = x_n) \overset{\triangle}{=} P[X_1 = x_1, \ldots, X_n = x_n].$$

(Note the abuse of the symbol P.) In a more compact notation,

$$P_{\mathbf{X}}(\mathbf{X} = \mathbf{x}) \overset{\triangle}{=} P[\mathbf{X} = \mathbf{x}].$$

Definition 4.64 implies certain properties (compare to Properties 3.4–3.7).

Lemma 4.13 *Let* $\mathbf{X} = (X_1, \ldots, X_n)$ *be a random vector with n elements. Then*

(1) $F_{\mathbf{X}}(\mathbf{x}) \ge 0, \quad \forall \mathbf{x} \in R^n.$

(2) $F_{X_1 X_2 \cdots X_n}(\infty, \infty, \ldots, \infty) = 1.$

(3) $F_{X_1 X_2 \cdots X_n}(x_1, x_2, \ldots, x_n) = 0,$ *when* $x_i = -\infty,$ *for some* $i = 1, 2, \ldots, n.$

(4) $F_{\mathbf{X}}(\mathbf{x})$ *is continuous from the right.*

(5) $F_{\mathbf{X}}(\mathbf{x})$ *is nondecreasing.*

4.11.2   Marginal CDF

The marginal cdf's of any random vector of dimensionality $n - 1$ or less can be computed from Definition 4.64. For example,

$$
\begin{aligned}
F_{X_1}(x_1) &= F_{X_1 X_2 \cdots X_n}(x_1, \infty, \ldots, \infty), \\
F_{X_i}(x_i) &= F_{X_1 X_2 \cdots X_n}(\infty, \ldots, x_i, \infty, \ldots, \infty), \\
F_{X_1, X_2}(x_1, x_2) &= F_{X_1 X_2 \cdots X_n}(x_1, x_2, \infty, \ldots, \infty).
\end{aligned}
$$

4.11.3   Conditional CDF

We can define a *family* of conditional cdf's since there are a variety of possible conditioning events defined in terms of random variables X_i. All the conditional cdf's are variations of the main idea in Chapter 3, where we defined the generic conditional cdf as

$$
F_X(x|A) = \frac{P[\{X \le x\} \cap A]}{P[A]}.
$$

Definition: The **conditional cdf** of the random variable X_i, given that $X_j = x_j$, is

$$
F_{X_i|X_j}(x_i|x_j) \triangleq \frac{P[\{X_i \le x_i\} \cap \{X_j = x_j\}]}{P[\{X_j = x_j\}]}.
$$

Definition: The conditional cdf of random variable X_i, given that $X_j = x_j, X_k = x_k$, is

$$
F_{X_i|X_j X_k}(x_i|x_j, x_k) \triangleq \frac{P[\{X_i \le x_i\} \cap \{X_j = x_j, X_k = x_k\}]}{P[\{X_j = x_j, X_k = x_k\}]}. \tag{4.65}
$$

Definition: The conditional cdf of the random vector (X_i, X_l), given that $X_j = x_j, X_k = x_k$, is defined as

$$
F_{X_i X_l|X_j X_k}(x_i, x_l|x_j, x_k) \triangleq \frac{P[\{X_i \le x_i, X_l \le x_l\} \cap \{X_j = x_j, X_k = x_k\}]}{P[\{X_j = x_j, X_k = x_k\}]}.
$$

The conditional densities are also very useful in practice; we show only one here.

Definition: The conditional pdf of the random vector (X_i, X_l), given that $X_j = x_j, X_k = x_k$, is defined as

$$
f_{X_i X_l|X_j X_k}(x_i, x_l|x_j, x_k) \triangleq \frac{f_{X_i X_l X_j X_k}(x_i, x_l, x_j, x_k)}{f_{X_j X_k}(x_j, x_k)}.
$$

4.11.4 Independence

The random variables $X_1, \ldots, X_n$ are called **independent** if and only if

$$P[X_1 \in A_1 \cap X_2 \in A_2 \cap \cdots \cap X_n \in A_n] = \prod_{k=1}^{n} P[X_k \in A_k],$$

for all such events $A_k, \ k = 1, \ldots, n$. The analogs of Theorems 4.3 and 4.4 are given below:

Theorem 4.12 *The random variables $X_1, \ldots, X_n$ are independent if and only if*

$$F_{X_1 X_2 \cdots X_n}(x_1, x_2, \ldots, x_n) = F_{X_1}(x_1) \cdot F_{X_2}(x_2) \cdots F_{X_n}(x_n), \ \forall x_1, \ldots, x_n \in R.$$

Theorem 4.13 *The random variables $X_1, \ldots, X_n$ are independent if and only if*

$$f_{X_1 X_2 \cdots X_n}(x_1, x_2, \ldots, x_n) = f_{X_1}(x_1) \cdot f_{X_2}(x_2) \cdots f_{X_n}(x_n), \ \forall x_1, \ldots, x_n \in R.$$

4.11.5 A Single Function of n Random Variables

Consider the function of n arguments $g(x_1, x_2, \ldots, x_n)$. The cdf of the random variable $Z = g(X_1, X_2, \ldots, X_n)$ can be found again easily (in principle) once the equivalent event of $\{Z \leq z\}$ is found in R^n.

Theorem 4.14 *Let $Z = g(X_1, X_2, \ldots, X_n)$ be a transformation of the random vector $(X_1, X_2, \ldots, X_n)$. Let*

$$R_z \overset{\triangle}{=} \{(x_1, \ldots, x_n) \in R^n : g(x_1, \ldots, x_n) \leq z\}.$$

Then the cdf of Z is given by

$$F_Z(z) = \int \int \cdots \int_{R_z} f_{X_1 X_2 \cdots X_n}(x_1, x_2, \ldots, x_n) \, dx_1 \cdots dx_n.$$

4.11.6 Expected Values

Equations 4.52 and 4.53 have fairly straightforward extensions:

$$E(X_1 + \cdots + X_n) = EX_1 + \cdots + EX_n.$$

Whenever the random variables are independent, we have

$$\text{var}(X_1 + \cdots + X_n) = \text{var}(X_1) + \cdots + \text{var}(X_n).$$

4.11.7   Gaussian and Poisson Models

Consider n random variables $\mathbf{X} \triangleq (X_1, \ldots, X_n)$. Denote their mean values as another vector $\mathbf{m}$:

$$\mathbf{m} \triangleq \begin{bmatrix} EX_1 \\ EX_2 \\ \vdots \\ EX_n \end{bmatrix}.$$

Define their covariance matrix C as

$$C \triangleq \begin{bmatrix} \text{var}(X_1) & \text{cov}(X_1, X_2) & \cdots & \text{cov}(X_1, X_n) \\ \text{cov}(X_2, X_1) & \text{var}(X_2) & \cdots & \text{cov}(X_2, X_n) \\ \cdots & \cdots & \cdots & \cdots \\ \text{cov}(X_n, X_1) & \text{cov}(X_n, X_2) & \cdots & \text{var}(X_n) \end{bmatrix}.$$

The vector is said to be jointly Gaussian if and only if the jpdf is given by

$$f_{\mathbf{X}}(\mathbf{x}) = \frac{1}{(2\pi)^{\frac{n}{2}} |C|^{\frac{1}{2}}} e^{-\frac{1}{2}(\mathbf{x}-\mathbf{m})^T C^{-1}(\mathbf{x}-\mathbf{m})}. \tag{4.66}$$

Consider a random vector $\mathbf{X} = (X_1, \ldots, X_n)$. Now $\mathbf{X}$ is an n-dimensional Poisson random vector if and only if its jpmf is given by

$$P[X_1 = k_1, \ldots, X_n = k_n] = e^{-(\lambda_1 + \cdots + \lambda_n)} \prod_{l=1}^{n} \frac{\lambda_l^{k_l}}{k_l!},$$

where $\lambda_l > 0$ are given parameters. Observe that the definition implies that the random variables $\{X_l\}$ are independent and Poisson-distributed.

4.11.8   Transformations

Consider a linear transformation $\mathbf{Z} = A\mathbf{X}$, where A is an $n \times n$ invertible matrix. Let $\mathbf{z} \triangleq (z_1, \ldots, z_n)$. The extension of Theorem 4.9 is the following:

Theorem 4.15 *Let* $\mathbf{Z} = A\mathbf{X}$. *Let* $(x_1, \ldots, x_n)^T = A^{-1}\mathbf{z}$. *Then*

$$f_{\mathbf{Z}}(\mathbf{z}) = \frac{f_{\mathbf{X}}(x_1, \ldots, x_n)}{|A|}.$$

The extension of Theorem 4.10 is given by the following:

Theorem 4.16 *Let* $Z_1 = g_1(X, Y), \ldots, Z_n = g_n(X, Y)$. *Then*

$$f_{Z_1 \cdots Z_n}(z_1, \ldots, z_n) = \frac{f_{\mathbf{X}}(h_1(z_1, \ldots, z_n), h_2(z_1, \ldots, z_n), \ldots, h_n(z_1, \ldots, z_n))}{|A|},$$

$$\tag{4.67}$$

with the obvious change in notation if there are multiple roots.

Let's see some examples next.

Example 4.60. *Linear Transformations of Gaussian Random Variables.* Consider the n-dimensional Gaussian vector $\mathbf{X}$ of Section 4.11.7. Let

$$\mathbf{Z} = A\mathbf{X} \tag{4.68}$$

be the linear transformation, where in order to avoid trivialities, we assume that the matrix A is invertible. From Equation 4.67 we have

$$f_{\mathbf{Z}}(\mathbf{z}) = \frac{f_{\mathbf{X}}(A^{-1}\mathbf{z})}{(2\pi)^{n/2}|C|^{0.5}|A|}. \tag{4.69}$$

Applying Equation 4.66 to the numerator of Equation 4.69, we see that the exponent is

$$e^{-\frac{1}{2}(A^{-1}\mathbf{z}-\mathbf{m})^T C^{-1}(A^{-1}\mathbf{z}-\mathbf{m})}.$$

We can simplify the expression in this exponent somewhat, using elementary matrix properties. Let I denote the identity matrix. We can rewrite C^{-1} as

$$C^{-1} = IC^{-1}I = A^T(A^T)^{-1}C^{-1}A^{-1}A,$$

and after some simple multiplications, we can rewrite the exponent in the simpler form

$$(\mathbf{z} - A\mathbf{m})^T D^{-1}(\mathbf{z} - A\mathbf{m}),$$

where for notational simplicity we have defined $D = ACA^T$. The denominator can also be rewritten in terms of the new matrix D as $(2\pi)^{n/2}|D|^{0.5}$. Therefore,

$$f_{\mathbf{Z}}(\mathbf{z}) = \frac{e^{-(\mathbf{z}-A\mathbf{m})^T D^{-1}(\mathbf{z}-A\mathbf{m})}}{(2\pi)^{n/2}|D|^{0.5}},$$

which we recognize as the jpdf of a Gaussian random vector, with mean $A\mathbf{m}$ and covariance matrix D.

We summarize the above as a theorem; we will extensively use this theorem later, e.g., in Chapter 7.

Theorem 4.17 *Let $\mathbf{X}$ be a vector of n jointly Gaussian random variables with mean m_X and covariance matrix C_X, and let $\mathbf{Z} = A\mathbf{X}$. Then $\mathbf{Z}$ is also jointly Gaussian, with mean $m_Z = A \cdot m_X$ and covariance matrix $C_Z = AC_X A^T$.*

Example 4.61. *Sums of Poisson Random Variables.* Let X_i, $i = 1, 2, \ldots, n$, be n independent Poisson random variables, each with parameter λ_i. What is the pmf of $Z = X_1 + \cdots + X_n$?

In the examples in Section 4.8.2, we saw that the sum of two independent Poisson random variables is again a Poisson random variable. Therefore, using a simple induction argument, we can see that the sum of n independent Poisson random variables, each with parameter λ_i, will be a Poisson random variable with parameter $\lambda = \sum_{i=1}^{n} \lambda_i$.

Example 4.62. *Sums of Gamma Random Variables.* Let X_i, $i = 1, 2, \ldots, n$, be n independent gamma random variables, each with parameters a_i, λ. What is the pdf of $Z = X_1 + \cdots + X_n$?

In Section 4.8.2 we also saw that the sum of two independent gamma random variables with a common parameter λ and arbitrary parameters a_i is another gamma random variable. Therefore, the sum of n independent gamma random variables with a common parameter λ will be again gamma-distributed with parameters λ and $a = \sum_{i=1}^{n} a_i$.

Example 4.63. *Sums of Exponential Random Variables.* Now let X_i, $i = 1, 2, \ldots, n$, be n independent exponential random variables, each with parameter λ_i. Let $Z = X_1 + \cdots + X_n$. Since the exponential random variable is a special case of the gamma random variable, Z is Erlang with parameters n and $\lambda = \sum_{i=1}^{n} \lambda_i$.

4.12   GENERATION OF VALUES OF A RANDOM VECTOR

In this section we present an algorithm for generating variates of $n \geq 2$, jointly distributed random variables $X_1, \ldots, X_n$. Of course, when the random variables are independent, the algorithm in Section 3.13 can be used. We simply generate independent uniform variates $u_1, u_2, \ldots$ and then generate the variates in question through

$$(x_1, x_2, \ldots) = (F_{X_1}^{-1}(u_1), F_{X_2}^{-1}(u_2), \ldots).$$

When the random variables follow an arbitrary joint cdf, the issue is far more complicated. In the sequel, we address only a special case: How do we generate jointly distributed random variables that have a given covariance matrix?

The basic idea is to start with variates $\mathbf{X}$ that have zero mean and unit variance and are uncorrelated, i.e.,

$$E\mathbf{X} = 0,$$

$$EX_i X_j = \begin{cases} 1, & i = j, \\ 0, & i \neq j. \end{cases} \tag{4.70}$$

(How can we obtain them?) Then we can transform the variates $\mathbf{X}$ with a *linear transformation*

$$\mathbf{Y} = A\mathbf{X}, \tag{4.71}$$

in order to achieve the desired covariance matrix. So the issue is how to determine matrix A from knowledge of the covariance matrix $\text{cov}(\mathbf{Y})$.

Let's write Equation 4.71 in its scalar form

$$Y_i = \sum_{k=1}^{n} a_{ik} X_k, \quad i = 1, \ldots, n,$$

where a_{ik} denotes the i, k element of A. From Equation 4.70 we have

$$EY_i = E \sum_{k=1}^{n} a_{ik} X_k = \sum_{k=1}^{n} a_{ik} EX_k = 0, \quad i = 1, \ldots, n,$$

and using Equation 4.70 one more time, we have

$$
\begin{aligned}
EY_i Y_j &= E \sum_{k=1}^{n} a_{ik} X_k \sum_{l=1}^{n} a_{jl} X_l = \sum_{k=1}^{n} \sum_{l=1}^{n} a_{ik} a_{jl} EX_k X_l \\
&= \sum_{k=1}^{n} a_{ik} a_{jk}, \quad i, j = 1, \ldots, n.
\end{aligned}
\tag{4.72}
$$

We recognize the last equality in Equation 4.72 as matrix multiplication, so

$$\text{cov}(\mathbf{Y}) = A \cdot A^T. \tag{4.73}$$

We must solve System 4.73 for A; this is a well-known problem in linear algebra, and it is solved via *diagonalization*. The *Cholesky factorization*, in particular, is one popular technique for doing that.

Briefly, the solution of System 4.73 proceeds as follows. Let λ_i denote the ith eigenvalue of the given matrix $\text{cov}(\mathbf{Y})$. Let e_i denote the ith eigenvector of $\text{cov}(\mathbf{Y})$.

1. Put all eigenvalues in a diagonal matrix Λ; i.e., set $\Lambda_{ii} = \lambda_i$, $i = 1, \ldots, n$.

2. Put all eigenvectors (as columns) in an $n \times n$ matrix M.

3. Then set $A = M\sqrt{\Lambda}$, where the square root of a diagonal matrix Λ is also diagonal, with elements the square roots of the elements of Λ.

Remark: Notice that

$$AA^T = M\sqrt{\Lambda}(M\sqrt{\Lambda})^T = M\sqrt{\Lambda}\sqrt{\Lambda}^T M^T = M\Lambda M^T = \text{cov}(\mathbf{Y}).$$

Therefore, the algorithm to generate zero-mean variates, with a given covariance matrix $\text{cov}(\mathbf{Y})$ is the following:

1. **Factorize** $\text{cov}(\mathbf{Y})$ as $A \cdot A^T$.

2. **Generate** zero-mean, uncorrelated, unit-variance variates $\mathbf{X}$.

3. **Output** $\mathbf{Y} = A\mathbf{X}$.

Since the mean and covariance matrix suffice to fully characterize the jpdf of the n-dimensional Gaussian model, this algorithm suffices to generate Gaussian variates that follow an arbitrary pdf.

Example 4.64. Let's consider the case $n = 4$; suppose that

$$\text{cov}(\mathbf{Y}) = \begin{bmatrix} 10 & 2 & 1 & 3 \\ 2 & 10 & 1 & 0 \\ 1 & 1 & 4 & 6 \\ 3 & 0 & 6 & 1 \end{bmatrix}.$$

The eigenvalues of this array are equal to $6.64, 8.93, 13.4, -3.97$. The corresponding eigenvectors are the columns in the matrix

$$\begin{bmatrix} 0.63 & 0.29 & 0.70 & 0.13 \\ -0.17 & -0.83 & 0.51 & -0.06 \\ -0.66 & 0.31 & 0.34 & 0.58 \\ -0.36 & 0.34 & 0.33 & -0.79 \end{bmatrix}.$$

Thus the matrix A is given by

$$\begin{bmatrix} 0.63 & 0.29 & 0.70 & 0.13 \\ -0.17 & -0.83 & 0.51 & -0.06 \\ -0.66 & 0.31 & 0.34 & 0.58 \\ -0.36 & 0.34 & 0.33 & -0.79 \end{bmatrix} \cdot \begin{bmatrix} \sqrt{6.64} & 0 & 0 & 0 \\ 0 & \sqrt{8.93} & 0 & 0 \\ 0 & 0 & \sqrt{13.4} & 0 \\ 0 & 0 & 0 & \sqrt{-3.97} \end{bmatrix}$$

$$= \begin{bmatrix} 4.19 & 2.60 & 9.45 & -0.54 \\ -1.19 & -7.44 & 6.93 & 0.24 \\ -4.38 & 2.79 & 4.63 & -2.33 \\ -2.43 & 3.10 & 4.53 & 3.14 \end{bmatrix}.$$

4.13   SUMMARY OF MAIN POINTS

- A **random vector** is a (measurable) function from S onto R^n. It is a convenient way to avoid dealing with the subtleties of Q.

- The **joint cumulative distribution function** of two random variables X, Y provides the conceptual link between probabilities in the original sample space and probabilities in R^2.

- The **joint probability density function** is the central concept of this chapter. It solves the questions of how to satisfy Axiom 3 and to which events we assign probabilities. It also enables us to calculate probabilities via **two-dimensional histograms** in practice and via integration over R^2 in theory.

- Random vectors are classified as **continuous, discrete, and mixed. A joint probability mass function** is used when the random variables are discrete.

- The main random vector **models** are still Poisson and Gaussian with or without correlations.

- It is easy to check for **independence** of two random variables; we can use the jpdf, jcdf, jpmf, or a number of **conditional probability functions** for that purpose.

- The jpdf of **transformations** of random vectors is given by Theorems 4.6, 4.7, 4.8, 4.9, 4.10, and 4.16.

- **Means, covariances,** and **correlations** summarize information about the jpdf of a random vector.

- A linear transformation and uncorrelated, unit-variance random numbers suffice to generate **variates** of a random vector with a desired covariance.

4.14   CHECKLIST OF IMPORTANT TOOLS

- Theorems 4.6, 4.7, 4.8, 4.9, 4.10, and 4.16

- The algorithm to generate variates of a random vector

4.15   PROBLEMS

Two-Dimensional Sets

4.1 Express the following sets as unions, intersections, complements, or limits of sets of the form $(-\infty, x] \times (-\infty, y]$. Note that the boundaries of the sets are included.
(a) A rectangle with (3,5) and (15,40) as the lower left and upper right corners.
(b) A square with side length 3 and the lower left corner at $(-3, -3)$.

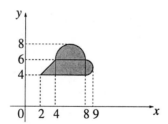

Figure 4.57   A nonrectangular event.

(c) A vertical infinite strip of width 2, centered around the y axis.

(d) A horizontal infinite strip that intersects the y axis at $y = 2$ and $y = -10$.

(e) $A = \{(x, y) \in R^2 : x^2 - \frac{y^2}{3} \leq 4\}$.

4.2 Derive an expression for a circle with arbitrary radius r and origin at (x_0, y_0) in terms of sets of the form $(-\infty, x] \times (-\infty, y]$. Discuss whether in your expression you have included the boundary of the circle.

4.3 Derive an expression for a circle, but using squares instead of rectangles. Is this more difficult than Problem 4.2?

4.4 Derive expressions for the set shown in Figure 4.57 in terms of rectangular sets. The boundary is included in this set.

4.5   Consider a random variable (X, Y) with range $S_{XY} = [0, 100] \times [0, 10]$. Identify and sketch the following events as subsets of S_{XY}.

(a) $\{X < Y\}$.

(b) $\{X \leq Y\}$.

(c) $\{|X| \leq Y\}$.

(d) $\{3X + 2Y = 0\}$.

(e) $\{3X + 2Y > -2\}$.

(f) $\{|X - Y| > 2\}$.

(g) $\{XY < 5\}$.

(h) $\{\max\{X, Y\} > 2\}$.

(i) $\{\min\{X, \frac{Y}{3}\} \leq 3\}$.

(j) $\{X^2 > 3\}$.

(k) $\{3 < X^2 + Y^2 \leq 6\}$.

(l) $\{X^2 + \frac{Y}{2} = 4\}$.

(m) $\{\sin(X) + \sin(Y) = 1\}$.

(n) $\{\sin(X) + \sin(Y) = 2\}$.

(o) $\{\sin(X) + \sin(Y) \leq 1\}$.

4.6 Repeat Problem 4.5 when the range is the circle $C = \{(x, y) : x^2 + y^2 \leq 10\}$.

4.7 Consider two random variables X and Y and the following events:

$$A = \{X \le x_1,\ Y \le y_1\}, \quad B = \{X \le x_2,\ Y \le y_2\},$$
$$C = \{X \le x_3\}, \quad D = \{Y \le y_3\}.$$

(a) Which of the following is/are true?
(1) $x_1 \le x_2$ implies that $A \subseteq B$,
(2) $x_1 \le x_3$ implies that $A \subseteq C$,
(3) $x_1 \le x_3$ and $y_1 \le y_3$ imply that $A \subseteq (C \cup D)$.
(b) Show the sets $A \cap C, A \cap D, A \cup B \cap (C \cap D)$ for a few values of x_1, x_2, x_3 and y_1, y_2, y_3.

The Joint CDF; Simple Skills

4.8 Are $F_{XY}(x, y)$ and $F_{YX}(x, y)$ the same function? How about $F_{XY}(x, y)$ and $F_{YX}(y, x)$?

4.9 Plot the sets $(x < n,\ y < n)$ for various n. What is the meaning of $F_{XY}(\infty, \infty)$? What is the value of $F_{XY}(-\infty, \infty)$?

4.10 Prove Lemmas 4.4–4.8.

4.11 Consider two random variables X, Y with jcdf given by

$$F_{XY}(x, y) = \begin{cases} 1, & x \ge 1,\ y \ge 1, \\ xy, & 0 \le x \le 1,\ 0 \le y \le 1, \\ x, & 0 \le x \le 1,\ y \ge 1, \\ y, & 0 \le y \le 1,\ x \ge 1, \\ 0, & \text{otherwise.} \end{cases}$$

Plot this function. Then evaluate, if possible, the following probabilities:
(a) $P[X \le 0.5,\ Y > 0.3]$,
(b) $P[0.1 < X < 0.9]$,
(c) $P[0.5 < X + Y \le 1]$,
(d) $P[X^2 + Y^2 < 0.1]$,
(e) $P[(X - 0.5)^2 + (Y - 0.5)^2 < 0.1]$,
(f) $P[\min\{X, \frac{Y^3}{10}\} > 005]$.

4.12 Determine the cdf's $F_X(x)$ and $F_Y(y)$ and the conditional cdf $F_X(x|X < 0.5)$ for the random variables in Problem 4.11.

4.13 *Valid CDF?* Consider the function

$$F(x, y) = \begin{cases} 1 - ae^{-\lambda(x+y)}, & x > 0,\ y > 0, \\ 0, & \text{otherwise.} \end{cases}$$

where $a \in R$, $\lambda > 0$ are given constants.
(a) Plot this function for various a, λ.
(b) Under what conditions on a and λ, if any, is F a valid jcdf?

4.14 Consider the function

$$F(x, y) = \begin{cases} ae^{-\lambda(x+y)}, & x > 0, \ y > 0, \\ 0, & \text{otherwise.} \end{cases}$$

where $a \in R$, $\lambda > 0$ are given constants.
(a) Plot this function for various a, λ.
(b) Under what conditions on a and λ, if any, is F a valid jcdf?

4.15 Consider the function

$$F(x, y) = \begin{cases} \left(1 - e^{-\lambda x}\right) u(y - x), & x > 0, \quad y > 0, \\ 0, & \text{otherwise,} \end{cases}$$

where $\lambda > 0$ is a given constant and $u(s)$ is the usual step function.
(a) Plot this function.
(b) Verify whether it is a valid jcdf for two random variables X, Y.

The Joint PDF; Clarify Concepts

4.16 Prove Lemma 4.10.

4.17 Prove Theorems 4.1 and 4.2.

4.18 Consider two random variables X, Y with joint pdf given by

$$f_{XY}(x, y) = \begin{cases} e^{-y}, & 0 < x < y < \infty, \\ 0, & \text{otherwise.} \end{cases}$$

(a) Plot this function.
(b) Find $P[X > 3 | Y \leq 2]$.
(c) Find $P[X > 1 | Y \leq 2]$.
(d) Find $P[X > 1, Y > 10]$.
(e) Can you find an event that has probability 0.5?

4.19 Consider two random variables X, Y with a joint pdf $f_{XY}(x, y)$ that takes nonzero values *only* in the square $[-1, 1] \times [-1, 1]$. Indicate the limits of integration for calculating the probability of the following events:
(a) $P[X < Y]$, (b) $P[X^2 < Y^2]$, (c) $P[|X| < Y]$.

4.20 Define the subset A as the region of the unit square included between the curves $y = x^2$ and $y = \sqrt{x}$. Consider the random variables X, Y with joint pdf

$$f_{XY}(x, y) = \begin{cases} 3, & (x, y) \in A, \\ 0, & \text{otherwise.} \end{cases}$$

(a) Plot the joint pdf.
(b) Find and plot the marginal densities of X and Y.

4.21 Consider two random variables X, Y with a joint pdf $f_{XY}(x, y)$ that is of the mixed type. The continuous part of $f_{XY}(x, y)$ takes nonzero values only in the circle with unit radius and center at the origin. The discrete part of $f_{XY}(x, y)$ contains three spikes at the points $(x, y) = (1, 0), (2, 0)$, and $(\frac{1}{\sqrt{2}}, \frac{1}{\sqrt{2}})$. All spikes have equal weight. Let

$$\begin{aligned} A &= P[1 \le X^2 + Y^2 \le 2], \quad B = P[1 < X^2 + Y^2 \le 2], \\ C &= P[1 \le X^2 + Y^2 < 2], \quad D = P[1 < X^2 + Y^2 < 2]. \end{aligned}$$

Arrange A, B, C, D in increasing order.

4.22 Consider the random variables X, Y with joint pdf

$$f_{XY}(x, y) = \begin{cases} \frac{1+xy}{3}, & 0 \le x \le 2, \ 0 \le y \le 1, \\ 0, & \text{otherwise.} \end{cases}$$

(a) Plot the joint pdf.
(b) Find and plot the marginal densities of X and Y.
(c) Find and plot the conditional densities $f_X(x|y)$ and $f_Y(y|x)$.

The Joint PDF; Apply Concepts

4.23 Let $S = [0, 1] \times [0, 1]$. Let $P[A]$ be the area of A, an arbitrary subset of S.
(a) Construct a random variable (X, Y) with jpdf

$$f_{XY}(x, y) = \begin{cases} 1, & 0 \le x, \ y \le 1, \\ 0, & \text{otherwise.} \end{cases}$$

(In other words, derive the mappings $X : S \to R$, $Y : S \to R$.)
(b) Construct two random variables (X, Y) and (X', Y') with a *common* jpmf $f(\cdot, \cdot)$ defined as follows:

$$f(z, w) = \begin{cases} 0.250, & (z, w) = (0, 0), \\ 0.250, & (z, w) = (0, 1), \\ 0.375, & (z, w) = (1, 1), \\ 0.125, & (z, w) = (1, 0). \end{cases}$$

The vector random variables should not be identical; in other words, $(X(\zeta), Y(\zeta)) \neq (X'(\zeta), Y'(\zeta))$ for at least one $\zeta \in S$.

4.24 Can you construct a discontinuous jpdf with continuous marginal pdf's?

4.25 *Valid Joint PDF?* Consider the function

$$f(x, y) = \begin{cases} ae^{\lambda(x^2+y^2)}, & x > 0, \ y > 0, \\ 0, & \text{otherwise.} \end{cases}$$

where $a, \lambda \in R$ are given constants.
(a) Plot this function for various a, λ.
(b) Under what conditions on a and λ, if any, is f a valid jpdf?
(c) Find the conditional pdf $f_X(x|y)$.

4.26 For given $a, \lambda > 0$, let

$$f(x, y) = \begin{cases} ae^{-\lambda y}, & y > x \geq 0, \\ 0, & \text{otherwise.} \end{cases}$$

(a) Plot this function for various a, λ.
(b) Under what conditions is $f(x, y)$ a valid two-dimensional pdf?

4.27 Let

$$f(x, y) \triangleq \begin{cases} a|x|, & -10 \leq x \leq 0, \ -10 \leq y < x, \\ 0, & \text{otherwise.} \end{cases}$$

(a) Plot this function for various a.
(b) Under what conditions is $f(x, y)$ a valid two-dimensional pdf?

4.28 Consider the function

$$f(x, y) = \begin{cases} 1 - |x + y|, & -c \leq x + y \leq c, \ -1 \leq x - y \leq 1, \\ 0, & \text{otherwise.} \end{cases}$$

(a) Plot this function for various c.
(b) Find the constant c for which $f(x, y)$ is the joint pdf of two random variables.
(c) Find the marginal pdf's $f_X(x), f_Y(y)$.

4.29 Consider the function

$$f(x, y) = \begin{cases} c(1 - x^2 - y^2), & 0 \leq x^2 + y^2 \leq 1, \ x \geq 0, \ y \geq 0, \\ 0, & \text{otherwise.} \end{cases}$$

(a) Plot this function for various c.
(b) Find the constant c for which $f(x, y)$ is the joint pdf of two random variables.

4.30 Consider the function

$$f(x,y) = \begin{cases} c, & 0 \leq x+y \leq 1, \ x \geq 0, \ y \geq 0, \\ 0, & \text{otherwise.} \end{cases}$$

(a) Plot this function for various c.
(b) Find the constant c for which $f(x,y)$ is the joint pdf of two random variables.
(c) Find the marginal pdf's $f_X(x), f_Y(y)$.

Independence; Clarify Concepts

4.31 Consider two random variables X_1, X_2 defined on two *different* probability spaces S_1, S_2. Are they independent?

4.32 Consider two random variables X_1, X_2 defined on the same space S with ranges S_{X_1}, S_{X_2} that have an empty intersection. Are they dependent or independent? Or can you not tell?

4.33 Consider the unit circle as a sample space S. Can you define two random variables that are independent?

4.34 Consider a sample space S. Construct two dependent random variables on S with different marginal cdf's.

4.35 Consider a sample space S. Construct two dependent random variables on S with the same marginal cdf's.

Independence; Simple Skills

4.36 Evaluate the probabilities of some of the events in Problem 4.5 for two independent random variables X, Y that are exponentially distributed with unit parameters. [Note that now $S_{XY} = [0, \infty) \times [0, \infty)$.]

4.37 Are the two random variables in Problem 4.25 independent?

4.38 Are the two random variables in Problem 4.28 independent?

4.39 Can you determine whether the two random variables in Problem 4.27 are independent, without resorting to any calculations?

4.40 Repeat Problem 4.39 for the random variables in Problem 4.30.

4.41 Consider two random variables X, Y such that $P[X = 1] = 1$, $P[Y = 1] = 1$. Are X, Y independent? Explain.

4.42 Consider two random variables X, Y with $P[X = c] = 1$ and Y arbitrary. Are X, Y independent? Explain.

4.43 Consider the region $A = \{(x, y) \in R^2 : 5 \le x^2 + y^2 \le 10\}$.
(a) Can you find a constant $k > 0$ such that the function

$$f(x, y) = \begin{cases} k, & (x, y) \in A, \\ 0, & \text{otherwise.} \end{cases}$$

is a valid jpdf?
(b) Can you find a constant $k > 0$ such that the function

$$f(x, y) = \begin{cases} ke^{-(x^2+y^2)}, & (x, y) \in A, \\ 0, & \text{otherwise.} \end{cases}$$

is a valid jpdf?
(c) Plot both functions.
(d) If you answered yes to (a) or (b), are the random variables independent?

Mathematical Skills

4.44 Consider two independent random variables X, Y such that for some constant c, $P[X + Y = c] = 1$. Show that both X and Y are constant.

4.45 Consider two random variables X and Y with jcdf given by

$$F_{XY}(x, y) = \begin{cases} 1 - e^{-\lambda x} - e^{-\lambda y} + e^{-\lambda(x+y)}, & x > 0, \ y > 0, \\ 0, & \text{otherwise,} \end{cases}$$

where $\lambda > 0$ is a given constant.
(a) Plot this function.
(b) Are the two random variables independent? Are they identically distributed?
(c) Calculate $P[X > 1, \ Y < 1]$.
(d) Calculate $P[X > 10 | Y > 10]$.
(e) Calculate $P[X = 3]$.

4.46 Consider the random variables in Problem 4.18.
(a) Find and plot the joint cdf.
(b) Are the two random variables independent?
(c) Calculate $P[X > 1, \ 1 < Y < 3]$.
(d) Calculate $f_X(x), f_Y(y)$.
(e) Calculate $F_X(x|Y > 2)$.
(f) Calculate $f_X(x|Y < 3)$.

4.47 Consider the random variables defined in Problem 4.28. Calculate

$$P[X < 0], \quad P[X > -0.5, \ -0.1 < Y < 0.3], \quad F_Y(y|X > 0), \quad f_Y(y|Y \le 0.2).$$

4.48 Consider two independent geometric random variables X_i, $i = 1, 2$. Find $P[X_1 = X_2]$.

4.49 Consider two independent random variables X, Y such that for all $x \in R$, $F_X(x) \le F_Y(x)$. Show that $P[X \ge Y] \ge 0.5$.

4.50 Let X and Y be two independent random variables. Show that

$$P[Y \le X] = \int_{-\infty}^{\infty} F_Y(x) f_X(x) dx, \quad P[Y \le X] = 1 - \int_{-\infty}^{\infty} F_X(y) f_Y(y) dy.$$

4.51 Consider two random variables X, Y. Prove or disprove the following statements.
(a) X, Y are independent if and only if X^2, Y^2 are independent.
(b) X, Y are independent if and only if X^3, Y^3 are independent.

Advanced Skills

4.52 Consider two random variables X and Y with sample space

$$S = \{(0,0), (0,1), (1,0), (1,1)\}.$$

The probabilities of these four points are p_1, p_2, p_3, and p_4, respectively.
(a) Are the two random variables Bernoulli?
(b) Under what conditions, if any, are they independent?
(c) Calculate $P[X > 0]$, $P[X > Y]$.
(d) Use an English statement to express an event whose probability is equal to $p_1 + p_3$.
(e) Plot the jcdf of X and Y.

4.53 Consider a sample space S with four equiprobable elements a, b, c, d.
(a) Construct two independent Bernoulli random variables on S.
(b) How many different Bernoulli random variables can you construct?
(c) Construct two dependent Bernoulli random variables on S. Plot their joint pmf and cdf.

4.54 Consider a sample space S with a finite number n of equiprobable elements.
(a) Can you construct two independent Bernoulli random variables on S?
(b) How many different Bernoulli random variables can you construct?
(c) Construct two dependent Bernoulli random variables on S. Plot their joint pmf and cdf.

4.55 Repeat Problem 4.54 when the outcomes are not equiprobable.

Models

4.56 Let X and Y be two independent Gaussian random variables with means m_X and m_Y and unit variance. Find

$$P[XY > 0 \text{ or } XY < 0].$$

4.57 Plot the conditional pdf in Equation 4.61 as a function of m_1.

4.58 Plot the conditional pdf in Equation 4.61 as a function of m_2.

4.59 Plot the conditional pdf in Equation 4.61 as a function of σ_1.

4.60 Plot the conditional pdf in Equation 4.61 as a function of ρ_{XY}.

4.61 Discuss how to modify the χ^2 test presented in Chapter 3 to fit a two-dimensional histogram.

Transformations; Generic Problems

4.62 Consider the function

$$g(x) = \begin{cases} g_1(x), & \text{with probability } 0.3, \\ g_2(x), & \text{with probability } 0.7. \end{cases}$$

Let $Y = g(X)$; is this a valid transformation? If yes, can you find $f_Y(y)$?

4.63 Suppose that $k \in R$ is a constant. Consider the transformation

$$Z_1 = aX + bY, \quad Z_2 = kaX + kbY,$$

for which the matrix A is singular.
(a) Evaluate the joint pdf of Z_1, Z_2.
(b) Plot this pdf when X, Y are independent, uniform random variables with parameters $0, 1$.

4.64 State and prove Theorems 4.6 and 4.7 for the case of two discrete random variables. Repeat for X continuous and Y discrete.

4.65 Consider a monotonic real-valued function $g(x)$. Let $Y = g(X)$ be a transformation of the random variable X. Find the jcdf $F_{XY}(x, y)$ in terms of $F_X(x)$ and $g(x)$.

4.66 Consider two independent, identically distributed random variables X, Y. Let

$$Z_1 = \frac{X}{Y}, \quad Z_2 = \frac{Y}{X}.$$

Based on the identical distribution of X, Y and on the symmetry in the definition of Z_1, Z_2, we can argue that $f_{Z_1}(z) = f_{Z_2}(z)$. On the other hand, since $Z_1 = 1/Z_2$, we can argue that $f_{Z_1}(z) \neq f_{Z_2}(z)$. Which argument is wrong, and why?

4.67 Consider two independent, identically distributed random variables X_1, X_2. Let

$$Z = \frac{X_1}{X_1 + X_2}, \quad Z' = \frac{X_2}{X_1 + X_2}.$$

Are the Z, Z' random variables identically distributed? Are they independent?

4.68 Let X and Y be independent random variables with respective probability generating functions $G_X(z)$ and $G_Y(z)$. Let $D = X - Y$. Show that

$$G_X(z)G_Y(\frac{1}{z}) = \sum_k P[D = k]z^k.$$

4.69 Let S be the unit circle, with a uniform probability assignment on it.

(a) Let X, Y be the Cartesian coordinates of a point selected at random. Show that X, Y are not independent random variables.
(b) Let V, W be the polar coordinates of the same point. Show that V, W are independent.
(c) Do you find these results surprising?

4.70 Consider two random variables X, Y. The conditional cdf, denoted by $F_X(x|y)$, is a function of x, with y a parameter. Consider the transformation $Z = g(X)$, where $g(x) = F_X(x|y)$. In Problem 3.177, we saw that the random variable $Z' = F_X(X)$ is always uniformly distributed in $[0, 1]$. What can you say about the random variable $Z = F_X(X|y)$?

4.71 Consider two random variables X_1, X_2. Let

$$M = \frac{X_1 + X_2}{2}, \quad V = \frac{(X_1 - M)^2 + (X_2 - M)^2}{2}.$$

(a) Find the jpdf of (M, V) if $f_{X_1 X_2}(x_1, x_2)$ is known.
(b) Evaluate your answer in part (a) when X_1, X_2 are independent, exponential random variables with parameter 1.
(c) Evaluate your answer in part (a) when X_1, X_2 are independent, Poisson random variables with parameters 1 and 2.

4.72 Consider two positive random variables X_1, X_2. Let

$$Z = \frac{X_1}{X_1 + X_2}.$$

(a) Find the pdf of Z in terms of the joint pdf of X_1, X_2.

(b) Evaluate your answer in part (a) when the X_i are independent Poisson random variables with parameters 1 and 2.

(c) Evaluate your answer in part (a) when the X_i are independent χ^2 random variables.

(d) Evaluate your answer in part (a) when the X_i are the dependent random variables in Example 4.21.

4.73 Let X and Y be independent Poisson random variables, with parameters λ_X and λ_Y. Verify that the conditional pmf of X, given the event $X + Y = n$, is binomial with parameters n and $p = \lambda_X/(\lambda_X + \lambda_Y)$.

4.74 Suppose that X is Poisson with parameter λ, which is a random variable.

(a) Find the pmf of X when λ is exponential with parameter a.

(b) Repeat part (a) when λ is a gamma random variable with parameters b, c.

4.75 Consider the joint, dependent Poisson model in Equation 4.59.

(a) Verify that the marginal pmf's are Poisson. Show that the parameter of X is equal to $\lambda_{XY} + \lambda_X$.

(b) Show that $\rho_{XY} \geq 0$.

(c) Show that X, Y are independent if and only if $\lambda_{XY} = 0$.

(d) Plot the joint pmf as a function of λ_{XY}.

4.76 Let X and Y be independent, geometric random variables. Let

$$Z = X - Y, \quad V = \min\{X, Y\}.$$

Find the joint pmf of Z, V. Are Z, V independent?

4.77 Show that the sum of two independent Pascal random variables with the same parameter p is also a Pascal random variable.

4.78 Let X, Y be two independent, uniform random variables with parameters a, b and c, d, respectively. Graph the pdf of $Z = X + Y$.

4.79 Let X, Y be two independent, uniform random variables with parameters $-a, a$. Find the pdf of $X - Y$.

4.80 Let X and Y be independent and uniformly distributed in $[0, 1]$. Find the pdf of the random variable $Z = |X - Y|$.

4.81 Consider two independent random variables X, Y with Y uniform in $[0, 1]$. Let $Z = X + Y$. Show that $f_Z(z) = F_X(z) - F_X(z - 1)$.

4.82 Consider two independent random variables X, Y with Y uniform in $[0, 1]$. Let $Z = (X + Y)\bmod 1$. Find the pdf of Z.

4.83 Consider two independent random variables X, Y. Let $t \in R$ be a constant. Define $Z = X \sin(2\pi t + Y)$. Show that when Y is uniformly distributed in $[0, 2\pi]$, we have

$$f_Z(z) = \frac{1}{\pi} \int_{-\infty}^{-|z|} \frac{f_X(x)}{\sqrt{x^2 - z^2}} dx + \frac{1}{\pi} \int_{|z|}^{\infty} \frac{f_X(x)}{\sqrt{x^2 - z^2}} dx.$$

4.84 Let A, B be two independent random variables that are uniformly distributed in $[-1, 1]$. Consider the quadratic equation $x^2 + 2Ax + B = 0$. Find the probability that the roots are (a) complex, (b) real, and (c) equal.

4.85 Repeat Problem 4.84 for the equation $Ax^2 + Bx + 1 = 0$.

4.86 Consider two dependent random variables X, Y with the following properties: X is positive and has continuous pdf, and Y is uniform on the set $(0, X)$. Show that if Y and $X - Y$ are independent, X follows a gamma distribution with parameter $a = 2$.

4.87 Let X and Y be independent and uniformly distributed in $[0, 1]$. Consider the transformation $Z = \log(X) \cos(2\pi Y)$.
(a) Determine the range of Z.
(b) Find the pdf of Z.

4.88 Let X and Y be independent and uniformly distributed in $[0, 1]$. Find the jpdf of the random variables

$$Z = X^2 \cos(2\pi Y), \quad V = X^2 \sin(2\pi Y).$$

Are Z, V independent?

4.89 Let Θ be a random variable, uniformly distributed in $(0, 2\pi]$. Let

$$X = \sin(\Theta), \quad Y = \cos(\Theta).$$

Evaluate $F_{XY}(x, y), f_X(x|y), f_Y(y|x)$.

4.90 Let X and Y be two independent, uniform random variables in $[0, 1]$. Show that the transformation

$$Z = \sqrt{-2 \log(X)} \cos(2\pi Y), \quad V = \sqrt{-2 \log(X)} \sin(2\pi Y)$$

results in two independent, Gaussian random variables with zero mean and unit variance.

Transformations of Gaussian Random Variables

4.91 Let X, Y be two independent random variables, with X zero-mean Gaussian. Suppose that $P[Y = -1] = P[Y = 1] = 0.5$. Let $Z = XY$.
(a) Find the pdf of Z.
(b) Are X, Z independent? Are Y, Z independent?

4.92 Let X, Y be independent Gaussian random variables with zero mean and unit variance. Show that the random variable $Z = X/Y$ follows a Cauchy distribution.

4.93 Consider two independent random variables X, Y. Variable X is Gaussian with zero mean and unit variance, while Y is uniform in $[0, \pi]$. Let $Z = X + \sin(Y)$. Show that

$$f_Z(z) = \frac{1}{\pi^{\frac{3}{2}} \sqrt{2}} \int_0^\pi e^{-\frac{1}{2}[z - \sin(y)]^2} \, dy.$$

4.94 Let X, Y be two dependent, Gaussian random variables. Let $Z = A(X, Y)$, where $Z = (V, W)$ and A is a 2×2 matrix. Determine A such that V, W are independent.

4.95 Let X and Y be two independent, zero-mean Gaussian random variables with variances σ_X^2 and σ_Y^2. Find the jpdf of the random variables

$$Z = 0.5(X^2 + Y^2), \quad V = 0.5(X^2 - Y^2).$$

4.96 Consider two zero-mean, independent Gaussian random variables X, Y. Let $Z_1 = aX + bY$, $Z_2 = X/Y$. Find the jpdf $f_{Z_1 Z_2}(z_1, z_2)$.

4.97 Consider two zero-mean, unit-variance, Gaussian random variables X_1, X_2 with correlation coefficient $\rho_{X_1 X_2} = \rho$. Is this enough information to:
(a) Find the jpdf of the random variables $Z = X_1^2$, $U = X_2^4$?
(b) Find the jpdf of the random variables $Z' = X_1^2$, $U' = X_1^4$?
(c) Are Z, Z' identically distributed?
(d) Are U, U' identically distributed? Independent?

Transformations of Exponential Random Variables

4.98 Consider two random variables X, Y with joint pdf

$$f_{XY}(x, y) = 2e^{-(x+y)}, \quad 0 \le y \le x < \infty.$$

Find the pdf of the random variable $Z = X - Y$.

4.99 Consider the random variables X, Y with joint pdf

$$f_{XY}(x, y) = \begin{cases} 2e^{-(x+2y)}, & x, y > 0, \\ 0, & \text{otherwise.} \end{cases}$$

Let $Z = X + Y$, $U = X/Y$. Find the joind pdf of Z, U.

4.100 Consider two independent, exponential random variables X_1 and X_2, with parameters λ_1 and λ_2, respectively. Let

$$Z = \max\{X_1, X_2\}, \quad Y = \min\{X_1, X_2\},$$

$$D = Y - Z, \quad X = \begin{cases} 1, & X_1 < X_2, \\ 2, & X_1 \geq X_2. \end{cases}$$

(a) Find $F_Y(y)$ and $f_Y(y)$.
(b) Find $F_Z(z)$ and $f_Z(z)$.
(c) Find $F_X(x)$ and $f_X(x)$.
(d) Are X and Y independent?
(e) Are Y and D independent?
(f) Evaluate $F_D(d|X = 1)$, $d \in R$.
(g) Find $P[Z < z, Y < y|Z < T]$, where $T > 0$ is a given constant.

Transformations; the Rest of the Models

4.101 Let X and Y be two independent Cauchy random variables. Find the jpdf of the random variables $Z = X^2 + Y^2$, $V = XY$.

4.102 Consider two independent Rayleigh random variables X, Y with the same parameter. Find the pdf of $Z = X/Y$.

4.103 Consider two independent χ^2 random variables X, Y with parameters n, k, respectively. Let

$$Z = \frac{X/n}{Y/k}.$$

Show that Z follows an F distribution.

4.104 Let X be an F-distributed random variable. Show that $Y = \sqrt{X}$ is also an F-distributed random variable.

4.105 Consider two random variables X, Y that are F-distributed, with parameters v, w and w, v respectively. Show that

$$F_X^{-1}(1 - a) = \frac{1}{F_Y^{-1}(a)},$$

where by definition, $F_X^{-1}(1-a)$ represents the $1-a$ percentile of X.

Expected Values; Simple Skills

4.106 Consider two random variables X, Y such that EX, EY exist.
(a) Show that $E(X \pm Y)$ exists as well.
(b) Show by counterexample that EXY may not exist.
(c) Show by counterexample that $E\frac{X}{Y}$ may not exist.

4.107 Suppose that $E(X+Y)$ exists. Show by counterexample that EX, EY may not exist.

4.108 Suppose that $E(X+Y), E(X-Y)$ exist. Do EX, EY exist?

4.109 Show that if EX^2, EY^2 exist, $E(X-Y)^2 \le 2EX^2 + 2EY^2$.

4.110 Consider two independent random variables X, Y such that for some $p > 0$, $E(X+Y)^p < \infty$. Show that $EX^p, EY^p < \infty$.

4.111 If X_1, X_2 are positive, identically distributed random variables with mean 1, show that $E \max\{X_1, X_2\} \le 2$.

4.112 Construct two random variables X, Y such that

$$EX > 0, \quad EY > 0, \quad EXY > EXEY.$$

Useful Inequalities

4.113 *Holder's Inequality.* Consider two random variables X, Y such that

$$E|X|^p < \infty, \quad E|Y|^q < \infty,$$

where $p > 1$, $q > 1$, and $p^{-1} + q^{-1} = 1$. Show that

$$E|XY| \le (E|X|^p)^{\frac{1}{p}} \cdot (E|Y|^q)^{\frac{1}{q}}.$$

The special case $p = q = 2$ is known as *Schwarz' inequality.*

4.114 Use Schwarz' inequality to show that

$$[\text{cov}(X, Y)]^2 \le \text{var}(X) \, \text{var}(Y).$$

4.115 *Minkowski's Inequality.* Consider two random variables X, Y such that, for some $p \ge 1$, $E|X|^p < \infty$, $E|Y|^p < \infty$. Use Holder's inequality to show that

$$E|X + Y|^p \le E|X|^p + E|Y|^p.$$

When does equality hold?

4.116 Consider an arbitrary random variable X with $E|X|^p < \infty$ for all $p \geq 0$. Define $f(p) \triangleq \log(EX^p)$. Use Holder's inequality to show that $f(p)$ is a convex function.

4.117 Consider two nonnegative random variables X, Y. Show that, for all real numbers $p > 0$,

$$E(X + Y)^p \leq 2^p(EX^p + EY^p).$$

4.118 Consider an arbitrary, nonnegative random variable X. Use Holder's inequality to show that

$$EX^b \leq (EX^{(}a)^{\frac{c-b}{c-a}} \cdot (EX^c)^{\frac{b-a}{c-a}},$$

for all real numbers $0 < a < b < c$.

Mathematical Skills

4.119 Show that $|\rho_{XY}| \leq 1$ for any two random variables X, Y.

4.120 Suppose we take the following measurements of two random variables X, Y: (1,0), (2,1), (2,0), (2,1), (1,1), (2,1), (2,0), (0,1). Suppose further that the resulting histogram is a very good approximation of the actual joint pdf. Calculate EXY.

4.121 Consider two random variables X and Y with jpdf given by

$$f_{XY}(x, y) = \begin{cases} \frac{1}{10}, & 0 < x < 5, \ 0 < y < 2, \\ 0, & \text{otherwise.} \end{cases}$$

Calculate EX, EY, EXY, EX^kY^l.

4.122 Consider two independent random variables X and Y, with means EX, EY and variances σ_X^2, σ_Y^2, respectively. Calculate

$$E(X + YX^2), \quad E(X^2 + Y^2).$$

4.123 Consider two discrete random variables X and Y with joint pdf

$$\begin{aligned} f_{XY}(x, y) = \ & \delta(x - (a))[0.1\delta(y - a) + 0.2\delta(y) + 0.1\delta(y + 1)] \\ & + \delta(x)[0.05\delta(y - a) + 0.1\delta(y) + 0.05\delta(y + 1)] \\ & + \delta(x + a)[0.15\delta(y - a) + 0.2\delta(y) + 0.05\delta(y + 1)]. \end{aligned}$$

For what value of a is the correlation between X and Y maximum? Minimum?

4.124 Consider two random variables X and Y with finite variances. Show that $\rho_{XY} = 1$ if and only if

$$\frac{X - EX}{\sigma_X} = \frac{Y - EY}{\sigma_Y}.$$

4.125 Consider the random variables in Problem 4.20.
(a) Evaluate $EX, \text{var}(X), EY, \text{var}(Y)$.
(b) Evaluate ρ_{XY}.

4.126 Consider two random variables X, Y with joint pdf given by

$$f_{XY}(x, y) = \begin{cases} 8xy, & 0 \le x \le y, \ 0 \le y \le 1, \\ 0, & \text{otherwise.} \end{cases}$$

(a) Find the pdf of X.
(b) Find the conditional pdf of X given Y.
(c) Find $E(X|Y = y)$; find $E(X|Y)$.

4.127 Consider the random variables in Problem 4.98. Calculate EX, EY, EXY, and $EX^k Y^l$.

4.128 Find, if possible, EX^k and EY^k for the random variables X, Y in Problem 4.26.

4.129 Consider the random variables in Problem 4.30. Calculate EX, EY, EXY, ρ_{XY}, $\text{var}(X)$, $\text{var}(Y)$.

4.130 Are the random variables X, Z in Problem 4.91 uncorrelated? How about Y, Z?

4.131 Repeat Problem 4.130 for the random variables X and Y in Problem 4.29.

4.132 Let $Y = X + N$, where X is a Bernoulli random variable with $p = 0.2$ and N is Gaussian with zero mean and unit variance. Variables X and N are independent. Find ρ_{XY}.

4.133 Consider two random variables X, Y with finite variances. Show that

$$\rho_{aX+b, cY+d} = \rho_{XY} \cdot \text{signum}(a, c),$$

where the signum function is defined as

$$\text{signum}(a, c) = \begin{cases} +1, & \text{if } ac > 0, \\ -1, & \text{if } ac < 0, \\ 0, & \text{if } ac = 0. \end{cases}$$

4.134 Consider a sample space S with n elements. Construct two random variables on S, with $EX = EY$ and $\rho_{XY} = +1$.

4.135 Consider a sample space S with $n \geq 2$ elements. Can you construct two random variables X and Y with a given ρ_{XY}?

4.136 Consider a random variable X with $P[X = 1] = P[X = -1] = 0.5$. Let $Y = \sin(X)$, $Z = \cos(X)$. Calculate ρ_{ZY}. Are Y, Z uncorrelated? Orthogonal?

4.137 (a) If X, Y are independent random variables, are X, XY independent?
(b) If X, Y are independent random variables, are X, XY orthogonal?
(c) If X, Y are uncorrelated random variables, are X^2, Y^2 uncorrelated?
(d) If X, Y are uncorrelated random variables, are X^3, Y^3 orthogonal?

4.138 Consider the linear transformation of two random variables X, Y

$$Z_1 = aX + bY, \quad Z_2 = cX + dY.$$

(a) Under what conditions, if any, are the random variables Z_1 and Z_2 independent? Uncorrelated?
(b) Can you find a linear transformation that will result in the two random variables Z_1 and Z_2 having prescribed means? Variances?

4.139 Consider two random variables X and Y with $\rho_{XY} = 0.1$, $EX = EY = 1$, and $\sigma_X^2 = 2, \sigma_Y^2 = 1$. Let $Z = (aX + bY)^2$.
(a) For what values of a, b is $EZ = 1$?
(b) For what values of a, b is EZ minimum?
(c) Find the range of values that EZ can possibly take when $a, b \in [-5, 1]$.

Conditional Expectations

4.140 Show that if $X - Y^{10}$ is independent of Y, then $E(X|Y) = Y^{10} + c$, where c is a constant. Evaluate c.

4.141 Consider two zero-mean random variables X, Y; let $Z = X + aY$. Suppose that Z is independent of Y. Show that $E(X|Y) = -aY$.

4.142 Consider the expression $E\{E[XY|Y, Z]|Z\}$.
(a) Is it a function of X? Y? Z? X, Y? Y, Z? None of the above?
(b) Can you simplify this expression?

4.143 Show that if $E(X|Y, Z) = E(X|Y)$, then $E(XZ|Y) = E(X|Y)E(Z|Y)$.

4.144 We know that $E(X^2|X) = X^2$. Is $E(X|X^2) = X$?

4.145 Is $E(Y|X^2) = E[E(Y|X)|X^2]$? Explain.

Useful Models; Mathematical Skills

4.146 Let X, Y be jointly Gaussian random variables, with jpdf

$$f_{XY}(x, y) = \frac{1}{4\pi 0.707} e^{-(x^2/8 - x/2 - 4.414xy/8 + 2.828y/8 + y^2/2 + 0.5)}.$$

Evaluate $EX, EY, \text{var}(X), \text{var}(Y), \rho_{XY}$.

4.147 Consider two independent, zero-mean Gaussian random variables X, Y with variance σ^2. Show that

$$E|X - Y| = \frac{2\sigma}{\sqrt{\pi}}, \quad E|X - Y|^2 = 2\sigma^2.$$

4.148 Let X_1, X_2 be zero-mean, jointly Gaussian random variables with covariance matrix

$$C = \begin{bmatrix} 1.0 & 0.1 \\ 0.1 & 2.0 \end{bmatrix}.$$

Find a transformation $Z = g(X_1, X_2)$ such that $EZ = 1, \text{var}(Z) = 1$. Can you argue, without resorting to calculations, that $g(X_1, X_2)$ *must* be nonlinear?

4.149 Let X_1, X_2 be zero-mean, jointly Gaussian random variables with covariance matrix

$$C = \begin{bmatrix} 3.0 & 0.2 \\ 0.2 & 2.0 \end{bmatrix}.$$

Find a 2×2 transformation matrix A such that the (linearly) transformed random variables are independent.

4.150 Calculate EZ^2 for a Rice random variable Z, using the property that $Z = \sqrt{X^2 + Y^2}$, with X, Y independent Gaussian random variables with means $0, m$ and equal variance σ^2.

4.151 Consider two independent, exponential random variables X, Y. Calculate $E \max\{X, Y\}$, $E \min\{X, Y\}$, and $E \max\{0, X - Y\}$.

4.152 Suppose that X is a binomial random variable with parameters N, p, where N itself is binomial with parameters M and p'. Show that X is binomial with parameters M, pp'.

4.153 Consider two independent, integer-valued, nonnegative random variables X, Y with $P[X + Y = 0] > 0$. Let n, m be nonnegative integers and M, N positive ones. Show that if

$$P[X = n | X + Y = n + m] = \frac{\binom{M}{m}\binom{N}{n}}{\binom{M+N}{m+n}}, \quad n \leq N, \ m \leq M,$$

then X, Y are binomial, with parameters N, p and M, p, respectively.

4.154 Consider two independent, integer-valued random variables X, Y with the property that, for some $p \in (0, 1)$,

$$P[X = n | X + Y = n + m] = \begin{cases} \binom{n+m}{n} p^n (1-p)^m, & n, m \geq 0, \\ 0, & \text{otherwise.} \end{cases}$$

Suppose further that $P[X = n], P[Y = m] > 0$, for all $n, m \geq 0$. Show that X, Y are Poisson-distributed.

4.155 Let X and Y be independent geometric random variables with parameter p. Verify that the conditional pmf of X, given the event $X + Y = n$, is discrete uniform with parameters $0, n$.

Transforms

4.156 Consider two random variables X and Y; suppose that

$$\Phi_{XY}(\omega_1, \omega_2) \triangleq E e^{j\omega_1 X} e^{j\omega_2 Y} = \frac{1}{(1 - j\omega_1)(1 - 2j\omega_2)}.$$

Here $\Phi_{XY}(\omega_1, \omega_2)$ is called the *joint characteristic function* of X, Y. Calculate EX, EY, EXY, $EX^k Y^l$, and ρ_{XY}.

4.157 Determine the pdf of the sum of two independent Poisson random variables, using characteristic functions.

4.158 Repeat Problem 4.157 for two independent exponential random variables that have the same parameter.

4.159 Repeat Problem 4.157 for two independent Gaussian random variables.

Multidimensional Random Variables; Mathematical Skills

4.160 Consider four independent random variables X_1, X_2, X_3, X_4. Calculate the following probabilities in terms of $f_{X_i}(x)$.
(a) $P[\max\{X_1, X_2, X_3, X_4\} < 1]$.
(b) $P[\min\{X_1, X_2, X_3, X_4\} > 2]$.
(c) $P[\max\{X_1, X_2, X_3, X_4\} > 2\min\{X_1, X_2, X_3, X_4\}]$.
(d) $P[\max\{X_1, X_2\} > \max\{X_3, X_4\}]$.
(e) $P[\max\{\min\{X_1, X_2\}, \max\{X_3, X_4\}\} > 2]$.

4.161 Consider n random variables $X_1, \ldots, X_n$ with means EX_i and covariances $\text{cov}(X_i, X_j) = ij$. Let $Y = a_1 X_1 + \cdots + a_n X_n$, where a_i are given constants. Evaluate EY and $\text{var}(Y)$.

4.162 The covariance matrix of three random variables X_1, X_2, X_3 with means 1, 2, and 3, respectively, is given by

$$C = \begin{bmatrix} 1.0 & 0.1 & 0.2 \\ 0.1 & 2.0 & 0.9 \\ 0.2 & 0.9 & 3.0 \end{bmatrix}.$$

Consider the linear transformation

$$\begin{aligned} Y_1 &= X_1 + X_2 - X_3, \\ Y_2 &= 2X_1 - X_2 + 3X_3, \\ Y_3 &= 3X_1 + X_2 + X_3. \end{aligned}$$

Find the means EY_i and the covariance matrix of Y_1, Y_2, Y_3.

4.163 Let

$$f_{XYZ}(x, y, z) = \begin{cases} 12x^2yz, & 0 < x < 1, \quad 0 < y < 1, \quad 0 < z < 1, \\ 0, & \text{otherwise.} \end{cases}$$

(a) Indicate the limits of integration for $P[X < Y < Z]$.
(b) Indicate the limits of integration for $P[X + Y + Z^2 < 1]$.

4.164 Consider three independent random variables X, Y, Z. Let $V = X + Y + Z$. Find $f_V(v|x), \quad f_V(v|x, y)$.

Multidimensional Random Variables; Discrete Models

4.165 Consider n independent Bernoulli random variables X_i, with parameter p_i. Let $S_n = X_1 + \cdots + X_n$. Evaluate $P[S_n = k]$, ES_n, and $\text{var}(S_n)$.

4.166 Suppose that you want to construct 16 Bernoulli random variables, with given parameters $p_1, \ldots, p_{16}$. *You can select* the number of elements n in the sample space S; let $S = \{\zeta_1, \ldots, \zeta_n\}$ be your choice. The probabilities $P[\{\zeta_k\}]$, $k = 1, \ldots, n$, are also your choice.
(a) Choose $n = 2$; that is, let $S = \{\zeta_1, \zeta_2\}$. Show that this choice for n is insufficient. In other words, show that you cannot specify all 16 required mappings.
(b) Choose $n = 16$. Show that this choice for n is sufficient. In other words, provide 16 such mappings.
(c) Can you find an $n < 16$ that is still sufficient?

4.167 Let X be a binomial random variable with parameters n, p. Consider the interpretation of X as the sum $\sum_k X_k$ of n independent Bernoulli random variables X_k. Suppose that $X = 1$. Find $P[X_k = 1|X = 1]$ for any $k \leq n$.

4.168 Show by example that the sum of *dependent* Bernoulli random variables is *not* a binomial random variable.

4.169 Consider three independent, identically distributed, geometric random variables X_i, $i = 1, 2, 3$. Find $P[X_1 + X_2 > X_3]$.

4.170 Can you construct a five-dimensional, discrete uniform random variable $X = (X_1, X_2, X_3, X_4, X_5)$ on the set $S = \{a, b, c, d, \ldots, y, z\}$? If you need any further information, make your own assumptions and state them clearly.

4.171 The random variables X_1, X_2, X_3, X_4 are independent and follow a discrete uniform distribution with parameters 0 and 9.
(a) Let $X = X_1 + X_2 + X_3 + X_4$. Evaluate (theoretically or experimentally) the pmf $P[X = k]$.
(b) Find the value(s) of k which will maximize this pmf. Find the value(s) of k which will minimize this pmf. Can you provide an intuitive argument for your answer?
(c) Can you generalize for the sum $X = \sum_{i=1}^{n} X_i$?

Multidimensional Random Variables; Uniform Model

4.172 The random variables $\{X_i\}_{i=1}^{n}$ are independent and uniformly distributed in $[0, 1]$. For a fixed $n > 0$, let $Z = \max_{1 \le i \le n}\{X_i\}$. Show that

$$F_Z(z) = \begin{cases} 0, & z < 0, \\ z^n, & 0 \le z \le 1, \\ 1, & z > 1. \end{cases}$$

Interpret intuitively the limiting form of $F_Z(z)$ as $n \to \infty$.

4.173 Consider three random variables X_1, X_2, X_3 with the following property:

$$F_{X_1}(x_1) = \begin{cases} 0, & x_1 < 0, \\ x_1, & 0 \le x_1 \le 1, \\ 1, & x_1 > 1. \end{cases} \qquad F_{X_2}(x_2|X_1 = x_1) = \begin{cases} 0, & x_2 < 0, \\ x_2, & 0 \le x_2 \le x_1, \\ 1, & x_2 > x_1. \end{cases}$$

$$F_{X_3}(x_3|X_2 = x_2, X_1 = x_1) = \begin{cases} 0, & x_3 < 0, \\ x_3, & 0 \le x_3 \le x_2, \\ 1, & x_3 > x_2. \end{cases}$$

(a) Is this enough information to find the joint cdf of the vector (X_1, X_2, X_3)? If no, what further information do you need?
(b) Find, if possible, $F_{X_i}(x_i)$, $i = 1, 2, 3$.
(c) Find, if possible, $F_{X_i}(x_i|X_j = x_j)$, $i = 1, 2, 3$, $j = 1, 2, 3$.
(d) Find, if possible, all three conditional cdf's

$$F_{X_i}(x_i|X_j = x_j, X_l = x_l), \quad i = 1, 2, 3, \quad j \ne i, \quad l \ne i.$$

(e) Find, if possible, all three joint cdf's

$$F_{X_i X_j}(x_i, x_j | X_l = x_l), \quad i = 1, 2, 3, \quad j \neq i, \quad l \neq i.$$

Multidimensional Random Variables; Gaussian Model

4.174 Consider a zero-mean Gaussian vector $(X_1, \ldots, X_n)$ with given covariance matrix $C = [c_{ij}]$.
(a) For a given $k > 0$ and indices j_i calculate $EX_{j_1} X_{j_2} \cdots X_{j_{2k}}$ in terms of the elements of C.
(b) Show that $EX_i X_j X_k X_l = c_{ij} c_{kl} + c_{ik} c_{jl} + c_{il} c_{jk}$.
(c) Let X be a zero-mean, Gaussian random variable, with variance v^2. Show that

$$EX^{2k} = 1 \cdot 3 \cdot 5 \cdot (2k - 1) v^{2k}.$$

4.175 Show that if three jointly Gaussian random variables are independent in pairs, they are independent.

4.176 Consider a sequence $X_1, \ldots, X_n$ of independent Gaussian random variables with zero mean and unit variance. Define

$$Y_n = \sum_{k=1}^{n} X_k, \quad Z_n = \sum_{k=1}^{n} (-1)^k X_k.$$

(a) Show that Y_n and Z_n have identical distributions.
(b) Compute $E(Y_k | Z_k)$ for a few values of k.
(c) Can you generalize for arbitrary k?

4.177 Let X_1, X_2, X_3 be three independent, jointly Gaussian random variables, with zero mean and unit variance. Find the pdf of $Y = \sqrt{X_1^2 + X_2^2 + X_3^2}$.

4.178 Consider n independent, identically distributed Gaussian random variables $\{X_i\}$ with mean m and variance σ^2. Let

$$M_n = \frac{X_1 + \cdots + X_n}{n}, \quad V_n = \frac{(X_1 - M_n)^2 + \cdots + (X_n - M_n)^2}{n}.$$

Show that M_n, V_n are independent.

4.179 Let $X_1, \ldots, X_n, M_n, V_n$ be defined as in Problem 4.178. Let

$$Y = \frac{1}{\sqrt{n}} \frac{M_n - m}{V_n}.$$

Show that Y follows the Student's-t distribution.

4.180 Let $X_1, \ldots, X_n$ be independent Gaussian random variables with zero mean and unit variance. Let $Y = X_1^2 + \cdots + X_n^2$. Show that Y is a χ^2 random variable with parameter n.

4.181 Consider n independent Gaussian random variables $X_1, \ldots, X_n$ with zero mean and unit variance. Consider the change from Cartesian to polar coordinates, defined by the transformation

$$
\begin{aligned}
X_1 &= R \sin \Theta_1, \\
X_2 &= R \sin \Theta_2 \cos \Theta_1, \\
X_3 &= R \sin \Theta_3 \cos \Theta_2 \cos \Theta_1, \\
&\vdots \\
X_{n-1} &= R \sin \Theta_{n-1} \cos \Theta_{n-2} \cdots \cos \Theta_1, \\
X_n &= R \cos \Theta_{n-1} \cos \Theta_{n-2} \cdots \cos \Theta_1.
\end{aligned}
$$

Find the joint pdf of $(R, \Theta_1, \Theta_2, \ldots, \Theta_{n-1})$.

Multidimensional Random Variables; Other Models

4.182 *Exponential.* Verify that the function

$$
f(x_1, \ldots, x_n) = \begin{cases} x_1 x_2 \cdots x_{n-1} e^{-(x_1 + x_1 x_2 + \cdots + x_{n-1} x_n)}, & x_k > 0, \quad k = 1, \ldots, n, \\ 0, & \text{otherwise,} \end{cases}
$$

is a valid multidimensional pdf for every integer $n > 0$.
(a) Find the marginal pdf's for X_1, X_2.
(b) Find the marginal two-dimensional pdf for the pair (X_1, X_2).
(c) Find the conditional pdf of X_1 given X_2.
(d) Find the correlation $E X_1 X_2$.

4.183 *Cauchy.* Find the pdf of the sum of n independent, identically distributed Cauchy random variables $X_1, \ldots, X_n$. Show that the random variable

$$
Z = \frac{1}{n} \sum_{k=1}^{n} X_k
$$

is also Cauchy-distributed.

4.184 *Gamma.* Consider n independent and identically distributed gamma random variables $X_1, \ldots, X_n$ with parameters a, λ. Let

$$
Y_k = \frac{X_k}{X_1 + \cdots + X_n}, \qquad 1 \le k \le n - 1.
$$

(a) Are the $n - 1$ random variables $Y_1, \ldots, Y_{n-1}$ independent? Identically distributed?

(b) Determine the marginal pdf of Y_1, experimentally or theoretically.

Multidimensional Random Variables; Generic Problems

4.185 Verify that the function

$$f(x_1, \ldots, x_n) = \begin{cases} \dfrac{1}{x_1 \cdot x_2 \cdots x_{n-1}}, & 0 < x_n < x_{n-1} < \cdots < x_1 < 1, \\ 0, & \text{otherwise}, \end{cases}$$

is the joint pdf of n random variables $X_1, \ldots, X_n$. When $n = 1$, we define

$$f(x_1) = \begin{cases} 1, & 0 < x_1 < 1, \\ 0, & \text{otherwise}. \end{cases}$$

(a) Find the marginal two-dimensional pdf of (X_1, X_2).

(b) Find the marginal pdf of the random variable X_2.

(c) Find the conditional pdf's $f_{X_1}(x_1 | X_2 = x_2)$ and $f_{X_2}(x_2 | X_1 = x_1)$.

(d) Find the correlation $E X_1 X_2$.

4.186 Consider three random variables X_1, X_2, X_3 with joint pdf

$$f_{X_1 X_2 X_3}(x_1, x_2, x_3) = \begin{cases} \dfrac{1}{\frac{4}{3}\pi}, & x_1^2 + x_2^2 + x_3^2 \leq 1, \\ 0, & \text{otherwise}. \end{cases}$$

(a) Are the random variables independent?

(b) Are they identically distributed?

(c) Are the pairs (X_1, X_2) and (X_2, X_3) identically distributed?

(d) Find the conditional pdf $f_{X_1}(x_1 | x_2)$.

(e) Find the conditional pdf $f_{X_1}(x_1 | x_2, x_3)$.

(f) Find the conditional pdf $f_{X_1 X_2}(x_1, x_2 | x_3)$.

Multidimensional Random Variables; Advanced Skills

4.187 Consider three random variables X, Y, Z. Prove or disprove the following statements.

(a) If X, Y are independent and Y, Z are independent, then X, Z are independent.

(b) If X is independent of the vector (Y, Z), then X is independent of Y.

(c) If X is independent of the vector (Y, Z), then X is independent of any real-valued function $g(Y, Z)$.

4.188 The random vector (X_1, X_2) is independent of the random vector (X_3, X_4). Are the random vectors (X_1, X_3), (X_2, X_4) independent?

4.189 Consider three random variables X, Y, Z and a constant $\epsilon > 0$.
(a) Prove the following "triangle inequality" for probabilities:

$$P[|X - Y| > \epsilon] \leq P\left[|X - Z| > \frac{\epsilon}{2}\right] + P\left[|Z - Y| > \frac{\epsilon}{2}\right].$$

(b) Does the triangle inequality hold if we replace the term $\frac{\epsilon}{2}$ on the right-hand side with α and β, where $\alpha + \beta = \epsilon$?

4.190 The random variables $\{X_i,\ i = 1, \ldots, n\}$ are independent and identically distributed, with pdf $f_X(x)$ and cdf $F_X(x)$. Let

$$Z_1 = \max_{1 \leq i \leq n} \{X_i\}, \quad Z_2 = \min_{1 \leq i \leq n} \{X_i\}.$$

(a) Find the joint cdf $F_{Z_1 Z_2}(z_1, z_2)$.
(b) Find the joint pdf $f_{Z_1 Z_2}(z_1, z_2)$.
(c) Find the pdf $f_{Z_1}(z_1)$.
(d) Find the pdf $f_{Z_2}(z_2)$.
(e) Find the conditional pdf's $f_{Z_1}(z_1 | z_2)$ and $f_{Z_2}(z_2 | z_1)$.

4.191 Consider n independent, identically distributed, continuous random variables $X_1, \ldots, X_n$ with common pdf $f_X(x)$. Let $Y_1, \ldots, Y_n$ be the ordered random variables (in increasing order). Show that

$$f_{Y_k}(y) = \frac{n!}{(n-k)!(k-1)!}[F_X(y)]^{k-1}[1 - F_X(y)]^{n-k} f_X(y), \quad k = 1 \ldots, n.$$

4.192 The random variables $\{X_i,\ i = 1, \ldots, n\}$ are independent and identically distributed. Consider the convex combination

$$\bar{X} = \sum_i a_i X_i, \quad a_i \geq 0, \quad \sum_i a_i = 1.$$

Show that the selection $a_i = 1/n$ minimizes the variance of $\bar{X}$.

4.193 Let $X_1, \ldots, X_n$ be n independent, nonnegative, integer-valued random variables with a common pmf. Let $R_n \triangleq \sum_{k=n}^{\infty} P[X_1 = k]$. Show that

$$E \min\{X_1, \ldots, X_m\} = \sum_{n=1}^{\infty} R_n^m.$$

4.194 Prove the chain rule (see Problem 2.90) for the joint pmf of a random vector:

$$\begin{aligned}
P[X_1 = x_1, \ldots, X_n = x_n] \;=\; & P[X_1 = x_1 | X_2 = x_2, \ldots, X_n = x_n] \\
& \cdot P[X_2 = x_2 | X_3 = x_3, \ldots, X_n = x_n] \\
& \cdots P[X_{n-1} = x_{n-1} | X_n = x_n] \cdot P[X_n = x_n].
\end{aligned}$$

4.195 Prove the chain rule for multidimensional pdf's.

4.196 Two sequences A, B of $\{0, 1\}$ digits, each of length n, are selected at random. A *match of length* k, $k \leq n$, occurs if a subsequence of k digits appears in both sequences A, B. Let X_k be the number of matches of length k. Find $EX_k, \mathrm{var}(X_k)$.

Miscellaneous

4.197 Let n, k be positive integers. Show that the binomial coefficients satisfy the relationship

$$\sum_{l=\max(0, k-n)}^{\min(n, k)} \binom{n}{l}\binom{n}{k-l} = \binom{2n}{k}$$

4.198 *A Modeling Problem. Are Communication Protocols 100 Percent Secure?* A submarine with nuclear missiles can fire its weapons only when its main computer receives a special order from the President of the United States, over a radio channel. To ensure that no irreversible mistakes are made, the Pentagon has hired BBC (Big Bucks Company), a well-known defense contractor, to design a *secure* communication protocol for delivering this special order. The only security threat this protocol has to overcome is removal of packets from the radio channel. In particular, every single order has a probability $p > 0$ of being removed from the channel, while every single acknowledgment has a probability $q > 0$ of being removed from the channel.

The big guns of BBC came up with the following protocol: The President sends $L \geq 1$ copies of the order, one after the other, to the submarine. The computer sends back to the President acknowledgments for every single order it has received. The rationale behind this order and acknowledgment duplication is obvious: The chances of getting an order or acknowledgment through must be better than the case of sending only one order.

You are hired as an independent consultant to evaluate the security features of this proposal.

(a) Can the President be 100 percent sure that her or his order will be executed?

(b) Can the submarine commander be 100 percent sure that the President knows that her or his order was executed?

(c) What is the probability of an ordered attack's being executed as a function of L?

(d) What is the probability of an ordered attack's being executed and acknowledged as a function of L?

(e) The President insists on making *absolutely sure* that her or his command will be executed, to avoid risk of American lives. What value of L should we set in this protocol, to fulfill this requirement?

(f) Can you design another protocol to give the President 100 percent assurance?

4.199 *Tough!* Let $A_1, \ldots, A_m$ be m events defined on some sample space S. Define the quantities P_i as in Problem 2.68. Let

$$a_{mk} \triangleq P[\text{exactly } k \text{ events occur}], \quad b_{mk} \triangleq P[\text{at least } k \text{ events occur}].$$

Find a_{mk}, b_{mk} in terms of P_i. More specifically, show that

$$a_{mk} = \sum_{n=k}^{m} (-1)^{n-k} \binom{n}{k} P_n, \quad b_{mk} = \sum_{n=k}^{m} (-1)^{n-k} \binom{n-1}{k-1} P_n.$$

4.200 *Generating Variates.* The algorithm presented in Section 4.12 generates variates with a desired *covariance*, not a desired *joint pdf*. Discuss the difficulty of obtaining variates with a desired jpdf.

4.201 *Experimental.* Perform n measurements of N and M, the respective number of packets coming into and out of a computer interface (e.g., using the *netstat 1* command on Unix workstations). Evaluate and plot the following:

(a) $P[N = k, M = m], \quad k, m = 0, 1, 2, \ldots$

(b) $P[N \leq k], \quad k = 0, 1, 2, \ldots$

(c) $P[M \leq m], \quad m = 0, 1, 2, \ldots$

(d) $P[N \leq k | M \leq m], \quad k, m = 0, 1, 2, \ldots$

(e) $E(N \cdot M), E(N^2 \cdot M^2)$.

(f) $\text{cov}(N, M)$.

5

INTRODUCTION TO ESTIMATION

Estimation refers to the general subject of making inferences about the value(s) of one (or more) random variable(s), based on measurements (observations) of one or more related random variable(s). For example, employers use the value of the GPA random variable in order to estimate the unknown true quality of a potential employee (another random variable). In a communication channel, we use the measured value of a received bit to estimate the unknown value of a transmitted bit (see Figure 1.2).

Do *not* confuse the subject of estimating the values of a random variable with a related subject from statistics, namely, estimating the value of an *unknown parameter*, such as a mean or variance, based on measurements of random variables.

We can put the random variable estimation problem in the following generic framework. There is a possibly vector-valued random variable Y that for various reasons we cannot measure directly. Instead, we measure another random variable X (also possibly vector-valued).[1] We then wish to make an inference about the value of Y given that the measured value of X was equal to x. In general, this inference process introduces an error, called the *estimation error*. Figure 5.1 depicts this framework pictorially. And $\hat{Y} \triangleq g(X)$ is the inferred value of the unknown Y.

---

[1] The case of infinite measurements is treated in Chapter 9.

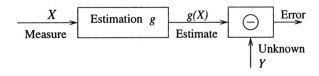

Figure 5.1 The general estimation problem.

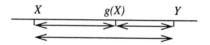

Figure 5.2   Processing of a random variable reduces the error.

The estimation problem can take a variety of specialized forms, depending on (1) the number of measurements involved (single-multiple measurements), (2) the way we process them (recursive or nonrecursive), and (3) the criteria for processing them (mean square, maximum-likelihood estimation, etc.). In the next section we present some of the choices for estimation criteria. In Section 5.2 we discuss in detail one very widely used criterion that is based on *mean square errors* and single measurements. In Section 5.3 we discuss how multiple measurements can be handled, based again on the mean square error criterion. In the following section we work out in detail a realistic example, based on the Dow Jones data. In Section 5.5 we present the *maximum-likelihood* criterion, which is used very widely in communication receivers and, in a slightly different context, in statistics. Finally, we conclude this chapter with an historical remark that depicts the close relationship between the Gaussian pdf and estimation.

5.1   CRITERIA TO CONSIDER

Consider a sample space S, on which there exist two random variables X and Y, with a known joint pdf. We take a measurement of X, and based on that, we want to estimate the value of Y. For example, X could be the value of a received bit in a communication channel, while Y could be the value of a transmitted bit. In another example, X could be the GPA of a prospective employee and Y his or her true quality.

In general, an *error* is made in this estimation. Ideally, we would like this error to be zero; in practice, we try to keep it as small as possible. The only hope for doing so is, of course, to *process* the raw measurements; we encountered this processing in Chapters 3 and 4, under the name *transformation* of random variables. Other commonly used terms in the estimation environment are *filtering* and *prediction*. As Figure 5.2 suggests, we hope that the transformation $g(\cdot)$ will bring the unknown and estimated values closer together (in a sense that we discuss shortly).

Definition: The random variable $Z \triangleq Y - g(X)$ is called the **estimation error**.

So we are looking now for a function $g(\cdot)$ such that the estimation error $Z = Y - g(X)$ is "small." There are a variety of ways to define what small means. One possibility is the following: Let $Z_h = Y - h(X)$, where $h(\cdot)$ is "any" function (in a set of functions H we want to consider). We may require that

$$Z(\zeta) \le Z_h(\zeta), \quad \forall h \in H, \quad \forall \zeta \in S,$$

or, equivalently,

$$g(X(\zeta)) \ge h(X(\zeta)), \quad \forall h \in H, \quad \forall \zeta \in S. \tag{5.1}$$

At first glance, this sounds like a tough requirement since we are asking for this inequality to hold for all outcomes of the sample space. A moment's thought should convince you that it is a silly requirement, too. (Think what happens with negative and positive errors.) A closer investigation of Inequality 5.1, say, with the functions $h(x) = g(x)$ and $h(x) = 1/g(x)$ should convince you that it is an impossible requirement after all! It is instructive to consider what Inequality 5.1 would actually mean in the channel transmission example of Figure 1.2.

An alternative, weaker requirement is to demand that

$$F_{Z_h}(x) \le F_Z(x), \quad \forall x \in R, \quad \forall h \in H,$$

where, as usual, $F_X(x)$ denotes the cdf of the random variable X. This still sounds like a tough requirement since it must hold for all values of $x \in R$. (Is it an impossible requirement, as Inequality 5.1?) An even less demanding requirement can be cast in terms of the means, namely,

$$Eg(X) \ge Eh(X), \quad \forall h \in H. \tag{5.2}$$

In the sequel, we will consider only criteria expressed in terms of means since in general these are the only tractable ones.

In Criterion 5.2, the *sign* of the error is *not* taken into account, and that may put us in some awkward situations, where "large" negative errors may lead us to believe that a function g is a good choice! For this reason, it makes sense to define the estimation error as $Z = |Y - g(X)|$ and to consider

$$EZ = E|Y - g(X)| \tag{5.3}$$

as the estimation criterion. In Equation 5.3, it is implicit that all errors count the same, no matter how small or large. But in many applications, some errors may not be as significant as others. For example, small noise values in a communication channel can actually be absorbed without any problems, while large ones will cause an incorrect detection. In this case, we may define the estimation error as in Figure 5.3(a) or 5.3(b). The idea behind the first error

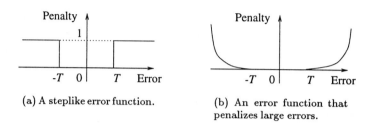

(a) A steplike error function.

(b) An error function that penalizes large errors.

Figure 5.3   Possible error functions.

choice is clear. We pay no price for small errors, and all big errors cost the same. The second error choice emphasizes the idea that errors larger than the threshold T cost us according to their size.

The error criteria in Equation 5.3 and Figures 5.3(a) and 5.3(b) are fine, but there is little we can do with them from a mathematical point of view (see Problems 5.1–5.3). As we will see in the next section, a "compromise" that combines more or less the salient features of all is to use the following criterion:

$$EZ = E(Y - g(X))^2. \qquad (5.4)$$

This criterion deals with mean square errors; it is called the *minimum mean square error (MMSE) criterion*. As we will see in the next section, it is very easy to handle Criterion 5.4 mathematically, and this will lead us to a rich theory (this will become even more apparent in Chapter 9).

Finally, another possible criterion is to *maximize the probability* of zero errors, given that the measured value was equal to x. More precisely, let $Z = Y - g(X)$ denote the estimation error; we then wish to choose $g(\cdot)$ to maximize

$$P[Z = 0|X = x] \triangleq P[Y = g(X)|X = x] = P[Y = g(x)|X = x]. \qquad (5.5)$$

This criterion is used a lot in communication channels and economics. We will say more about it in Section 5.5.

For historical reference, let us mention that D. Bernoulli was the first to study an estimation problem, in the eighteenth century. More specifically, he tried to estimate the value of a constant c in terms of n "noisy" measurements $c + e_i$, $i = 1, \ldots, n$. See also the historical reference in Section 5.6.

To summarize, a variety of criteria exist. In the next section we focus on the mean square error criterion introduced in Equation 5.4 because it is easy to handle mathematically. The maximum-likelihood criterion in Equation 5.5, another criterion that can be handled analytically, is discussed in Section 5.5.

The rest of the criteria are briefly considered in the problems, designed to show the difficulties of dealing with them.

5.2   MMSE ESTIMATION, SINGLE MEASUREMENT

We can pose this problem in its simplest form as follows: Consider a random experiment E with model (S, Q, P) and two random variables X and Y defined on it. Suppose that we perform a trial of the experiment and the value $X = x$ is observed. [Equivalently, an outcome $\zeta \in S$ is selected, and $X(\zeta) = x$.] We wish to find a transformation $\hat{Y} = g(X)$ such that the estimate $\hat{Y}$ minimizes the error criterion:

$$\min E(Y - \hat{Y})^2 = \min E[Y - g(X)]^2. \qquad (5.6)$$

Of course, we must specify what are the allowable choices for the class of functions G, from which $g(\cdot)$ may be selected.[2] We will now solve the problem in Equation 5.6 in three steps; we start with a very simple set G and progressively enrich it. More specifically, we consider three cases:

- The set of all constant functions, i.e.,

$$G_c = \{g(x) = a : a \in R\}$$

- The set of all linear, real-valued functions, i.e.,

$$G_l = \{g(x) = ax + b : a, b \in R\}$$

- The set G_{nl} of all nonlinear, real-valued functions

The reason for this staggered approach should become apparent when you reach Section 5.2.3.

5.2.1   Estimation with a Constant Value

Consider the following problem:

$$\min_{\hat{Y} \in G_c} E(Y - \hat{Y})^2 \triangleq \min_{a \in R} E(Y - a)^2. \qquad (5.7)$$

In other words, we want to choose a constant a, *irrespective of the measured value x*, for an estimate of Y. This is equivalent to taking no measurements at

---

[2]Do not forget that all functions in G must be such that $g(X)$ is a random variable!

all, or throwing any taken measurements away! It sounds like an unwarranted restriction, but recall the cola example in Section 3.9.1. There may be situations in which we do not want to handle a lot of information! At any rate, the constant a in Equation 5.7 must be chosen as follows: Let

$$e(a) \triangleq E(Y - a)^2$$

denote the mean square error in estimating the random variable Y with constant a. Expanding $e(a)$, we get

$$e(a) = a^2 - 2aE(Y) + E(Y^2). \tag{5.8}$$

Equation 5.8 is a simple quadratic in a. There is no constraint on a, so the minimum can be found by simple differentiation. The optimum value a^* is then given by

$$a^* = EY,$$

and indeed it produces a minimum, since the second derivative of Equation 5.8 is positive. The average square error is equal to $E(Y - EY)^2$, that is, var(Y). Let's state this result as a lemma:

Lemma 5.1 *The minimum mean square, constant estimate of a random variable Y is*

$$a^* = EY.$$

The minimum average square error is given by

$$e_{\min} = \text{var}(Y).$$

Implementation issues

From Lemma 5.1 we see that in order to implement such an estimator, we need to take no measurements of X at all; all we have to know is the average value of the random variable Y. (This, of course, we could know theoretically; or, it would require some a priori measurements of Y. See Section 6.6 in that respect.)

5.2.2   Estimation with a Linear Function

Consider now the set G_l of all linear functions. Since such functions can be represented by two real numbers a, b, mathematically the problem now is the following. Let

$$e(a, b) \triangleq E(Y - \hat{Y})^2 \triangleq E(Y - aX - b)^2. \tag{5.9}$$

We wish to solve the problem

$$\min_{a, b \in R} e(a, b). \tag{5.10}$$

Note that in Expression 5.10, the free variables are both a and b. Since a, b are not constrained at all, we can solve this problem in a straightforward manner, by setting

$$\frac{\partial e}{\partial a} = 0, \quad \frac{\partial e}{\partial b} = 0. \tag{5.11}$$

We ask you to complete the remaining steps of this approach in Problem 5.8. We can solve the same problem with a little trick that enables us to use the approach in Section 5.2.1. Indeed, we can view Equation 5.9 as estimation of the variable $Y - aX$ with a constant b. Then, for any choice of a, the best b, call it $b(a)$, is given by

$$b(a) = E(Y - aX) = EY - aEX. \tag{5.12}$$

The best a can be determined as follows. Substituting Equation 5.12 in Equation 5.9, we get

$$e(a) \overset{\triangle}{=} E[Y - aX - (EY - aEX)]^2 = E[\{Y - EY\} - a\{X - EX\}]^2.$$

Expanding the square inside the expectation, we get a second-order equation in the unknown a:

$$e(a) = E[a^2\{X - EX\}^2 - 2a\{Y - EY\}\{X - EX\} + \{Y - EY\}^2]. \tag{5.13}$$

We can differentiate now Equation 5.13 with respect to a; after some elementary algebra, the optimum value a^* for the coefficient a is given by

$$a^* = \frac{\text{cov}(X, Y)}{\text{var}(X)} = \rho_{XY} \frac{\sigma_Y}{\sigma_X}. \tag{5.14}$$

We can easily calculate the average value of the minimum square error. Substituting Equations 5.14 and 5.12 into Equation 5.9, we get

$$\begin{aligned} E(Y - \hat{Y})^2 &= E[Y - a^*X - b(a^*)]^2 = E[(Y - EY) - a^*(X - EX)]^2 \\ &= \text{var}(Y)(1 - \rho_{XY}^2). \end{aligned}$$

We summarize this result as a theorem since it is widely used in practice:

Theorem 5.1 *Consider two random variables X and Y with mean values EX and EY, correlation coefficient ρ_{XY}, and standard deviations σ_X and σ_Y. Then the minimum mean square error linear estimator of the random variable Y, in terms of measured values of X, is given by*

$$\hat{Y} = \rho_{XY} \cdot \sigma_Y \left(\frac{X - EX}{\sigma_X} \right) + EY. \tag{5.15}$$

The minimum mean square error is given by

$$e_{\min} = \text{var}(Y)(1 - \rho_{XY}^2).$$

Since the correlation coefficient is at most 1 (see Problem 4.119), we see that the error with a linear estimator is no greater than the error with a constant estimator. This should be expected, since the class of constant estimators is a subset of the class of linear ones. Observe that the minimum errors, described in Lemma 5.1 and Theorem 5.1, are equal when $\rho_{XY} = 0$, that is, when the two random variables are *uncorrelated*. Moreover, the error is zero when $\rho_{XY} = \pm 1$, that is, when the two random variables are linearly correlated. Of course, this does not mean that *always* (i.e., for all $\zeta \in S$, or all measurements) the error is zero. We discuss this further in Chapter 6.

Finally, it is easy to calculate that the average value of the estimator $\hat{Y}$ is equal to EY. Estimators with such a property are called *unbiased*.

1. Implementation issues

Equation 5.15 provides the processing algorithm for the best function $g(x)$: When a measurement x of the random variable X is taken, our estimated value for the random variable Y is simply

$$g(x) = \rho_{XY}\sigma_Y\left(\frac{x - EX}{\sigma_X}\right) + EY.$$

Since this is a linear transformation of x into a value $g(x)$, implementing the best linear estimator is easy: It takes only one multiplication (with the constant $\rho_{XY}\sigma_Y/\sigma_X$) and one addition (with the constant $EY - \rho_{XY}\sigma_Y EX/\sigma_X$).

2. The famous orthogonality principle

We derived the optimal value a^* in Equation 5.14 by differentiating the mean square error $e(a)$ in Equation 5.13. Therefore, we see that a necessary condition that the best linear estimate should satisfy is

$$\frac{d}{da}\left\{E[Y - EY - a(X - EX)]^2\right\} = 0. \tag{5.16}$$

Rearranging Equation 5.16, we see that a^* must satisfy the equation

$$E\{[(Y - EY) - a^*(X - EX)]\cdot(X - EX)\} = 0. \tag{5.17}$$

Equation 5.17 has a very intuitive "geometric" explanation in terms of the zero-mean random variables $Y - EY$ and $X - EX$. The quantity inside square brackets in Equation 5.17 is simply the best error in our estimation of $Y - EY$ in terms of $X - EX$. Then what Equation 5.17 says, in a geometric sense, is that *the error vector is orthogonal to the measurement vector* (hence the name *orthogonality principle* for this necessary condition). We use this principle directly in Chapter 9 to derive the form of MMSE estimators for random processes.

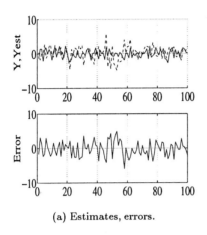

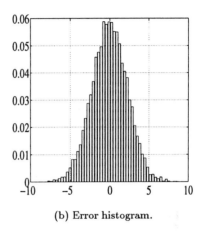

(a) Estimates, errors.   (b) Error histogram.

Figure 5.4 Linear estimation of dependent Gaussian variables. The dashed line is the estimate.

Let's see some examples next.

Example 5.1. *The Gaussian Random Variables in Section 4.10.* Consider two random variables X, Y that are jointly Gaussian with their joint density function given by (see Figure 4.55)

$$f_{XY}(x,y) = \frac{e^{-\frac{1}{2(1-\rho_{XY}^2)}\left\{\left(\frac{x-m_X}{\sigma_X}\right)^2 - 2\rho_{XY}\frac{x-m_X}{\sigma_X}\frac{y-m_Y}{\sigma_Y} + \left(\frac{y-m_Y}{\sigma_Y}\right)^2\right\}}}{2\pi\sigma_X\sigma_Y\sqrt{1-\rho_{XY}^2}} \tag{5.18}$$

We have generated 10,000 samples of X and Y values, using the algorithm in Section 4.12 for the density function in Equation 5.18. Based on those values, we have calculated the errors in linear estimation, shown in Figure 5.4(a). Note that only 100 out of 10,000 values are depicted.

Example 5.2. *The ftp Transfer in Section 1.9.* Let Y denote the throughput of the file transfer and X denote the measured transfer delay. We wish to estimate the throughput values, based on the measured delays. Using the same data we used to produce Figure 4.20 in Section 4.4.5, we calculated the best linear estimator. In Figure 5.5 we show some of the actually measured values for the throughput Y, some estimated values based on delay measures, and the resultant estimation errors. It is instructive to compare these graphs with the corresponding ones for the best nonlinear estimator of the next section (Figure 5.6).

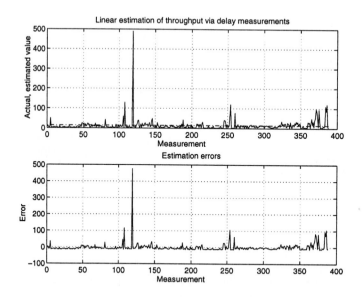

Figure 5.5   The *ftp* throughput, best linear estimator.

5.2.3   Estimation with a Nonlinear Function

This estimator is the most general case we consider. As expected, the estimation errors must be the smallest since the previous two estimators are special cases of the class of nonlinear functions. Consider therefore the problem

$$\min_{g(\cdot) \in G_{nl}} E[Y - g(X)]^2, \tag{5.19}$$

where G_{nl} is the set of all nonlinear (measurable, of course) functions on R. Observe that the minimization in Problem 5.19 is taken over *a set of functions;* therefore, taking derivatives will not work here, as it did with the sets G_c (actually R) and G_l (actually R^2). Instead, we can use a property of conditional expectations (see Section 4.9.1), which will allow us to recast the problem into the constant estimator case. Applying Equation 4.54 in Equation 5.19, we get

$$
\begin{aligned}
E[Y - g(X)]^2 &= E[E[[Y - g(X)]^2 | X = x]] \\
&= \int_{-\infty}^{\infty} E[[Y - g(X)]^2 | X = x] f_X(x) \, dx \\
&= \int_{-\infty}^{\infty} E[[Y - g(x)]^2 | X = x] f_X(x) \, dx. \tag{5.20}
\end{aligned}
$$

In the last step, we can use the constant $g(x)$ instead of the random variable $g(X)$ since we have conditioned on $X = x$ already. Observe from Equation 5.20 that for all $x \in R$, the integrand is nonnegative. Therefore, the function in 5.19 will be minimized if the following simpler problem is solved:

$$\min_{g(x) \in R} E[[Y - g(x)]^2 | X = x]. \tag{5.21}$$

But in Problem 5.21, $g(x)$ is a constant since we have already conditioned on the event $\{X = x\}$. Observe that the expectation in 5.21 is taken with respect to the conditional density $f_Y(y|X = x)$. Therefore, the optimal solution $g^*(x)$ is simply the conditional mean $g^*(x) = E[Y|X = x]$. We summarize these findings as a theorem:

Theorem 5.2 *Consider two random variables X and Y. The minimum mean square error, nonlinear estimator of the random variable Y, in terms of measured values of X, is given by*

$$\hat{Y} = E(Y|X). \tag{5.22}$$

In general, the form of the estimator will depend on the form of the joint distribution of X and Y. We cannot, therefore, calculate the minimum error in a closed form, as we have done for the linear estimator.

Implementation issues

Let's see now what it takes to implement the best nonlinear estimator of Equation 5.22. We have

$$\hat{Y}(x) = \int_{-\infty}^{\infty} y f_Y(y|X = x) \, dy.$$

The measurement x is processed in a rather peculiar way: Essentially, it determines which conditional pdf (from the class of all conditional pdf's) we choose to integrate. In most situations, exact knowledge of this pdf may not be possible. Even when it is, integration is another problem. Compare these difficulties to the implementation aspects of the constant and linear estimators of the two previous subsections! These difficulties reduce the interest in the subject of best nonlinear estimators to rather theoretical aspects only.

Let's look at some examples now.

Example 5.3. *X, Y Independent.* Consider two random variables X, Y that are independent. The best nonlinear estimate of Y, given that a measurement $X = x$ was taken, is in this case given by

$$\hat{Y}(x) = E(Y|X = x) = EY.$$

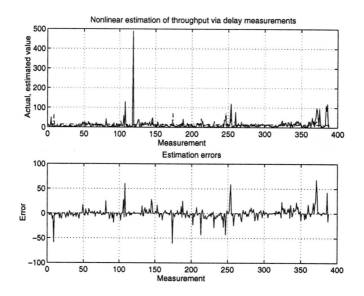

Figure 5.6   The best nonlinear estimator, *ftp* example.

Therefore, regardless of the measured value, it is optimal to always estimate Y with a constant, independent of measurement x.

Example 5.4. $Y = h(X)$ *Is a Known Function of X.* Consider two random variables X, Y that have a strong dependence [e.g., see the examples in Figures 4.21 and 4.22(a), in Section 4.4.5, for some special types of dependence or forms of the function $h(\cdot)$]. The best nonlinear estimate of Y, given that a measurement $X = x$ was taken, is in this case

$$\hat{Y}(x) = E[Y|X = x] = E[h(x)|X = x] = h(x).$$

Therefore, processing of the measured value is straightforward in this case because the nature of the dependence between X and Y is *explicitly* known.

Example 5.5. *The ftp Transfer in Section 1.9.* Let's investigate now a more realistic example, in which the random variables X, Y are dependent, but the form of dependence is not known, as in the previous example. Let Y denote the throughput of the file transfer and X denote the measured transfer delay. In Figure 5.6 we show $E[Y|X = x]$ for various values of x and the resulting errors; this conditional expectation was evaluated based on approximately 400 measured file transfers.

5.3 LINEAR PREDICTION, MULTIPLE MEASUREMENTS

We saw that the best nonlinear estimator is quite difficult to implement. This difficulty only increases when *multiple* measurements are taken since in that case we have to evaluate conditional means of joint distributions of multidimensional vectors. Focusing then on linear estimators is quite natural when we have to process multiple measurements. Then the problem of using multiple measurements in estimating the value of a single random variable can be formulated as follows.

We are given n measurements, which we regard as values of random variables $X_1, \ldots, X_n$. We wish to find a linear estimate of a random variable Y, which we denote now as X_{n+1}, in terms of all the measurements.[3] We first present the problem and its solution for the special case of $n = 2$ since notation is much simpler in this case. We then address the general case.

5.3.1 The Case of $n = 2$

Let X_1, X_2, and X_3 be random variables with zero mean. (The zero-mean assumption simplifies presentation, but is not essential.) We wish to estimate X_3 as a linear function $aX_1 + bX_2$ so as to minimize the error

$$E[X_3 - (aX_1 + bX_2)]^2. \tag{5.23}$$

Applying the principle of orthogonality, or taking derivatives in 5.23 with respect to a and b, we get

$$E[(X_3 - aX_1 - bX_2)X_1] = 0,$$
$$E[(X_3 - aX_1 - bX_2)X_2] = 0.$$

Rearranging the terms in these equations, we get

$$aEX_1^2 + bEX_1X_2 = EX_1X_3,$$
$$aEX_1X_2 + bEX_2^2 = EX_2X_3, \tag{5.24}$$

or in matrix form (which will be helpful in the general case),

$$\begin{bmatrix} EX_1^2 & EX_1X_2 \\ EX_1X_2 & EX_2^2 \end{bmatrix} \begin{bmatrix} a \\ b \end{bmatrix} = \begin{bmatrix} EX_1X_3 \\ EX_2X_3 \end{bmatrix}.$$

---

[3] If we consider $X_1, \ldots, X_{n+1}$ as the same family of random variables, we can understand the term linear *prediction* that has traditionally been used to describe the problem of this section: X_{n+1} can be considered as the *future*, X_n is the *present*, and $X_1, \ldots, X_{n-1}$ are the *past*.

The solution of this 2×2 system is easily shown to be

$$a = \frac{\text{var}(X_2)\text{cov}(X_1, X_3) - \text{cov}(X_1, X_2)\text{cov}(X_2, X_3)}{\text{var}(X_1)\text{var}(X_2) - [\text{cov}(X_1, X_2)]^2},$$

$$b = \frac{\text{var}(X_1)\text{cov}(X_2, X_3) - \text{cov}(X_1, X_2)\text{cov}(X_1, X_3)}{\text{var}(X_1)\text{var}(X_2) - [\text{cov}(X_1, X_2)]^2}.$$

The minimum error is now laborious but straightforward to calculate. We leave that as an exercise (see Problem 5.18).

5.3.2   The Case of $n > 2$

Now let $X_1, \ldots, X_n$ be a sequence of $n > 2$ zero-mean random variables. We wish to estimate another zero-mean random variable X_{n+1} in terms of a linear combination of the random variables $X_1, \ldots, X_n$ so as to minimize

$$E[X_{n+1} - a_1 X_1 - a_2 X_2 - \cdots - a_n X_n]^2.$$

From the orthogonality principle, we get

$$E[(X_{n+1} - a_1 X_1 - a_2 X_2 - \cdots - a_n X_n)X_i] = 0, \quad i = 1, \ldots, n. \qquad (5.25)$$

Equation 5.25 is a linear system of n equations in the n unknowns $a_1, a_2, \ldots, a_n$. In matrix form,

$$\begin{bmatrix} EX_1^2 & EX_1X_2 & \cdots & EX_1X_n \\ EX_1X_2 & EX_2^2 & \cdots & EX_2X_n \\ \cdots & \cdots & \cdots & \cdots \\ EX_1X_n & EX_2X_n & \cdots & EX_n^2 \end{bmatrix} \begin{bmatrix} a_1 \\ a_2 \\ \vdots \\ a_n \end{bmatrix} = \begin{bmatrix} EX_1X_{n+1} \\ EX_2X_{n+1} \\ \vdots \\ EX_nX_{n+1} \end{bmatrix}.$$

The solution of this system is conceptually simple: Any standard method of solving linear systems would do. However, as n increases, standard methods (like the Gauss-Seidel) will become less and less efficient. A *recursive* method of solving this system is known as *Levinson's algorithm*, which we discuss in Section 9.4.4.

5.3.3   Examples

In the following examples, we adjusted the random variables so that their means are equal to zero, as the theory of this section requires.

Example 5.6. *The Random Variables X and $\sin(X)$ in Section 4.4.5.* Consider the two random variables $X_1 \triangleq X$ and $X_2 \triangleq \sin(X)$, for which we have evaluated an experimental histogram (see Figure 4.21 in Section 4.4.5). Suppose

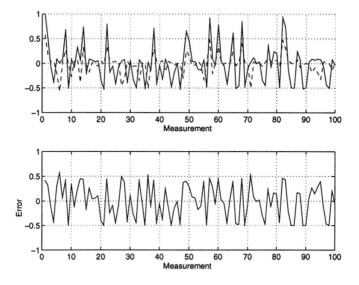

Figure 5.7 Estimation errors.

we wish to estimate the random variable $X_3 = X\cos(X)$ from those measurements. In order to apply the solution of System 5.24, we must calculate the values EX_iX_j from the measured data. Based on these values, we get

$$a = 1.17749, \quad b = 0.29742.$$

Figure 5.7 shows a few sample errors from this estimation. The horizontal axis represents measurement id; the vertical axis represents the estimated value (dashed line) and the actual value (solid line).

Example 5.7. *The Random Variables $\bar{X}$ and $\bar{X}/\bar{Y}$ in Section 4.4.5.* Consider now the random variables $X_1 \triangleq \bar{X}$, $X_2 \triangleq \bar{X}/\bar{Y}$ in Figure 4.22(a). Suppose we wish to estimate the random variable $X_3 = 2(X_1 + X_2)$, for which we suspect a zero error since it is already a linear combination of the measurements. As Figure 5.8 depicts, the processing of this data set *does* confirm our expectation (as it should). The axes have a meaning similar to that in Figure 5.7; the small deviations from 0 are due to numerical inaccuracies.

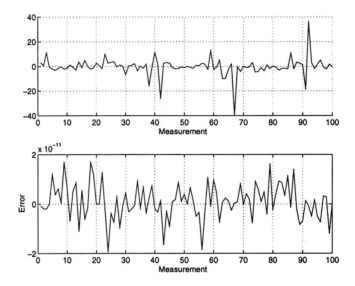

Figure 5.8   An exact prediction.

5.4   DOW JONES EXAMPLE

Let's see now how the ideas of the previous sections apply to a realistic example, namely, the Dow Jones index discussed in Section 1.9. In order to make the discussion more interesting, consider the following "game."

You graduate and find a job with a hefty salary. You start thinking about investments. You learn about investment brokers, and you visit one. Here is a strategy that the broker proposes to you: Every business day, when the New York Stock Exchange opens, you buy stock at the opening value; this is the random variable O. Suppose that you spend x dollars every day, and you distribute this amount proportionally to all 30 stocks of the Dow Jones index. You always sell your stock at closing time, getting back revenue described by the closing random variable C. The broker (not knowing you had taken a course in probability theory) claims that this strategy is a "sure bet." In 5 years you can retire and think about Bermuda, Tahiti, etc. How can you tell whether this is a good investment strategy? Can you find a better one?

Let's evaluate the proposed strategy with our Dow Jones data for years 1987 to 1993. The expected gain per day is of course $EC - EO = \$2789.10 - \$2788.20 = \$0.90$ per $\$2788.20$ (the amount you invest every day). Over the 1589 (business) days of that period, you can expect a total return from your $\$2788.20$ investment of less than $\$1500$. In about 5 years, this strategy gave you

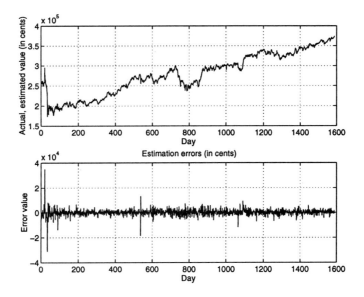

Figure 5.9 Dow Jones example, linear estimation.

a total return of about 50 percent, or about 10 percent per year, not counting the broker's fees.

How can we improve the strategy? Suppose we can predict somehow that on a given day, the closing value will be higher or lower than the opening one. Then investing only on days with higher estimated closing values will result in a better gain. What we need then is an estimator of the random variable C given measurements of the random variable O.

Let's start with an estimator that utilizes only one such measurement. Let's consider an MMSE estimator.[4] The best linear estimator of C given O is

$$\hat{C} = aO + b,$$

where from the given data we have

$$a = 1.1456, \quad b = 1192.10.$$

The estimation errors are shown in Figure 5.9.

The best nonlinear estimator can be calculated as $E(C|O)$, and the errors are shown in Figure 5.10. The new gain is \$0.93, for the same amount of investment.

---

[4] Even though a negative error is quite different from a positive error here.

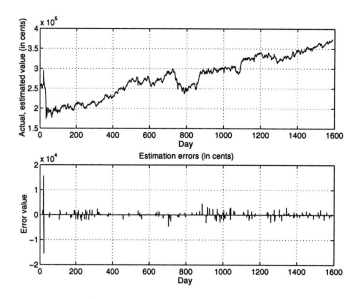

Figure 5.10   Dow Jones example, nonlinear estimation.

Consider now a linear predictor that utilizes two random variables, namely, the opening and high values (variables O and H) to predict the value of the random variable C. (Do not forget that in developing the theory, we had to assume all variables had zero means. So in deriving these results, we have to subtract the means from O, H, and C.) From the given data, again, the parameters for the best predictor, when two measurements are used, are

$$a = 0.0205, \quad b = 0.9752$$

and the errors are shown in Figure 5.11. Is this predictor implementable as an investment strategy?

Will we get any better estimates if we use the random variable $(O + H)/2$, instead of O, to estimate the value of C? (Can you think of any reason why we should consider such a choice for measurements?) What if we use the *vector* random variable (O, H)? Or the random variable O' representing O values on a given day only, say, Friday? Or the random variable O'' taken from half the data (i.e., from every other value)? Figures 5.12 and 5.13 show some of the errors when these new estimators are used. See also Problems 5.23–5.35 for some additional suggestions.

Let's summarize our discussions concerning this example. Among the proposed estimation criteria and the relevant estimators, the best one (i.e., the

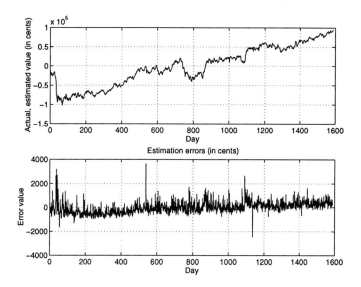

Figure 5.11 Dow Jones example, two measurements.

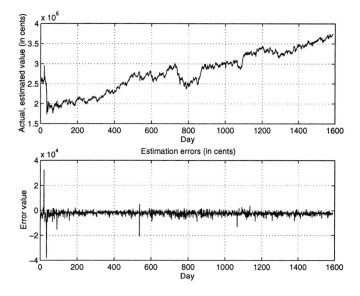

Figure 5.12 Dow Jones example, estimator $(O + H)/2$.

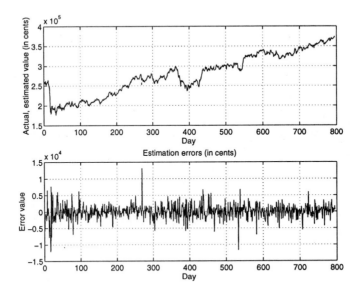

Figure 5.13   Dow Jones example, estimator O''.

one which results in the highest income at the end of the 5-year period) is the nonlinear estimator; however, all our estimators came close, and none was extremely profitable.

5.5   MAXIMUM-LIKELIHOOD ESTIMATION

This estimation criterion is very heavily used in communication channels. Here we present an abstract channel model only since the emphasis is on the new estimation criterion, rather than the channel application itself.

Consider the *noisy communication channel* shown in Figure 2.3. The input to the channel is a random sequence of 0 and 1 bits. Because of noise inside the channel, these bits are not received correctly all the time. Occasionally a bit inversion and a consequent detection error occur. The function of the receiver is to decide what bit was actually sent by the transmitter, based on the received bit.

More precisely, the estimation problem at the receiver is the following. Let X, Y denote the values of a specific received and transmitted bit, respectively. Estimate the value of the random variable Y based on the value of the received bit $X = x$. The estimator function $\hat{Y}(x)$ is chosen (among a set of allowable functions) so that the probability of zero error is maximized:

$$\max_{\hat{Y}(x)} P[Y - \hat{Y}(x) = 0 | X = x]. \tag{5.26}$$

The estimator that maximizes 5.26 is known as a *maximum-likelihood estimator (MLE)*. (If the random variables are continuous, pdf's instead of probabilities are used.) A criterion similar to 5.26 is also used in statistics, in the problem of processing multiple measurements for estimating an unknown parameter (such as a mean or variance), instead of a random variable.

Let's see some examples next.

Example 5.8. *Two Independent Random Variables.* Let X and Y be two independent random variables. We saw that the MMSE estimate of Y given a measurement $X = x$ is simply the mean value of Y, or EY. From 5.26, since X and Y are independent, we can see immediately that

$$P[Y - \hat{Y}(X) = 0 | X = x] = P[Y - \hat{Y}(x) = 0 | X = x] = P[Y = \hat{Y}(x)]. \tag{5.27}$$

The form of the best (in the MLE sense) function, and at the same time an algorithm for processing the measurement x, should now be obvious: Given any measurement x, map it *always* to the same value y^* which maximizes the marginal pmf or pdf of Y. As an example, let X and Y be Poisson-distributed with parameters $\lambda_1 = 1$ and $\lambda_2 = 10$, respectively. Since the mass function $P[Y = k]$ of Y attains its maximum at the point $k = 10$, the best MLE for all x is

$$\hat{Y}(x) = 10.$$

Using the random number generator of Chapter 3 for two independent Poisson random variables, we generated sample values for X and Y with the above-mentioned parameters. Figure 5.14 shows some of the errors involved in this estimation. It is instructive to compare the performance of this estimator to the performance of the MMSE estimator.

Example 5.9. *Binary Symmetric Channel (BSC).* Let's work out in detail the MLE for the channel in Figure 2.3. Based on our notation, $P[X = 0 | Y = 1] = P[X = 1 | Y = 0] = \epsilon$. Suppose that $x = 0$; that is, a 0 bit was received. Let $A \triangleq \max_{\hat{Y}(0)} P[Y - \hat{Y}(0) = 0 | X = 0]$. Since $\hat{Y}(0)$ can take the values of 0 and 1 only, from 5.26 we have

$$
\begin{aligned}
A &= \max\{P[Y - 0 = 0 | X = 0], P[Y - 1 = 0 | X = 0]\} \\
&= \max\{P[Y = 0 | X = 0], P[Y = 1 | X = 0]\} = \max\{1 - \epsilon, \epsilon\}.
\end{aligned}
$$

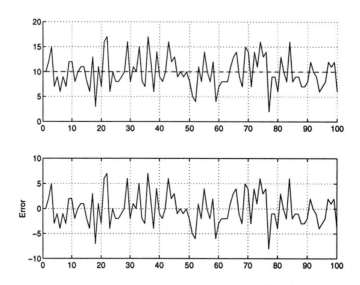

Figure 5.14   MLE estimation, independent Poisson random variables.

(The last equality follows from Bayes' rule.) Therefore, when $x = 0$, the optimal MLE is given by

$$\hat{Y}(0) = \begin{cases} 0, & 0.5 \geq \epsilon, \\ 1, & 0.5 < \epsilon. \end{cases}$$

Similarly, when $x = 1$,

$$\hat{Y}(1) = \begin{cases} 1, & 0.5 \geq \epsilon, \\ 0, & 0.5 < \epsilon. \end{cases}$$

5.6   AN HISTORICAL REMARK

The MMSE approach was heralded as one of the most significant advances in statistics and probability in the nineteenth century. As we see in Chapter 9, its extension to random processes has made it a cornerstone of the development of the theory in the twentieth century as well. It is worthwhile then to trace how the approach was originally conceived and developed.

The first appearance of the method dates back to 1805, when Legendre[5] used it to solve a problem in cometary orbits. The problem, in a more abstract setting, is the following: Suppose that we take (noisy) observations $\{x_i, y_i\}_{i=1}^n$.

---

[5] Adrien Marie Legendre, 1752–1833, a French mathematician.

We want to fit a straight line $y = ax + b$ to those observations, where a, b are unknown constants, the best value of which we have to determine. In general, the linear system

$$y_i = ax_i + b, \quad i = 1, \ldots, n,$$

is overdetermined, since $n \gg 2$, in general. Legendre realized that any solution $(\bar{a}, \bar{b})$ would inevitably introduce an error; let

$$E_i \overset{\triangle}{=} y_i - \bar{a}x_i - \bar{b}, \quad i = 1, \ldots, n,$$

denote the value of this error when the experimental observation (x_i, y_i) is used. Legendre chose (without any justification) to consider the sum of the squared errors

$$E(\bar{a}, \bar{b}) \overset{\triangle}{=} \sum_{i=1}^{n} E_i^2$$

and then determine the sought-after values (a^*, b^*) so that $E(a^*, b^*)$ is a minimum. Legendre's approach was dubbed the *method of least squares*. In view of the modern exposition of the problem, he tacitly assumed that all errors were equiprobable, and he dealt with the best linear estimator only.

As early as the middle eighteenth century, scientists were studying models for capturing the essential properties of errors in observations (the primary force behind this being astronomical observations). In 1755, Simpson[6] proposed what we would call today a *discrete triangular pmf* (see Problem 3.116); he also gave an example in which taking the mean of n observations would reduce the error. A few years later, in the 1770s, Laplace spelled out the two most desirable properties of an error pdf: It should be symmetric around zero since both positive and negative error values of magnitude x are equally likely; and it should approach zero as $|x| \to \infty$ since extremely large errors are not likely. Then he presented what we call today the *Laplace pdf*, as a model for error functions.

In the early nineteenth century, Gauss had the idea of combining Laplace's error function properties with Legendre's least squares approach. His crack at the problem went as follows.

Let $\{X_i\}_{i=1}^{n}$ denote measurements of a quantity Y. There is an error $E_i = Y - X_i$ associated with measurement X_i. Gauss assumed that the errors were identically distributed and independent, with pdf $\phi(x)$; the pdf should satisfy Laplace's postulated properties. The probability $P[E]$ of the error's assuming values $\{E_i\}_{i=1}^{n}$ is thus

$$P[E] = \prod_{i=1}^{n} \phi(E_i).$$

---

[6] Thomas Simpson, 1710–1761, a British mathematician.

Gauss looked for estimates that would maximize $P[E]$, called the *likelihood function*, thus providing the justification missing from Legendre's approach to estimation. By that time, belief in taking the arithmetic mean of n observations was quite strong, so Gauss assumed that the quantity $1/n \sum_i X_i$ would achieve the maximum $P[E]$. He then proceeded as follows: Let $\bar{x} = \frac{1}{n} \sum_{i=1}^{n} X_i$ be the arithmetic mean of the measurements. The maximum occurs of course when

$$\frac{\partial \log(P[E])}{\partial \bar{x}} = 0,$$

since $P[E]$ is a product. Thus, the necessary condition is

$$\frac{\partial \phi(E_1)/\partial \bar{x}}{\phi(E_1)} + \frac{\partial \phi(E_2)/\partial \bar{x}}{\phi(E_2)} + \cdots + \frac{\partial \phi(E_n)/\partial \bar{x}}{\phi(E_n)} = 0.$$

Now, from the chain rule for derivatives, for all i,

$$\frac{\partial \phi(E_i)}{\partial \bar{x}} = \frac{\partial \phi(E_i)}{\partial E_i} \frac{\partial E_i}{\partial \bar{x}} = \phi'(E_i) \frac{\partial E_i}{\partial \bar{x}}.$$

Since $\frac{\partial E_i}{\partial \bar{x}} = -\frac{1}{n}$, we get

$$\frac{\phi'(E_1)}{\phi(E_1)} + \frac{\phi'(E_2)}{\phi(E_2)} + \cdots + \frac{\phi'(E_n)}{\phi(E_n)} = 0.$$

Since the measurements $X_i, i = 2, \ldots, n$, appear only as the sum $\sum_{i=2}^{n} X_i$, Gauss supposed that each one was equal to $X_1 - nd$, where d was some constant. Then $E_1 = (n-1)d$, $E_i = -d$, $i = 2, \ldots, n$, and we have

$$\frac{\phi'((n-1)d)}{\phi((n-1)d)} = (1-n)\frac{\phi'(-d)}{\phi(-d)}.$$

Then Gauss deduced that for all $x \in R$,

$$\frac{\phi'(x)}{x\phi(x)} = c,$$

for some constant c. Thus,

$$\phi(x) = ae^{\frac{1}{2}cx^2}$$

for some constants a, c. The symmetry condition of Laplace is obviously satisfied. The asymptotic condition is satisfied by choosing a negative c, say, $c = -1/\sigma^2$, in today's notation. Finally the constant a is determined from the normalizing condition

$$\int_{-\infty}^{\infty} \phi(x)\,dx = 1,$$

and finally

$$\phi(x) = \frac{1}{\sqrt{2\pi}\sigma} e^{-\frac{x^2}{2\sigma^2}}.$$

Now,

$$P[E] = \frac{1}{(\sqrt{2\pi}\sigma)^n} e^{-\frac{E_1^2 + \cdots + E_n^2}{2\sigma^2}}$$

and thus the "natural" choice of maximizing the likelihood of errors leads to Legendre's minimization of the sum of the square errors.

5.7 SUMMARY OF MAIN POINTS

- **Minimum mean square error** (MMSE) is a tractable criterion for estimating the values of one random variable in terms of measured values of another.

- The **orthogonality principle** provides us with necessary conditions that the best estimator parameters must satisfy.

- **Linear estimators** are very easy to implement.

- The best (MMSE) **nonlinear estimator** is the **conditional expectation** $E(Y|X)$. In general, it is very difficult to implement.

- **Maximum likelihood** is another criterion for estimation, which yields easy-to-implement estimators for some problems.

5.8 CHECKLIST OF IMPORTANT TOOLS

- The orthogonality principle

- The linear and nonlinear estimators

- The maximum-likelihood estimator

5.9 PROBLEMS

Estimation Criteria

5.1 *What If?* Consider the criterion shown in Figure 5.3(a). We will investigate the mathematical difficulties associated with this criterion.

(a) Write down a mathematical expression for the average value of the penalty function. (*Hint:* Use step functions.)

(b) For simplicity, consider estimators in the class G_l of linear functions only, so take $g(X) = aX$, with $a \in R$. Let $T = 0.5$. Assume that the random variables X, Y are independent Bernoulli with parameter p. Evaluate the average error.

(c) The average error is a nonlinear function of a. Solve this function for the best value of a.

(d) Identify precisely where the difficulty comes from.

(e) Discuss how this difficulty will be compounded when nonlinear functions $g(X)$ and general jpdf's $f_{XY}(x, y)$ enter the picture.

5.2 Consider the criterion shown in Figure 5.3(b). Repeat Problem 5.1.

5.3 Consider the criterion in Equation 5.3. Discuss the problems that arise from the fact that the function is not differentiable at zero.

5.4 Consider the criterion in Inequality 5.1 and the binary symmetric channel in Section 5.5. Show that $g(X(\zeta))$ must be equal to 1 for all ζ, in order to satisfy Inequality 5.1. Thus this criterion does not make sense.

5.5 Consider the following estimation criterion. The random variable Y must be estimated as a linear function of another random variable X, in the form $Y = aX$. We wish to use the error criterion $E(Y - aX)^3$. What value of the real number a minimizes this error? Explain.

5.6 *Why?* In many situations, credit companies use the mortgage of a person to determine his or her income. This is an estimation procedure. Explain why companies would prefer to rely on mortgages and not incomes.

5.7 *Hmmm!* Is stereotyping like estimating a random variable with a constant?

Minimum Mean Square Estimation; Mathematical Skills

5.8 Complete the proof of Equation 5.11.

5.9 Show that $E\hat{Y} = EY$, where $E\hat{Y}$ is given by Equation 5.15. Calculate the variance of $\hat{Y}$ and the covariance $\text{cov}(X, \hat{Y})$.

5.10 Let $Y = X^k$, where k is a fixed positive integer. Find the MMSE linear and nonlinear estimates of Y in terms of moments of X.

5.11 Repeat Problem 5.10 for the random variable $Y = \sin(X)$.

5.12 Repeat Problem 5.10 for the random variable $Y = e^{j\omega X}$, where ω is a fixed constant.

5.13 Let X, Y be independent exponential random variables with parameter $\lambda = 1$. Let

$$Z = \frac{X^2 - E(Y|X)}{\text{var}(Y)}.$$

(a) What is the best linear estimator of Y given Z?
(b) What is the best nonlinear estimator of Y given Z?

5.14 Consider a nonlinear estimator of the form $\hat{Y} = aX^2 + bX + c$. Find the best MMSE values for the constants a, b, c.

5.15 *Do It!* Compare the performance of the estimator of Problem 5.14 to the performance of the best linear estimator and the best nonlinear estimator, using some actual data.

5.16 *Do It!* Calculate the pdf of the best linear estimator experimentally, using your own data.

5.17 *Do It!* Consider the random variables in Example 5.1, Section 5.2.2. Calculate the minimum variance of the best linear estimate, both theoretically and by generating 10,000 samples of the random variables involved.

Multiple Measurements

5.18 *Mathematical Skills.* Calculate the minimum error variance in the linear prediction, two measurements case, in Section 5.3.1, either theoretically or experimentally. Compare this variance to the single-measurement case.

5.19 We want to estimate a random variable X based on multiple measurements $Y_1, \ldots, Y_n$, of another random variable Y. The two random variables X, Y_i have a known given joint pdf that is independent of i.
(a) Evaluate the best estimator of X based on $n = 2, 3, 5$ measurements of Y. Assume a joint pdf for X, Y_i. Then:
(b) Plot the resulting errors.
(c) Plot histograms of the errors.
(d) Calculate the error variances.

5.20 *Concepts.* Can you design an experiment based on the Unix command *ping* to estimate the round-trip propagation delay between your campus and the east or west coast? (*Hint:* The total delay a packet will experience in a network is the sum of four distinct delays: propagation, transmission, queuing, and processing delays.)
(a) Discuss the difficulties of your approach. If you decide that the theory we developed does not apply, suggest how it should be modified to accommodate the needs of this problem.

(b) If you had control over the network operation, how would you change things to improve your estimate?

5.21 *Huh?* Is it better to get one house appraisal or more than one? Formulate this question as an estimation problem. Carefully define the criterion of error.

5.22 Suppose we take two house appraisals X_1 and X_2. We form two estimates of the actual house value X_1 and $(X_1 + X_2)/2$. Under what conditions on X_1 and X_2 and what criterion, is $(X_1 + X_2)/2$ a better estimate than X_1 alone?

Projects

5.23 Consider the Dow Jones data in the Web site. We want to estimate C, the closing value, using only half the data, say, the first half, of the opening value.
(a) Plot the resulting errors.
(b) Plot the histogram of the error random variable.
(c) Calculate the mean square error. Explain any differences you may have from the case where you use all the data points.

5.24 Repeat Problem 5.23 with the second half of the data.

5.25 Repeat Problems 5.23 and 5.24 for estimating the closing random variable with measurements of the volume random variable.

5.26 Repeat Problem 5.23, using only Monday's measurements for the opening value.

5.27 Model the Dow Jones data (say, the closing value), using five random variables $X_1, \ldots, X_5$, one for each day of the week. Answer the questions of Problem 5.23 for estimating X_1 from measurements of X_5.

5.28 Let $X_1, \ldots, X_5$ be as in Problem 5.27. Let

$$X = \frac{X_1 + X_2 + X_3 + X_4 + X_5}{5}.$$

Answer the questions in Problem 5.23 for estimating X from measurements of X_i, $i = 1, 2, 3, 4, 5$.

5.29 Let Y be the average value of the month for, say, the closing value of the Dow Jones index. Let X, $X_1, \ldots, X_5$ be defined as in Problem 5.28. Which measurements give a better estimate of Y, the weekly ones X or the daily ones X_i? Make a guess and prove or disprove it, using the actual data.

5.30 Consider the following investment strategy. We buy only on days for which the opening value was less than the opening value of the previous day.

We sell only when we predict that the closing value of the day will be higher than the closing value of the previous day. Evaluate the potential gain of this strategy, using the actual data.

5.31 Define your own investment strategy and evaluate its performance, using the actual data.

5.32 Estimate the Monday opening value based on the average weekly opening value of the previous week.

5.33 Suppose you measure both the opening (O) and high (H) values of the Dow Jones index.
(a) Find the best linear MMSE estimate of the closing value C, based on these two measurements.
(b) Plot the resulting errors.
(c) Plot a histogram of the errors.
(d) Calculate the minimum mean square error.

5.34 Repeat Problem 5.33, using measurements of the $(O+H)/2$ and O values. Is the new estimate better than the previous one?

5.35 Consider a new criterion for evaluating estimation of Dow Jones data. Let $X, \hat{X}$ be the estimated random variable and its estimate. Define the average error as
$$E\max\{X - \hat{X}, 0\}.$$
In other words, negative errors are reset to zero. Find experimentally, using the actual data, a good estimator according to this criterion. Can you find the optimal estimator theoretically?

5.36 Consider the buffer project data in the Web site. Find the best constant, linear, and nonlinear estimators of the delay random variable, using measurements of the transmission times.

5.37 Using a suitable compression command (e.g., *compress* on Unix workstations), compress a number N of files in your system. Let X, Y denote the size of a file before and after compression. For the MMSE criterion, derive and plot:
(a) The best constant estimator of Y, based on a single measurement of X.
(b) The best linear estimator of Y, based on X.
(c) The best nonlinear estimator of Y, based on X.
(d) Calculate the variances of all estimation errors.
(e) Repeat questions (a)–(d) for $2N$ files.

5.38 Consider the channel project data in the Web site. Let Y, X denote the value of the transmitted and received bit, respectively. Repeat questions (a)–(d) of Problem 5.37 for the MMSE criterion.

5.39 Let X, Y be the random variables in Problem 5.38.
(a) Find the best estimator of X according to the MLE criterion.
(b) Find the best estimator of Y according to the MLE criterion.

5.40 Let X, Y be the random variables in Problem 5.38.
(a) Find the best estimator of X according to the criterion shown in Figure 5.3(a).
(b) Find the best estimator of Y according to the same criterion.
(c) Which of the three estimation criteria gives the "smallest" errors? (You have to define what *smallest* means to you.)

5.41 Consider the *ftp* project data in the Web site. Let X, Y denote the throughput and delay random variables, respectively. Repeat Problem 5.37.

5.42 Consider the *ftp* project data in the Web site. Let X, Y, Z denote the throughput, delay, and file size random variables respectively.
(a) Find $\hat{X}$, the best linear estimator of X, based on Y, Z. Plot the estimation errors. Plot a histogram of the errors. Calculate the error variance.
(b) Set $W = Z/Y$, where W is a nonlinear estimate of X. Plot the estimation errors. Plot a histogram of the errors.
(c) Determine which throughput estimate is better, and explain why.

SEQUENCES OF (IID) RANDOM VARIABLES

6.1 EXPERIMENTS WITH AN UNBOUNDED NUMBER OF MEASUREMENTS

In a variety of experiments, the number of measurements concerning a random variable cannot be known in advance. For example, in the *netstat* experiment, we may not know how many 1-s intervals we will measure. In the Dow Jones example, we may be interested in observing opening, closing, and high values for the indefinite future. In such cases, we cannot utilize a random vector model with a *known* dimensionality N since we may exceed that bound in our measurement-taking process.

In essence, in such cases we have an experiment in which an unbounded number of measurements and therefore random variables must be defined. The random vector model must be now augmented to that of an infinite sequence $\{X_1, \ldots, X_n, \ldots\}$; in this chapter, we introduce this notion and discuss some fundamental theorems concerning the behavior of sequences of random variables.

In Section 6.2 we discuss some preliminary notions about product spaces and independent, identically distributed (IID) random variables, the simplest but also most important case of sequences of random variables. In Section 6.3 we introduce sums of IID random variables, a very important topic, with many applications, mainly in statistics and queuing theory. In Section 6.4 we introduce the concept of random sums. In Section 6.5 we present the Weak Law of Large Numbers, a result of historical significance. In Section 6.6 we introduce the Strong Law of Large Numbers, one of the masterpieces of Kolmogorov's work and a result that is very widely used in practice. In Section 6.7 we discuss the central limit theorem, the cornerstone of statistics. In Section 6.8

we introduce various modes of convergence for sequences of random variables, not necessarily IID. Finally, in Section 6.9, we present two widely used results about infinite sequences of events.

6.2   IID RANDOM VARIABLES

In a large number of experiments, the next trial does not depend on the current trial or the previous ones. Based on our discussion in Section 4.7.3, a *product space* is a natural sample space for such a case. Since the probabilistic behavior of the experiment does not change from trial to trial, the random variable X_i is independent of the random variable X_j, for all $j \neq i$. Moreover, the cdf's of all random variables are identical. This observation leads to the following definition:

Definition: A sequence of random variables $\{X_i\}_{i=1}^{\infty}$ is called a sequence of **independent, identically distributed (IID)** random variables if the following conditions are met:

- $F_{X_i}(x) = F(x), \quad \forall x \in R, \ i = 1, 2, \ldots.$

- A collection of random variables $\{X_i\}$ is independent of any collection $\{X_j\}$ as long as there are no common elements in the two collections.

The important thing about IID random variables is that we need only one cdf to describe the statistical behavior of any collection of random variables. Indeed, let $C \subset \{1, 2, \ldots\}$ be an arbitrary (i.e., finite or infinite) collection of integers. Let $\mathbf{X} \triangleq \{X_i\}_{i \in C}$ be an arbitrary family of IID random variables. Then the joint cdf of this family is given by

$$F_{\mathbf{X}}(\ldots, x_i, \ldots) = \prod_{i \in C} F(x_i), \tag{6.1}$$

and (at least conceptually) all probabilities involving random variables in C can be evaluated from Equation 6.1.

Remark: Note that no dependence is allowed between members of an IID sequence. So such a model cannot describe our familiar "systems with memory" where future system states depend on the past. We discuss other models that can capture such dependencies (namely, Markov Chains and random processes) in later chapters.

6.2.1   Useful Models of IID Random Variables

What are some useful models for IID random variables? Even though an infinite sequence of random variables is involved, for such models we need only specify a *single* pdf! Any pdf from Chapter 3 will do; for this reason, we do not repeat any specific model here. As expected, the Gaussian, Poisson, Bernoulli, χ^2, etc., random variables in Chapter 3 are still among the most important models.

6.3   SUMS OF IID RANDOM VARIABLES

Sums of random variables arise very naturally in a large number of applications. For example, in the Dow Jones game, the total gain from an investment strategy is the sum of the daily gains. In the *netstat* experiment, the cumulative number of errors or collisions in a time period is the sum of the errors or collisions over all the 1-s intervals that comprise the total time period. In statistics, the most widely used parameter estimators for means and variances involve summations of random variables. Finally, we mention the very fundamental model of *random walks* (see Section 7.5.1), which is again defined in terms of sums of random variables. A lot can be said about such sums when the random variables involved are IID. We focus primarily on such sums in this and the next section.

More specifically, consider a sequence of independent, identically distributed random variables $\{X_i\}_{i=1}^{\infty}$. In the following two definitions, n is a fixed integer.

Definition: The random variable

$$S_n \triangleq \sum_{i=1}^{n} X_i$$

is called the **sum** of the n IID random variables $\{X_i\}_{i=1}^{n}$.

Definition: The random variable

$$M_n \triangleq \frac{1}{n} \sum_{i=1}^{n} X_i = \frac{1}{n} S_n$$

is called the **normalized sum** of the n IID random variables $\{X_i\}_{i=1}^{n}$.

Remark: Variable M_n is also known as the arithmetic mean of the values X_i. Lagrange used it in estimation problems, even though he did not give any rationale for doing so. As we saw in the previous chapter, Gauss also used it in deriving the Gaussian pdf.

In this section we determine the characteristic function, pdf, mean, and variance of these two random variables. In Section 6.4 we do the same for the case where n itself is a random variable. In the next two sections we evaluate the limiting behavior of M_n as $n \to \infty$.

6.3.1   Characteristic Function of S_n, M_n

We can easily determine the characteristic function. From its definition, we have

$$
\begin{aligned}
\Phi_{S_n}(\omega) \quad &\triangleq \quad Ee^{j\omega S_n} = Ee^{j\omega(X_1+\cdots+X_n)} = E(e^{j\omega X_1})E(e^{j\omega X_2})\cdots E(e^{j\omega X_n}) \\
&= \quad \Phi_{X_1}(\omega)\Phi_{X_2}(\omega)\cdots\Phi_{X_n}(\omega) \tag{6.2}
\end{aligned}
$$

Equation 6.2 holds true for independent random variables, not necessarily identically distributed. Of course, since the random variables are identically distributed, we have

$$
\Phi_{S_n}(\omega) = [\Phi_X(\omega)]^n, \tag{6.3}
$$

where $\Phi_X(\omega)$ denotes the common characteristic function. The characteristic function of the random variable M_n is found easily from Theorem 3.2 in Section 3.8:

$$
\Phi_{M_n}(\omega) = \left[\Phi_X\left(\frac{\omega}{n}\right)\right]^n. \tag{6.4}
$$

6.3.2   Probability Density Function of S_n, M_n

The pdf of the random variables S_n, M_n is in principle determined by inverting the characteristic functions in Equations 6.3 and 6.4. Therefore,

$$
f_{S_n}(x) = \mathcal{F}^{-1}\{[\Phi_X(\omega)]^n\} = \frac{1}{2\pi}\int_{-\infty}^{\infty} e^{-j\omega x}[\Phi_X(\omega)]^n\, d\omega, \tag{6.5}
$$

where $\mathcal{F}^{-1}$ denotes the inverse transform. Similarly,

$$
f_{M_n}(x) = \mathcal{F}^{-1}\{[\Phi_X(\frac{\omega}{n})]^n\} = \frac{1}{2\pi}\int_{-\infty}^{\infty} e^{-j\omega x}[\Phi_X(\frac{\omega}{n})]^n\, d\omega. \tag{6.6}
$$

Of course, in the general case, evaluating the integrals in Equations 6.5 and 6.6 is not trivial. So most often we settle for the partial information we discuss in the next section.

6.3.3   Mean and Variance of S_n, M_n

The mean of S_n is easily calculated as

$$
ES_n = EX_1 + EX_2 + \cdots + EX_n = nEX, \tag{6.7}
$$

where EX denotes the common mean. Equation 6.7 holds true even for dependent random variables, as long as they are identically distributed (see Equation 4.52). The mean of M_n is given by

$$
EM_n = \frac{1}{n}ES_n = EX. \tag{6.8}
$$

The variance of S_n is easily calculated from Equation 4.53 since the random variables are independent:

$$\text{var}(S_n) = \sum_{i=1}^{n} \text{var}(X_i) = n \cdot \text{var}(X). \qquad (6.9)$$

From Property 3.55 we immediately get

$$\text{var}(M_n) = \frac{\text{var}(X)}{n}. \qquad (6.10)$$

6.3.4 Examples

Let's look at some examples for specific distributions. In all of them, X represents a generic random variable that has the same pdf as the sequence $\{X_i\}$.

Example 6.1. *Gaussian Random Variables.* In this case,

$$\Phi_X(\omega) = e^{jm\omega - \frac{\sigma^2 \omega^2}{2}},$$

so

$$\Phi_{S_n}(\omega) = e^{nmj\omega - \frac{n\sigma^2\omega^2}{2}}, \quad \Phi_{M_n}(\omega) = e^{mj\omega - \frac{\sigma^2\omega^2}{2n}},$$

$$f_{S_n}(x) = \frac{1}{\sqrt{2\pi n}\sigma} e^{-\frac{(x-nm)^2}{2n\sigma^2}}, \quad f_{M_n}(x) = \frac{\sqrt{n}}{\sqrt{2\pi}\sigma} e^{-\frac{n(x-m)^2}{2\sigma^2}},$$

$$ES_n = nm, \quad EM_n = m, \quad \text{var}(S_n) = n\sigma^2, \quad \text{var}(M_n) = \frac{\sigma^2}{n}.$$

From the above expressions we recognize both S_n and M_n as Gaussian random variables, as expected (see Table 3.4).

Example 6.2. *Poisson Random Variables.* In this case,

$$\Phi_X(\omega) = e^{\lambda(e^{j\omega} - 1)},$$

so

$$\Phi_{S_n}(\omega) = e^{n\lambda(e^{j\omega} - 1)}, \quad \Phi_{M_n}(\omega) = e^{n\lambda(e^{j\frac{\omega}{n}} - 1)},$$

$$P[S_n = k] = \frac{(n\lambda)^k}{k!} e^{-n\lambda}, \quad k = 0, 1, 2, \ldots,$$

$$ES_n = n\lambda, \quad EM_n = \lambda, \quad \text{var}(S_n) = n\lambda, \quad \text{var}(M_n) = \frac{\lambda}{n}.$$

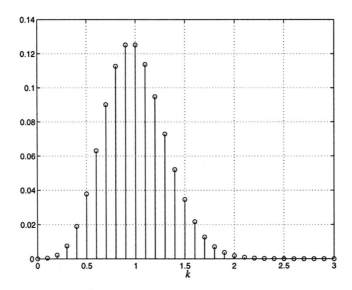

Figure 6.1   The pmf of M_n, $n = 10$, $\lambda = 1$, Poisson case.

We cannot easily invert $\Phi_{M_n}(\omega)$, in order to derive the pmf of M_n. However, from the definition of M_n,

$$P[M_n = k] = P[S_n = k \cdot n] = \begin{cases} e^{-n\lambda} \frac{(n\lambda)^{kn}}{(kn)!}, & k \cdot n \text{ is integer,} \\ 0, & \text{otherwise.} \end{cases}$$

In Figure 6.1 we plot this pmf for $\lambda = 1$ and $n = 10$.

Example 6.3. *Exponential Random Variables.* The characteristic function of an exponential random variable is given by

$$\Phi_X(\omega) = \frac{\lambda}{\lambda - j\omega},$$

so

$$\Phi_{S_n}(\omega) = \left(\frac{\lambda}{\lambda - j\omega}\right)^n, \quad \Phi_{M_n}(\omega) = \left(\frac{\lambda}{\lambda - j\omega/n}\right)^n,$$

$$f_{S_n}(x) = \frac{\lambda(\lambda x)^{n-1}e^{-\lambda x}}{(n-1)!},$$

$$ES_n = \frac{n}{\lambda}, \quad EM_n = \frac{1}{\lambda}, \quad \text{var}(S_n) = \frac{n}{\lambda^2}, \quad \text{var}(M_n) = \frac{1}{n\lambda^2}.$$

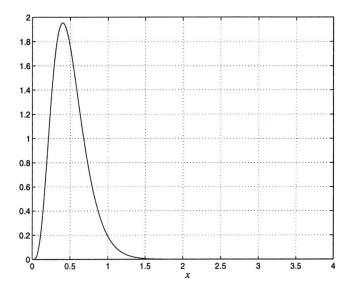

Figure 6.2 The pdf of M_n, $n = 5, \lambda = 2$, exponential case.

We can determine the pdf of M_n from that of S_n since the two random variables are linearly related. From Theorem 3.2

$$f_{M_n}(x) = n\frac{\lambda(\lambda x n)^{n-1}e^{-\lambda x n}}{(n-1)!}.$$

In Figure 6.2 we plot this pdf for $\lambda = 2$ and $n = 5$.

Example 6.4. *Bernoulli Random Variables.* We have

$$P[X = 0] = 1 - p, \quad P[X = 1] = p,$$

so

$$\Phi_{S_n}(\omega) = (1 - p + pe^{j\omega})^n, \quad \Phi_{M_n}(\omega) = (1 - p + pe^{j\omega/n})^n,$$

$$P[S_n = k] = \binom{n}{k}p^k(1 - p)^{n-k}, \quad k = 0, 1, \ldots, n,$$

$$ES_n = np, \quad EM_n = p, \quad \text{var}(S_n) = np(1 - p), \quad \text{var}(M_n) = \frac{p(1 - p)}{n}.$$

The pmf of M_n is given by

$$P[M_n = k] = P[S_n = nk] = \begin{cases} \binom{n}{nk}p^{nk}(1 - p)^{n-nk}, & n \cdot k = 0, 1, \ldots, n, \\ 0, & \text{otherwise.} \end{cases}$$

This pmf is shown in Figure 6.3 for the special case $p = 0.5, n = 10$.

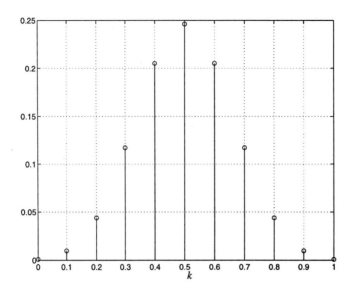

Figure 6.3   The pdf of M_n, $n = 10, p = 0.5$, Bernoulli case.

6.4   RANDOM SUMS OF IID RANDOM VARIABLES

In the preceding section, the number of random variables we were adding in S_n or M_n was fixed. What happens when this number is itself random? Situations in which *random sums* are of interest arise quite naturally in telecommunications, economics, etc. For example, the total number of errors in the transmission of a random size file is a random sum. The total number of packets that arrive in a router, while a random size packet is transmitted, is another random sum.

Mathematically, let N denote a positive integer-valued random variable with generating function $G_N(z)$, mean EN, and variance var(N). The quantity

$$S_N \overset{\triangle}{=} \sum_{i=1}^{N} X_i$$

is called a **random sum**. We evaluate next the characteristic function, mean, and variance of S_N.

Suppose at first that N is independent of $\{X_i\}_{i=1}^{\infty}$. The trick that simplifies our calculations is conditional expectations (Property 4.54). Indeed,

$$
\begin{aligned}
\Phi_{S_N}(\omega) \;\triangleq\; & Ee^{j\omega S_N} = E[E\{e^{j\omega S_N}|N=n\}] \\
=\; & E[E\{e^{j\omega(X_1+\cdots+X_n)}|N=n\}] = E[E\{e^{j\omega(X_1+\cdots+X_n)}\}] \\
=\; & E[Ee^{j\omega X_1}]\cdots E[Ee^{j\omega X_n}] = E[[\Phi_X(\omega)]^n] \\
=\; & E[z^N]\big|_{z=\Phi_X(\omega)} = G_N(\Phi_X(\omega))
\end{aligned}
\tag{6.11}
$$

In the first equality we simply use Property 4.54; in the second equality we write out the definition of S_n; in the third we use the fact that the X_i random variables are independent of N; in the fourth we use the independence property of the sequence; in the fifth we use the identical distribution property of the sequence; and finally, in the last two equalities we use the definition of the generating function of N.

We can similarly obtain the mean and variance:

$$
\begin{aligned}
ES_N \;=\; & E[E[S_N|N=n]] = E[E[S_n|N=n]] \\
=\; & E[nE(X)|N=n] = E[NE(X)] = E(N)E(X).
\end{aligned}
\tag{6.12}
$$

$$
\begin{aligned}
\operatorname{var}(S_N) \;\triangleq\; & E[(S_N - ES_N)^2] = E[(S_N - ENEX)^2] \\
=\; & E[E\{(S_N - ENEX)^2|N=n\}] \\
=\; & E[E\{(S_n - ENEX)^2|N=n\}] \\
=\; & E[E[(S_n - nEX + nEX - ENEX)^2|N=n]] \\
=\; & E[E(S_n - nEX)^2 - 2E[(S_n - nEX)(n - EN)EX] \\
& + E[(n - EN)^2(EX)^2]] \\
=\; & E[n \cdot \operatorname{var}(X)] + (EX)^2\operatorname{var}(N) \\
=\; & EN \cdot \operatorname{var}(X) + (EX)^2\operatorname{var}(N).
\end{aligned}
$$

The first equality for the variance follows immediately from Equation 6.12; the second and third come from our conditioning trick; in the fourth we simply add and subtract nEX; in the fifth we expand the square. In the sixth we use independence of N and S_n and the fact that $E(S_n - nEX) = 0$.

We summarize these findings as a theorem:

Theorem 6.1 *Let N denote a positive integer-valued random variable with probability generating function $G_N(z)$, mean EN, and variance $\operatorname{var}(N)$. Let $\{X_i\}_{i=1}^{\infty}$ be a sequence of IID random variables, independent of N, with characteristic function $\Phi_X(\omega)$, mean EX, and variance $\operatorname{var}(X)$. Let $S_N = \sum_{i=1}^{N} X_i$. Then*

$$
\Phi_{S_N}(\omega) \;=\; G_N(\Phi_X(\omega)),
$$

$$ES_N = E(N)E(X),$$
$$\text{var}(S_N) = EN\text{var}(X) + \text{var}(N)(EX)^2.$$

Let's see an example next.

Example 6.5. *Geometric Sums of Exponential Random Variables.* In this case, N is geometrically distributed, and X is exponential with parameter λ. Therefore,

$$G_N(z) = \frac{p}{1 - (1-p)z},$$

$$\Phi_X(\omega) = \frac{\lambda}{\lambda - j\omega}.$$

Therefore, from Equation 6.11

$$\Phi_{S_N}(\omega) = \frac{p}{1 - (1-p)\frac{\lambda}{\lambda - j\omega}} = p + (1-p)\frac{p\lambda}{p\lambda - j\omega}. \tag{6.13}$$

The inverse Fourier transform of Equation 6.13 is rather easy:

$$f_{S_N}(x) = p\delta(x) + (1-p)p\lambda e^{-p\lambda x} u(x).$$

Can you recognize S_N as a mixture random variable? What are the components of the mix?

6.4.1   Wald's Equalities

In Theorem 6.1, N was assumed independent of the sequence. The next two theorems address the case where N depends on the sequence. It should be clear that when this dependence is arbitrary, there is no hope for developing generic formulas like those in Theorem 6.1. Counterexamples can be constructed easily. A very broad and quite useful form of dependence, for which general results can be obtained, is described in the definition of a *stopping time*.

Consider a sequence of IID random variables $\{X_i\}_{i=1}^{\infty}$. Let σ_n denote the σ algebra generated by $\{X_1, \ldots, X_n\}$. (This is the set of all events that involve the random variables $X_1, \ldots, X_n$.) From Problem A.42, we know that $\{\sigma_i\}_{i=1}^{\infty}$ is an increasing sequence of sets. Consider a random variable T, taking values in the set $\{1, 2, 3, \ldots, \infty\}$.

Definition: The random variable T with values in $\{1, 2, 3, \ldots, \infty\}$ is called a **stopping time** (with respect to the sequence $\{\sigma_n\}_{n=1}^{\infty}$) if, for all positive integers k, the event $\{T = k\} \in \sigma_k$.

In other words, the event $\{T = k\}$ depends only on the first k variables from the sequence $\{X_i\}_{i=1}^{\infty}$. We explore some further properties of stopping times in Problems 6.22 and 6.23. The following two theorems are known as *Wald's equality for means* and *Wald's equality for second moments*.[1] Their proof is omitted since it is beyond the scope of this book. The interested reader may consult, for example, [11], in which more general versions are presented.

Theorem 6.2 *Consider a sequence of IID random variables $\{X_i\}_{i=1}^{\infty}$ with finite mean EX. Let T be a stopping time with $ET < \infty$. Then*

$$ES_T = EX \cdot ET.$$

Theorem 6.3 *Consider a sequence of IID random variables $\{X_i\}_{i=1}^{\infty}$ with mean $EX = 0$ and finite variance $\mathrm{var}(X) = \sigma^2$. Let T be a stopping time with $ET < \infty$. Then*

$$ES_T^2 = \sigma^2 \cdot ET.$$

Let's see an example next.

Example 6.6. *Busy Cycles in Queuing Systems.* Queues (or waiting lines) arise anywhere: in ticket booths for movie theaters and football games; in professors' offices; inside computer network nodes and computer memories. Specific examples of technical interest include buffers inside routers in computer networks and CPU schedulers.

Consider a packet buffer inside a router (see Example 1.27). The queue inside the router's memory stores packets that arrive unpredictably at input lines and, after processing, are forwarded to appropriate output lines. The buffer contents, measured in units of packets, change randomly as packets come and go. A typical scenario is depicted in Figure 3.17(a).

Loosely speaking, the *busy cycle B* is defined as the time interval during which the queue is not empty. The time interval during which the queue is empty is defined as the *idle cycle* (or idle period). Busy and idle cycles alternate in a queuing system. In Figure 6.4, the first busy cycle has a duration of $t_2 - t_1$ time units. During this busy cycle, the maximum queue size was 3. The first idle cycle started at time 0 and ended at time t_1, while the second had a duration of $t_3 - t_2$ time units.

The lengths of busy cycles are clearly random variables. Their properties (e.g., mean value, maximum value, variance) are of extreme interest in performance studies of queuing systems. For example, if the busy cycle is infinite, intuitively the queue will never be empty. This property can be interpreted as a sign of system instability in the router example.

---

[1] After Abraham Wald, 1902–1950, the father of sequential analysis and decision theory in statistics. He discovered these equalities in 1944.

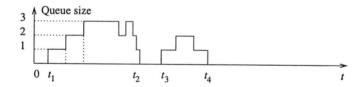

Figure 6.4   Busy and idle cycles.

To study the properties of busy cycles, we need a formal definition. Mathematically, we can express a busy cycle as a random sum. Let $\{X_i\}_{i=1}^{\infty}$ be a sequence of random variables (in the router example, X_i denotes the transmission time of the ith forwarded packet). Then we can express the length of a busy cycle as

$$B = \sum_{i=1}^{N} X_i,$$

where N is a random variable. The properties of B can be derived from the theorems of this section.

6.5   WEAK LAW OF LARGE NUMBERS

The two Laws of Large Numbers (Weak and Strong) deal with the behavior of the random variable M_n as n becomes arbitrarily large. The intuition behind them is the following. From Equation 6.10, note that the variance of M_n approaches 0 as n increases, which suggests that the density function of M_n becomes "narrower" and "narrower," and intuitively, we expect it to approach a delta function as $n \to \infty$. Then M_n should approach a constant as $n \to \infty$.

The Weak Law of Large Numbers (WLLN) is an easy consequence of Chebyshev's inequality. This law was first proved by James Bernoulli, in the late seventeenth century. It appeared in his seminal book *Ars Conjectandi* in 1713. Bernoulli's notation was not the one we use below since means and variances of random variables were not formally defined at the time; the scope of Bernoulli's result was also a bit more limited. Nevertheless, this result was remarkable since it was the *first theorem* of the young theory of probability!

Theorem 6.4 *Consider a sequence of IID random variables* $\{X_i\}_{i=1}^{\infty}$ *with finite mean* $EX = m$ *and finite variance* $\text{var}(X) = \sigma^2$. *Then for any fixed* $\epsilon > 0$

$$\lim_{n \to \infty} P[|M_n - m| < \epsilon] = 1. \qquad (6.14)$$

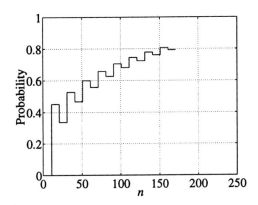

Figure 6.5   The WLLN for Bernoulli random variable.

Proof: Apply Chebyshev's inequality (see Inequality 3.84, in Chapter 3) to the random variable M_n. Since

$$EM_n = m, \quad \text{var}(M_n) = \frac{\sigma^2}{n},$$

we have

$$P[|M_n - m| \geq \epsilon] \leq \frac{\sigma^2}{n\epsilon^2}.$$

Therefore,

$$P[|M_n - m| < \epsilon] \geq 1 - \frac{\sigma^2}{n\epsilon^2},$$

and the result follows immediately, by letting $n \to \infty$ in the last inequality.[2] □

6.5.1   Examples

Example 6.7.   Consider the case of Bernoulli IID random variables X_i with parameter $p = 0.5$. In this case, the pmf of M_n was calculated in Example 6.4. Figure 6.5 shows how fast the convergence in Equation 6.14 takes place as a function of n. We used $\epsilon = 0.05, p = 0.5$ in this example.

Example 6.8.   In Example 6.7 we were able to evaluate the probability in Equation 6.14 in a closed form. When the random variables X_i are, say, Weibull, such a closed form does not exist. In Figure 6.6 we show the full pdf of the simulated random variable M_n for various values of n. We used

---

[2] As a testament to the effort that pioneers must devote to advance the frontiers of knowledge, it took Bernoulli about 20 years to formulate this theorem and finalize its proof. At that time, Chebyshev's inequality was not known.

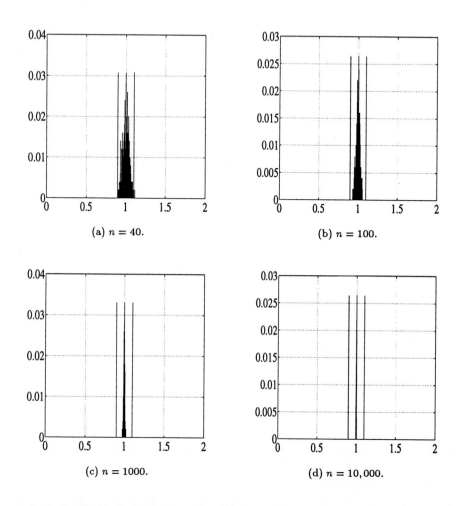

Figure 6.6   The WLLN for Weibull random variable.

$\epsilon = 0.1, EX_i = 1$. The three vertical bars at $x = 0.9, 1, 1.1$ show the "spread" of the pdf. In Figure 6.6 the calculated percentages of the pdf mass inside the interval $(0.9, 1.1)$ were 53, 70, 99.4, and 100 percent, respectively.

6.6   STRONG LAW OF LARGE NUMBERS

Theorem 6.4 can be loosely interpreted as follows: Since ϵ is arbitrary, in the limit the density of M_n should approach a delta function at $x = m$, because the entire density should be in a strip narrower than 2ϵ, as, for example, the sequence of graphs in Figure 6.6 suggests.

The "catch" here (which results in such an easy proof for the Weak Law and an extremely more elaborate one for the Strong Law) is that ϵ is fixed in Equation 6.14. In order to make the density argument of the previous paragraph rigorous, we essentially need to let $\epsilon \to 0$ in Equation 6.14 *before* we let $n \to \infty$.

Bernoulli conjectured that this law should be true; in his own words, one should be able to apply the "inverse" of his theorem as well. He was not able to provide a proof, though, and in fact infuriated Leibnitz, a great mathematician of that time, by his belief. This conjecture was finally proved by Emil Borel in the beginning of the twentieth century, for the special case of Bernoulli random variables with $p = 0.5$. Such an astonishing result attracted the attention and scrutiny of a large number of mathematicians; the proof was found incomplete a few years later. Cantelli gave a proof for the case $0 < p < 1$ in 1917. The general case of an arbitrary random variable was proved by Kolmogorov a few years later. Formally, the Strong Law of Large Numbers (SLLN) can be stated as follows:

Theorem 6.5 *Consider a sequence of IID random variables $\{X_i\}_{i=1}^{\infty}$ with finite mean $EX = m$ and finite variance σ^2. Then*

$$P[\lim_{n \to \infty} M_n = m] = 1. \tag{6.15}$$

A rigorous proof of this theorem is rather deep and certainly beyond the scope of this introductory book. Various mathematicians presented extensions of Theorem 6.5 for the cases of infinite variances and non-IID random variables [19]. Reference [13] contains Kolmogorov's original proof, a masterpiece of mathematical elegance; Borel's proof is in reference [9]. In Problems 6.31–6.39 and 6.41–6.44, we ask you to consider non-IID cases for both the WLLN and the SLLN.

6.6.1   Practical Significance of SLLN

Relationship 6.15 is the basis for justifying *simulations* and analysis of all experimental results. Consider, for example, the case where throughput of a protocol such as *ftp* must be evaluated with computer simulations or measured directly, through repeated trials of an experiment, as we did in Chapters 1 and 3.

Typically, we repeat a computer simulation a large number of times (that is, for all practical purposes infinite), and we simulate the unknown throughput

T as

$$T = \frac{1}{n} \sum_{i=1}^{n} T_i, \tag{6.16}$$

where T_i is the observed throughput in simulation run i. What Theorem 6.5 guarantees, then, is that *if we make sure that our simulation produces IID random variables T_i*, we can trust that the result of Equation 6.16 will not change significantly if we repeat the simulation another time, since $T \to ET_i$.

The same is true when we measure the throughput T directly, in repeated trials of the experiment. Equation 6.16 still holds true, with T_i now being the measurement in trial i. Again, if we make sure that our trials result in IID random variables T_i, we can trust that any other repetition of the same trials will not provide different answers.

In Equation 6.15, only the expected value m appears; this may lead us to the hasty conclusion that we can only measure or simulate means. What about *probabilities* of events? We saw in Chapter 2 how difficult it can be to evaluate them analytically.

The SLLN also suggests a way to evaluate the probability of an event A based on n repeated, independent trials of an experiment. Indeed, let

$$X_i \triangleq \begin{cases} 1, & \text{event } A \text{ occurs in } i\text{th trial,} \\ 0, & \text{otherwise,} \end{cases}$$

be the indicator function of the event A during the ith trial. The sequence $\{X_i\}_{i=1}^{\infty}$ is an IID sequence of Bernoulli random variables, since the trials are independent; moreover, $EX_i = P[A]$ and $\text{var}(X_i) = (1 - P[A])P[A]$. Therefore, the conditions of Theorem 6.5 are satisfied, and thus

$$M_n \to P[A],$$

which explains why M_n is so widely used as a good estimator of the unknown probability.

6.6.2   Examples

Let's look at some examples now, to get a feeling for this convergence to the mean EX.

Example 6.9. Figure 6.7(a) was produced with exponential random variables and a mean $EX = 1$. The sample values of the random variables X_i are shown, along with the values of $M_n, n = 1, \ldots, 1000$. And M_n is the almost "horizontal" line, around the value 1. The figure shows how fast the convergence in Equation 6.15 takes place in this case: After $n \approx 200$, the value of M_n "stabilizes" around $EX = 1$. Figure 6.7(b) shows the oscillations of M_n for the first 50 values of n.

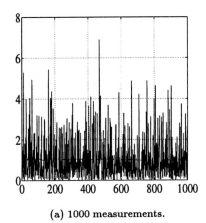

(a) 1000 measurements.

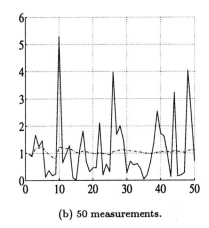

(b) 50 measurements.

Figure 6.7   SLLN for the exponential distribution.

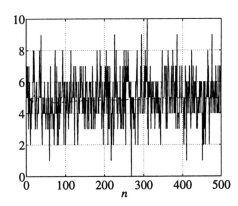

Figure 6.8   SLLN for the binomial distribution.

Example 6.10. Figure 6.8 depicts the same convergence results for a discrete-valued random variable that is binomially distributed with parameters $n = 10$ and $p = 0.5$ (do not confuse the parameter n of the distribution with the index n in the SLLN). Again M_n is the "horizontal" line around the value 5.

6.7   CENTRAL LIMIT THEOREM

The central limit theorem[3] (CLT) is the cornerstone of statistics; it explains why the Gaussian random variable and pdf are so important and common in seemingly different application areas. The theorem describes the limiting behavior of the distribution function of a normalized sum of IID random variables. As our experimental results with finite and small values of n will show, the theorem is a rather good approximation even when n is away from ∞. Formally, for each $n \geq 1$, let's define the following normalized sum:

$$Z_n \triangleq \frac{S_n - nm}{\sigma\sqrt{n}}, \tag{6.17}$$

where $m = EX_i, \sigma^2 = \text{var}(EX_i)$. We can easily check that, for all n, the new random variable Z_n has $EZ_n = 0$ and $\text{var}(Z_n) = 1$. Therefore, even when $n \to \infty$, the mean and variance of the limiting random variable will not change. A natural question then is What is the shape of the limiting cdf or pdf? The CLT provides the answer.

Theorem 6.6 *Consider a sequence of IID random variables $\{X_i\}_{i=1}^{\infty}$ with finite mean m and finite variance σ^2. Let Z_n be defined as in Equation 6.17. Then*

$$\lim_{n \to \infty} P[Z_n \leq z] = \frac{1}{\sqrt{2\pi}} \int_{-\infty}^{z} e^{-\frac{x^2}{2}} \, dx.$$

Sketch of the Proof: A rigorous proof of the theorem is rather involved and is beyond the scope of this book. The interested reader can consult [13]. Laplace was the first to supply a proof, based on expanding binomials through Stirling's approximation (see Equation B.2, Appendix B). Our sketch is based on characteristic functions. Let

$$Z_n \triangleq \frac{1}{\sigma\sqrt{n}}(S_n - nm) = \frac{1}{\sigma\sqrt{n}} \sum_{i=1}^{n}(X_i - m). \tag{6.18}$$

Taking characteristic functions in Equation 6.18, we have

$$\begin{aligned}
\Phi_{Z_n}(\omega) &= Ee^{j\omega Z_n} = E\left(e^{\frac{j\omega}{\sigma\sqrt{n}}\sum_{i=1}^{n}(X_i - m)}\right) \\
&= E\left(\prod_{i=1}^{n} e^{\frac{j\omega}{\sigma\sqrt{n}}(X_i - m)}\right) = \prod_{i=1}^{n} E\left(e^{\frac{j\omega}{\sigma\sqrt{n}}(X_i - m)}\right) \\
&= \left[E\left(e^{\frac{j\omega}{\sigma\sqrt{n}}(X - m)}\right)\right]^n.
\end{aligned}$$

---

[3]Named as such by G. Pólya, 1887–1985, a Hungarian mathematician, in 1920.

The first equality above is the definition of the characteristic function of the random variable Z_n; we used Equation 6.18 to derive the second equality; we used the property $e^{a+b} = e^a e^b$ to derive the third one; and we used the independence of the X_i random variables for the fourth and their identical distribution for the fifth equality. And X is a random variable that has the same pdf as the sequence $\{X_i\}$.

Let's use now a Taylor series expansion for the exponential function in the last equality, with all the third- and higher-order terms combined into a single function $h_n(j\omega)$ (the reason should be clear soon). We have

$$
\begin{aligned}
E\{e^{\frac{j\omega}{\sigma\sqrt{n}}(X-m)}\} &= E\left[1 + \frac{j\omega}{\sigma\sqrt{n}}(X - m) + \frac{(j\omega)^2}{2n\sigma^2}(X - m)^2 + h_n(j\omega)\right] \\
&= 1 + \frac{j\omega}{\sigma\sqrt{n}}E(X - m) + \frac{(j\omega)^2}{2n\sigma^2}E(X - m)^2 + Eh_n(j\omega) \\
&= 1 - \frac{\omega^2}{2n} + Eh_n(j\omega).
\end{aligned}
$$

The first equality in the above equation is the definition of the Taylor series expansion; in the third we use the definitions of the mean and variance of the random variable X. The term $Eh_n(j\omega)$ is a function of n that grows as n^{-2}. Therefore,

$$
\Phi_{Z_n}(\omega) = \left[1 - \frac{\omega^2}{2n} + Eh_n(j\omega)\right]^n \to e^{-\frac{\omega^2}{2}}. \tag{6.19}
$$

The right-hand side of Equation 6.19 is the characteristic function of a Gaussian random variable with zero mean and unit variance. Therefore, the cdf of Z_n approaches in the limit the cdf of a zero-mean, unit-variance Gaussian random variable. □

Many theoreticians had given incomplete or flawed proofs since the first appearance of the theorem. In his famous challenge to axiomatize probability theory, in 1900, Hilbert specifically suggested that a stronger, cleaner proof of this theorem be given. Lindeberg responded to this challenge a few years later in 1922, generalizing and strengthening Laplace's original proof [49]. The list of people who worked on this theorem includes Chebyshev, Markov, Lyapunov, and more recently, Lévy and Feller.[4] Here is Lindeberg and Feller's generalization:

Theorem 6.7 *(Lindeberg). Consider a sequence of zero-mean random variables $\{X_i\}_{i=1}^{\infty}$ with variances σ_n^2 and cdf's $F_i(x)$; let $s_n^2 \triangleq \sum_{i=1}^{n} \sigma_n^2$. Suppose*

---

[4] William Feller, 1906–1970, a prominent U.S. mathematician.

that the cdf's satisfy the condition

$$\sum_{i=1}^{n} \int_{|x| > \epsilon s_n} x^2 \, dF_i(x) = o(s_n^2), \quad \forall \epsilon > 0. \tag{6.20}$$

Let $Z_n \triangleq \sum_{i=1}^{n} X_i / s_n$. *Then*

$$\lim_{n \to \infty} P[Z_n \leq z] = \frac{1}{\sqrt{2\pi}} \int_{-\infty}^{z} e^{-\frac{x^2}{2}} \, dx. \tag{6.21}$$

(Feller). Conversely, if Equation 6.21 holds and

$$\lim_{n \to \infty} \frac{\sigma_n}{s_n} = 0, \quad \lim_{n \to \infty} s_n = \infty,$$

then Equation 6.20 holds true.

Remark 1: There is no result similar to Theorem 6.6 for the pdf in the general case. Only when the random variables are continuous can it be shown that the pdf's $f_{Z_n}(x)$ will converge to the Gaussian pdf

$$\frac{1}{\sqrt{2\pi}} e^{-\frac{x^2}{2}}.$$

Remark 2: Notice that the sequence $\{Z_i\}_{i=1}^{\infty}$ is *not* an IID sequence, in contrast to the two Laws of Large Numbers, which require the IID assumption.

6.7.1   Importance of CLT in Practice; Examples

In practice, $n \approx 30$ is a rule-of-thumb value for Theorem 6.6 to hold. This approximation is the cornerstone of statistics. All the fundamental distributions and tests in the theory and practice of statistical inference use such approximations.

In the following examples, we used small values for n on purpose. We generated 1500 samples for the random variables X_i. In each diagram, we show these samples in the upper right corner. The histogram of the normalized sum is shown in the upper left corner, to see how close to a Gaussian it becomes. Finally, the limiting Gaussian and experimental cdf's are shown in the lower part of each diagram, to see how close the approximation is, for such low values of n.

Example 6.11. Figure 6.9 depicts the CLT for some discrete-valued random variables.

Example 6.12. Figure 6.10 depicts the CLT for some continuous-valued random variables. Figure 6.10(c) shows a remarkable matching for the two distributions, even though n is only equal to 3. But that should come as no surprise since linear transformations of Gaussian random variables are also Gaussian!

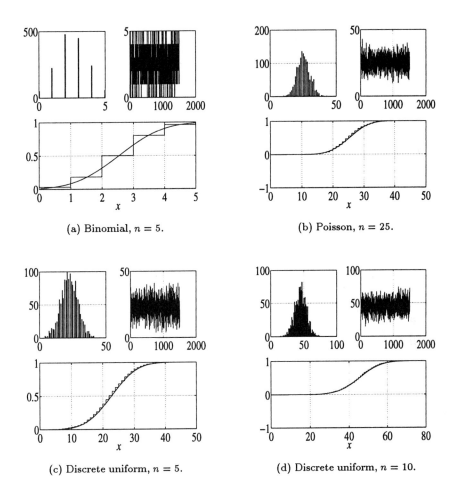

(a) Binomial, $n = 5$.

(b) Poisson, $n = 25$.

(c) Discrete uniform, $n = 5$.

(d) Discrete uniform, $n = 10$.

Figure 6.9 The CLT in practice, discrete random variables.

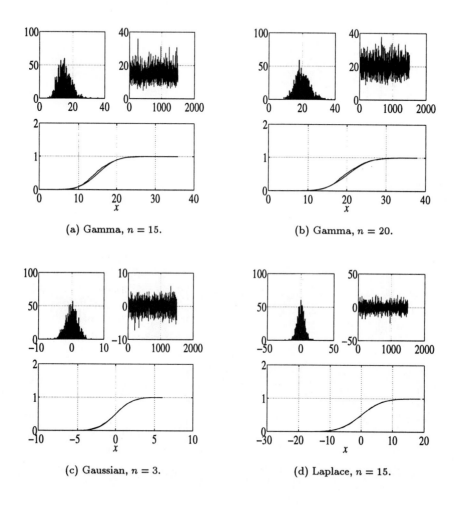

(a) Gamma, $n = 15$.

(b) Gamma, $n = 20$.

(c) Gaussian, $n = 3$.

(d) Laplace, $n = 15$.

Figure 6.10   The CLT in practice, continuous random variables.

6.8 CONVERGENCE OF SEQUENCES OF RANDOM VARIABLES

Many engineering systems (from the simplest circuit to the most complex computer networks) are dynamic, in the sense that their behavior changes with time. In linear systems courses we talk about transient and steady-state system behavior. Random systems exhibit similar characteristics. For example, as a result of a sudden overload in packet arrivals, a computer network node may go through a transient, congested phase; then (hopefully) it may enter a noncongested, steady-state phase.

Example 6.13. A quite publicized example is the early 1990 "meltdown" of the AT&T long-distance telephone network. Because of a software bug, a central telephone switch in Manhattan, New York, took itself out of operation and started a series of shutdowns in other nearby switches. As a result, network capacity was reduced to about 50 percent for a few hours, increasing the probability of calls being blocked quite a lot. After the network engineers fixed the problem, the network entered again its "steady-state" operation. The network controls were subsequently modified to decrease the probability of similar transients in the future.

It is of interest, therefore, to study convergence of *random variables* since in the presence of randomness the system state can only be described through random variables. In this section we discuss various definitions of convergence and present some theorems for convergence of (not necessarily IID) random variables. The alert student may quickly remark that IID random variables do not suffice to describe dynamic systems with memory, as the majority of real systems are. Convergence for Markov Chains is discussed in Chapter 10; martingale convergence results are beyond the scope of this book. The interested student should consult [38] or [11]. Convergence of non-IID sequences (to constant values) is also possible, and numerous results are summarized in the proceedings of a workshop [19].

We saw three convergence results: the two Laws of Large Numbers and the central limit theorem. We introduce next some modes of convergence formally and comment where they are useful. For the rest of this section, $\{X_n\}_{n=1}^{\infty}$ is a sequence of random variables, not necessarily IID, unless specifically stated otherwise.

6.8.1 Sure Convergence

Recall that a random variable is a function on the sample space S. Suppose that for a fixed element $\zeta \in S$,

$$X_n(\zeta) \to X(\zeta) \tag{6.22}$$

as $n \to \infty$; that is, the sequence of real numbers $X_n(\zeta)$ converges to another real number $X(\zeta)$. The function $X : S \to R$ may or may not be a random variable itself.

Definition: The sequence $\{X_n\}_{n=1}^{\infty}$ **converges surely** if the convergence in 6.22 is true *for all $\zeta \in S$.*

Note that the definition allows for the sequence $\{X_n\}$ to converge to a function X that may not be a random variable! On the other hand, $X(\zeta)$ may be a degenerate function of ζ, that is, a constant.

Let's see some examples next.

Example 6.14. *Convergence to a Random Variable X.*   Let $S = [-1, 1]$. Define the sequence of random variables

$$X_n(\zeta) = \zeta^2 \left(1 - \frac{1}{n} \right).$$

For any fixed $\zeta \in S$, we can easily check that

$$X_n(\zeta) \to \zeta^2 \quad \text{as } n \to \infty.$$

The function $X(\zeta) = \zeta^2$ *is* a random variable (why?), and thus the sequence converges surely to a random variable.

Example 6.15. *Convergence to a Constant.* Let $S = [-1, 1]$ again. Define the sequence of random variables

$$X_n(\zeta) = \frac{\zeta^2}{n}.$$

For any fixed $\zeta \in S$, we can easily check that $X_n(\zeta) \to 0$ as $n \to \infty$, and thus the sequence converges surely to a constant.

Example 6.16. *A Divergent Sequence.* Let $S = [-1, 1]$. Define the sequence of random variables

$$X_n(\zeta) = (-1)^n \zeta + 1, \quad n = 1, 2, \dots.$$

For any fixed $\zeta \in S$, the sequence of real numbers $\{X_n(\zeta)\}$ diverges (it oscillates between two values).

Example 6.17. *X Not a Random Variable.* Let $S = \{a, b, c\}$, with $a, b, c \in R$; suppose that all outcomes are equiprobable. Let

$$X_n(\zeta) = \begin{cases} 1 + \frac{1}{n}, & \zeta = a, \\ n, & \zeta = b, \\ 3(c + \frac{1}{n}), & \zeta = c. \end{cases}$$

Then $\{X_n\}$ converges to $X(\zeta)$, where

$$X(\zeta) = \begin{cases} 1, & \zeta = a, \\ \infty, & \zeta = b, \\ 3c, & \zeta = c. \end{cases}$$

Clearly, X is not a random variable since $P[X = \infty] \neq 0$.

Cauchy criterion for sure convergence

How can we check whether a given sequence of random variables converges surely? When we know (or suspect) the limit $X(\zeta)$, this task is straightforward. The *Cauchy criterion* is helpful when this limit is not known since the criterion is not stated in terms of the limit. We state this criterion as a lemma; its proof is standard and therefore omitted.

Lemma 6.1 *The sequence $\{X_n\}_{n=1}^{\infty}$ converges surely to some limit X if and only if for every $\zeta \in S$*

$$|X_{n+k}(\zeta) - X_n(\zeta)| \to 0, \quad \text{as } n \to \infty \text{ uniformly in } k.$$

6.8.2 Almost Sure Convergence

Suppose now that the convergence property in 6.22 holds true only for a *proper subset A of S*, such that $P[A] = 1$. In other words,

$$X_n(\zeta) \to X(\zeta), \quad \forall \zeta \in A, \tag{6.23}$$

as $n \to \infty$.

Definition: The sequence $\{X_n\}_{n=1}^{\infty}$ **converges almost surely**, or **converges with probability 1**, when there is a proper subset $A \subset S$, with $P[A] = 1$, such that 6.23 holds true.

Symbolically, we write $X_n \xrightarrow{\text{a.s.}} X$ or $X_n \xrightarrow{\text{w.p.1}} X$. There may be elements in S (since $A^c \neq \emptyset$) for which 6.22 does not hold. However, the probability of those elements is zero. You may wonder why and how something like that can happen. An example that justifies the need for this definition is the GPA example in Section 3.2: When we artificially augment a sample space, it is not a priori certain that 6.22 will hold true for *all* $\zeta \in S$.

Remark: The SLLN is an example of almost sure convergence to a *constant*, i.e., a degenerate random variable X.

Let's see some examples next.

Example 6.18. *Almost Sure Convergence to a Random Variable.* Let $S = [0, \infty)$. Suppose that $P[\{\zeta = 0\}] = 0$. Define the sequence of random variables

$$X_n(\zeta) = \frac{1}{\zeta}\left(1 - \frac{1}{n}\right).$$

For any fixed $\zeta \in S$, except for $\zeta = 0$, we can easily check that

$$X_n(\zeta) \to \frac{1}{\zeta} \quad \text{as } n \to \infty$$

and thus the sequence converges almost surely to a random variable.

Example 6.19. *Almost Sure Convergence to a Constant.* Let $S = [0, \infty)$ again, with $P[\{\zeta = 0\}] = 0$. Define the sequence of random variables

$$X_n(\zeta) = \frac{1}{n\zeta}.$$

For any fixed $\zeta \in S$, except for $\zeta = 0$, we have

$$X_n(\zeta) \to 0 \quad \text{as } n \to \infty,$$

and thus the sequence converges almost surely to a constant.

Example 6.20. *A Divergent Sequence.* Let $S = [-1, 1]$. Assume a uniform probability distribution on the outcomes of S. Define the sequence of random variables

$$X_n(\zeta) = \begin{cases} 0, & \text{if } \zeta \text{ is rational,} \\ (-1)^n, & \text{if } \zeta \text{ is irrational.} \end{cases}$$

Then the sequence diverges almost surely since irrational numbers in $[-1, 1]$ have probability 1 (check that).

Cauchy criterion for almost sure convergence

Lemma 6.2 *The sequence $\{X_n\}_{n=1}^{\infty}$ converges almost surely to some random variable X if and only if for every $\zeta \in A$, with $P[A] = 1$ and $A \subset S$, we have that*

$$|X_{n+k}(\zeta) - X_n(\zeta)| \to 0 \quad \text{as } n \to \infty \quad \text{uniformly in } k.$$

This criterion is not applicable when we do not know the form of the function X_n. Criteria for almost sure convergence that are based on the cdf of X_n are easier to use. We present such criteria in Section 6.9.

6.8.3   Convergence in Probability

The first two modes of convergence involved the *actual values* of random variables. Convergence in probability deals with probability values instead. Let $\epsilon > 0$ be a fixed real number.

Definition: The sequence $\{X_n\}_{n=1}^{\infty}$ **converges in probability** to the random variable X if there exists a random variable X such that

$$P[|X_n - X| > \epsilon] \to 0, \tag{6.24}$$

for any $\epsilon > 0$ as $n \to \infty$.

We denote that by $X_n \xrightarrow{P} X$.

Remark: The WLLN is an example of convergence in probability to a *constant*, i.e., a degenerate random variable X.

Example 6.21. *Convergence.* Let $S = [0,1]$. Assume a uniform probability distribution on the outcomes of S. Define the sequence of dependent, not identically distributed random variables

$$X_n(\zeta) = \zeta(1 + \frac{1}{n}), \tag{6.25}$$

and the random variable $X(\zeta) = \zeta$. We can easily see that X is uniformly distributed, with parameters $a = 0, b = 1$, while for any fixed n, the random variable X_n is also uniformly distributed, but with parameters $a = 0, b = 1 + (1/n)$. The two random variables are dependent, with the joint pdf given by

$$f_{XX_n}(x, y) = \tilde{\delta}\left(x, \left(1 + \frac{1}{n}\right)x\right), \tag{6.26}$$

where $\tilde{\delta}(x, y) = \delta(x - y)$. The pdf in Equation 6.26 is shown pictorially in Figure 6.11 for the case $n = 3$. Observe that the pdf is nonzero only across the diagonal $x = 4/3y$.

For any fixed $\epsilon > 0$, we can easily write for the probability in Equation 6.24

$$P[|X_n - X| > \epsilon] = P\left[\zeta \in [0,1] : \frac{\zeta}{n} > \epsilon\right] = P[\zeta \in [0,1] : \zeta > n\epsilon]$$

It is clear from the above equation that, for any fixed $\epsilon > 0$, if we choose $n > 1/\epsilon$,

$$P[|X_n - X| > \epsilon] = 0,$$

and thus

$$\lim_{n \to \infty} P[|X_n - X| > \epsilon] = 0.$$

Example 6.22. *Divergence.* Consider $S = [0,1]$, with a uniform probability distribution on its outcomes. Define the sequence

$$X_n(\zeta) = n\zeta.$$

We can easily see that X_n is uniformly distributed, with parameters $a = 0, b = n$. Consider a set A, with $P[A] \neq 0$, and any random variable X, with $X(\zeta) < \infty$, for all $\zeta \in A$. Then, for any $\epsilon > 0$,

$$\lim_{n \to \infty} P[|X_n - X| > \epsilon] \geq P[A] > 0,$$

and the sequence cannot converge in probability.

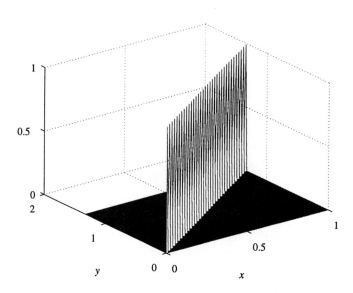

Figure 6.11   The joint pdf of X, X_n.

Cauchy criterion for convergence in probability

Lemma 6.3 *The sequence $\{X_n\}_{n=1}^{\infty}$ converges in probability to some random variable X if and only if for every $\epsilon > 0$,*

$$P[|X_{n+k} - X_n| \geq \epsilon] \to 0 \quad \text{as } n \to \infty \quad \text{uniformly in } k.$$

6.8.4   Mean Square Convergence

This mode of convergence involves second moments only.

Definition: The sequence $\{X_n\}_{n=1}^{\infty}$ **converges in the mean square (m.s.) sense** to the random variable X if there exists a random variable X with $EX^2 < \infty$ such that

$$\lim_{n \to \infty} E(X_n - X)^2 = 0. \tag{6.27}$$

We denote mean square convergence by $X_n \xrightarrow{\text{m.s.}} X$ or l.i.m. $X_n = X$ (*limit in the mean*).

Remark: For any random variable Y, we have $EY^2 = \text{var}(Y) + (EY)^2$. Then from Equation 6.27 we can conclude that $\text{var}(X_n - X) \to 0$ and $EX_n \to EX$. Can we further conclude that $X_n \xrightarrow{\text{a.s.}} X$ as well since the limiting density is a delta function? As we will see later, not in general!

Example 6.23. *Convergence to a Random Variable.* Let $S = [0, 1]$, with a uniform probability distribution for all elements. Define the sequence

$$X_n(\zeta) = \zeta \left(1 + \frac{1}{n}\right).$$

Since for all $\zeta \in S$, $X_n(\zeta) \to \zeta$, we can take $X(\zeta) = \zeta$ for the random variable required in the definition. Now

$$X_n(\zeta) - X(\zeta) = \frac{\zeta}{n},$$

and therefore the random variable $X_n - X$ is uniformly distributed with parameters $a = 0, b = n^{-1}$. Thus

$$E(X_n - X)^2 = \frac{1}{3n^2},$$

and the sequence converges in the mean square sense to a uniform random variable.

Example 6.24. *Convergence to a Constant.* Consider the same S as in Example 6.14, and define now the sequence

$$X_n(\zeta) = 1 + \frac{\zeta}{n}.$$

We can easily check that this sequence converges to the constant 1, in the mean square sense, since

$$E(X_n - 1)^2 = \frac{1}{3n^2}.$$

Example 6.25. *Divergence.* Consider the same S again, as in Example 6.14. Define now

$$X_n(\zeta) = \begin{cases} 2^n, & \text{if } 0 \le \zeta \le \frac{1}{n}, \\ 0, & \text{otherwise.} \end{cases}$$

We can easily compute that

$$EX_n^2 = \frac{2^{2n}}{2n},$$

and thus the sequence does not converge in the mean square sense.

Cauchy criterion for mean square convergence

Lemma 6.4 *The sequence* $\{X_n\}_{n=1}^{\infty}$ *converges in the mean square sense to some random variable* X *if and only if*

$$E(X_{n+k} - X_n)^2 \to 0 \quad \text{as } n \to \infty \text{ uniformly in } k.$$

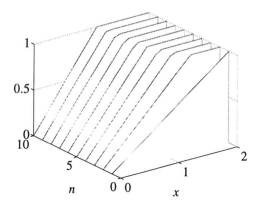

Figure 6.12   Convergence of $F_{X_n}(x)$.

6.8.5   Convergence in Distribution

Consider a sequence of random variables $\{X_n\}_{n=1}^{\infty}$ with cdf's $F_{X_n}(x)$ and a random variable X with cdf $F_X(x)$.

Definition: The sequence $\{X_n\}_{n=1}^{\infty}$ **converges in distribution** to the random variable X if

$$\lim_{n \to \infty} F_{X_n}(x) = F_X(x)$$

for every point $x \in R$ at which $F_X(x)$ is continuous.

We denote that by $X_n \xrightarrow{D} X$. The central limit theorem is the most famous example of convergence in distribution.

Example 6.26. *Convergence to a Random Variable.* Consider the sample space and sequence in Example 6.21. We can easily see that

$$F_{X_n}(x) = \begin{cases} \frac{x}{1+\frac{1}{n}}, & x \in [0, 1 + \frac{1}{n}], \\ 0, & \text{otherwise.} \end{cases}$$

Now, $\lim_{n \to \infty} F_{X_n}(x) = x$, and therefore the sequence converges in distribution to a uniformly distributed random variable X, with range $[0, 1]$. Figure 6.12 gives an idea of how fast this convergence is.

Example 6.27. *Convergence to a Constant.* Let $S = [0, 1]$, with the usual uniform distribution. Define the sequence as

$$X_n(\zeta) = \frac{\zeta}{n^2}.$$

The random variable X_n is uniform, with parameters $a = 0, b = n^{-2}$; we can easily check that its cdf converges to zero.

6.8.6 Relationship Between Modes of Convergence

As might be expected, the different modes of convergence are related to each other. We saw in some examples that the same sequence converges in more than one sense. The following theorems and examples explain the relationship between the various modes.

Theorem 6.8 *If the sequence $\{X_n\}_{n=1}^{\infty}$ converges almost surely, it must also converge in probability.*

Proof: Since $\{X_n\}$ converges almost surely, for every $\epsilon > 0$, there exists an integer $N(\zeta, \epsilon) \geq 1$ such that

$$|X_n(\zeta) - X(\zeta)| < \epsilon \qquad (6.28)$$

for all $n \geq N(\zeta, \epsilon)$ and for all $\zeta \in A$, where $P[A] = 1$. Intuitively, Inequality 6.28 states that for sufficiently large n, for "all" ζ, the sample values of X_n are close to those of X.

Consider the event $B_n(\epsilon) = \{|X_n - X| \geq \epsilon\}$. From Inequality 6.28 we can see that

$$P[B_n(\epsilon)] = P[\{\zeta \in S : n < N(\zeta, \epsilon)\}],$$

i.e., the only ζ values that contribute to the event $B_n(\epsilon)$ are those for which n has not been "large enough."

The set $B_n(\epsilon)$ is a nonincreasing function of n. The limit set $B = B_1(\epsilon) \cap B_2(\epsilon) \cap B_3(\epsilon) \cap \cdots$ cannot be a subset of A. Therefore, $P[B] = 0$, and

$$\lim_{n \to \infty} P[\{|X_n - X| \geq \epsilon\}] = P[B] = 0.$$

$\square$

Theorem 6.9 *If the sequence $\{X_n\}_{n=1}^{\infty}$ converges in the mean square sense, it must also converge in probability.*

Proof: The theorem is an immediate consequence of Markov's inequality (see Inequality 3.83) since the random variable $|X_n - X|$ is nonnegative. Indeed, from Inequality 3.83 we have, for every fixed $\epsilon > 0$,

$$P[|X_n - X| \geq \epsilon] = P[|X_n - X|^2 \geq \epsilon^2] \leq \frac{E(X_n - X)^2}{\epsilon^2} \to 0 \quad \text{as } n \to \infty.$$

$\square$

Theorem 6.10 *If the sequence $\{X_n\}_{n=1}^{\infty}$ converges in probability, it must also converge in distribution.*

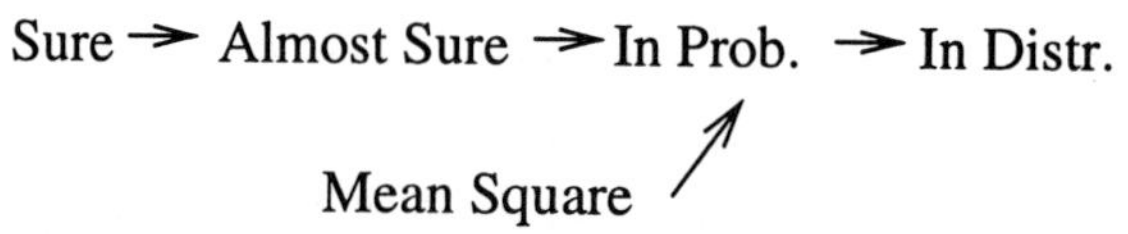

$$\text{Sure} \rightarrow \text{Almost Sure} \rightarrow \text{In Prob.} \rightarrow \text{In Distr.}$$

$$\text{Mean Square} \nearrow$$

Figure 6.13   The various convergence modes.

Proof: Let $F_{X_n}(x), F_X(x)$ denote the cdf's of the random variables X_n, X, respectively. Fix an $\epsilon > 0$, and consider a point of continuity of $F_X(x)$. We can write

$$
\begin{aligned}
F_{X_n}(x) &= P[X_n \le x, \ X \le x + \epsilon] + P[X_n \le x, \ X > x + \epsilon], \\
F_X(x + \epsilon) &= P[X \le x + \epsilon, \ X_n \le x] + P[X \le x + \epsilon, \ X_n > x].
\end{aligned}
$$

Subtracting the two probabilities on the left-hand side, we have

$$
\begin{aligned}
|F_X(x + \epsilon) - F_{X_n}(x)| &\le P[X_n \le x, \ X > x + \epsilon] + P[X \le x + \epsilon, \ X_n > x] \\
&= P[X_n \le x, \ X > x + \epsilon] + P[X \le x - \epsilon, \ X_n > x] \\
&\quad + P[X_n > x, \ x - \epsilon < X \le x + \epsilon].
\end{aligned}
$$

Since the event $\{X_n \le x, \ X > x + \epsilon\} \cup \{X \le x - \epsilon, \ X_n > x\}$ implies that $\{|X_n - X| > \epsilon\}$, we have

$$|F_X(x + \epsilon) - F_{X_n}(x)| \le P[|X_n - X| > \epsilon] + P[x - \epsilon < X \le x + \epsilon].$$

Since the sequence converges in probability, the first term in the last inequality vanishes as $n \to \infty$. Since $F_X(x)$ is continuous at the point x, we have

$$\lim_{n \to \infty} |F_X(x) - F_{X_n}(x)| \le \lim_{\epsilon \to 0} P[x - \epsilon \le X < x + \epsilon] = 0.$$

$\square$

Figure 6.13 summarizes graphically the relationship between the various convergence modes.

In the special case where the sample space S is countable, and $\mathcal{Q} = 2^{|S|}$, we can prove the following theorem:

Theorem 6.11 *The sequence $\{X_n\}_{n=1}^{\infty}$ converges almost surely if and only if it converges in probability.*

The Borel-Cantelli lemmas in Section 6.9 are another way to establish almost sure convergence from knowledge of convergence in probability. Let's see some examples.

Example 6.28. *Almost Sure Convergence Need Not Imply Convergence in the Mean Square Sense.* Let $S = [0, 1]$, with a uniform probability distribution for all elements. For a fixed n, define

$$X_n(\zeta) = e^{-n^2(n\zeta - 1)}.$$

For any fixed $\zeta > 0$, the sequence $X_n(\zeta) \to 0$, so it converges almost surely. On the other hand,

$$E(X_n - 0)^2 = \int_0^1 e^{-2n^2(n\zeta - 1)} \, d\zeta = \frac{e^{2n^2}}{2n^3}(1 - e^{-2n^3}) \to \infty,$$

and the sequence does not converge in the mean square sense.

Example 6.29. *Convergence in the Mean Square Sense Need Not Imply Almost Sure Convergence.* Let $S = [0, 1]$, with a uniform probability distribution for all elements. For a fixed integer $l \geq 0$, define a sequence of 2^l random variables

$$Y_{l,k}(\zeta) = \begin{cases} 1, & \text{if } \frac{k-1}{2^l} \leq \zeta \leq \frac{k}{2^l}, \\ 0, & \text{otherwise}, \end{cases} \tag{6.29}$$

for $k = 1, \ldots, 2^l$. Arrange the $Y_{l,k}$ variables in an array, the rows of which correspond to a fixed value of l. Define now a sequence of random variables $\{X_n\}$ by enumerating the elements of the array rowwise. In other words,

$$X_1 = Y_{0,1}, \quad X_2 = Y_{1,1}, \quad X_3 = Y_{1,2}, \quad X_4 = Y_{2,1},$$

etc. We can easily see from Equation 6.29 that

$$E[|Y_{l,k} - 0|^2] = \frac{1}{2^l},$$

and therefore X_n converges to 0 in the mean square sense.

Now, fix a $\zeta \in S$. From Equation 6.29 we can again easily see that $Y_{l,k}(\zeta) = 1$ for an infinite number of pairs l, k. Therefore, X_n cannot converge to 0 almost surely.

Example 6.30. *Convergence in Probability Need Not Imply Convergence in the Mean Square Sense.* Let $S = [0, 1]$, with a uniform probability distribution for all elements. For a fixed n, define

$$X_n(\zeta) = \begin{cases} 3^n, & \text{if } 0 \leq \zeta \leq \frac{1}{n}, \\ 0, & \text{otherwise}. \end{cases}$$

Fix any $\epsilon > 0$. The set $\{|X_n - 0| > \epsilon\}$ is a subset of the interval $[0, 1/n]$. Therefore,

$$P[|X_n - 0| > \epsilon] \to 0$$

and the sequence converges in probability. On the other hand,

$$E(X_n - 0)^2 = \frac{1}{n}3^{2n} \to \infty,$$

and the sequence does not converge in the mean square sense.

Example 6.31. *Convergence in Probability Need Not Imply Almost Sure Convergence.* Consider the sequence of $Y_{l,k}$ random variables introduced in Example 6.29. We can easily see that

$$P[|Y_{l,k} - 0| > 0] = \frac{1}{2^l}$$

and therefore X_n converges to 0 in probability. However, X_n does not converge almost surely to 0, as we have seen.

6.9   BOREL-CANTELLI LEMMAS

The next two theorems are very useful, generic tools for dealing with infinite sequences and events of the form "something happens infinitely often" or "something happens eventually." Such events arise very frequently in the study of dynamic systems (for example, queuing systems). The results were developed in the beginning of the twentieth century, by Emil Borel and Fransesco Paolo Cantelli. We present them here without proof.

Consider a sequence of events $\{A_n\}_{n=1}^{\infty}$. The notation A_n, i.o. will be used as shorthand for the event {events in the family A_n occur infinitely often}. Mathematically, we can write

$$\{A_n, \text{i.o.}\} = \limsup_{n \to \infty} A_n = \lim_{n \to \infty} \cup_{k=n}^{\infty} A_k.$$

The proof is not difficult and is left as an exercise.

Theorem 6.12 *Direct Borel-Cantelli Lemma. Consider a sequence of events* $\{A_n\}_{n=1}^{\infty}$ *such that*

$$\sum_{n=1}^{\infty} P[A_n] < \infty.$$

Then

$$P[A_n, \text{i.o.}] = 0.$$

Note that the events in Theorem 6.12 need not be independent. When, however, they are, we can show the following.

Theorem 6.13 *Inverse Borel-Cantelli Lemma. Consider a sequence of independent events* $\{A_n\}_{n=1}^{\infty}$ *such that* $\sum_{n=1}^{\infty} P[A_n] = \infty$. *Then*

$$P[A_n, \text{i.o.}] = 1.$$

The following results are very useful in establishing almost sure convergence. We leave the proof of the first one as an exercise.

Theorem 6.14 *Consider a sequence of random variables* $\{X_n\}$. *Then* $X_n \xrightarrow{\text{a.s.}} 0$ *if and only if for every integer* $k > 0$,

$$P\left[|X_n| \geq \frac{1}{k}, \text{ i.o.}\right] = 0.$$

Theorem 6.15 *Consider a sequence of random variables* $\{X_n\}$. *Then* $X_n \xrightarrow{\text{a.s.}} X$ *if and only if* $\sup_{k \geq n} |X_k - X| \xrightarrow{P} 0$.

Proof: From Theorem 6.14, $X_n \xrightarrow{\text{a.s.}} X$, or equivalently $X_n - X \xrightarrow{\text{a.s.}} 0$, if and only if for every integer $k > 0$,

$$P\left[|X_n - X| \geq \frac{1}{k}, \text{ i.o.}\right] = 0.$$

The above equation holds true if and only if for every integer $k > 0$

$$\lim_{n \to \infty} P\left[\sup_{k \geq n} |X_k - X| \geq \frac{1}{k}\right] = \lim_{n \to \infty} P\left[\cup_{k=n}^{\infty} |X_k - X| \geq \frac{1}{k}\right]$$

$$= P\left[\cap_{n=1}^{\infty} \cup_{k=n}^{\infty} |X_k - X| \geq \frac{1}{k}\right] = 0,$$

which is tantamount to the condition $\sup_{k \geq n} |X_k - X| \xrightarrow{P} 0$. $\square$

Let's see some examples next.

Example 6.32. *All Teams Will Be Beaten Someday, Somehow.* Is it possible to ever have a team in the NCAA that will go undefeated forever? Consider the sequence of events $\{A_n(i)\}$, where $A_n(i)$ is defined as

$$A_n(i) = \{\text{team } i \text{ is undefeated during season } n\}.$$

Then we claim that for any team i, we must have $P[A_n(i), \text{i.o.}] = 0$. Can you argue for or against this claim?

Example 6.33. Consider the "experiment" of dialing a given telephone number (e.g., an 800 number). Suppose that the probability of getting a busy signal during the nth trial is $p_n = p$, where $0 < p < 1$ (that is, it does not change with time). Intuitively, the probability that you will *eventually get through* must be 1, no matter how close to 1 the value of p is. How can you *formally justify* such intuition? Consider the events

$$A_n \triangleq \{n\text{th trial is the first successful one}\}.$$

Then, assuming independence (can you say of which events?),

$$P[A_n] = p^{n-1}(1-p).$$

Therefore $\sum_{n=1}^{\infty} P[A_n] = p < \infty$ and $P[A_n, \text{i.o.}] = 0$.

Example 6.34. *Kolmogorov's 0-1 Law for Tail Events.* Consider a sequence of random variables $\{X_n\}_{n=1}^{\infty}$. Define the σ algebra generated by the random variables $(X_j, X_{j+1}, X_{j+2}, \ldots)$ as σ_j, $j = 1, 2, \ldots$. Let $\sigma_X \triangleq \cap_{j=1}^{\infty}\sigma_j$. The sets in σ_X are called *tail events*. Intuitively, these events do not depend on any finite number of random variables. A typical example of a tail event is $\{X_n \text{ converges}\}$ or $\{1/n \sum_{i=1}^{n} X_i \text{ converges}\}$ since convergence depends only on the tail of the sequence $\{X_n\}_{n=1}^{\infty}$.

Kolmogorov's 0-1 law states that probabilities of tail events can have only two values.

Theorem 6.16 *The tail events of a sequence $\{X_n\}_{n=1}^{\infty}$ of independent random variables have probability either 0 or 1.*

Is it an intuitive result? For a proof, see [11].

Example 6.35. Consider a sequence of independent events A_n, with $P[A_n] = 1/n$. What is the probability of the event $\{A_n, \text{i.o.}\}$? Since

$$\sum_{n=1}^{\infty} P[A_n] = \sum_{n=1}^{\infty} \frac{1}{n} = \infty,$$

we immediately get $P[A_n, \text{i.o.}] = 1$.

6.10   SUMMARY OF MAIN POINTS

- **Independent, identically distributed (IID)** random variables are a very useful model for describing dynamic systems in discrete time or repeated trials of the same experiment.

- **Sums** (and **normalized sums**) of IID random variables are easy to characterize; they arise quite frequently in practice.

- The **Weak** and **Strong Law of Large Numbers** and the **central limit theorem** describe the behavior of sums of random variables with an arbitrarily large number of summands. The laws can be used to obtain approximations of the behavior of finite sums.

- A sequence of random variables (not necessarily IID) can **converge** in a variety of modes: **surely, almost surely, in the mean square sense, in probability, and in distribution.**

6.11   CHECKLIST OF IMPORTANT TOOLS

- Equations 6.5, 6.7, 6.9, and 6.3 for calculating the pdf, mean, variance, and characteristic function of the sum of IID random variables, respectively

- Equations 6.6, 6.8, 6.10, and 6.4 for calculating the pdf, mean, variance, and characteristic function of the normalized sum of IID random variables, respectively

- The Strong Law of Large Numbers

- The central limit theorem

- All modes of convergence

- The Borel-Cantelli lemmas

6.12   PROBLEMS

In the following problems, if the range of the index n of a sequence $\{X_n\}$ is not defined, it is assumed that it is the set of all positive integers. As usual, S_n denotes the sum, and M_n denotes the normalized sum associated with $\{X_n\}$.

Sequences of (IID) Random Variables; Simple Skills

6.1 Consider an IID sequence $\{X_n\}$. Define a new sequence of random variables $\{Y_n\}$ by $Y_n = f(X_n)$, where $f(x)$ is a (measurable) function on R. Show that the sequence $\{Y_n\}$ is also IID.

6.2 Consider an integer-valued, IID sequence $\{X_n\}$ with the property that $P[X_i = X_j] = 0$, for all $j \neq i$. Let

$$Y_n \triangleq \sum_{i=1}^{n} I(X_i \leq X_n),$$

where $I(A)$ is the indicator function of the event A.
(a) Express Y_n in words.
(b) Find the pmf of Y_n.
(c) Is the sequence $\{Y_n\}$ IID?

6.3 Consider an IID Bernoulli sequence $\{X_n\}$. Define a new random variable Y as follows:

$$Y = \begin{cases} 1, & X_1 = 1, \\ 2, & X_1 = 0, X_2 = 1, \\ \vdots & \vdots \\ k, & X_1 = 0, X_2 = 0, \ldots, X_k = 1, \\ \vdots & \vdots \end{cases}$$

In other words, Y marks the first index for which $X_n = 1$. Find the pmf of Y.

6.4 Consider an IID Bernoulli sequence $\{X_n\}$. Define a new sequence $\{Y_n\}$ as follows: Fix a positive integer m, and let $Y_n = X_n + X_m$. Is the sequence $\{Y_n\}$ IID?

6.5 Consider an IID Bernoulli sequence $\{X_n\}$. Let $Y_n = \max\{X_1, \ldots, X_n\}$.
(a) Find the pmf of Y_n.
(b) Is the sequence $\{Y_n\}$ IID?

6.6   Consider three independent IID Bernoulli sequences $\{X_n\}$, $\{Z_n\}$, and $\{Y_n\}$, with parameters p_X, p_Z, p_Y. Define a new sequence $\{V_n\}$ as follows:

$$V_n = \begin{cases} X_n, & \text{if } Y_n = 1, \\ Z_n, & \text{if } Y_n = 0. \end{cases}$$

Is $\{V_n\}$ IID?

6.7   Consider two independent IID Bernoulli sequences $\{X_n\}$, $\{Z_n\}$, with parameters p_X, p_Z. Define a new sequence $\{V_n\}$ via $V_n = X_n Z_n$.
(a) Is $\{V_n\}$ IID?
(b) Is $\{V_n\}$ a Bernoulli sequence? If yes, find its parameter.

6.8 Consider an IID Poisson sequence $\{X_n\}$. Define two new sequences $\{Y_n\}$ and $\{V_n\}$ as follows:

$$Y_n = \max\{X_n, X_{n-1}\}, \quad V_n = \min\{X_n, X_{n-1}\}.$$

(a) Is $\{Y_n\}$ IID?
(b) Is $\{V_n\}$ IID?
(c) Is either of the two sequences Poisson?

Sequences of (IID) Random Variables; Advanced Skills

6.9 Consider an IID sequence $\{X_n\}$ of continuous random variables. Define the random variable N as the smallest integer n such that

$$X_1 < X_2 < X_3 < \cdots < X_{n+1} > X_{n+2}.$$

If $X_1 > X_2$, let $N = 0$. Express N in words and find $P[N = k]$.

6.10 Let $\{X_n\}$ be a sequence of random variables, with $E|X_n| < \infty$. Define the sequence $\{Y_n\}$ as follows: Set $Y_0 = X_0$ and let

$$Y_n = X_n - E(X_n|X_{n-1}, \ldots, X_1, X_0), \quad n = 1, 2, \ldots.$$

(a) Find $E(Y_n|X_{n-1}, \ldots, X_1, X_0), \quad n = 1, 2, \ldots.$
(b) Find $E(Y_n|Y_{n-1}, \ldots, Y_1, Y_0), \quad n = 1, 2, \ldots.$
(c) Find $E(Y_n), \quad n = 1, 2, \ldots.$

Sums; Simple Skills

6.11 Consider an IID sequence $\{X_n\}$.
(a) Find the pdf of the following random variables.

$$X_1 + X_2, \quad X_1 + X_3, \quad X_j + X_l, \quad l \neq j.$$

(b) Suppose that X_n is Bernoulli, with parameter $p > 0.5$. Define

$$Y = \begin{cases} 1, & \text{if } X_l = 1, \text{ some } l, \\ 2, & \text{if } X_l = 0, \text{ some } l. \end{cases}$$

Is $P[Y = 1] > P[Y = 2]$? (*Hint:* This is a trick question.)

6.12 Consider a zero-mean sequence $\{X_n\}$ with the property that $EX_iX_j = EX_i^2$, for all $j > i$. Evaluate ES_n and $\text{var}(S_n)$.

6.13 Consider a sequence of IID discrete uniform random variables $\{X_n\}$, with parameters $1, b$. Calculate the probability generating function of S_n.

6.14 Consider the sequence $\{V_n\}$ in Problem 6.7 and its sum process $\{S_n\}$. Find the joint pdf of S_n, S_m.

6.15 Consider an IID Poisson sequence $\{X_n\}$. Find the joint pmf of the sum random variables S_k, S_l.

6.16 Consider an IID sequence $\{X_n\}$ of Cauchy random variables and its sum process $\{S_n\}$. Find the joint pdf of S_n, S_m.

6.17 Show that sums of IID gamma random variables are also gamma random variables.

6.18 Consider a sequence of IID Gaussian random variables with zero mean and unit variance. Define

$$Y_n = \sum_{k=1}^{n} X_k, \quad Z_n = \sum_{k=1}^{n} (-1)^k X_k.$$

Show that Y_n and Z_n have identical distributions.

Sums; Advanced Skills

6.19 Consider an IID sequence $\{X_n\}$ with variance σ^2. Show that

$$\frac{1}{n-1} E \left[\sum_{j=1}^{n} (X_j - M_n)^2 \right] = \sigma^2.$$

6.20 Consider a sequence of IID random variables $\{X_n\}$ with the property that $P[X_i = 0] < 1$. Show that for every positive constant k there exists an integer n_k such that $P[|S_{n_k}| > k] > 0$. Explain intuitively why you should expect such a result to be true.

6.21 Consider a sequence of IID random variables $\{X_n\}$. Show that for any given $\epsilon > 0$ and for any given positive integers m, N,

$$\sum_{n=0}^{N} P[|S_n| < m\epsilon] \leq 2m \sum_{n=0}^{N} P[|S_n| < \epsilon].$$

6.22 Let T be a stopping time and k a positive, given constant. Which of the following random variables are stopping times?

$$T \pm k, \quad T \cdot k, \quad \frac{T}{k}.$$

6.23 Let T_1, T_2 be two stopping times. Which of the following random variables are stopping times?

$$T \triangleq \min\{T_1, T_2\}, \quad U \triangleq \max\{T_1, T_2\}, \quad V \triangleq T_1 \pm T_2, \quad W \triangleq \max\{T_1, 9\}.$$

6.24 Consider a sequence of exponential IID random variables $\{X_n\}$. Consider a sequence of Bernoulli IID random variables $\{I_n\}$ that are independent of

$\{X_n\}$. For any integer $k \geq 1$, let

$$N_k = \sum_{n=1}^{k} I_n, \quad Y_k = \sum_{n=1}^{N_k} X_n.$$

(a) Find the pdf of Y_k, EY_k, and $\text{var}(Y_k)$.
(b) Find the joint pdf of Y_k and Y_l.
(c) Is the sequence $\{Y_n\}$ IID?

6.25 *Kolmogorov's Inequality for Sums.* Consider a sequence of independent random variables $\{X_n\}$. For any $a > 0$, show that

$$P[\max_{1 \leq k \leq n} |S_k - ES_k| < a \cdot \text{var}(S_n)] \geq 1 - a^{-2}.$$

How does that differ from Chebyshev's inequality?

6.26 Consider a sequence of positive IID random variables $\{X_n\}$ with $EX_n < \infty$. Suppose that EX_n^{-1} exists for all n. Prove the following:
(a) $ES_n^{-1} < \infty$.
(b) For all $k = 1, \ldots, n$, $EX_k S_n^{-1} = n^{-1}$, $ES_k S_n^{-1} = kn^{-1}$.
(c) For all $k > n$, $ES_k S_n^{-1} = 1 + (k - n)EX_n ES_n^{-1}$.

6.27 Consider a sequence of integer-valued, nonnegative IID random variables $\{X_{nk}, n \geq 0, k \geq 0\}$. Set $Y_0 = 1$ and recursively define the sequence $\{Y_n\}$ via

$$Y_{n+1} = \sum_{k=1}^{Y_n} X_{nk}.$$

(If $Y_n = 0$, the sum is interpreted as 0.) Show that $G_{Y_{n+1}}(z) = G_{Y_n}(G_X(z))$.

6.28 Construct a counterexample to show that Theorem 6.1 may not hold when N depends arbitrarily on the sequence $\{X_i\}$.

Laws of Large Numbers; Simple Skills

6.29 Consider a sequence of IID random variables $\{X_n\}$.
(a) Let $Y_n = aX_n + b$, where a, b are real-valued constants. Does the WLLN hold for the sequence $\{Y_n\}$?
(b) Repeat part (a) for the nonlinear transformation $Y_n = f(X_n)$.

6.30 The WLLN for IID Bernoulli random variables with parameter p asserts that $P[|M_n - p| \geq \epsilon] \to 0$. Prove the stronger result

$$\sum_{n=1}^{\infty} P[|M_n - p| \geq \epsilon] < \infty$$

for all $\epsilon > 0$. (*Hint:* Use the fact that S_n is a binomial random variable; bound the *fourth* central moment of S_n.)

6.31 Consider a sequence of random variables $\{X_n\}$ with $EX_i = EX$, but not necessarily IID. Show that a sufficient condition for having

$$P[|M_n - EX| \geq \epsilon] \to 0$$

is

$$\lim_{n \to \infty} \frac{\text{var}(S_n)}{n} = 0.$$

6.32 Consider a sequence of independent random variables $\{X_n\}$, defined by

$$X_n = \begin{cases} +n^a, & \text{with probability } 0.5, \\ -n^a, & \text{with probability } 0.5. \end{cases}$$

Here $a < 0.5$ is a given constant.
(a) Is $\{X_n\}$ an IID sequence?
(b) Show that $P[|M_n| \geq \epsilon] \to 0$.

6.33 Consider a sequence of independent random variables $\{X_n\}$ with

$$P[X_n = 0] = 1 - \frac{1}{n^a}, \quad P[X_n = \pm n] = \frac{1}{2n^a},$$

where a is a given constant.
(a) Is $\{X_n\}$ an IID sequence?
(b) Determine values for the constant a for which $P[|M_n| \geq \epsilon] \to 0$.

6.34 Consider a sequence of independent random variables $\{X_n\}$ with

$$P[X_n = 0] = 1 - \frac{1}{2^{2n}}, \quad P[X_n = \pm 2^n] = 2^{-(2n+1)}.$$

Does $P[|M_n| \geq \epsilon] \to 0$?

6.35 Consider a sequence of independent, discrete uniform random variables $\{X_n\}$. The range of X_n is the set

$$S_{X_n} = \{-na_n, -(n-1)a_n, \ldots, 0, \ldots, (n-1)a_n, na_n\},$$

and thus the random variables are not identically distributed. Determine, if possible, values for the real constants a_k for which $P[|M_n| \geq \epsilon] \to 0$.

6.36 Consider a sequence of dependent random variables $\{X_n\}$ with mean $EX_n = m$ and covariance given by

$$\text{cov}(X_i, X_j) = \begin{cases} \sigma^2, & i = j, \\ a\sigma^2, & i = j \pm 1, \\ 0, & \text{otherwise}, \end{cases}$$

where $|a| < 1, \sigma^2 > 0$ are given constants. Show that the result of the WLLN holds.

6.37 (a) Would the results in Problem 6.36 change if $\text{cov}(X_i, X_j) = a\sigma^2$ for $i = j \pm 2$, instead of $i = j \pm 1$?
(b) What if $i = j \pm M$, where M is a given positive integer?

6.38 Consider a sequence of dependent random variables $\{X_n\}$ with covariance $\text{cov}(X_i, X_j) = a^{|i-j|}$, where $0 < a < 1$ is a given constant. Show that the result of the WLLN holds.

6.39 Consider a sequence of dependent random variables $\{X_n\}$ with covariance

$$\text{cov}(X_i, X_j) \begin{cases} = \sigma^2, & i = j, \\ \leq 0, & i \neq j, \end{cases}$$

where $a > 0$ is a given constant. Show that the result of the WLLN holds.

6.40 Let X be an F-distributed random variable, with parameters a and $b = \infty$. Show that the random variable $Y = aX$ is χ^2-distributed, with parameter a.

6.41 Consider a sequence of IID uniform random variables $\{X_n\}$ with parameters 0, 1.
(a) Let

$$Y_i = \begin{cases} X_i^2, & \text{with probability } 0.1, \\ X_i, & \text{with probability } 0.9. \end{cases}$$

Let $M_n = \frac{1}{n} \sum_{i=1}^n Y_i$. What is $\lim_{n \to \infty} M_n$?
(b) Let now

$$Z_i = \begin{cases} 0, & i = 1, 3, 5, 7, 9, \ldots, \\ X_i, & i = 2, 4, 6, 8, 10, \ldots. \end{cases}$$

Let $M_n = \frac{1}{n} \sum_{i=1}^n Z_i$. Is $\lim_{n \to \infty} M_n$ equal to 0, 0.25, or 0.5? Explain.

Laws of Large Numbers; Advanced Skills

6.42 Consider again a sequence of random variables $\{X_n\}$ with $EX_i = EX$, but not necessarily IID. Show that if $|X_n| \leq c < \infty$ (i.e., the random variables are uniformly bounded), then $\lim_{n \to \infty} P[|M_n - EX| \geq \epsilon] = 0$.

6.43 Construct a sequence of independent, not identically distributed random variables $\{X_n\}$ for which the WLLN holds.

6.44 Construct a sequence of independent, not identically distributed random variables $\{X_n\}$ for which $\lim_{n \to \infty} P[|M_n| \geq \epsilon] \neq 0$.

6.45 Consider a sequence of IID Bernoulli random variables with parameter p. Prove the SLLN.

6.46 Consider a sequence of IID Poisson random variables with parameter λ. Prove the SLLN. (*Hint:* Evaluate $P[|M_n - \lambda|^4 > \epsilon^4]$.)

Central Limit Theorem; Simple Skills

6.47 Verify the CLT *experimentally*, using IID variates drawn from a distribution of your own choice. Observe how fast you approach the Gaussian cdf.

6.48 Consider a sequence of IID random variables $\{X_n\}$.
(a) Let $Y_n = aX_n + b$, where a, b are real-valued constants. Does the CLT hold for the sequence $\{Y_n\}$?
(b) Does it hold for the nonlinear transformation $Y_n = f(X_n)$?

6.49 Use the CLT and a suitable sequence of random variables to show that

$$\lim_{n \to \infty} e^{-n} \sum_{i=0}^{n} \frac{n^i}{i!} = 0.5.$$

6.50 Consider a sequence of independent random variables $\{X_n\}$ such that

$$P[X_n = 0] = \frac{1}{4} - \frac{2}{n^3}, \quad P[X_n = \pm 1] = \frac{3}{8}, \quad P[X_n = \pm n] = \frac{1}{n^3}.$$

(a) Does the sequence satisfy the CLT?
(b) Repeat part (a) if the term n^3 is replaced by n^k, for k a positive real constant.

Central Limit Theorem; Advanced Skills

6.51 Verify Lindeberg's condition for an IID sequence of your choice.

6.52 Consider a sequence of independent random variables $\{X_n\}$ such that

$$|X_n| \leq c_n, \quad n \geq 1.$$

Here $c_n > 0$ are given constants. Let $s_n^2 \triangleq \sum_{i=1}^{n} E(X_i - EX_i)^2$. Suppose that $s_n \to \infty$ and $c_n = o(s_n)$. Show that the random variable $(S_n - ES_n)/s_n$ satisfies the CLT.

6.53 Consider two independent, identically distributed random variables X, Y with zero mean and unit variance and cdf $F(x)$. Use the CLT to show that if

$(X + Y)/\sqrt{2}$ has the same cdf, then

$$F(x) = \frac{1}{\sqrt{2\pi}} \int_{-\infty}^{x} e^{-\frac{t^2}{2}} dt.$$

Convergence in Probability; Simple Skills

6.54 Calculate $P[|X_n - X| > \epsilon]$ for all n, ϵ for the example in Equation 6.25.

6.55 Consider a sequence of random variables $\{X_n\}$. Show that if $X_n \xrightarrow{P} X$ and $X_n \xrightarrow{P} Y$, then $P[X = Y] = 1$.

6.56 Consider two sequences $\{X_n\}, \{Y_n\}$. Suppose that $X_n \xrightarrow{P} X$ and $Y_n \xrightarrow{P} Y$. Show that $X_n + Y_n \xrightarrow{P} (X + Y)$. Is the opposite statement also true?

6.57 Consider a sequence of random variables $\{X_n\}$ that converges in probability to a random variable X. Evaluate the convergence of the sequence $\{g(X_n)\}$, where $g(x)$ is a continuous function.

6.58 Consider a sequence $\{X_n\}$ of random variables with $X_n \xrightarrow{P} 0$. Show that the medians $m_{X_n} \to 0$ as well.

Convergence in Probability; Advanced Skills

6.59 Consider a sequence $\{X_n\}$ of identically distributed random variables with the property that $nP[|X_i| > n] \to 0$. Show that

$$\frac{1}{n} \max_{1 \le i \le n} |X_i| \xrightarrow{P} 0.$$

6.60 Consider two sequences $\{X_n\}$ and $\{Y_n\}$. Suppose that

$$|X_n - X| \le Y_n, \quad \text{almost surely}$$

for some random variable X. Show that if $EY_n \to 0$, then $EX_n \to EX$ and $X_n \xrightarrow{P} X$.

6.61 Consider a sequence of independent random variables $\{X_n\}$ that converges in probability to a random variable X. Show that X is a constant.

6.62 Consider a sequence of random variables $\{X_n\}$ such that for all n and for some positive constant c, $P[|X_n| \ge c] > \epsilon > 0$. Show that if the sequence of real numbers $\{a_n\}$ is such that $a_n X_n \xrightarrow{P} 0$, then $a_n \to 0$.

6.63 Construct a sequence of discrete random variables that converge to zero in probability, but EX_n does not approach zero.

6.64 Consider a sequence of random variables $\{X_n\}$. Show that

$$E\left(\frac{|X_n|}{1+|X_n|}\right) \to 0$$

implies that $X_n \xrightarrow{P} 0$.

6.65 Show that $X_n \xrightarrow{P} X$ if and only if

$$\lim_{n \to \infty} E\frac{|X_n - X|}{1+|X_n - X|} = 0.$$

Almost Sure Convergence; Simple Skills

In some of the following (almost sure convergence) problems, you may need the Borel-Cantelli lemma of Section 6.9.

6.66 Let $S = [0,1]$. Let P assign uniform probabilities. Define

$$X_n(\zeta) = ne^{-n\zeta}.$$

Show that $X_n \to 0$ almost surely.

6.67 For an arbitrary random variable X, let

$$X^+ \triangleq \max(X,0), \quad X^- \triangleq \max(-X,0).$$

Show that the following inequalities hold true surely

$$(X+Y)^+ \le X^+ + Y^+, \quad (X+Y)^- \le X^- + Y^-,$$

$$(X+Y)^+ \ge \max\{(X^+ - Y^-),(Y^+ - X^-)\}.$$

6.68 Consider a sequence of positive real numbers $a_n \to \infty$ and a sequence of finite random variables $\{X_n\}$. Can you construct the sequence on a suitable sample space such that almost surely $\limsup_n \frac{S_n}{a_n} = 0$? 1? ∞?

6.69 Consider a sequence of real numbers $a_n \to 0$ and an arbitrary random variable X. Define $X_n = a_n X$. Show that $\{X_n\}$ converges almost surely.

Almost Sure Convergence; Advanced Skills

6.70 Prove Theorem 6.14.

6.71 Consider a sequence of (possibly dependent) random variables $\{X_n\}$. Show that if for every $\epsilon > 0$

$$\sum_{n=1}^{\infty} P[|X_n| > \epsilon] < \infty,$$

then $X_n \to 0$ almost surely.

6.72 Consider a sequence of independent, finite-valued random variables $\{X_n\}$. Let $\{a_n\}$ denote a sequence of positive real numbers with $a_n \to \infty$. Show that

$$\liminf_{n\to\infty} \frac{S_n}{a_n} = c, \quad \limsup_{n\to\infty} \frac{S_n}{a_n} = d, \quad \text{almost surely,}$$

where c, d are real constants.

6.73 Consider a sequence $\{X_n\}$ of uncorrelated, identically distributed random variables, with finite variance σ^2 and zero mean. Show that

$$\frac{S_n}{n} \to 0, \quad \text{almost surely.}$$

6.74 Consider an IID sequence $\{X_n\}$ of Bernoulli random variables, with parameter $p = 0.5$. Consider a given pattern $p_0, p_1, \ldots, p_m$ of 1s and 0s that has length $m + 1$. Define a sequence $\{Y_n\}$:

$$Y_n = \begin{cases} 1, & X_n = p_0, X_{n+1} = p_1, \ldots, X_{n+m} = p_m, \\ 0, & \text{otherwise.} \end{cases}$$

(a) Is the sequence $\{Y_n\}$ IID?
(b) Show that

$$\frac{1}{n} \sum_{k=1}^{n} Y_k \to (0.5)^{m+1}, \quad \text{almost surely.}$$

6.75 Consider a sequence of IID random variables $\{X_n\}$. Show that

$$\limsup_{n\to\infty} \frac{|X_n|}{n} \le 1 \quad \text{almost surely,}$$

if and only if $\sum_{n=1}^{\infty} P[|X_n| > n] < \infty$.

6.76 Consider a sequence of IID random variables $\{X_n\}$. Show that

$$\limsup_{n \to \infty} X_n = \infty \text{ almost surely,}$$

if and only if for every $c < \infty$, $P[X_1 < c] < 1$.

6.77 Let $\{X_n\}$ be an IID sequence of random variables. Show that

$$\frac{1}{n} \max_{1 \le i \le n} \{X_i\} \to 0 \text{ almost surely, if and only if } E|X_1| < \infty.$$

Mean Square Convergence; Simple Skills

6.78 Let $S = [0, 1]$. Let P assign uniform probabilities. Define

$$X_n(\zeta) = \begin{cases} n, & 0 \le \zeta \le \frac{1}{n^3}, \\ 0, & \text{otherwise.} \end{cases}$$

Show that $X_n \xrightarrow{\text{m.s.}} 0$.

6.79 Show that l.i.m. X_n exists if and only if EX_nX_m exists as $n, m \to \infty$. Use the Cauchy criterion.

6.80 The random variables $\{X_i\}$ are independent, with zero mean and variance $\text{var}(X_i) = \sigma_i^2$. Show that l.i.m. X_n exists if and only if $\sum_{i=1}^{\infty} \sigma_i^2 < \infty$.

6.81 Consider a sequence $\{X_n\}$ with $X_n \xrightarrow{\text{m.s.}} X$. Show that $EX_n \to EX$ and $EX_n^2 \to EX^2$.

6.82 Consider a sequence $\{X_n\}$ of Gaussian random variables, with $X_n \xrightarrow{\text{m.s.}} X$. Show that X is a Gaussian random variable.

6.83 Show that if $X_n \xrightarrow{\text{m.s.}} X$ and $Y_n \xrightarrow{\text{m.s.}} Y$, then

$$\lim_{m, n \to \infty} EX_nY_m = EXY.$$

6.84 Consider a sequence of independent random variables $\{X_n\}$ such that

$$EX_n = 0, \quad \sum_{n=1}^{\infty} EX_n^2 < \infty.$$

Show that the random variable X converges in the mean square sense. Find EX^2.

6.85 Consider two sequences of random variables $\{X_n\}$ and $\{Y_n\}$ which have limits X and Y, in the mean square sense. Let a, b be given constants. Define a new sequence $\{Z_n\}$ via $Z_n = aX_n + bY_n$. Does this sequence also converge in the mean square sense? If yes, what is its limit?

Convergence in Distribution; Simple Skills

6.86 Consider an IID sequence $\{X_n\}$ of uniform random variables with parameters $0, 1$. Define a sequence $\{Y_n\}$ via $Y_n = \min\{X_1, \ldots, X_n\}$.
(a) Is the sequence $\{Y_n\}$ IID?
(b) Show that $nY_n \xrightarrow{D} Y$, where Y is an exponential random variable with parameter 1.

6.87 Consider an independent sequence $\{X_n\}$ of Bernoulli random variables with parameters p_n such that $np_n \to \lambda > 0$. Show that $S_n \xrightarrow{D} X$, where X is Poisson with parameter λ.

6.88 Consider a sequence $\{X_n\}$ of discrete uniform random variables with parameters $1, n$. Show that X_n/n converges in distribution to a uniform random variable with parameters $0, 1$.

6.89 Consider an IID sequence $\{X_n\}$ of Cauchy random variables with parameters $a = 0, b > 0$. Define the sequence $\{Y_n\}$, via

$$Y_n = \frac{1}{n^2} \sum_{k=1}^{n} X_k.$$

(a) Show that Y_n does not converge in distribution.
(b) Examine the convergence in distribution of the sequence $\{Y_n\}$, defined via

$$Y_n = \frac{1}{n^d} \sum_{k=1}^{n} X_k,$$

where $d \neq 0$ is a real number.

6.90 Consider an IID sequence of Gaussian random variables $\{X_n\}$ with zero mean and unit variance. Show that

$$\frac{1}{n\sqrt{n}} \sum_{k=1}^{n} S_k \xrightarrow{D} Y,$$

where Y is a zero-mean Gaussian random variable with variance $1/3$.

6.91 Consider a Gaussian random variable X with zero mean and variance σ^2. Define a sequence of (Gaussian) random variables $\{X_n\}$, via

$$X_n = \frac{X}{\sqrt{n}}.$$

Does $\{X_n\}$ converge in distribution? If so, to what?

Convergence in Distribution; Advanced Skills

6.92 Let $\{Y_n\}$ be an IID sequence of zero-mean, unit-variance Gaussian random variables. Let $X_0 = Y_0$; define the sequence $\{X_n\}$ recursively via $X_{n+1} = aX_n + Y_{n+1}$. Show that

$$X_n \xrightarrow{D} X,$$

where X is a zero-mean Gaussian random variable. Find its variance.

6.93 Consider a sequence of random variables $\{X_{nk}, \ n \geq 1, \ k \geq 1\}$ with the property that for all $\epsilon > 0$,

$$\sum_{k=1}^{\infty} P[|X_{nk}| > \epsilon] \to 0 \ \text{ as } n \to \infty.$$

Show that $\sup_{k \geq 1} |X_{nk}| \xrightarrow{D} 0$.

General Convergence Problems; Simple Skills

6.94 Let X_p be a geometric random variable with parameter p. Define $Y_p = pX_p$. Show that for every $x > 0$,

$$\lim_{p \to 0} P[Y_p < x] = 1 - e^{-x}.$$

6.95 Consider a sequence of independent random variables $\{X_n\}$ such that

$$P[X_n = (-1)^{n+1}\sqrt{n}] = \frac{n^2}{n^2 + \sqrt{n}}, \quad P[X_n = (-1)^n n^2] = 1 - \frac{n^2}{n^2 + \sqrt{n}}.$$

(a) Does this sequence converge in any sense?
(b) Does your answer change if the term $\sqrt{n}$ in the denominator is replaced by n? By n^a, where $a > 2$?

6.96 Consider a sequence $\{X_n\}$ that converges almost surely to the random variable X. Suppose that $|X_n| < Y$, with $EY^2 < \infty$. Show that $X_n \xrightarrow{\text{m.s.}} X$.

6.97 Consider a sequence $\{X_n\}$ with $X_n \xrightarrow{D} c$, a constant. Show that $X_n \xrightarrow{P} c$ as well.

General Convergence Problems; Advanced Skills

6.98 Consider a sequence of IID random variables $\{X_n\}$. Let

$$Y_n \triangleq \max_{1 \le i \le n} |X_i|.$$

Under what conditions on $E|X_i|$ can
(a) $\frac{Y_n}{n}$ converge in probability? Can you find the limit?
(b) $\frac{Y_n}{n}$ converge almost surely? Can you find the limit?

6.99 Consider a sequence of random variables $\{X_n\}$, such that $X_n \xrightarrow{a.s.} X$ and $X_n \xrightarrow{m.s.} Y$, where X, Y are two random variables. Evaluate $P[X = Y]$.

6.100 Consider an IID sequence $\{X_n\}$ of Cauchy random variables.
(a) Does the random variable M_n converge? In what sense(s)?
(b) Does the random variable $\sqrt{n}M_n$ converge? In what sense(s)?
(c) Does the random variable M_n/n converge? In what sense(s)?

6.101 Show that if $E(X_n - X)^2 \le n^{-2}$, then

$$P\left[|X_n - X| > \frac{1}{\sqrt{n}}\right] \le \frac{1}{n}.$$

General Convergence Problems; Experimental

6.102 Verify convergence in the mean square sense, using variates drawn from a distribution of your own choice. Observe how fast convergence takes place.

6.103 Verify convergence in distribution, using variates drawn from a distribution of your own choice. Observe how fast convergence takes place.

Borel-Cantelli Lemmas; Simple Skills

6.104 Can you find a sequence of dependent events $\{A_n\}$ such that $\sum P[A_n] = \infty$, but $P[A_n, \text{i.o.}] < 1$?

6.105 Consider a sequence of IID random variables $\{X_n\}$. Evaluate the probability of the event $\{X_n \text{ converges to a constant}\}$.

6.106 Consider a sequence $\{X_n\}$ of positive, integer-valued, IID random variables. What is the value of $P[\sum_{n=1}^{\infty} X_i < \infty]$?

6.107 Consider a sequence of events $\{A_n\}$ such that

$$\sum_{n=1}^{\infty} P[A_n \cap A_{n+1}^c] < \infty, \quad \lim_{n \to \infty} P[A_n] = 0.$$

Show that $P[A_n, \text{i.o.}] = 0$.

6.108 Let $\{X_n\}$, $\{Y_n\}$ be two independent sequences, each one being IID. Suppose that $P[X_1 \geq Y_1] > 0$. Show that $P[X_n \geq Y_n, \text{i.o.}] = 1$.

6.109 Consider a sequence of random variables $\{X_n\}$. Show that $X_n \xrightarrow{\text{a.s.}} 0$ if and only if $P[|X_n| > k^{-1}, \text{i.o.}] = 0$, $k = 1, 2, \ldots$.

6.110 Consider a sequence $\{X_n\}$ of independent Bernoulli random variables. Show that $X_n \xrightarrow{\text{a.s.}} 0$ if and only if $\sum_{n=1}^{\infty} P[X_n = 1] < \infty$.

Borel-Cantelli Lemmas; *Advanced Skills*

6.111 Consider a Bernoulli sequence $\{X_n\}$ with the property that

$$P[X_n = 1 | X_0, X_1, \ldots, X_{n-1}] \geq c > 0$$

for some constant c. Show that (a) $P[X_n = 1, \text{i.o.}] = 1$ and (b) $P[X_n = 1, \text{at least one } n] = 1$. Are the two events the same?

6.112 Consider a sequence $\{X_n\}$ of random variables. Let X be a random variable. Suppose that the sequences of nonnegative numbers $\{\epsilon_n\}$ and $\{c_n\}$ are such that $P[|X_n - X| \geq \epsilon_n] \leq c_n$. Suppose further that

$$\lim_{n \to \infty} \epsilon_n = 0, \quad \sum_{n=1}^{\infty} c_n < \infty.$$

Show that $X_n \xrightarrow{\text{a.s.}} X$.

6.113 Consider a sequence of ± 1-valued, independent random variables $\{X_n\}$ such that $P[X_n = +1] = 0.5$. Set $S_0 = 0$. Show that

$$P[\sup_n S_n = \infty] = 1, \quad P[\inf_n S_n = -\infty] = 1.$$

6.114 Consider a sequence of random variables $\{X_n\}$, defined on a countable sample space S, with $\mathcal{Q} = 2^{|S|}$. Show that $\{X_n\}$ converges almost surely if and only if it converges in probability.

6.115 Consider a sequence of random variables $\{X_n\}$ with the property

$$\sum_{n=1}^{\infty} E|X_n - X|^r < \infty$$

for some random variable X and constant $r > 0$. Show that $X_n \xrightarrow{\text{a.s.}} X$.

6.116 Consider a sequence of random variables $\{X_n\}$ with the property

$$\sum_{n=1}^{\infty} EX_n < \infty, \quad X_n \geq 0 \text{ almost surely.}$$

Suppose that $ES_n > 0$. Show that S_n/ES_n converges almost surely.

Computer Problems

6.117 Simulate the operation of a buffer in a computer network node. Assume that the packet interarrival times form an IID sequence $\{X_n\}$ and that the packet transmission times form another IID sequence $\{Y_n\}$, independent of $\{X_n\}$. Calculate histograms of the busy cycle as a function of the means EX and EY. Calculate the mean and variance of the busy cycle.

6.118 Plot the functions $\Phi_{S_n}(\omega)$, $\Phi_{M_n}(\omega)$ for the Pascal random variable.

6.119 Consider a random sum of N Poisson IID random variables, where N is a geometric random variable. Generate a large number of variates, and verify experimentally Wald's equalities.

6.120 Generate n variates of an IID sequence of exponential random variables with parameter λ.
(a) Evaluate the WLLN and observe how fast $P[|M_n - \lambda^{-1}| < \epsilon]$ approaches its limiting value.
(b) Study experimentally and plot the convergence rate as a function of λ.

6.121 Generate n variates of an IID sequence of Bernoulli random variables with parameter p.
(a) Evaluate the SLLN and observe how fast $|M_n - p|$ approaches its limiting value.
(b) Study experimentally and plot the convergence rate as a function of p.

6.122 Generate n variates of an IID sequence of discrete uniform random variables with parameters $a = 1$, b.
(a) Evaluate the CLT and observe how fast $P[Z_n \leq z]$ approaches the Gaussian cdf.
(b) Study experimentally and plot the convergence rate as a function of b.

6.123 Consider the sample space $S = [0, 1]$, with the uniform distribution on it. Define the sequence of random variables $\{X_n\}$ as follows:

$$X_n(\zeta) = \left[\zeta + \frac{(-1)^n}{n^2}\right]\zeta^2.$$

Generate a large number of X_n variates, and describe its modes of convergence experimentally.

7

RANDOM PROCESSES

Until the beginning of the twentieth century, the random processes studied were almost exclusively discrete in time (i.e., sequences of random variables). Random processes in continuous time were first studied by Einstein (1905, Brownian motion) and Wiener (1921, Brownian motion). Bruno de Finetti[1] started a systematic study of general continuous-time random processes in 1929. Kolmogorov put together a systematic study of Markov processes in 1931. Finally, Kolmogorov presented the framework that is accepted today for the general study of continuous-time random processes in 1933, in his seminal book [45].

Section 7.1 provides the basic definitions and some examples. Section 7.2 introduces the joint pdf and cdf of a random process; they suffice to completely characterize a random process. Partial characterization is provided by the mean and autocorrelation functions, which we introduce in Section 7.3. Special but extremely useful cases of random processes are presented in Section 7.4; two widely used models are presented in Section 7.5. The notions of continuity, derivatives, and integrals, presented in Section 7.6, are the foundation for the mathematical manipulation of random processes, which we use in Chapter 9. The notion of ergodicity, a subtle concept and a widely used tool, is presented in Section 7.7. The Karhunen-Loève expansion in Section 7.8 is another useful, theoretical tool. Finally, in Section 7.9 we discuss how values of a random process can be generated for simulations.

---

[1] Bruno de Finetti, 1907–1985, a contemporary Italian mathematician, the father of subjective probabilities (see Section C.2.2).

7.1   DEFINITION OF A RANDOM PROCESS AND EXAMPLES

Consider a probability space (S, Q, P) and an infinite (countable or uncountable) set I, called the *index set*. It is usual (even though not necessary) to consider I as a set of *time* indices, such as $I = (-\infty, \infty)$ or $I = (0, \infty)$ or $I = [0, 1]$. Consider a family of random variables defined on (S, Q, P) and indexed by the set I. Consider first the case where I is uncountable.

Definition: The (uncountable) collection of random variables $\{X(t), t \in I\}$ is called a **continuous-time random process**.

Consider now the case where I is countable, i.e., finite or countably infinite. Typical examples are the set of all integers, the set of nonnegative integers, and the set of rational numbers. Since all infinite countable sets can be mapped one-to-one onto the set of nonnegative integers, it is usual to consider $I = \{1, 2, \ldots\}$ or $I = \{0, 1, 2, \ldots\}$.

Definition: The (countable) collection of random variables $\{X(t), t \in I\}$ is called a **discrete-time random process**.

We saw a discrete-time random process in Chapter 6, namely, an IID sequence. Moreover, the very special case of an index set with a finite number of elements was covered in Section 4.11. We use the notation $X(t)$, $X(n)$, or X_n to denote a single discrete-time random variable.

We can consider a random process as a function $X(t, \zeta)$ of two arguments, where $t \in I$ and $\zeta \in S$. When ζ is fixed, the deterministic function $X(t, \cdot)$ is called a **sample path** or **realization** or **sample function** of the random process.

Remark 1:   A rigorous formulation for continuous-time random processes is much more difficult than that for discrete-time ones. Handling an uncountable infinity of random variables is not an easy game! The main difficulty arises from the fact that limits are much more difficult to handle in the continuous-time case (see Section 7.6).

Remark 2:   A word of caution about notation is appropriate here. The symbol $X(t)$ is very frequently used to denote the entire collection of random variables; i.e., it replaces the more accurate notation $\{X(t), t \in I\}$ or the slightly vague $\{X(t)\}$. It is also used to represent a generic, single random variable from the same collection (for which, of course, t is fixed). Finally, it is also used to denote a sample path itself (for which, of course, ζ is fixed). The exact meaning associated with the symbol $X(t)$ will become clear from the context. Make sure that you fully understand the intended use of this symbol before you attempt any of the problems. Experience has shown that this notational vagueness is the single most important source of confusion that makes working with random processes difficult.

7.1.1 Classification and Examples

Random processes can be classified in numerous ways, based on a variety of criteria. We can have **discrete-** or **continuous-time** random processes, based on whether the index set I is discrete or continuous. We can also have **discrete-** or **continuous-space** random processes, based on whether the sample paths of the random process can take discrete or continuous values. We will see more classifications in Section 7.4. Let's look at some examples of random processes now. The first five describe random processes with continuous spaces; the latter five describe random processes with discrete spaces.

Example 7.1. *ECG.* Electrocardiograms (ECGs) are "pictures" of the human heart, taken with ultrasounds. There is a lot of "noise" in such pictures since the ultrasound signal can rebound from a lot of places inside the chest cavity.

Example 7.2. *Noisy Channel Output.* The output of a communication channel is usually corrupted by electromagnetic noise. Figure 1.2 shows typical input and output signals of a such a channel, corrupted by "weak" noise. Note that both the input and the modulated signal have discrete spaces; only the output signal has continuous space. All three signals have continuous-time index sets.

Example 7.3. *Sound Signals.* Sound signals can be considered as random since the sequence of phonems that produces them is not deterministic. Figure 7.1(a) is the first few phonems from Handel's *Messiah*. This random process has continuous space and index set.

Example 7.4. *FM Signal with Noise.* A typical FM (frequency-modulated) signal has the form

$$X(t) = \cos\left(2\pi f_c t + 2\pi k_f \int_0^t a(s)\, ds\right),$$

where f_c is the frequency of the carrier (e.g., something in the band from 88 to 108 MHz), k_f is the so-called sensitivity gain, and $a(s)$ is the sound signal we wish to transmit [e.g., something like the signal in Figure 7.1(a) if you are listening to classical music stations]. Even without any noise, the signal $X(t)$ is random since $a(s)$ is random.

Example 7.5. *A Random Sinusoid.* Let $I = (-\infty, \infty)$. Consider a probability space (S, Q, P), with $S = [0, 1]$ and a probability function P that assigns uniform probabilities. Define the random process $\{X(t),\ t \in I\}$ via the sample path equation

$$X(t, \zeta) = \zeta \sin(2\pi t), \quad \zeta \in [0, 1], \quad t \in I. \tag{7.1}$$

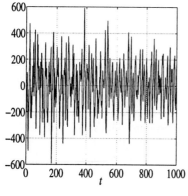

(a) Sound signal, Handel's *Messiah*.

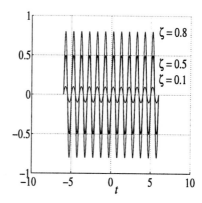

(b) A random amplitude sinusoid.

Figure 7.1   Typical continuous-space random processes.

The sample paths are, in other words, sinusoidal functions of time with random amplitudes. Figure 7.1(b) shows a typical case, for three values of ζ. We will use this random process a lot in subsequent sections to illustrate a number of concepts.

Example 7.6.   *Buffer Contents in Network Routers.*   Network routers are devices (typically small specialized computers) that switch packets from input links to output links. When a packet is received over an input link, it is temporarily stored in some buffer (memory) until protocol processing, such as checking for errors and finding addresses, is done. When a packet is ready for transmission over an output link and that link is free, the packet is removed from the buffer. Therefore, the number of packets waiting in buffer for transmission varies with time. This variation is random since errors and arrivals on the input links are themselves random. Figure 7.2 depicts typical buffer contents.

Example 7.7.   *Network Link Utilization.*   Computer and telephone network operators such as IBM, AT&T, and MCI are very interested in how much the links of their networks are utilized since typically higher utilization translates to higher revenues. We say that a link in a computer network is utilized at time t if and only if a packet transmission takes place at time t. Mathematically, let $U(t)$ denote the link utilization at time t, where we can take $t \in [0, 24]$ to

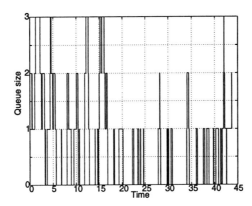

Figure 7.2 Typical buffer contents at a network node.

denote a 1-day interval. Clearly,

$$U(t) = \begin{cases} 1, & \text{if packet transmission is going on at time } t, \\ 0, & \text{otherwise.} \end{cases}$$

Now $U(t)$ *is* a random process since its value on the same day instant in two different days can be different depending on, say, how many users are logged onto the network. (Can you think of other reasons for this randomness? What could be a suitable S for this random process?) Figure 7.3(a) depicts a typical utilization sample path (of a busy link).

Example 7.8. *ftp Throughput.* The familiar *ftp* example can give rise to a random process in discrete time; consider the throughput of an *ftp* transfer. Suppose we are interested in the throughput of the nth file transfer during a given day. Then $\{X_n\}$ is a discrete-time and discrete-space random process. Figure 7.3(b) shows a typical sample path of such a process.

Example 7.9. *Discrete Modulated Signals.* Data signals that are transmitted over channels are random since the sequence of bits which generates them is itself random. Figure 7.4 represents a typical Manchester-modulated signal of 20 bits.

Example 7.10. *Random Binary Sequences.* Let ζ be a randomly selected number in $S = [0, 1]$. The binary expansion of ζ can be represented as

$$\zeta = \sum_{i=1}^{\infty} \frac{b_i}{2^i},$$

where b_i can take the value 1 or 0 only. (Can you determine the b_i for $\zeta = 0.5, 0.3333, \pi/4$?) A discrete-time random process $X(n, \zeta)$ can then be defined

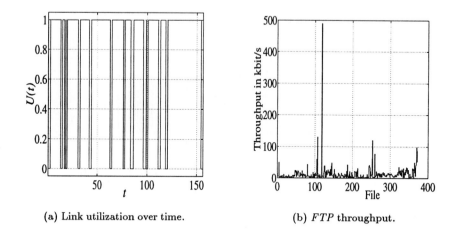

(a) Link utilization over time.   (b) *FTP* throughput.

Figure 7.3   Typical discrete-space random processes.

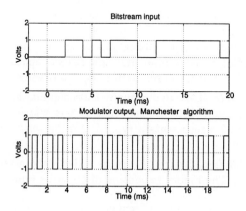

Figure 7.4   A Manchester-modulated data signal.

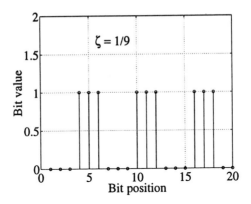

Figure 7.5   Binary expansions of numbers in $[0, 1]$.

as follows:

$$X(n, \zeta) = b_n.$$

Figure 7.5 shows a sample path for the number $\zeta = 1/9$.

7.2   JOINT CDF AND PDF

How do we characterize the probabilistic behavior of a random process? In analogy to the vector random variable or the random sequence cases in Chapters 4 and 6, we must specify now the joint cdf or pdf for an infinity of random variables, and even worse, for an uncountable infinity. It should be obvious right at the outset that doing so for the general case requires a prohibitive amount of work. Questions should also be raised regarding whether we can *actually* do so.

The following approach is due to Kolmogorov and (expectedly) answers the above-mentioned question. The basic idea is that events of interest regarding the random process do not necessarily involve *all* the random variables of the collection. When a finite number n are involved, the jcdf of the n random variables suffices to calculate any probability of interest.

More precisely, consider a fixed integer n, and let $t_1, \ldots, t_n$, be a given, fixed set of n time indices ($t_k \in I$). Let's define, for ease of later notation, the random variables

$$X_k \triangleq X(t_k), \quad k = 1, \ldots, n. \tag{7.2}$$

The n random variables in Equation 7.2 are obviously an n-dimensional random vector; therefore, to describe its behavior, the n-dimensional jcdf[2]

$$F_{X_1 X_2 \cdots X_n}(x_1, x_2, \ldots, x_n) = P[X_1 \leq x_1, X_2 \leq x_2, \ldots, X_n \leq x_n] \qquad (7.3)$$

must be specified[3] *for any value of* $n = 1, 2, \ldots$ *and any choice of indices* $t_1, \ldots, t_n$! Similarly, an n-dimensional jpdf, $f_{X_1 X_2 \cdots X_n}(x_1, x_2, \ldots, x_n)$, would suffice as well. This is easier said than done, of course. Note that even for $n = 1$, this task still involves specification of an infinite, possibly uncountable number of cdf's, since we must specify one for each $t_1 \in I$.

There is of course a question of *consistency* associated with Equation 7.3: The distributions in 7.3 for $n{+}1$ must provide the specified distributions for n as marginals, as we saw in Section 4.11.2. So this raises a fundamental theoretical question: *Can we specify* an infinity of jcdf's to describe a random process?

An affirmative answer to that question was given by Kolmogorov, in the beginning of this century, as part of his effort to provide a solid foundation for the theory of stochastic processes. The opposite question (the obvious extension to those in Sections 3.3.3 and 4.3.1) also merits attention: Given a family of functions

$$F(x_1, x_2, \ldots, x_n), \quad x_1, \ldots, x_n \in R,$$

for all n, is there a random process with those functions as jcdf's of the random variables $\{X_{t_k}\}$ for all choices $\{t_k \in I, \ k = 1, \ldots, n\}$, where I is a given index set?

The answer was given by Kolmogorov, also in the affirmative, in the following theorem, known as *Kolmogorov's existence theorem.*

Theorem 7.1 *Given a sequence of consistent finite-dimensional joint cdf's, there exists a random process described by these joint cdf's.*

Kolmogorov actually *constructed* a random process with the given jcdf's in a fashion reminiscent of the algorithm in Section 3.13.2. See [45] for more details. Let's see now some simple examples where specifying the jcdf's is not difficult, at least conceptually.

Example 7.11. *IID Families in Continuous Time.* An IID family of random variables in continuous time is a random process $\{X(t)\}$, such that the one-dimensional cdf $F_{X(t_1)}(x_1)$ is independent of t_1 and any two collections of random variables (with noncommon elements) are independent. This is a direct analog of the IID *sequences* in the previous chapter. Again, a *single* pdf or cdf

---

[2]Sometimes we write $F_{X(t_1)X(t_2)\cdots X(t_n)}(x_1, x_2, \ldots, x_n)$ to denote the jcdf.

[3]There are random processes, called *nonseparable*, for which specifying the n-dimensional jcdf does not suffice [8]. However, such processes are of pure mathematical interest and we do not deal with them at all.

suffices to describe the entire process. Indeed, for any integer n and any choice of indices $t_k \in I, k = 1, \ldots, n$, we easily have

$$F_{X_1 X_2 \cdots X_n}(x_1, x_2, \ldots, x_n) = \prod_{i=1}^{n} F(x_i),$$

where $F(x)$ is the common cdf.

Example 7.12. *Sinusoid Random Process of Equation 7.1.* Consider the random process whose sample paths are described in Equation 7.1. We evaluate next the one-dimensional cdf $F_{X(t_1)}(x_1)$:

$$F_{X(t_1)}(x_1) \triangleq P[\zeta \in [0,1] : \zeta \sin(2\pi t_1) \leq x_1], \quad x_1 \in R.$$

Consider the case $\sin(2\pi t_1) > 0$ first. We have

$$F_{X(t_1)}(x_1) = P[\zeta \in [0,1] : \zeta \leq \frac{x_1}{\sin(2\pi t_1)}],$$

and thus

$$F_{X(t_1)}(x_1) = \begin{cases} 0, & x_1 \leq 0, \\ \frac{x_1}{\sin(2\pi t_1)}, & 0 < x_1 \leq \sin(2\pi t_1), \\ 1, & \sin(2\pi t_1) < x_1. \end{cases}$$

Consider next the case $\sin(2\pi t_1) < 0$. We have

$$F_{X(t_1)}(x_1) = P[\zeta \in [0,1] : \zeta \geq \frac{x_1}{-|\sin(2\pi t_1)|}],$$

and thus

$$F_{X(t_1)}(x_1) = \begin{cases} 1, & x_1 \geq 0, \\ 1 - \frac{x_1}{-|\sin(2\pi t_1)|}, & 0 > x_1 \geq -|\sin(2\pi t_1)|, \\ 0, & -|\sin(2\pi t_1)| > x_1. \end{cases} \qquad (7.4)$$

Finally, when $\sin(2\pi t_1) = 0$, we trivially have

$$F_{X(t_1)}(x_1) = \begin{cases} 1, & x_1 \geq 0, \\ 0, & x_1 < 0. \end{cases}$$

This function is shown in Figure 7.6, for $t_1 \in [0,2]$ and $x_1 \in [-1,2]$.

Fix next $t_1, t_2 \in I$ and consider the two random variables $X_1 = X(t_1)$, $X_2 = X(t_2)$. Let's evaluate next the two-dimensional jcdf $F_{X_1 X_2}(x_1, x_2)$. We have

$$F_{X_1 X_2}(x_1, x_2) = P[\zeta \in [0,1] : \zeta \sin(2\pi t_1) \leq x_1, \ \zeta \sin(2\pi t_2) \leq x_2]. \qquad (7.5)$$

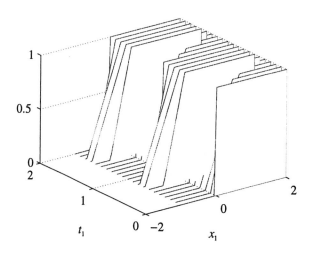

Figure 7.6   The one-dimensional distribution for $X(t_1)$.

Observe that the range for X_i, $i = 1, 2$, is

$$[A_i, B_i] \triangleq [\min\{0, \sin(2\pi t_i)\}, \max\{0, \sin(2\pi t_i)\}].$$

Consider the case $\sin(2\pi t_1) > 0$, $\sin(2\pi t_2) > 0$. (The other cases can be treated similarly, and we do not work them out here.) We have from Equation 7.5

$$F_{X_1 X_2}(x_1, x_2) = P\left[\zeta \in [0, 1] : \zeta \leq \min\left\{\frac{x_1}{\sin(2\pi t_1)}, \frac{x_2}{\sin(2\pi t_2)}\right\}\right].$$

After a little algebra, we get

$$F_{X_1 X_2}(x_1, x_2) = \begin{cases} 0, & x_1 < A_1 \text{ or } x_2 < A_2, \\ \min\{1, \frac{x_1}{\sin(2\pi t_1)}, \frac{x_2}{\sin(2\pi t_2)}\}, & A_1 \leq x_1 \text{ and } A_2 \leq x_2. \end{cases} \quad (7.6)$$

Figure 7.7 shows the cdf in Equation 7.6 for $t_1 = 0.3, t_2 = 0.1$ and $x_1 \in [0, 1]$, $x_2 \in [0, 0.6]$. The pattern for general n should be obvious now, but the number of cases we have to consider grows out of hand as n increases.

The point of this example is twofold: When we know the exact functional form of the sample paths $X(t, \zeta)$, it is possible, in principle, to evaluate the jcdf's. But as this example shows, this can be rather cumbersome, even in simple cases.

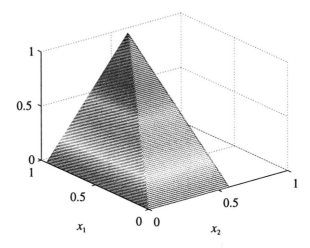

Figure 7.7 The two-dimensional cdf, $t_1 = 0.3, t_2 = 0.1$.

7.2.1 Multiple Random Processes

Consider two random processes $\{X(t)\}, \{Y(t)\}$ defined on the same sample space S and having the same index set[4] I. The probabilistic behavior of the two random processes can be specified through the collection of the joint cdf's

$$F_{X_1 X_2 \cdots X_n; Y_1 Y_2 \cdots Y_l}(x_1, x_2, \ldots, x_n, y_1, y_2, \ldots, y_l), \qquad (7.7)$$

for all $n, l = 1, 2, \ldots$ and all choices of time indices $t_1, t_2, \ldots \in I$ that define the above random variables. Multidimensional random processes with more than two element processes can be defined in a similar way, but we do not even attempt to establish the notation here since we are not going to use them in this book.

Clearly, to develop a theory with a chance for applications, we must confine our attention to special cases where the jpdf's can be measured in practice, in a rather easy fashion. We present a few such cases in Section 7.4, after introducing fundamental notions in the next section.

---

[4] The index set need not be the same; this restriction simplifies our notation somewhat.

7.3   EXPECTATION, AUTOCOVARIANCE, AND CORRELATION FUNCTIONS

Partial information about the random process can be provided in a fashion similar to that in Section 3.9.1. In the following definitions, it is assumed that the expectations (single or double integrals) exist.

Definition: The **mean** of a random process $X(t)$ is defined as[5]

$$m_X(t) \stackrel{\triangle}{=} EX(t) = \int_{-\infty}^{\infty} x f_{X(t)}(x)\,dx, \quad t \in I. \tag{7.8}$$

Other names frequently used for $m_X(t)$ are **expectation, average,** and **expectation function.**

Definition: The **autocovariance function** of a random process $X(t)$ is defined for all $t_1, t_2 \in I$ as

$$
\begin{aligned}
C_X(t_1, t_2) &\stackrel{\triangle}{=} E\left[[X(t_1) - m_X(t_1)][X(t_2) - m_X(t_2)]\right] \\
&= \int_{-\infty}^{\infty} \int_{-\infty}^{\infty} [x - m_X(t_1)][y - m_X(t_2)] f_{X(t_1)X(t_2)}(x, y)\,dx\,dy.
\end{aligned}
\tag{7.9}
$$

Another name frequently used for $C_X(t_1, t_2)$ is **covariance function.**

Definition: The **autocorrelation function** of a random process $X(t)$ is

$$R_X(t_1, t_2) \stackrel{\triangle}{=} EX(t_1)X(t_2), \quad t_1, t_2 \in I. \tag{7.10}$$

The autocorrelation and autocovariance functions of a random process are related to each other in the following way:

$$C_X(t_1, t_2) = R_X(t_1, t_2) - m_X(t_1)m_X(t_2).$$

This property can be easily derived from Equations 7.8, 7.9, and 7.10.

Definition: The **cross-correlation** of two random processes $X(t)$ and $Y(t)$ is

$$R_{XY}(t_1, t_2) \stackrel{\triangle}{=} EX(t_1)Y(t_2), \quad t_1, t_2 \in I.$$

Definition: The **correlation coefficient** of a random process $X(t)$ is

---

[5]Notice the double use of the symbol $X(t)$ here: In the text of the definition, it denotes the entire random process, so t ranges over the entire set I. In Equation 7.8, t is fixed, so $X(t)$ denotes a single random variable, and $f_{X(t)}(x)$ denotes the pdf of this random variable.

$$\rho_X(t_1, t_2) \triangleq \frac{C_X(t_1, t_2)}{\sqrt{C_X(t_1, t_1)C_X(t_2, t_2)}}. \tag{7.11}$$

Remark: Definition 7.11 specifies a *function* of two arguments t_1, t_2. When these two arguments take a *fixed* value, we have the correlation coefficient of two random variables, as we saw in Chapter 4.

How do we calculate all these characteristics of a random process in practice? In general, this is difficult. Specifying them for all $t_i \in I$ requires an infinite amount of work, so at best we can only hope for approximations. This difficulty is another good reason for developing mathematical models! We will see in Section 7.7 a special case where the calculation is very simple.

Example 7.13. Let's calculate now $m_X(t), C_X(t_1, t_2)$, and $R_X(t_1, t_2)$ for the sinusoid random process in Equation 7.1. Instead of calculating those parameters through their definitions (which we leave as an exercise), we can use a little trick. Let Z denote a uniform random variable with parameters $a = 0, b = 1$. Then we can write $X(t) = \sin(2\pi t) Z$, and thus

$$m_X(t) = \sin(2\pi t) \ EZ = \frac{\sin(2\pi t)}{2}.$$

Similarly,

$$\begin{aligned}
R_X(t_1, t_2) &= EX(t_1)X(t_2) = \sin(2\pi t_1) \ \sin(2\pi t_2) \ EZ^2 \\
&= \frac{\sin(2\pi t_1) \ \sin(2\pi t_2)}{3}.
\end{aligned} \tag{7.12}$$

A plot of this function is given in Figure 7.8. Finally,

$$C_X(t_1, t_2) = R_X(t_1, t_2) - m_X(t_1)m_X(t_2) = \frac{\sin(2\pi t_1) \ \sin(2\pi t_2)}{12}.$$

7.4   SOME IMPORTANT SPECIAL CASES

We already saw one of the simplest possible examples of a random process, the IID process in discrete and continuous time. Here are some more special cases for which the jcdf's can be specified with relative ease.

1. Independent increment process. Consider a random process $\{X(t)\}$ that takes discrete values (we make this assumption for notational simplicity only). For any integer n, fix the indices $t_1, \ldots, t_n \in I$ such that $t_1 < t_2 < \cdots < t_n$. Consider the *increments*

$$Y_1 \triangleq X(t_2) - X(t_1), Y_2 \triangleq X(t_3) - X(t_2), \ldots, Y_{n-1} \triangleq X(t_n) - X(t_{n-1}),$$

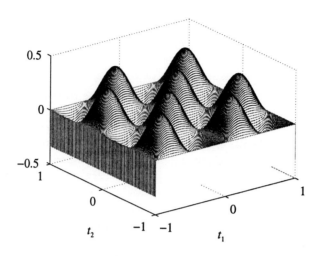

Figure 7.8   The autocorrelation function of the sinusoid process.

and define $Y_0 = X(t_1)$.

Definition: The random process $\{X(t)\}$ is called an **independent increment process** if the random variables $Y_k, k = 0, 1, 2, \ldots, n - 1$, are independent for any choice of n and indices t_k.

Independent increment processes were studied systematically by B. de Finetti in 1929. For such a process, the jcdf in Equation 7.3 (actually, the joint pmf) can be easily calculated as follows. Let $A = P[X_1 = x_1, X_2 = x_2, \ldots, X_n = x_n]$. We have

$$
\begin{aligned}
A &= P[Y_{n-1} = x_n - x_{n-1}, \ldots, Y_1 = x_2 - x_1, Y_0 = x_1] \\
&= P[Y_{n-1} = x_n - x_{n-1}] \cdots P[Y_1 = x_2 - x_1] \cdot P[Y_0 = x_1]. \quad (7.13)
\end{aligned}
$$

Therefore, to determine $F_{X_1 X_2 \cdots X_n}(x_1, x_2, \ldots, x_n)$, we need to know only the one-dimensional pmf's for the increment random variables. In the special case where all the increments are identically distributed, Equation 7.13 reduces to

$$
P[X_1 = x_1, X_2 = x_2, \ldots, X_n = x_n] = P[Y_0 = x_1] \cdot \prod_{i=1}^{n-1} P[Y_1 = x_{i+1} - x_i].
$$

$$(7.14)$$

In this case, the random process is called a *independent, stationary increment process*. Make sure you understand the difference between Equations 7.13 and 7.14!

Example 7.14. The discrete-time process $\{S_n, n = 1, 2, \ldots\}$, defined by sums of IID random variables $\{X_i\}$ in Section 6.3, is a very special but important case of an independent increment process. Let's calculate the pdf's for this example. Denote the common pdf of the $\{X_i\}$ variables by $f(x)$. Then

$$
\begin{aligned}
f_{S_1}(x_1) &= f(x_1), \\
f_{S_1 S_2}(x_1, x_2) &= f(x_1)f(x_2 - x_1), \\
f_{S_1 S_2 \cdots S_n}(x_1, x_2, \ldots, x_n) &= f(x_1)f(x_2 - x_1) \cdots f(x_n - x_{n-1}).
\end{aligned}
$$

In Section 8.2 we will encounter the Wiener process, a continuous-time analog to the process $\{S_n\}$.

2. Stationary random processes.[6] Intuitively, a stationary process has a "behavior" that does not depend on time. More precisely, consider a random process $\{X(t), t \in I\}$; for any integer $n > 0$, let $t_1, \ldots, t_n$ be an increasing sequence of time indices.

Definition: The random process $\{X(t)\}$ is called **stationary** if

$$
F_{X(t_1)X(t_2)\cdots X(t_n)}(x_1, x_2, \ldots, x_n) = F_{X(t_1+s)X(t_2+s)\cdots X(t_n+s)}(x_1, x_2, \ldots, x_n)
\tag{7.15}
$$

for any choice of s (provided of course that $t_k + s \in I, \ \forall k$).

In words, the jcdf does not depend on "where the time origin is placed." Take some time to really understand what is behind this definition. As a check, is a random vector of N elements stationary? How would you apply Definition 7.15 to a random vector?

Definition 7.15 has certain immediate implications. To avoid trivialities, let $I = R$. When $n = 1$, we get

$$
F_{X(t_1)}(x_1) = F_{X(t_1+s)}(x_1), \quad \forall s \in R, t_1 \in R, x_1 \in R.
\tag{7.16}
$$

Applying Equation 7.16 with $s = -t_1$, we get

$$
F_{X(t_1)}(x_1) = F_{X(0)}(x_1), \quad \forall t_1 \in R,
$$

and thus the one-dimensional cdf must be independent of time. To emphasize this independence, we usually denote the distribution function as $F_X(x)$. When $n = 2$, we get

$$
F_{X(t_1)X(t_2)}(x_1, x_2) = F_{X(t_1+s)X(t_2+s)}(x_1, x_2)
\tag{7.17}
$$

(where, as usual, $s \in R$, $t_1, t_2 \in R$, and $x_1, x_2 \in R$). When $s = -t_2$, we get from the above equation

$$
F_{X(t_1)X(t_2)}(x_1, x_2) = F_{X(t_2-t_1)X(0)}(x_1, x_2),
$$

---

[6] Introduced by A. Khinchine, 1894–1959, a Russian mathematician.

and thus we conclude that the two-dimensional cdf must be a function of only the difference $t_2 - t_1$. Again, to emphasize this fact, we may write

$$F_{X(\tau)X(0)}(x_1, x_2)$$

for the jcdf, where by definition $\tau = t_1 - t_2$.

From Equations 7.16 and 7.17 we can immediately deduce that the mean, covariance, and correlation of a stationary random process $\{X(t)\}$ enjoy certain special properties. More specifically, for any $t, t_1, t_2 \in I$, we have

$$
\begin{aligned}
m_X(t) &= EX(t) = m, \\
C_X(t_1, t_2) &= C_X(t_1 - t_2) = C_X(t_2 - t_1), \\
R_X(t_1, t_2) &= R_X(t_1 - t_2) = R_X(t_2 - t_1).
\end{aligned}
$$

The proofs of the above equalities are left as exercises.

The notion of stationarity can be easily extended to that of **joint stationarity** for two random processes $X(t)$ and $Y(t)$: $X(t)$ and $Y(t)$ will be jointly stationary if their joint cdf's (as defined in Equation 7.7) do not depend on the placement of the time origin. The notion of joint stationarity is used extensively in Section 9.4.

How do we check stationarity in practice? If we know all the cdf's as a function of $t_1, t_2, \ldots$, then we may be able to check stationarity directly by applying the definitions. Otherwise, this is very difficult to do, if not impossible. Note that no theorem that establishes stationarity has been reported in the literature. Instead, we must assume this property and justify it for a given situation by using first principles.

Example 7.15. Consider the sinusoid random process in Equation 7.1. Is it stationary? Since the one-dimensional cdf in Equation 7.4 is a function of time, the answer is easily no.

Example 7.16. Consider the independent increment random process. Is it stationary? Since the n-dimensional pmf in Equation 7.13 depends in general on time, the answer is no.

Example 7.17. Consider the IID random process. Is it stationary? Since the n-dimensional pmf in Equation 7.14 does not depend on time, the answer is yes.

Example 7.18. Consider a sample space $S = [0, 1]$ with a uniform distribution for its outcomes ζ. Let $I = [1, \infty)$. Consider the random process $\{X(t)\}$ with sample paths given by $X(t, \zeta) = \zeta/t^2$. Is it stationary?

We can easily check that $0 \le X(t, \zeta) \le 1$. Fix a $t_1 \in I$. We will evaluate $F_{X(t_1)}(x)$. Clearly, for $x \le 0$, $F_{X(t_1)}(x) = 0$; for $x \ge 1$, $F_{X(t_1)}(x) = 1$. Consider therefore $0 < x < 1$. We have

$$F_{X(t_1)}(x) = P[X(t_1) \le x] = P[\zeta \in [0, 1] : \zeta \le xt_1^2],$$

and thus

$$F_{X(t_1)}(x) = \begin{cases} xt_1^2, & xt_1^2 \le 1, \\ 1, & xt_1^2 > 1. \end{cases}$$

Therefore, the process is not stationary.

3. Wide-sense stationary random processes. As we alluded in the previous case, it is rather impractical to determine whether a random process is stationary. It certainly seems that determining whether

$$m_X(t) \overset{?}{=} m \tag{7.18}$$

(i.e., the mean is a constant) and

$$C_X(t_1, t_2) \overset{?}{=} C_X(t_2 - t_1) \tag{7.19}$$

[i.e., the autocovariance (or, equivalently, the autocorrelation)] are a function of $t_2 - t_1$ only, is much less work.[7]

Definition: A random process $\{X(t)\}$ is called a **wide-sense stationary** (WSS) process if Equations 7.18 and 7.19 hold true.

Definition: Two random processes $\{X(t)\}$ and $\{Y(t)\}$ are called **jointly wide-sense stationary** (JWSS) if they are individually wide-sense stationary and

$$R_{XY}(t_1, t_2) = R_{XY}(t_1 - t_2).$$

Wide-sense stationary processes were also introduced by A. Khinchine, in 1938 [39]. They will play a central role in Chapter 9. For this reason, we list here some general properties of their autocorrelation function. For simplicity, let's write $R_X(\tau) = EX(t + \tau)X(t)$.

Theorem 7.2 *The autocorrelation function of a wide-sense stationary random process satisfies the following properties:*
(a) $R_X(\tau) = R_X(-\tau)$, that is, it is an even function,
(b) $R_X(\tau) \le R_X(0)$, that is, the function is maximum at $\tau = 0$,
(c) If there is a constant $T > 0$ such that $R_X(0) = R_X(T)$, then $R_X(\tau)$ is periodic.

Proof: (a) We have

$$R_X(\tau) = EX(t + \tau)X(t) = EX(t)X(t + \tau) = R_X(-\tau).$$

(b) This property follows directly from Schwarz' inequality (see Problem 4.113) by setting $X = X(t)$ and $Y = X(t + \tau)$.

---

[7]But how much less? What does it take to verify Equation 7.19?

(c) From Schwarz' inequality, we have

$$\{EX(t)[X(t+\tau+T) - X(t+\tau)]\}^2$$
$$\le \quad EX^2(t)E[X(t+\tau+T) - X(t+\tau)]^2,$$

which implies that

$$[R_X(\tau+T) - R_X(\tau)]^2 \le 2\,[R_X(0) - R_X(T)]\,R_X(0).$$

Therefore, if $R_X(0) = R_X(T)$, $R_X(\tau+T) = R_X(\tau)$, for all τ.   □

We can ask the "inverse" question as well. Suppose we are given a function $R(\tau)$; under what conditions can it be the autocorrelation function of a WSS random process $\{X(t)\}$? The following theorem, which we state without proof (see [23]), provides the answer.

Theorem 7.3 *The function $R(\tau)$, $\tau \in R$, is the autocorrelation function of a real-valued, WSS, continuous-time random process $\{X(t)\}$ if and only if $R(\tau)$ is a positive definite function.*

Checking the property of positive definiteness is not easy in general. The following criterion, which we state without proof, is easier to check. Note that the next theorem provides a sufficient condition, whereas the previous one provided a necessary and sufficient condition.

Theorem 7.4 *The even function $R(\tau)$, $\tau \in R$, is the autocorrelation function of a real-valued, WSS, continuous-time random process $\{X(t)\}$ if*
(a) $R(\tau)$ is concave for $\tau > 0$ and
(b) $\lim_{\tau \to \infty} R(\tau) = c$, where c is a finite constant.

Example 7.19.   Are stationary processes wide-sense stationary? From the definition of a stationary process we can easily verify that Equation 7.19 holds true, so the answer is yes.

Example 7.20.   Consider the sinusoid process in Equation 7.1. Is it wide-sense stationary? Since

$$m_X(t) = \frac{\sin(2\pi t)}{2},$$

from Example 7.13 the process is not WSS.

4. Cyclostationary random processes.   Another useful class of random processes is the following:

Definition: A random process $\{X(t)\}$ is called **cyclostationary** if Equation 7.15 holds true only for values of s that are integer multiples of a basic period $T > 0$, more precisely, if

$$F_{X(t_1)\cdots X(t_n)}(x_1, x_3, \ldots, x_n) = F_{X(t_1+lT)\cdots X(t_n+lT)}(x_1, x_2, \ldots, x_n)$$

for every integer l and a fixed constant T.

Definition: A random process $\{X(t)\}$ is called **wide-sense cyclostationary** if for all integers l and a fixed constant $T > 0$, we have

$$
\begin{aligned}
m_X(t_1 + l \cdot T) &= m_X(t_1), \quad \forall t_1 \in I, \\
C_X(t_1 + l \cdot T, \, t_2 + l \cdot T) &= C_X(t_1, t_2) \quad \forall t_1, t_2 \in I.
\end{aligned}
$$

It is a trivial conclusion that a cyclostationary process is also wide-sense cyclostationary.

The sinusoid random process of Equation 7.1 is cyclostationary. The process with sample paths given by $X(t, \zeta) = \zeta/t^2$ is not. (Prove that.)

5. Discrete-time Markov Chains. Consider a process $\{X_n, n = 0, 1, 2, \ldots\}$ which takes values in a discrete set A.

Definition: The process $\{X_n, n = 0, 1, 2, \ldots\}$ is called a **Markov Chain** if for all positive integers n and all $j_{n+1}, j_n, \ldots, j_1, j_0 \in A$, its conditional pmf's satisfy the property

$$
P[X_{n+1} = j_{n+1} | X_n = j_n, \ldots, X_0 = j_0] = P[X_{n+1} = j_{n+1} | X_n = j_n]. \quad (7.20)
$$

If we consider the event $\{X_n = j_n, \ldots, X_1 = j_1, X_0 = j_0\}$ to be a representation of the "past up to time n," n to be the "present," and $n + 1$ the "future," then according to Equation 7.20, a process is a Markov Chain if its "future is independent of the past, given the present."

We study this important class of random processes in great detail in Chapter 10. We will now show only that the joint pmf can be easily evaluated. Let $A = P[X_n = j_n, \ldots, X_1 = j_1, X_0 = j_0]$. We have

$$
\begin{aligned}
A &= P[X_n = j_n | X_{n-1} = j_{n-1}, \ldots, X_0 = j_0] \cdot P[X_{n-1} = j_{n-1}, \ldots, X_0 = j_0] \\
&= P[X_n = j_n | X_{n-1} = j_{n-1}] \cdot P[X_{n-1} = j_{n-1}, \ldots, X_0 = j_0] \\
&= P[X_n = j_n | X_{n-1} = j_{n-1}] \cdot P[X_{n-1} = j_{n-1} | X_{n-2} = j_{n-2}, \ldots, X_0 = j_0] \\
&= \cdot P[X_{n-2} = j_{n-2}, \ldots, X_0 = j_0] \\
&= P[X_n = j_n | X_{n-1} = j_{n-1}] \cdot P[X_{n-1} = j_{n-1} | X_{n-2} = j_{n-2}] \\
&\quad \cdots P[X_0 = j_0].
\end{aligned}
$$

Therefore, to evaluate the jpmf of the process $\{X_n\}$, we need to determine *only* (1) the conditional pmf's $P[X_n = j_n | X_{n-1} = j_{n-1}]$ and (2) the initial pmf $P[X_0 = j_0]$. This requires a lot less information, especially when the conditional pmf's do not depend on n. The Markov Chain model has seen tremendous success in modeling systems with memory.

7.5   USEFUL RANDOM PROCESS MODELS

We will see now in greater detail some basic but useful random process models. The two most widely used models, namely, the Poisson and Gaussian, are studied separately in the next chapter.

7.5.1   Random Walk

The physical problems motivating the random walk model derive from diverse applications. The model has been used to study, among other applications, the walking patterns of drunk persons (hence the name), the fortune of a bettor after n bets in a game, and the queue size in a network buffer as a function of time.

1. Classification

A number of variations and extensions of the basic model have appeared. The model has *discrete-* and *continuous-*time versions; *symmetric* and *asymmetric* versions; *discrete-* and *continuous-*space versions; *one-* or *multi*dimensional random variables; the range of the random variables can be bounded or not, in which case we talk of *restricted* or *unrestricted* models; finally, restricted random walks can have *reflecting* or *absorbing* barriers (to be defined shortly).

2. Unrestricted random walk

Random walks were studied extensively by G. Pólya in the beginning of the twentieth century; he also coined the term. We describe next a discrete-time, discrete-space, one-dimensional, unrestricted model.

Let $I = \{1, 2, \ldots\}$ be the index set for this process. Define a sequence of IID random variables $\{Y_i\}_{i=1}^{\infty}$ on some space $(S, \mathcal{Q}, P)$, such that

$$P[Y_i = +1] = p, \quad P[Y_i = -1] = 1 - p,$$

where $0 < p < 1$ is a given parameter.

Definition: The random process $\{X_n\}_{n=1}^{\infty}$, defined by

$$X_n \triangleq \sum_{i=1}^{n} Y_i,$$

is called an **unrestricted random walk**.

The range of the random variable X_n is $\{-n, -n+1, \ldots, -1, 0, 1, \ldots, n\}$. Thus the range of the process is the set of all integers.

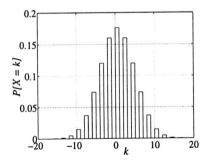

Figure 7.9   The pmf of a random walk, $n = 20, p = 0.5$.

3. Probabilistic description

We can immediately recognize a random walk as a sum process. Therefore, we can apply the theory in Section 6.3 to calculate its n-dimensional pmf, mean, variance, and generating function.

The one-dimensional pmf of the random variable X_n can be found in a simple way as follows. For any n, there will be k variables Y_i that have a value of $+1$ and $n - k$ variables Y_i that have a value of -1. Here the possible values for k are $k = 0, 1, \ldots, n$. Therefore, $X_n = (+1) \cdot k + (-1) \cdot (n - k) = 2k - n$. Thus,

$$P[X_n = 2k - n] = \binom{n}{k} p^k (1 - p)^{n-k}, \quad k \in \{0, 1, 2, \ldots, n\}. \tag{7.21}$$

Observe that the right-hand side of Equation 7.21 is identical to the binomial distribution; keep in mind, however, that X_n is *not* a binomial random variable since its range is not the set $\{0, 1, \ldots, n\}$. The pmf in Equation 7.21 is shown in Figure 7.9 for $n = 20, p = 0.5$.

The mean of the random variable X_n is given by

$$EX_n = E \sum_{i=1}^{n} Y_i = n(2p - 1).$$

Its variance is

$$\text{var}(X_n) = \text{var}(\sum_{i=1}^{n} Y_i) = 4np(1 - p).$$

4. Restricted random walk

Consider a finite subset of the integers $\{0, 1, \ldots, a\}$, where $0 < a$. Consider a sequence of random variables $\{X_k\}_{k=0}^{\infty}$ such that

$$P[X_{k+1} = i - 1 | X_k = i] = 1 - p, \quad P[X_{k+1} = i + 1 | X_k = i] = p,$$

whenever $i \neq 0, a$, and

$$P[X_{k+1} = 1 | X_k = 0] = 1, \quad P[X_{k+1} = a | X_k = a] = 1.$$

We say that 0 is a *reflecting barrier* and a is an *absorbing barrier*. Suppose for simplicity of presentation that, at time $k = 0$, $X_0 = 0$.

Definition: The sequence $\{X_k\}$ is called a **restricted random walk** (with both reflecting and absorbing) barriers.

The *time to absorption* is a random variable T, defined as

$$T = \min\{k > 0 : X_k = a\}.$$

We evaluate next the mean ET. Toward this end, we define the random variables $T_i, i = 1, \ldots, a$, as follows:

$$T_i = \min\{n > 0 : X_{k+n} = i, \text{ given that } X_k = i - 1\}.$$

In other words, T_i is the "time it takes" the random walk to move from a value $i - 1$ to value i. (Why doesn't T_i depend on k?) Then

$$ET = \sum_{i=1}^{a} ET_i.$$

We can evaluate ET_i using the total probability theorem for expected values (see Problem 3.194). Conditioning on the random variable $X_{k+1} - X_k$, we have

$$
\begin{aligned}
ET_i &= E[E(T_i | X_{k+1} - X_k)] \\
&= p \cdot E(T_i | X_{k+1} - X_k = +1) + (1 - p) \cdot E(T_i | X_{k+1} - X_k = -1).
\end{aligned}
$$

Observe that, given the event $X_{k+1} - X_k = -1$ (i.e., a transition to the left), we have $T_i = 1 + T_{i-1} + \tilde{T}_i$, where T_i and $\tilde{T}_i$ have identical distributions. Therefore,

$$ET_i = p \cdot 1 + (1 - p) \cdot E(1 + T_{i-1} + T_i),$$

and thus

$$ET_i = \frac{1}{p} + \frac{1-p}{p} ET_{i-1},$$

with initial condition $ET_1 = 1$ since 0 is a reflecting barrier. Using standard techniques from difference equations, we can solve for ET_i; then

$$ET = \begin{cases} a^2, & p = 0.5, \\ \frac{a}{2p-1} - \frac{2(1-p)}{2p-1} \frac{1-(\frac{1-p}{p})^a}{1-\frac{1-p}{p}}, & p \neq 0.5. \end{cases}$$

Remark: The random walk is related to a number of other models. It is a special case of a sum process and of a process with stationary increments. As we will see, it is a special case of a Markov Chain (Chapter 10); the Wiener process (Chapter 8) is a random walk in continuous time.

7.5.2 ARMA Processes

The *autoregressive moving-average (ARMA)* process is a widely used model that describes physical systems with memory. Consider a sequence of random variables $\{Y_i\}_{i=1}^\infty$ not necessarily IID. The ARMA process $\{X_i\}_{i=1}^\infty$ is defined via the recursion

$$X_n = \sum_{k=1}^{M} a_k X_{n-k} + \sum_{l=0}^{L} b_l Y_{n-l}, \quad n = 1, 2, \ldots, \tag{7.22}$$

where M, L are given nonnegative integers, $\{a_k\}, \{b_l\}$ are given real numbers, and $Y_0, \ldots, Y_{1-L}$ are random variables with known jpdf. When all $b_l = 0$, we have the *autoregressive* model; when all $a_k = 0$, we have the *moving-average* model.

The model has found wide applicability in signal processing. Since the random variables $\{X_i\}$ are in general dependent, as Equation 7.22 suggests, even when the $\{Y_i\}$ *are* IID, it is not possible to derive the n-dimensional distributions of the model in closed form. We see in Chapter 9 a more suitable description of this process, via the notion of power spectral densities.

7.6 CONTINUITY, DERIVATIVES, AND INTEGRALS

In Chapter 9 we see that random processes are commonly used as inputs to (linear) systems, such as integrators, delay systems, and differentiators. Such systems are commonly described via differential equations, which relate the output to the input of the system. The notions of continuity, derivatives, and integrals are central to the study of system behavior in the deterministic case, where the input signal is a deterministic function of time. We need to develop their probabilistic counterparts for a more rigorous study of systems with random processes as inputs. Moreover, in linear systems theory, it is known that the various transforms (Laplace, Fourier, Z transform) and signal expansions (trigonometric, Fourier expansion) play a fundamental role in the development of the theory. We treat transforms of random processes in Chapter 9; we study the Karhunen-Loève expansion of a random process in Section 7.8. We develop the three basic notions of continuity, derivatives, and integrals in this section. Recall from the deterministic case that all three notions involve some form of limits, so we can develop probabilistic counterparts using any of the convergence forms developed in Chapter 6. However, in practice, mean square convergence has proved to be the most useful, so we confine our attention to this case only.

7.6.1   Continuity

Definition: The random process $X(t)$ is **mean square continuous at the point** $t_0 \in I$ if

$$E[X(t) - X(t_0)]^2 \to 0 \quad \text{as } t \to t_0. \tag{7.23}$$

If Equation 7.23 holds true for all $t_0 \in I$, the random process is called **mean square continuous**.

How do we check whether a given random process is mean square continuous at a given point t_0? The following theorem provides a sufficient condition in terms of the autocorrelation function, which is easy to verify in practice.

Theorem 7.5 *Let $X(t)$ be a random process with autocorrelation $R_X(t, s)$. If $R_X(t, s)$ is continuous at the point (t_0, t_0), then $X(t)$ is mean square continuous at t_0.*

Proof: From the definition of continuity, we can easily get

$$E[X(t) - X(t_0)]^2 = R_X(t, t) - R_X(t, t_0) - R_X(t_0, t) + R_X(t_0, t_0).$$

Let $t \to t_0$ in the above expression; if $R_X(t, s)$ is continuous at the point (t_0, t_0), the right-hand side approaches 0. □

Example 7.21. Is the sinusoid process of Example 7.1 mean square continuous at any point?

The autocorrelation function for this process was calculated in Equation 7.12. We can immediately see that $R_X(t, s)$ is continuous, so the sinusoid process is mean square continuous at all points.

Example 7.22. As we will see in Problem 8.7, the Poisson random process is mean square continuous; however, all its sample paths are steplike functions, and thus none is a continuous function of time.

Remark: How do we check continuity in practice? To apply Theorem 7.5, we should know the form of $R_X(t, s)$. In most practical situations, we calculate $R_X(t, s)$ *approximately*, through a histogram. Therefore, it will never be continuous. So, often we must stipulate that the random process be mean square continuous.

7.6.2 Derivatives

Definition: The random variable $X'(t_0)$ is called the **derivative** of the random process $X(t)$, at the point $t_0 \in I$ if

$$E\left[\frac{X(t_0 + \epsilon) - X(t_0)}{\epsilon} - X'(t_0)\right]^2 \to 0 \quad \text{as } \epsilon \to 0. \tag{7.24}$$

If Equation 7.24 holds true for all $t_0 \in I$, the random process $X'(t)$ is called the derivative of the random process $X(t)$.

How do we check whether a given random process has a derivative at some given point t_0? A useful criterion, which does *not* involve the (unknown) derivative, can be given in terms of the autocorrelation function.

Theorem 7.6 *Let $X(t)$ be a random process with autocorrelation $R_X(t, s)$. If $R_X(t, s)$ has a second-order derivative at the point (t_0, t_0), then the random process $X(t)$ has a derivative at the point t_0.*

Proof: Define

$$A \triangleq \lim_{\epsilon, \delta \to 0} E\left[\frac{X(t + \epsilon) - X(t)}{\epsilon} - \frac{X(t + \delta) - X(t)}{\delta}\right]^2.$$

Let's apply Cauchy's criterion for convergence to the derivative in Equation 7.24:

$$A = \lim_{\epsilon \to 0} E\left[\frac{X(t_0 + \epsilon) - X(t_0)}{\epsilon}\right]^2 + \lim_{\delta \to 0} E\left[\frac{X(t_0 + \delta) - X(t_0)}{\delta}\right]^2$$

$$- 2 \lim_{\epsilon, \delta \to 0} E\left[\frac{X(t_0 + \epsilon) - X(t_0)}{\epsilon} \cdot \frac{X(t_0 + \delta) - X(t_0)}{\delta}\right]$$

$$= \frac{\partial^2 R_X(t_0, t_0)}{\partial t \, \partial s} + \frac{\partial^2 R_X(t_0, t_0)}{\partial t \, \partial s}$$

$$- 2 \lim_{\epsilon, \delta \to 0} \frac{R_X(t_0 + \epsilon, t_0 + \delta) - R_X(t_0 + \epsilon, t_0) - R_X(t_0, t_0 + \delta) + R_X(t_0, t_0)}{\epsilon \delta}$$

$$= \frac{\partial^2 R_X(t_0, t_0)}{\partial t \, \partial s} + \frac{\partial^2 R_X(t_0, t_0)}{\partial t \, \partial s} - 2\frac{\partial^2 R_X(t_0, t_0)}{\partial t \, \partial s} = 0.$$

QED.   □

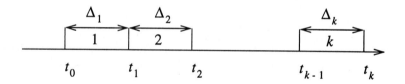

Figure 7.10   Riemann approximation of an integral.

Example 7.23. The process in Example 7.1 has a derivative at all points t since from Equation 7.12 we can see that $R_X(t, s)$ is differentiable at all points.

7.6.3   Integrals

Consider a time interval $(T_0, T_1) \subset I$. We are interested in defining the integral of the random process $X(t)$ over such an interval. Following the Riemann[8] notion of an integral in the deterministic case, let's consider a partition of the interval (T_0, T_1) into $k < \infty$ subintervals of length $\Delta_1, \ldots, \Delta_k$ (see Figure 7.10) such that $t_0 = T_0$, $t_k = T_1$, and

$$t_1 = t_0 + \Delta_1, \ t_2 = t_1 + \Delta_2, \ t_3 = t_2 + \Delta_3,$$

and so on.

For any finite value of k, the random variable

$$Y \triangleq \sum_{i=1}^{k} X(t_i) \Delta_i$$

is a well-defined sum. Now if we let $k \to \infty$ (so that $\Delta_i \to 0$), we can define the integral of a random process $X(t)$ in the mean square sense:

Definition: The random variable Y is called the **integral** of the random process $X(t)$ [over the time interval (T_0, T_1)] if

$$E \left[Y - \sum_{i=1}^{k} X(t_i) \Delta_i \right]^2 \to 0 \ \text{ as } k \to \infty.$$

We can easily check whether the integral exists, through a Cauchy-type criterion that again involves the autocorrelation function $R_X(t, s)$ only.

Theorem 7.7 *Let $X(t)$ be a random process with autocorrelation $R_X(t, s)$. If $R_X(t, s)$ is such that the integral*

$$\int_{T_0}^{T_1} \int_{T_0}^{T_1} R_X(t, s) \, dt \, ds \tag{7.25}$$

---

[8] After Georg Bernhard Riemann, 1826–1866, a German mathematician.

exists, then the integral of $X(t)$ over (T_0, T_1) exists.

Proof: From Cauchy's criterion, we know that the integral of $X(t)$ exists when

$$E\left[\sum_{i=1}^{k} X(t_i)\,\Delta_i - \sum_{j=1}^{l} X(t_j)\,\Delta_j\right]^2 \to 0, \qquad (7.26)$$

when $k, l \to \infty$. We expand now the square in Equation 7.26; we can easily see that the expansion will give us terms of the form

$$E\left[\sum_{i=1}^{k}\sum_{j=1}^{l} X(t_i)X(t_j)\,\Delta_i\,\Delta_j\right] = \sum_{i=1}^{k}\sum_{j=1}^{l} R_X(t_i, t_j)\,\Delta_i\,\Delta_j. \qquad (7.27)$$

The right-hand side of Equation 7.27 approaches the integral in 7.25 as $k, l \to \infty$. Therefore, if the integral in 7.25 exists, the integral of the random process itself exists. QED. $\qquad\qquad\square$

We state the following theorem without proof; it provides a sufficient condition for integrability.

Theorem 7.8 *Suppose that $X(t)$ is a mean square continuous process. Then its integral exists.*

7.7   ERGODICITY

Ergodicity is a subtle concept, difficult to grasp at first, but nevertheless very useful and widely used. Ergodicity is difficult partly because it can be defined in a number of ways (e.g., in the mean square sense, in the almost-sure sense) and can be applied to stationary or nonstationary processes. The first ergodic theorems (in the almost-sure sense) were derived by J. von Neumann and G. Birkoff in the first half of the twentieth century. A survey of the early results is given in [29].

Loosely speaking, ergodicity enables us to calculate statistical parameters of the random process, such as means and correlations, through time averages of *a single sample path*. We saw in Chapter 3 that histograms can take a lot of repeated measurements of a random variable. Since sample paths are *functions* of time, storing multiple sample paths in order to calculate a mean could be a nontrivial problem. Operating on *just one* can imply significant storage savings. Besides, time averaging (to be defined shortly) is a fairly simple and easily implementable procedure.

Let's elaborate on the time-averaging concept a little. Consider the *netstat* experiment in Section 1.9. Define a random process $\{N(t)\}$ for this experiment as follows: Choose $I = \{0.00, 0.01, \ldots, 23.59\}$; i.e., the index set is all the 1-min intervals in a day. Suppose we are interested in measuring the mean number of packets $EN(t)$, for $t = 10.00$ (i.e., mean traffic at 10 AM). We can perform, say, 100 experiments and approximate this mean as

$$EN(10.00) \approx \frac{1}{100} \sum_{i=1}^{100} N_i(10.00), \tag{7.28}$$

where $N_i(10.00)$ is the ith measurement. (Of course, we appeal to the Strong Law of Large Numbers here.)

This would require a rather lengthy experiment since we must perform it over 100 different days! And except for the 1-min interval from 10:00 to 10:01, our equipment would be idle! It would be nice if we could use the equipment from, say, 10:00 to 11:40, obtain 100 values again, and use those to approximate $EN(10.00)$ as

$$EN(10.00) \approx \frac{1}{100} \sum_{i=1}^{100} \tilde{N}_i, \tag{7.29}$$

where $\tilde{N}_i$ is again the ith measurement. The two experiments are of course drastically different: In Equation 7.28 we are using 100 variates of the *same* random variable, namely, $N(10.00)$. In the "continuous" measurement approach in Equation 7.29, we are using a single sample path of the process, namely, its portion from time 10:00 until time 11:40. In more mathematical jargon, in Equation 7.28 we are varying ζ 100 times, while we keep t fixed at 10.00. In Equation 7.29, we are varying t 100 times, while we keep ζ fixed.

How do we make sure that we can come up with the same value for $EN(10.00)$ in both experiments? The concept of ergodicity addresses precisely this question, and it allows us to use time averaging instead of statistical averaging to calculate parameters such as the mean or variance. As a matter of fact, the savings of time averaging over statistical averaging are so great that, in practice, we are willing to take advantage of ergodicity even when the process is known to be nonergodic!

Ergodic theorems were stated in the past under a variety of conditions regarding the random processes to which they apply; earlier versions were preoccupied with stationary or WSS processes only. It was thought for a while that stationarity was *needed* for ergodicity. Even though we do not present any theorems to that effect in this book, this is not the case at all.

Consider a random process $X(t)$, possibly nonstationary.

Definition: The random process $X(t)$ is mean ergodic if and only if

$$\lim_{T \to \infty} \frac{1}{2T} \int_{-T}^{T} X(t) \, dt = m \tag{7.30}$$

in the mean square sense. Note that the quantity m in Equation 7.30 need not be the mean of $X(t)$! It could, in general, be a random variable.

The following theorems provide conditions for mean ergodicity, in discrete and continuous time. The proof of the first one is left as an exercise (see Problem 7.60).

Theorem 7.9 *Let $\{X_n\}$ be a discrete-time WSS random process with finite mean m and covariance $C_X(n)$. A sufficient condition for $\{X_n\}$ to be mean ergodic is*

$$C_X(0) < \infty, \quad \lim_{n \to \infty} C_X(n) = 0. \tag{7.31}$$

Remark: An intuitive explanation of the conditions in Theorem 7.9 is that the random variables in the process become asymptotically uncorrelated. So we can think of them as "almost" independent when n is sufficiently large. Then Equation 7.30 can be thought of as an extension of the Strong Law of Large Numbers.

Theorem 7.10 *Let $\{X(t)\}$ be a continuous-time WSS random process with finite mean m and covariance $C_X(s)$. A necessary and sufficient condition for $\{X(t)\}$ to be mean ergodic is*

$$\lim_{T \to \infty} \frac{1}{2T} \int_{-2T}^{2T} \left(1 - \frac{|s|}{2T}\right) C_X(s) \, ds = 0.$$

Proof: Fix $T > 0$. For simplicity, let's define

$$X_T \triangleq \frac{1}{2T} \int_{-T}^{T} X(t) \, dt.$$

We can easily check that $EX_T = m$ for all T. Moreover,

$$
\begin{aligned}
\text{var}(X_T) &= E(X_T - m)^2 \\
&= E\left[\frac{1}{2T} \int_{-T}^{T} X(t) \, dt - m\right]\left[\frac{1}{2T} \int_{-T}^{T} X(s) \, ds - m\right] \\
&= \frac{1}{4T^2} \int_{-T}^{T} \int_{-T}^{T} E[X(t) - m)(X(s) - m] \, dt \, ds \\
&= \frac{1}{4T^2} \int_{-T}^{T} \int_{-T}^{T} C_X(t - s) \, dt \, ds.
\end{aligned}
$$

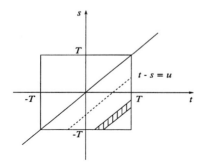
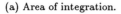
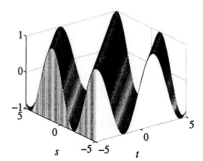

(a) Area of integration.   (b) A typical covariance function.

Figure 7.11

The integration in the above equation can be done easily over the diagonal lines $y = t - s$ [see Figure 7.11(a)] since the function $C_X(t - s)$ is constant over such lines [see Figure 7.11(b)].

After a little algebra, we can see that the area of an infinitesimal strip around the $y = t - s$ line is equal to $(2T - |y|)\, dy$. We can thus simplify the double integral in the previous expression, and therefore

$$
\begin{aligned}
\mathrm{var}(X_T) &= \frac{1}{4T^2} \int_{-2T}^{2T} (2T - |y|) C_X(y)\, dy \\
&= \frac{1}{2T} \int_{-2T}^{2T} \left(1 - \frac{|y|}{2T}\right) C_X(y)\, dy.
\end{aligned}
$$

The condition of the theorem now guarantees that $\mathrm{var}(X_T) \to 0$, and thus mean ergodicity is ensured. This proves the sufficiency part of the theorem. The necessary part is obvious. □

Example 7.24. Consider a WSS random process with covariance $C_X(\tau) = e^{-2a|\tau|}$. Is this process mean ergodic? Let

$$
A \triangleq \frac{1}{2T} \int_{-2T}^{2T} \left(1 - \frac{|s|}{2T}\right) e^{-2a|s|}\, ds.
$$

Applying Theorem 7.10, we have

$$
A = \frac{1}{T} \int_{0}^{2T} e^{-2as}\, ds - \frac{1}{2T^2} \int_{0}^{2T} s e^{-2as}\, ds
$$

$$= \frac{1}{2aT}\left(1 - e^{-4aT}\right) - \frac{1}{8a^2T^2}\left(1 - 4aTe^{-4aT} - e^{-4aT}\right) \to 0,$$

and thus the process is mean ergodic.

Example 7.25. Consider the trivial random process $X(t) = Y$, where Y is a random variable with a given mean and variance. Is this process mean ergodic?

We can easily check that $EX(t) = EY$ and $C_X(\tau) = \mathrm{var}(Y)$. Since C_X does not approach 0 as $\tau \to \infty$, we suspect trouble. Indeed, by applying Theorem 7.10 again, we can see that

$$\lim_{T\to\infty} \frac{1}{2T} \int_{-2T}^{2T} \left(1 - \frac{|s|}{2T}\right) \mathrm{var}(Y)\,ds = \mathrm{var}(Y),$$

and the process cannot be mean ergodic.

7.8   KARHUNEN-LOÈVE EXPANSIONS

The random process $X(t)$ is in general a collection of an uncountably infinite number of random variables. The Karhunen-Loève expansion of a random process is an alternative way of representing the random process with *only* a countably infinite number of random variables. Even though this does not sound like significant savings, it simplifies things a lot, especially in theoretical studies. The theory of this section is based on the work of Karhunen [36] and Cramér [14], in the middle of the twentieth century.

We present a special case of WSS processes first, to clarify the main ideas. Define the random process $\{X(t)\}$ as *mean square periodic*, with period T, if $E[X(t+T) - X(t)]^2 = 0$ for some $T > 0$.

Theorem 7.11 *Let $\{X(t),\ t \in [0,\infty)\}$ be a zero-mean, mean square periodic, WSS random process with period $T > 0$. Then for any fixed $t > 0$,*

$$X(t) = \sum_{k=-\infty}^{\infty} X_k e^{j2\pi kt/T},$$

in the mean square sense, where

$$X_k \triangleq \frac{1}{T}\int_0^T X(s) e^{-j2\pi ks/T}\,ds.$$

Proof: In the following, the superscript $*$ denotes conjugates of complex numbers, and all sums extend from $-\infty$ to ∞. Let $A \triangleq E\left|X(t) - \sum X_k e^{j2\pi kt/T}\right|^2$.

We have

$$
\begin{aligned}
A &= E\left[X(t) - \sum X_k e^{j2\pi kt/T}\right]\left[X(t) - \sum X_l e^{j2\pi lt/T}\right]^* \\
&= E|X(t)|^2 + E\left(\sum\sum X_k X_l^* e^{j2\pi(k-l)t/T}\right) \\
&\quad - E\left[X(t)\sum X_k^* e^{-j2\pi kt/T}\right] - E\left[X^*(t)\sum X_k e^{j2\pi kt/T}\right] \quad (7.32)
\end{aligned}
$$

It is easy to check that when the random process $\{X(t)\}$ is mean square periodic, its autocorrelation function $R_X(\tau)$ is periodic with the same period T. Therefore, $R_X(\tau)$ itself can be expanded in a Fourier series with coefficients a_k. We can also easily check that $R_X(\tau)$ is an even function; therefore, a_k is a real number. Thus,

$$
R_X(\tau) = \sum a_k e^{j2\pi k\tau/T}.
$$

Consider now the quantity $EX_k X_m^*$. We have

$$
\begin{aligned}
EX_k X_m^* &= \frac{1}{T}E\left[X_k \int_0^T X^*(s)e^{j2\pi ms/T}\,ds\right] \\
&= \frac{1}{T}\int_0^T EX_k X^*(s)e^{j2\pi ms/T}\,ds.
\end{aligned}
$$

We can evaluate the quantity $EX_k X^*(s)$ as follows:

$$
\begin{aligned}
EX_k X^*(s) &= \frac{1}{T}E\left[\int_0^T X(v)e^{-j2\pi kv/T}\,dv \cdot X^*(s)\right] \\
&= \frac{1}{T}\int_0^T R_X(v-s)e^{-j2\pi kv/T}\,dv \\
&= \frac{1}{T}\int_{-s}^{T-s} R_X(v)e^{-j2\pi kv/T}\,dv \cdot e^{-j2\pi ks/T} \\
&= a_k e^{-j2\pi ks/T}.
\end{aligned}
$$

The last equality follows from the periodicity of $R_X(v)$. Therefore,

$$
EX_k X_m^* = a_k \delta_{km},
$$

where δ_{km} denotes the usual Kronecker symbol. Substituting in Equation 7.32, we have

$$
E\left|X(t) - \sum X_k e^{j2\pi kt/T}\right|^2 = R_X(0) + \sum a_k - \sum a_k + \sum a_k^* = 0.
$$

The last equality follows since the a_k are all real and $R_X(0) = \sum a_k < \infty$. $\square$

In the above theorem, $X(t)$ was restricted to being mean square periodic. The Karhunen-Loève expansion allows us to handle nonstationary random processes. In the sequel, T is a fixed positive real number. (If $T = \infty$, too many problems arise, so we do not treat this case here.) Let $\phi_k(t)$ be orthonormal functions in $[0, T]$, that is,

$$\int_0^T \phi_k(t)\phi_m(t)\, dt = \delta_{km}, \quad k, m = 1, 2, \ldots.$$

Let

$$X_k \triangleq \int_0^T X(t)\phi_k^*(t)\, dt, \quad k = 1, 2, \ldots \tag{7.33}$$

in the mean square sense. (Why does the integral exist?) The Karhunen-Loève expansion of a random process is the following:

Theorem 7.12 *Consider a zero-mean random process $\{X(t),\ t \in [0, \infty)\}$ with autocorrelation function $C_X(t, s)$. Let $T > 0$ be a fixed real constant. Then*

$$X(t) = \sum_{k=1}^{\infty} X_k \phi_k(t), \quad 0 \leq t \leq T, \tag{7.34}$$

in the mean square sense. The random variables X_k are defined as in Equation 7.33, and $\phi_k(t)$ are the eigenfunctions of $C_X(t, s)$.

Proof: The orthonormal functions $\phi_k(t)$ are the solutions to the eigenvalue equation

$$\int_0^T C_X(t, s)\phi_k(t)\, dt = \lambda_k \phi_k(s), \quad 0 \leq s \leq T, \tag{7.35}$$

where λ_k and $\phi_k(t)$ are the eigenvalues and eigenfunctions of the covariance matrix $C_X(t, s)$. It can be shown that when $C_X(t, s)$ is continuous, the eigenfunctions are orthonormal and satisfy the so-called Mercer equation:

$$C_X(t, s) = \sum_{k=1}^{\infty} \lambda_k \phi_k(t)\phi_k^*(s), \quad t, s \in [0, T].$$

Consider now Equation 7.33, applied to the eigenfunctions $\phi_k(t)$. We have

$$
\begin{aligned}
EX_m^* X_k &= EX_m^* \int_0^T X(s)\phi_k^*(s)\, ds = \int_0^T EX(s)X_m^* \phi_k^*(s)\, ds \\
&= \int_0^T E\left[X(s) \int_0^T X^*(v)\phi_m(v)\, dv \right] \phi_k^*(s)\, ds
\end{aligned}
$$

$$= \int_0^T \int_0^T C_X(s,v)\phi_m(v)\,dv \cdot \phi_k^*(s)\,ds$$

$$= \int_0^T \lambda_m \phi_m(s)\phi_k^*(s)\,ds = \lambda_k \delta_{km},$$

and thus the random variables X_k, X_m are orthonormal. To complete the proof, we must show that Equation 7.34 holds true in the mean square sense. Proceeding as in the proof of Theorem 7.11, we can show that

$$E\left|X(t) - \sum_{k=1}^{\infty} X_k \phi_k(t)\right|^2 = C_X(t,t) + \sum_{k=1}^{\infty} \lambda_k |\phi_k(t)|^2$$

$$- \sum_{k=1}^{\infty} \lambda_k |\phi_k(t)|^2 - \sum_{k=1}^{\infty} \lambda_k^* |\phi_k(t)|^2$$

$$= 0$$

since the eigenvalues are real.   □

We give an example of the Karhunen-Loève expansion for the Wiener random process in Section 8.2.3.

7.9   GENERATION OF VALUES OF A RANDOM PROCESS

How do we generate values for the sample paths of a random process, as we did for variates of a single random variable in Chapter 3 and of a vector random variable in Chapter 4?

For an arbitrary process, generation of sample paths is clearly out of the question. We would have to generate an (uncountable) infinity of random numbers to simulate *just a single outcome* ζ! The Karhunen-Loève expansion could be brought to bear here since it requires a countable infinity of random variables. Conceptually, we could use it with a finite number, as an approximation. But usually we have no way to judge what is a good approximation. So we can only hope to answer this question for special cases. We investigate some of these cases next.

1. IID sequences in discrete time

This case is simple, and the algorithm in Section 3.13 suffices since we have to generate variates from a *single* cdf.

2. Independent increment random processes

This case can be handled with the algorithm in Section 4.12 since we have to generate a variate from an n-dimensional cdf. An alternative is to use the algorithm in Section 3.13, to generate variates of the increment random variable, and simply add the generated variates to produce a variate for the random process.

3. WSS random processes

We will see a "clever" way of generating variates of WSS random processes in Section 9.3.4. Since stationary processes are also WSS, the method described there can be used for generating stationary processes as well. When you study Section 9.3.4, be sure to compare the difficulty of implementing the required filter to that of implementing the algorithm in Section 3.13.

7.10 SUMMARY OF MAIN POINTS

- A **continuous-time random process** is an uncountable family of random variables.

- **Finite-dimensional joint distributions** suffice to characterize a random process completely, but specifying the joint cdf's at all time instants requires much work.

- Special random process **models** that are tractable include IID, processes with independent increments, stationary, wide-sense stationary, and cyclostationary processes and Markov Chains.

- The **mean, autocovariance,** and **autocorrelation** functions provide quite useful partial information about a random process.

- **Ergodicity** of a random process enables us to calculate the mean and autocorrelation function via a single sample path of the process.

- **Derivatives** and **integrals** of random processes arise quite frequently in applications. Simple conditions for their existence in the mean square sense can be given through the autocorrelation function.

- The **Karhunen-Loève** expansion simplifies the theoretical treatment of random processes in the same sense as the Fourier transform does for deterministic signals.

- Generating **variates** of a random process is not simple; it can be done easily only for some special models.

7.11   CHECKLIST OF IMPORTANT TOOLS

- The autocorrelation function, especially that of a WSS process

- The Karhunen-Loève expansion

- Ergodicity

- The random walk model

- The ARMA model

7.12   PROBLEMS

In the following problems, if the index set I of the random process is not specified, it is assumed that $I = R$ for a continuous-time process, or I is the set of all integers for a discrete-time process. As a reminder, in all problems, if you are asked to evaluate the pdf, mean, etc., of $X(t)$, the index t is *fixed*, and thus we consider a *single* random variable. Similarly, if you are asked to evaluate the joint pdf, autocorrelation, etc., of $X(t), X(s)$, the indices t, s are (arbitrary but) *fixed*, so we consider a *two-dimensional* pdf.

Random Processes; Simple Skills

7.1 Plot a few sample paths of the sinusoid process in Example 7.5. Generate histograms for the marginal pdf's of the random variables $X(-1)$, $X(10)$, and $X(0)$.

7.2 Consider the sample space $S = [0, 1]$ and a uniform probability assignment on it. Define a random process $\{X(t), \; t \geq 0\}$ with sample paths given by $X(t, \zeta) = t^k \zeta$, where $k > 0$ is a fixed constant.
(a) How many sample paths are there? Plot a few.
(b) Is there a sample path which is constant?
(c) Find and plot the pdf and cdf of $X(t)$.
(d) Find and plot the joint pdf and cdf of $X(t)$ and $X(s)$.

7.3 Consider the sample space $S = [0, 1]$ and a uniform probability assignment on it. Let I be the set of all nonnegative integers. Define a random process $\{X_n\}$, with sample paths given by $X(n, \zeta) = \zeta^n$.
(a) Plot a few sample paths. How many sample paths does this random process have?
(b) Find and plot the pdf and cdf of X_n.
(c) Find and plot the jpdf of X_n and X_k.

7.4 Repeat Problem 7.3 when the sample space is $S = [0, \infty)$ with the outcomes distributed as an exponential random variable, with parameter 1.

7.5 Consider the random process $\{X(t)\}$ defined by $X(t) = Ae^{-Bt}$, $t \geq 0$, where A is a constant and B is a positive random variable with known pdf $f_B(x)$.
(a) Calculate the density $f_{X(t)}(x)$, $x \in R$.
(b) Calculate the density $f_{X(t)X(s)}(x, y)$, $x, y \in R$.
(c) Suppose that B is a uniform random variable with parameters $0, 1$. What is the maximum value that the random variable $X(t)$ can take?
(d) For the same assumption as in part (c), what are the maximum and minimum values that the random process can take?

7.6 Consider the random process $\{X(t)\}$ defined by $X(t) = At + Bt^2$, where A, B are random variables.
(a) Sketch a few sample paths.
What is the minimal amount of information you must know about A, B in order to do the following?
(b) Find the pdf $f_{X(t)}(x)$.
(c) Find the cdf $F_{X(t)}(x)$.
(d) Find the jcdf $F_{X(t)X(s)}(x, y)$.
(e) Find the n-dimensional pdf.

7.7 Consider the sample space $S = (-1, 1)$ with a uniform probability assignment. Define a random process $\{X(t), \ t \in [0, \infty)\}$ as follows:

$$X(t, \zeta) = \begin{cases} \cos 2\pi t, & 0 \geq \zeta \geq -1, \\ 2t, & 0 < \zeta < 1. \end{cases}$$

(a) What is the set of values that $X(t)$ can take, for a fixed t?
(b) Find $F_{X(t)}(1)$ for all t.
(c) Find $F_{X(3), X(30)}(x, y)$ for all $x, y \in R$.

7.8 Consider the random process $\{X(t)\}$ defined by

$$X(t) = A \cos(2\pi f_0 t) + B \sin(2\pi f_0 t),$$

where f_0 is a given constant and A, B are random variables with a known joint pdf.
(a) Calculate the density $f_{X(t)}(x)$, $x \in R$.
(b) Calculate the density $f_{X(t)X(s)}(x, y)$, $x, y \in R$.
(c) Calculate the n-dimensional pdf.

7.9 Consider the random process $\{X(t)\}$ defined by

$$X(t) = A \cos(2\pi f_0 t + \Theta),$$

where f_0 is a given constant, A is a random variable with a Rayleigh distribution, and Θ is a random variable with a uniform distribution in $[0, 2\pi)$. And A and Θ are independent.

(a) Calculate the density $f_{X(t)}(x)$, $x \in R$.

(b) Calculate the density $f_{X(t)X(s)}(x, y)$, $x, y \in R$.

(c) What is the density $f_{X(t)X(t)}(x, y)$, $x, y \in R$?

7.10 Consider the sample space $S = \{-1, 1\}$, with $P[\{-1\}] = p$. Define a discrete-time random process with sample paths described by

$$X(n, \zeta) = \zeta(-1)^n, \quad n = 1, 2, \ldots.$$

(a) Draw all the sample paths of this process.

(b) Find the pmf, pdf, and cdf of the random variable $X(n)$.

(c) Find the joint pmf of the random variables $X(n)$ and $X(k)$.

7.11 Suppose that Y is a Gaussian random variable with zero mean and unit variance. Define a random process $\{X(t),\ t > 0\}$, with sample paths described by

$$X(t, \zeta) = h(t) + Y(\zeta),$$

where $h(t)$ is a given function.

(a) Consider a countable set of indices T. Show that

$$P[X(t) = 0, \text{ for at least one } t \in T] = 0.$$

(b) Consider an uncountable set of indices T, say, $T = [1, 2]$. What can you say about $P[X(t) = 0, \text{ for at least one } t \in T]$?

(c) Would your answers in parts (a) and (b) change if a random variable other than a Gaussian were used? Explain.

7.12 Consider a random process $\{X(t)\}$ with $t \in I$, an uncountable set of your choice. Can you define the sample paths of this process so that

(a) $P[X(t) = 0, \text{ for all } t \in I] = 1$?

(b) $P[X(t) \le 1, \text{ for all } t \in I] = 1$?

7.13 *Huh?* When we deal with a random process, do we know "what time it is"? In other words, when we condition on $X(1)$, can we use the fact that $t = 1$?

Means and Correlations; Simple Skills

7.14 Show how you could define the joint, multidimensional cdf of two random processes when their index sets are different.

7.15 Calculate and plot the mean $m_X(t)$ and autocorrelation $R_X(t, s)$ of the process in Problem 7.3. If you cannot find them analytically, do it experimentally.

7.16 Repeat Problem 7.15 for the process in Problem 7.2.

7.17 Plot $EX(t), R_X(t, s)$ for the random process in Problem 7.5. Choose your own pdf for the random variable B.

7.18 Find $EX_k X_l X_m$ for the random process in Problem 7.3, analytically or experimentally. Plot it as a function of k, l, with m fixed.

7.19 Consider the following transformation of a random process $\{X(t)\}$:

$$Y(t) = \begin{cases} 1, & X(t) \geq 0, \\ 0, & X(t) < 0. \end{cases}$$

(a) Find the pdf and cdf of $Y(t)$.
(b) Find the mean and autocovariance of $Y(t)$.

7.20 Consider the random process $\{X(t), \; t \geq 1\}$ with sample paths given by

$$X(t, \zeta) = \frac{\zeta}{t^2}.$$

The sample space on which the process is defined has a Gaussian probability assignment. Calculate $m_X(t)$ and $C_X(t_1, t_2)$ for this process.

7.21 Consider a zero-mean random process $\{X(t)\}$ with autocorrelation function

$$R_X(\tau) = \begin{cases} e^{-|\tau|}, & -10 < \tau < 10, \\ 0, & \text{otherwise.} \end{cases}$$

Based on that information only, what property can you identify for the random variables $X(-11), X(1), X(11)$? Explain.

Means and Correlations; Advanced Skills

7.22 Consider a zero-mean Gaussian random process $\{X(t)\}$ with autocorrelation function $R_X(t, s) = 0.5(|t| + |s| - |t - s|)$. Define the transformation process $\{Y(t), \; t \geq 0\}$ via $Y(t) = X(-t)$. Show that $\{Y(t), \; t \geq 0\}$ and $\{X(t), \; t \geq 0\}$ are independent processes.

7.23 Consider the three sequences of Problem 6.6.
(a) Find EV_n and the autocovariance function $C_V(n, k)$.
(b) Find the cross-correlation $R_{XV}(n, k)$ and $R_{VY}(n, k)$.

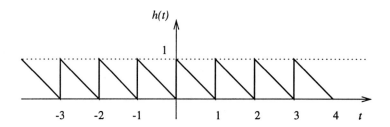

Figure 7.12   A sawlike periodic function.

7.24 Consider an IID sequence $\{X_n\}$ of Gaussian random variables with zero mean and variance σ^2. Let $\{Y_n\}$ be another IID sequence of Bernoulli random variables, independent of $\{X_n\}$. Define $Z_n = X_n Y_n$. Find the autocorrelation function $R_Z(n, k)$.

7.25 Let Y denote a uniform random variable defined on the sample space $S = [0, 1]$. Let $h(t)$ be the periodic function in Figure 7.12. Consider a random process $\{X(t)\}$ with sample paths described by $X(t, \zeta) = h(t - Y(\zeta))$.
(a) Draw the sample paths corresponding to $\zeta = 0, 0.3, 0.5, 1$.
(b) Find the pdf and cdf of $X(t)$.
(c) Find the jpdf of $X(t)$, $X(s)$.
(d) Find $EX(t)$, $R_X(t, s)$.

7.26 Define a cyclostationary process on the sample space $S = [0, 1]$. Plot a few sample paths. Plot the one- and two-dimensional pdf and cdf of this process.

7.27 Repeat Problem 7.26 for an ARMA process.

7.28 Plot the autocorrelation function of the processes you defined in Problems 7.26 and 7.27.

7.29 Consider a random process $\{X(t),\ t \geq 0\}$ with independent increments. Suppose that the random variable $X(t) - X(s)$, where $t > s$, is gamma-distributed with parameters λ and $t - s$. Suppose that $X(0) = 0$.
(a) Can you tell whether the function $X(t, \zeta)$ is monotonic for a fixed ζ? Determine and plot the following:
(b) The pdf of $X(t)$, $EX(t)$ and $\text{var}(X(t))$.
(c) The joint pdf $f_{X(t)X(s)}(x, y)$.
(d) The autocorrelation and autocovariance functions of $X(t)$.

7.30 Consider a zero-mean random process $\{X(t)\}$ with orthogonal increments; suppose that $X(0) = 0$ and $E[X(t) - X(s)]^2 = |t - s|$. Find $R_X(t, s)$.

7.31 Consider a zero-mean random process $\{X(t)\}$ with a continuous autocorrelation function $R_X(t, s)$. Suppose that $R_X(t, s)$ satisfies the property

$$R_X(t, s) = \frac{R_X(t, u) R_X(u, s)}{R_X(u, u)}, \quad s < u < t,$$

for all t, s in an interval I, and furthermore, $R_X(u, u) > 0$ for all $u \in I$. Show that the correlation coefficients satisfy the property

$$\rho_{X(t)X(s)} = \rho_{X(t)X(u)} \cdot \rho_{X(u)X(s)}.$$

7.32 Consider a zero-mean random process $\{X(t), \ t \geq 0\}$ with orthogonal increments and $X(0) = 0$.
(a) Show that $R_X(t, t)$ is nondecreasing.
(b) Let $\tau = \min\{t, s\}$. Show that $R_X(t, s) = R_X(\tau, \tau)$.

7.33 Consider a random walk $\{X_n\}$, with $P[Y_i = +1] = P[Y_i = -1] = 0.5$. Let N be a given positive integer. For any $n \geq 1$, define the events

$$A_n \triangleq \{\max_{1 \leq i \leq n} \{X_i\} \geq N\}, \quad B_n \triangleq \{\max_{1 \leq i \leq n} \{X_i\} = N\}.$$

Show that

$$
\begin{aligned}
P[A_n, X_n < N] &= P[X_n > N], \\
P[A_n] &= 2P[X_n \geq N] - P[X_n = N], \\
P[B_n] &= P[X_n = N] + P[X_n = N + 1], \\
P[X_1 \neq 0, \ldots, X_{n+1} \neq 0] &= P[X_n = 0] + P[X_n = 1].
\end{aligned}
$$

7.34 Consider the same random walk as in Problem 7.33 with $X_0 = 0$. Show that

$$
\begin{aligned}
P[\liminf_n X_n = -\infty] &= P[\limsup_n X_n = \infty] = 1, \\
P[X_n = k, \ \text{i.o.}] &= 1, \quad k = 0, \pm 1, \pm 2, \ldots.
\end{aligned}
$$

7.35 Consider the same random walk as in Problem 7.33 with $X_0 = 0$. Let $a > 0$, $b < 0$ be fixed integers. Show that

$$P[X_n \text{ reaches } a \text{ before } b] = \frac{a}{a - b}.$$

Special Cases; Simple Skills

7.36 Show that the condition $F_{X(t_1)X(t_2)}(x, y) = F_{X(0)X(t_2-t_1)}(x, y)$ implies that $F_{X(t_1)}(x) = F_{X(0)}(x)$.

7.37 Consider a zero-mean Gaussian random process $\{X(t)\}$ with autocorrelation function

$$R_X(\tau) = 10\frac{\sin^2(6\tau)}{(6\tau)^2}.$$

(a) Evaluate $\text{cov}(X(t), X(s))$ for arbitrary t, s.
(b) For what values of t, s are the random variables $X(t), X(s)$ independent?

7.38 Consider a stationary random process $\{X(t)\}$. Show that

$$\text{var}(X(t + s) - X(t)) = 2[R_X(0) - R_X(s)],$$
$$P|X(t + s) - X(t)| \geq \epsilon] \leq \frac{2}{\epsilon^2}[R_X(0) - R_X(s)], \quad \forall \epsilon > 0.$$

7.39 Consider two jointly WSS random processes $\{X(t), Y(t)\}$. Show that

$$|R_{XY}(\tau)| \leq 0.5[R_X(0) + R_Y(0)].$$

7.40 Consider the random process in Problem 7.5. Is it stationary?

7.41 Consider the ARMA process

$$Y_n = \frac{2}{3}X_n + \frac{1}{3}X_{n-1}, \quad n \geq 1.$$

(a) If the sequence $\{X_n\}$ is IID, is the process $\{Y_n\}$ stationary?
(b) If the sequence $\{X_n\}$ is stationary, is the process $\{Y_n\}$ stationary?

7.42 Consider the random process $\{X(t)\}$ defined by

$$X(t) = A\,\cos(t) + B\,\sin(t),$$

where A and B are random variables. Under what conditions (if any) is $\{X(t)\}$ a stationary process? WSS? Cyclostationary? WSS cyclostationary?

7.43 Consider a cyclostationary random process $\{X(t)\}$ with period T and a random variable Θ, independent of $\{X(t)\}$ and uniformly distributed in the interval $(-T, T)$. Show that the random process $\{Y(t)\}$ defined via $Y(t) = X(t - \Theta)$ is stationary. Find its n-dimensional cdf.

7.44 Consider the WSS cyclostationary random process $\{X(t)\}$ with period T and a uniform random variable Θ, with parameters $0, T$, independent of $\{X(t)\}$. Show that the random process $\{Y(t)\}$ defined via $Y(t) = X(t - \Theta)$ is WSS.

7.45 Consider the random process $\{X(t)\}$ defined by

$$X(t) = A\cos(2\pi f_0 t + \Theta),$$

where Θ is a uniformly distributed random variable in $[0, 2\pi)$ and A, f_0 are constants. Define the rectified process $\{Y(t)\}$ by $Y(t) = |X(t)|$.
(a) Find $EY(t)$.
(b) Determine the autocorrelation function $R_Y(t, s)$.
(c) Determine the cross-correlation function $R_{XY}(t, s)$.
(d) Is $\{Y(t)\}$ WSS?

7.46 Consider a WSS random process $\{X(t)\}$. Define the new process

$$Y(t) = X(t)\cos(2\pi f_0 t + \Theta),$$

where f_0 is a constant and Θ is a uniformly distributed random variable in $[0, 2\pi)$, independent of the process $\{X(t)\}$.
(a) Find $EY(t)$.
(b) Determine the autocorrelation function $R_Y(t, s)$.
(c) Determine the cross-correlation function $R_{XY}(t, s)$.
(d) Is $\{Y(t)\}$ WSS?

7.47 Let A be a Gaussian random variable; define the random process $\{X(t)\}$ via $X(t) = A\cos(2\pi f_0 t)$, where f_0 is a constant.
(a) Find $EX(t)$.
(b) Determine the autocorrelation function $R_X(t, s)$.
(c) Is $\{X(t)\}$ WSS?

7.48 Consider two independent random variables Θ_1 and Θ_2 uniformly distributed in $[0, 2\pi)$. Define two new random processes as

$$X(t) = \cos(2\pi f_1 t + \Theta_1), \quad Y(t) = \cos(2\pi f_2 t + \Theta_2),$$

where f_1, f_2 are given constants. Determine $R_{XY}(t, s)$. Under what conditions, if any, are the two random processes jointly WSS?

7.49 Is the random process in Problem 7.9 stationary?

7.50 Determine which of the periodic functions in Figure 7.13 can be autocorrelation functions for a WSS random process $\{X(t)\}$.

7.51 Determine which of the following functions can be autocorrelation functions for a WSS random process $\{X(t)\}$:

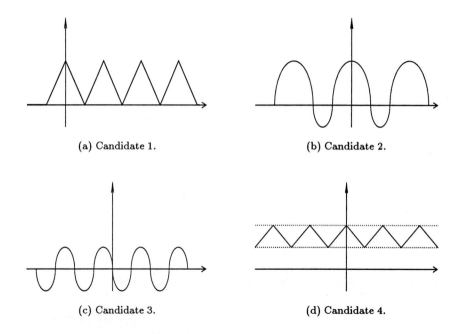

(a) Candidate 1.

(b) Candidate 2.

(c) Candidate 3.

(d) Candidate 4.

Figure 7.13   Candidate autocorrelation functions.

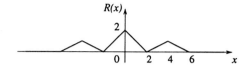

Figure 7.14 Autocorrelation function.

$$\cos(\tau), \quad \frac{\cos(\tau)}{\tau}, \quad e^{-\tau}\cos(\tau), \quad 10+\sin^2(\tau), \quad 10+\sin(\tau^2).$$

7.52 Consider a WSS random process $\{X(t)\}$ with the autocorrelation function shown in Figure 7.14. Which of the following quantities can you evaluate from the figure?
(a) $EX(0)$, (b) $EX(t)$, $t \neq 0$, (c) $EX^2(0)$, (d) $EX^2(t)$, $t \neq 0$.

7.53 Consider two jointly WSS random processes $\{X(t)\}$ and $\{Y(t)\}$, with known $R_X(\tau), R_Y(\tau)$, and $R_{XY}(\tau)$. Let $S(t) = X(t)+Y(t)$ and $D(t) = X(t)-Y(t)$.
(a) Find the autocorrelation functions of $S(t)$, $D(t)$.
(b) Find the cross-correlation function $R_{XS}(t,s)$.
(c) Find the cross-correlation function $R_{SD}(t,s)$.

7.54 Consider the product $P(t) = X(t)Y(t)$ and ratio $R(t) = X(t)/Y(t)$ of two jointly WSS random processes. Under what conditions are $\{P(t)\}$ and $\{R(t)\}$ WSS random processes? Jointly WSS?

7.55 Consider a WSS random process $\{X(t)\}$ with autocorrelation function $R_X(\tau)$. Define a new random process $\{Y(t)\}$ via $Y(t) = X(t)+X(t-T)$, where $T > 0$ is a fixed constant.
(a) Find $EY(t)$.
(b) Determine the autocorrelation function $R_Y(t,s)$.
(c) Determine the cross-correlation function $R_{XY}(t,s)$.
(d) Is $\{Y(t)\}$ WSS?

7.56 Repeat Problem 7.55 for $Y(t) = X(t+T) - X(t-T)$.

Special Cases; Advanced Skills

7.57 Consider a stationary random process $\{X(t)\}$ that takes on the values ± 1 with equal probability; that is, $P[X(t) = +1] = P[X(t) = -1] = 0.5$. Let $N(t)$ be the number of changes in the value of $X(t)$ in the interval $[0,t)$. Suppose

$$P[N(t) = k] = \frac{1}{1+\lambda t}\left(\frac{\lambda t}{1+\lambda t}\right)^k,$$

where $\lambda > 0$ is a given constant.
(a) Calculate $EN(t), \text{var}(N(t))$.
(b) Is $\{N(t)\}$ a stationary process? WSS?
(c) Calculate the probability that $\{X(t)\}$ will change value at a given time t. (*Hint:* Consider $P[X(t + \Delta t) + X(t) = 0]$, for small Δt.)

7.58 Consider a stationary sequence $\{X_n\}_{-\infty}^{\infty}$. Let $Y_n = f(X_k, \; k \leq n)$; in other words, the system described by $\{Y_n\}$ has infinite memory. Show that the sequence $\{Y_n\}$ is also stationary.

7.59 Consider a stationary random process $\{X(t)\}$ and a function $g(x)$. Show that $\{g(X(t))\}$ is a stationary random process.

Ergodicity

7.60 Prove Theorem 7.9.

7.61 Consider a random process $\{X_n, \; n = 1, 2, \ldots\}$ with $EX_n = 0$ and $EX_n X_k = 1$.
(a) Show that X_n converges in the mean square sense.
(b) Let $g(x)$ be a function on R. Show by example that

$$\lim_{n \to \infty} \frac{1}{n} \sum_{k=1}^{n} g(X_k)$$

may not converge in the mean square sense.

7.62 Construct a random process which is (a) stationary and ergodic, (b) stationary and nonergodic, (c) nonstationary and ergodic, and (d) nonstationary and nonergodic.

7.63 Consider the following (trivial) model for a computer network node. Packets, all of equal size, arrive at the node's buffer in a periodic fashion every $T > 0$ s. Each packet takes $X < T$ s to transmit. Let $N(t)$ denote the number of packets at the buffer at time t. Is $\{N(t), \; t \geq 0\}$ a stationary random process? Is it ergodic?

7.64 Consider the trivial random process $\{X(t)\}$ defined by $X(t) = A$, where A is a random variable. Find the mean and variance of the time average of $X(t)$.

7.65 Consider a zero-mean WSS random process $\{X(t)\}$ with autocorrelation function

$$R_X(\tau) = \begin{cases} 1 - |\tau|, & |\tau| \leq 1, \\ 0, & |\tau| > 1. \end{cases}$$

Prove that $\{X(t)\}$ is mean ergodic.

7.66 Consider a WSS random process $\{X_n\}$ with autocorrelation function $R_X(m) = 0.3^{|m|}$. Prove that $\{X_n\}$ is mean ergodic.

7.67 Consider two mean ergodic random processes $\{X(t)\}$ and $\{Y(t)\}$, with means m_X and m_Y. Let A, B be two random variables, independent of the two processes. Define the random process $\{Z(t)\}$ via $Z(t) = AX(t) + BY(t)$. Is the process $\{Z(t)\}$ mean ergodic?

7.68 Consider the random process $\{X(t)\}$ defined by $X(t) = A\cos(2\pi f_0 t)$, where A is a random variable and f_0 is a given constant. Find the mean and variance of the time average of $X(t)$.

7.69 Define the random process $\{X(t)\}$ via $X(t) = A\cos(2\pi f_0 t + \Theta)$, where f_0 is a given constant and A and Θ are independent random variables, with Θ being uniform in $[0, 2\pi)$. Find the mean and variance of the time average of $\{X(t)\}$.

7.70 Consider the random process in Problem 7.69, but now A is a nonnegative random variable. Does

$$X_T \triangleq \frac{1}{2T} \int_{-T}^{T} X(t)\, dt \to EX(t)$$

in the mean square sense? In probability?

7.71 If $\{X(t)\}$ is a WSS random process, is $\{X^k(t)\}$, where k is an integer, also a WSS random process?

7.72 Consider a zero-mean WSS random process $\{X(t)\}$. For a given $T > 0$ and integer $n > 1$, let

$$X_T \triangleq \frac{1}{2T} \int_{-T}^{T} X^n(t)\, dt$$

be an estimate of the nth moment of $X(t)$. State under what conditions X_T converges to $EX^n(t)$.

7.73 Consider a zero-mean WSS random process $\{X(t)\}$. For a given $T > 0$, let

$$R_T(s) \triangleq \frac{1}{2T} \int_{-T}^{T} X(t)X(t+s)\, dt$$

be an estimate of the autocorrelation of $X(t)$. State under what conditions $R_T(s)$ converges to $R_X(s)$.

7.74 Consider a random process $\{X(t)\}$. For a set $A \subseteq R$, define

$$I_A(t) \triangleq \begin{cases} 1, & X(t) \in A, \\ 0, & \text{otherwise}, \end{cases} \qquad X_T \triangleq \frac{1}{2T} \int_{-T}^{T} I_A(t)\, dt.$$

Find EX_T. Does $EX_T \to P[A]$? Does $X_T \xrightarrow{\text{a.s.}} P[A]$?

Continuity and Derivatives; Simple Skills

7.75 Let Y be a continuous random variable; let $u(t)$ denote the usual step function. Define the random process $\{X(t)\}$ via $X(t) = u(t + Y)$.
(a) Are the sample paths of $\{X(t)\}$ continuous?
(b) Is $\{X(t)\}$ mean square continuous?
(c) Does $\{X(t)\}$ have a mean square derivative?
(d) Does $\{X(t)\}$ have an integral?

7.76 Repeat Problem 7.75 when Y is a discrete random variable.

7.77 Consider a WSS random process $\{X(t)\}$ with autocorrelation function

$$R_X(\tau) = 10e^{-3\tau^2}.$$

(a) Is $\{X(t)\}$ mean square continuous?
(b) Does $\{X(t)\}$ have a mean square derivative?
(c) Does $\{X(t)\}$ have an integral?

7.78 Give an example of two mean square continuous processes whose product is not mean square continuous.

7.79 Show that if a random process $\{X(t)\}$ is mean square continuous, it has an integral.

7.80 Show that if $\{Y(t)\}$ is the integral process of the random process $\{X(t)\}$, the derivative of $Y(t)$ is equal to $X(t)$ in the mean square sense.

7.81 Consider a random process $\{X(t)\}$ with derivative process $\{X'(t)\}$. Show that

$$EX'(t) = \frac{dEX(t)}{dt}.$$

7.82 Consider a WSS random process $\{X(t)\}$ with a derivative process $\{X'(t)\}$. Show that $EX(t)X'(t) = 0$.

7.83 Consider a WSS random process $\{X(t)\}$ with autocorrelation function

$$R_X(\tau) = e^{-|\tau|} + e^{-5|\tau|}.$$

Does the kth derivative $d^k X(t)/dt^k$ exist? If you said yes, evaluate it.

7.84 Consider a zero-mean, continuous (in the mean square sense) random process $\{X(t)\}$. Show that the process has stationary orthogonal increments if and only if $R_X(t, s) = a^2 \min(t, s)$ for some constant a.

7.85 Consider a differentiable random process $\{X(t)\}$ with autocorrelation function $R_X(t, s)$. Let $\{X'(t)\}$ denote the derivative process. Show that

$$R_{X'}(t, s) = \frac{\partial^2}{\partial t \, \partial s} R_X(t, s), \quad R_{XX'}(t, s) = \frac{\partial R_X(t, s)}{\partial s}.$$

7.86 Consider a twice differentiable random process $\{X(t)\}$. Let $\{X'(t), X''(t)\}$ denote the first and second derivative processes. Find the following:
(a) $R_{X''}(t, s)$, (b) $R_{XX''}(t, s)$, (c) $R_{X'X''}(t, s)$.

7.87 Consider a random process $\{X(t)\}$ with nth derivative process $\{X^{(n)}(t)\}$ and a random process $\{Y(t)\}$ with kth derivative process $\{Y^{(k)}(t)\}$. Show that

$$R_{X^{(n)}Y^{(k)}}(t, s) = \frac{\partial^{n+k}}{\partial t^n \, \partial s^k} R_{XY}(t, s).$$

7.88 Consider a WSS random process $\{X(t)\}$ with nth derivative process $\{X^{(n)}(t)\}$. Show that

$$R_{X^{(n)}}(\tau) = (-1)^n \frac{\partial^{2n}}{\partial \tau^{2n}} R_X(\tau).$$

7.89 Evaluate and plot the derivative and integral of the sinusoid process in Example 7.1.

Continuity and Derivatives; Advanced Skills

7.90 Consider a zero-mean WSS random process $\{X(t)\}$ with autocorrelation function $R_X(\tau) = \sigma^2 e^{-c|\tau|}$, where $\sigma^2 > 0$ and $c > 0$ are given constants.
(a) Set up the equations for the Karhunen-Loève expansion of $X(t)$.
(b) Show that the eigenfunctions $\phi(t)$ satisfy the equation

$$\frac{d^2 \phi(t)}{dt^2} = \frac{c^2(\lambda - 2\sigma^2/c)}{\lambda} \phi(t).$$

(c) Verify that the solutions of the above equation are of the form

$$\phi(t) = A_1 \sin at, \quad \phi(t) = A_2 \cos at,$$

where A_i are constants. Find the values of A_i that normalize the eigenfunctions.
(d) Verify that the constant a is the solution of the equation

$$\tan aT = \pm \frac{a}{c}.$$

7.91 Consider a zero-mean WSS random process $\{X(t)\}$ with autocorrelation function

$$R_X(\tau) = \sum_{k=0}^{\infty} \frac{1}{1+k^2} \cos(2\pi k \tau).$$

(a) Set up the equations for the Karhunen-Loève expansion of $X(t)$.
(b) Find the eigenfunctions $\phi(t)$.

7.92 Let $T = 0.5$. Repeat Problem 7.91 for a zero-mean WSS random process $\{X(t)\}$ with autocorrelation function

$$R_X(\tau) = \begin{cases} 1 - |\tau|, & |\tau| \le 1, \\ 0, & \text{otherwise.} \end{cases}$$

7.93 Consider a random process $\{X(t),\ t \ge 0\}$ defined on a sample space $S = [0, 1]$. Let $X(t, \zeta) = te^{\zeta}$, $\zeta \in S$. Is this enough information to determine the following?
(a) $F_{X(t)}(x)$
(b) $F_{X_1 X_2 \cdots X_n}(x_1, x_2, \ldots, x_n)$
(c) $R_X(t_1, t_2)$
(d) $\frac{\partial X(t)}{\partial t}$
(e) $\int_0^t X(t, \zeta)\, dt$
(f) Stationarity of $\{X(t)\}$
(g) Wide-sense stationarity of $\{X(t)\}$

7.94 Consider an IID sequence of Bernoulli random variables $\{X_n,\ n = 0, 1, \ldots\}$, with parameter p. Define the sequence $\{Y_n,\ n \ge 1\}$ via $Y_n = 0.5(X_n + X_{n-1})$.
(a) Is $\{Y_n\}$ an ARMA process?
(b) Find the pmf of Y_n.
(c) Find the autocorrelation $R_Y(n, k)$.
(d) Does $\{Y_n\}$ have an integral?

7.95 Consider an IID sequence of Bernoulli random variables $\{X_n,\ n = 0, 1, \ldots\}$, with parameter p. Define the sequence $\{W_n,\ n \ge 1\}$, by

$$W_n = \frac{1}{2}X_n + \frac{1}{3}X_{n-1}.$$

(a) Find the pmf of W_n.
(b) Find the cross-correlation $R_{YW}(n, k)$.
(c) Does $\{W_n\}$ have an integral?

7.96 Consider an IID sequence of Bernoulli random variables $\{X_n,\ n = 0, 1, \ldots\}$, with parameter p. Define the sequences

$$Y_n = \frac{1}{3}Y_{n-1} + X_n, \quad Z_n = 3Z_{n-1} + X_n, \quad n = 1, 2, \ldots$$

with $Z_0 = Y_0 = 0$.
(a) Are $\{Y_n, \; n = 1, 2, \ldots\}$ and $\{Z_n, \; n = 1, 2, \ldots\}$ ARMA processes?
(b) Find the pmf of Z_n, Y_n.
(c) Find the autocorrelations $R_Z(n, k)$, $R_Y(n, k)$.
(d) Find the cross-correlation $R_{YZ}(n, k)$.
(e) Does $\{Y_n\}$ have an integral?
(f) Does $\{Z_n\}$ have an integral?

Computer Problems

7.97 Generate a large number of sample paths of a random walk with $p = P[Y_i = +1] = 0.5$. Repeat for another random walk, with $p = 0.51$. What do you observe for large values of i, for example, $i \geq 100$?

7.98 Generate a sample path of a random walk with $p = P[Y_i = +1] = 0.5$.
(a) What do you observe for large values of i, for example, $i \geq 100,000$?
(b) Repeat part (a) for $p = 0.51$ and $p = 0.49$.

7.99 Evaluate experimentally and plot the integral of a random walk process.

8

POISSON AND GAUSSIAN
RANDOM PROCESSES

The Poisson and Gaussian processes (and their variations) are undoubtedly the most celebrated examples of random processes. The former is the most widely used process in queuing theory, a discipline with applications in delay and throughput analysis of computer systems and networks; we study it in Section 8.1. The latter is widely used in studies of noise, with applications in telecommunication systems and signal processing; we study it in Section 8.2. In their basic forms, the processes were introduced a long time ago; in recent years, numerous extensions and modifications have been proposed and analyzed. The main reason for studying these processes is their *mathematical tractability*.

8.1 POISSON PROCESS

Numerous (but of course equivalent) definitions of the Poisson process have appeared in the past, each serving its own purpose. We present some of those definitions in Section 8.1.4. Consider a continuous-time random process $\{N(t), t \in [0, \infty)\}$ that takes nonnegative integer values. It is convenient to think of the random variable $N(t)$ as one that "counts events occurring in the interval $[0, t)$" or as one that "counts arrivals in the interval $[0, t)$."

Definition: The random process $\{N(t),\ t \in [0, \infty)\}$ is called a **Poisson random process** if the following properties are satisfied:

1. $N(0) = 0$, with probability 1.

2. $\{N(t)\}$ is a process with IID increments.

3. For all $t > s$, the random variable $I(t, s) \triangleq N(t) - N(s)$ is Poisson-distributed; i.e., its pmf is given by

$$P[I(t, s) = k] = e^{-\lambda(t-s)} \frac{[\lambda(t - s)]^k}{k!}, \quad k = 0, 1, 2, \ldots. \qquad (8.1)$$

The constant $\lambda > 0$ is called the **parameter** of the process. We see in Section 8.1.6 that λ has a natural interpretation as a rate parameter. The random variable $I(t, s)$ describes the "jumps" of the Poisson process in the time interval $[t, s)$. Since all the probabilistic information about the Poisson process is given in terms of the pmf of $I(t, s)$, it is important to understand that when we have to calculate probabilities, pmf's, and pdf's of other random variables associated with the Poisson process, we must reduce them to events involving $I(t, s)$. We apply this observation repeatedly in the following sections.

What do the sample paths of this process look like? Consider a time interval $[t, \, t + \Delta t)$ with a small length Δt. From Equation 8.1 we see that

$$\begin{aligned}
P[I(t + \Delta t, \, t) = 0] &= e^{-\lambda \, \Delta t}, \\
P[I(t + \Delta t, \, t) = 1] &= \lambda \, \Delta t \, e^{-\lambda \, \Delta t}, \\
P[I(t + \Delta t, \, t) = k] &= \frac{(\lambda \, \Delta t)^k}{k!} e^{-\lambda \, \Delta t}, \quad k > 1.
\end{aligned}$$

We can expand the exponential function around 0. We can rewrite the above expressions, dropping the expansion terms that are of order higher than $o(\Delta t)$, as follows:

$$\begin{aligned}
P[I(t + \Delta t, \, t) = 0] &\approx 1 - \lambda \, \Delta t \\
P[I(t + \Delta t, \, t) = 1] &\approx \lambda \, \Delta t \\
P[I(t + \Delta t, \, t) = k] &\approx 0, \quad k > 1.
\end{aligned}$$

The intuitive interpretation of the above relationships is that "events" or "arrivals" described by the Poisson process cannot occur in "batches"; in other words, the instantaneous jumps of the process must be equal to 1. It is traditional to refer to the time instants where the jumps occur as the *arrival instants* of the Poisson process.

From the definition of $I(t, s)$, we have $N(t+s, \, \zeta) \geq N(t, \zeta)$ for any positive s and all $t \geq 0$, $\zeta \in S$. Therefore, the sample paths of the Poisson process are nondecreasing functions of time. A typical path therefore looks like that in Figure 8.1. Notice that all the jumps are equal to 1; since the interval $[0, t)$, over which arrivals are counted, is open at the right, the sample paths are right-continuous functions of t.

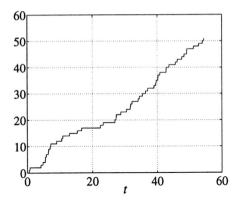

Figure 8.1 A typical sample path of the Poisson process.

Remark 1: Consider a time interval $[0, t)$; there are an infinite number of ways we can partition it in nonoverlapping intervals. For arbitrary $0 = t_0 < t_1 < \cdots < t_n = t$, consider the n IID random variables $\{N(t_i) - N(t_{i-1})\}_{i=1}^n$. We have

$$N(t) = \sum_{i=1}^n [N(t_i) - N(t_{i-1})] = \sum_{i=1}^n I(t_i, t_{i-1}).$$

It is not a priori certain that two different partitions will give rise to the same pmf for $N(t)$. However, since sums of Poisson random variables are still Poisson, we can immediately see that $N(t)$ will have a parameter

$$p = \sum_{i=1}^n \lambda(t_i - t_{i-1}) = \lambda \sum_{i=1}^n (t_i - t_{i-1}) = \lambda t,$$

verifying that, irrespective of the partition, the definition makes sense.

Remark 2: In the context of an arrival model, the assumptions in the definition of the Poisson process have the following consequences: Since $N(0) = 0$, we cannot model systems that do not start from an "empty" state. More importantly, since the process has IID increments, we cannot model nonstationary systems or systems with memory.

8.1.1 n-Dimensional PMFs

We can easily evaluate the n-dimensional pmf of the Poisson process. For $n = 1$, $P[N(t_1) = k]$ is given by Equation 8.1 (take $s = 0, t = t_1$). Consider the case $n = 2$ next. Fix $t_2 > t_1 \geq 0$ and nonnegative integers $k \geq l$. We have

$$P[N(t_2) = k, N(t_1) = l] \quad = \quad P[N(t_2) - N(t_1) = k - l, \ N(t_1) = l]$$

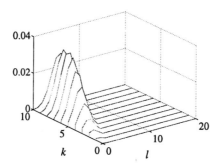

Figure 8.2   Two-dimensional distribution, Poisson process.

$$
\begin{aligned}
&= \quad P[I(t_2, t_1) = k - l] \cdot P[I(t_1, 0) = l] \\
&= \quad e^{-\lambda(t_2 - t_1)} \frac{[\lambda(t_2 - t_1)]^{k-l}}{(k - l)!} e^{-\lambda t_1} \frac{(\lambda t_1)^l}{l!} \quad (8.2)
\end{aligned}
$$

Equation 8.2 should be expected since the Poisson process has IID increments. In Equation 8.2, we must have $k \geq l$ since $N(t_2) \geq N(t_1)$. Therefore,

$$
P[N(t_2) = k, N(t_1) = l] = 0, \quad k < l.
$$

The two-dimensional pmf is depicted in Figure 8.2, where we have used $\lambda = 2$ and $t_1 = 3, t_2 = 5$.

For the case $n > 2$, let $i_n \geq i_{n-1} \geq \cdots \geq i_2 \geq i_1 \geq i_0 = 0$ be given integer values. Working as in Equation 8.2, we can show easily that

$$
P[N(t_n) = i_n, \ldots, N(t_1) = i_1] = e^{-\lambda t_n} \prod_{l=1}^{n} \frac{[\lambda(t_l - t_{l-1})]^{i_l - i_{l-1}}}{(i_l - i_{l-1})!}. \quad (8.3)
$$

From Equation 8.2 we can easily calculate the mean, autocovariance, and autocorrelation functions $m_N(t), C_N(t, s), R_N(t, s)$. We have

$$
\begin{aligned}
m_N(t) &= \lambda t, \\
C_N(t, s) &= \lambda \min(t, s), \\
R_N(t, s) &= C_N(t, s) + m_N(t) m_N(s) = \lambda \min(t, s) + \lambda t \lambda s.
\end{aligned}
$$

8.1.2   Random Variables and Processes Associated with the Poisson Process

Now let's investigate some random variables and processes associated with the Poisson random process.

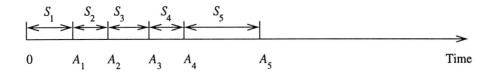

Figure 8.3 Interarrival times S_n.

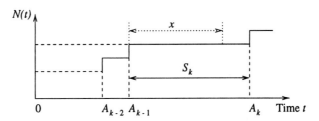

Figure 8.4 Equivalent events in the Poisson process.

1. Sequence of interarrival times

The time instants at which the jumps of the Poisson process occur are called *arrival instants*. Let $\{A_n\}_{n=0}^{\infty}$ denote the arrival instants of a Poisson process with parameter λ, where $A_0 \triangleq 0$. (Why do we have an infinite number of these random variables? See Problem 8.2.) Define the sequence of random variables $\{S_n\}_{n=1}^{\infty}$ as follows (see Figure 8.3):

$$S_n = A_n - A_{n-1}, \quad n = 1, 2, 3, \dots. \tag{8.4}$$

The random variable S_n represents the nth *interarrival time*. Clearly, S_n is a nonnegative-valued random variable. We show now the following:

Theorem 8.1 *The sequence $\{S_n, \; n = 1, 2, \dots\}$ is an IID sequence of exponential random variables, each distributed with parameter λ.*

Proof: We determine the distribution of the random variable S_k first. From Figure 8.4, we can see that the event $\{S_k > x\}$ is equivalent to the event $\{N(A_{k-1} + x) - N(A_{k-1}) = 0\}$. In other words, the kth interarrival time is at least x units of time if and only if no arrivals occur in the interval $[A_{k-1}, A_{k-1} + x)$. Therefore,

$$P[S_k > x] = 1 - P[S_k \le x] = P[N(A_{k-1} + x) - N(A_{k-1}) = 0] = e^{-\lambda x},$$

or

$$P[S_k \le x] = 1 - e^{-\lambda x},$$

which shows that for any $k = 1, 2, \ldots$, the random variable S_k is exponentially distributed with parameter λ, the same as that of the Poisson process. Consider next two random variables S_k and S_l, with $k \neq l$. From Figure 8.4 again, we can see that the event $\{S_k > x, \; S_l > y\}$ is equivalent to the event

$$\{N(A_{k-1} + x) - N(A_{k-1}) = 0, N(A_{l-1} + y) - N(A_{l-1}) = 0\}.$$

Let $B \triangleq P[S_k > x, S_l > y]$. We have

$$
\begin{aligned}
B &= P[N(A_{k-1} + x) - N(A_{k-1}) = 0, N(A_{l-1} + y) - N(A_{l-1}) = 0] \\
&= e^{-\lambda x} e^{-\lambda y},
\end{aligned}
$$

which shows that the two random variables S_k, S_l are independent. Independence of $n > 2$ random variables can be similarly established. $\square$

2. Time of nth arrival

The time of the nth arrival $A_n, n = 1, 2, \ldots$, can be defined as follows (see also Definition 8.4):

$$A_n \triangleq \sum_{i=1}^{n} S_i,$$

where $\{S_i\}$ is the sequence of interarrival times. We saw in Section 4.11.8 that the sum of n IID exponential random variables with parameter λ is an Erlang random variable with parameters n, λ. Therefore, the pdf of A_n is given by

$$f_{A_n}(x) = \frac{(\lambda x)^{n-1}}{(n-1)!} \lambda e^{-\lambda x} u(x), \quad x \in R.$$

Observe that $\{A_n\}_{n=1}^{\infty}$ is a sequence of sums of IID random variables, and thus its pdf and mean could be alternatively found by the theory in Chapter 6.

3. Conditional arrival times in $[T_1, T_2]$

Consider a fixed time interval $[T_1, T_2]$, with $T_2 > T_1 \geq 0$. Suppose that exactly $M > 0$ arrivals are known to have occurred in this interval, where M is a given integer constant. Denote by $B_i, i = 1, \ldots, M$, the random variable that represents the ith arrival instant (given that M arrivals have occurred). Of course, $0 < B_1 < B_2 < \cdots < B_M$ with probability 1. At first glance, one might be tempted to say that the random variables $\{B_n\}_{i=1}^{M}$ have the same pdf as the arrival instants $\{A_n\}_{i=1}^{M}$; we will see, however, that this is not so!

Since the Poisson process has IID increments, we can assume without loss of generality that $T_1 = 0$; let $T = T_2 - T_1$. We can consider the interval $[0, T]$ instead. We study the case $M = 1$ first.

Let B_1 denote the first arrival instant. We want to evaluate the function (conditional probability) $F(x)$, defined as

$$F(x) \triangleq P[B_1 \leq x | N(T) = 1], \quad x \in R.$$

Of course, when $x < 0$, $F(x) = 0$ since B_1 is nonnegative; moreover, when $x > T$, $F(x) = 1$ since $B_1 \leq T$ by definition. Consider, therefore, the case $0 \leq x \leq T$. Observe that the event $\{B_1 \leq x\}$ is equivalent to the event $\{N(x) = 1\}$. We then have

$$
\begin{aligned}
P[B_1 \leq x | N(T) = 1] &= \frac{P[B_1 \leq x \cap N(T) = 1]}{P[N(T) = 1]} \\
&= \frac{P[N(x) = 1 \cap N(T) - N(x) = 0]}{P[N(T) = 1]} \\
&= \frac{P[N(x) = 1]P[N(T) - N(x) = 0]}{P[N(T) = 1]} \\
&= \frac{\lambda x e^{-\lambda x} e^{-\lambda(T-x)}}{\lambda T e^{-\lambda T}} = \frac{x}{T},
\end{aligned}
$$

which shows that the conditional distribution of the arrival time is uniform in $[0, T]$. We just proved the following:

Lemma 8.1 *Suppose that exactly one arrival has occurred in the time interval* $[T_1, T_2]$. *Then the arrival instant is uniformly distributed, with parameters* T_1 *and* T_2.

This lemma expresses in a precise form the frequently used statement that, in a Poisson process, "arrivals are completely random."

Consider next the case $M = 2$; let

$$F(x, y) \triangleq P[B_1 \leq x, B_2 \leq y | N(T) = 2], \quad x, y \in R.$$

From the definition of B_1, B_2, observe that $F(x, y) = 0$ when $x < 0$, $x \geq y$, or when $y > T$. Consider therefore the case $0 \leq x < y \leq T$. We evaluate the density

$$f(x, y) \, dx \, dy \approx P[x < B_1 \leq x + dx, \ y < B_2 \leq y + dy | N(T) = 2].$$

Let $C = P[x < B_1 \leq x + dx, y < B_2 \leq y + dy | N(T) = 2]$. From Figure 8.5, we see that

$$
\begin{aligned}
C &= \frac{(e^{-\lambda x})(\lambda \, dx \, e^{-\lambda dx})(e^{-\lambda(y-x-dx)})(\lambda \, dy \, e^{-\lambda dy})(e^{-\lambda(T-y-dy)})}{e^{-\lambda T} \frac{(\lambda T)^2}{2}} \\
&= \frac{2 \, dx \, dy}{T^2}.
\end{aligned}
$$

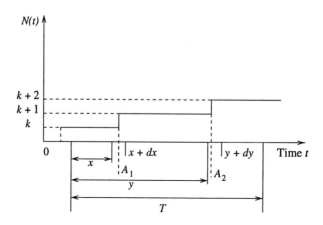

Figure 8.5   Two arrivals in $[T_1, T_2]$.

In Problem 8.3, we ask you to show that the above relationship implies that the two random variables B_1, B_2 are uniformly distributed in $[0, T]$. The general case $M > 2$ can be shown in a similar fashion, so we can state the following:

Lemma 8.2 *Suppose that exactly M arrivals have occurred in the interval* $[T_1, T_2]$. *Then the arrival instants are uniformly distributed.*

8.1.3   Two Remarkable Properties

1. Memoryless property

Since the exponential random variable enjoys the memoryless property (see Problem 3.118), the Poisson process enjoys it, too. In the context of an arrival process, this property can be intuitively interpreted as follows: The Poisson arrival process has no memory; i.e., given that the time from the last arrival is T, the time until the next arrival is still exponentially distributed, with parameter λ, independent of T.

This property is the main reason for the popularity of the Poisson process as a modeling tool. Tremendous model simplifications can be attributed to this property, as can be seen in any book on queuing theory [41].

2. PASTA property

In Section 7.7, we saw that there are tremendous practical advantages in using time averages instead of statistical ones. There is an additional form of averaging, namely, conditional averaging, that has similar practical (and theoretical) advantages. In the context of this section, the conditioning event of interest will always be an arrival.

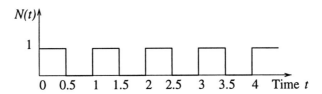

Figure 8.6 Buffer contents.

We can best introduce the notion of conditional averaging with an example.

Example 8.1. Consider the following problem. Packets arrive at a computer network node according to a Poisson process. We want to evaluate, for example, the mean buffer size, *as seen by the arriving packet.* This mean need not be the same as the time average of the buffer size. The following counterexample should convince you about the truth of this statement.

Counterexample. Consider a network node with arrivals that are periodic, with period 1 s. Suppose it takes 0.5 s to transmit a packet. Then, $N(t)$, the buffer contents as a function of time, will be as shown in Figure 8.6. From that figure, we can immediately see that an arriving packet will always see a zero buffer size; therefore, the average buffer size seen by an arriving packet will also be 0. However, the time average of the buffer size is 0.5. □

In a more general setting, let A_n denote, as usual, the arrival instants associated with the Poisson process. Let $\{X(t)\}$ denote a random process. The Poisson process has the remarkable property described in the following theorem, which we present without proof; for more details, see [67].

Theorem 8.2 *Let $\{A_n\}$ be the arrival instants associated with a Poisson process. Let $\{X(t), t \in [0, \infty)\}$ be a random process defined on the same sample space. Assume that future jumps of the Poisson process do not depend on past values of $\{X(t)\}$. Then, with probability 1,*

$$\lim_{n \to \infty} \frac{1}{n} \sum_{i=1}^{n} X(A_i) = \lim_{T \to \infty} \frac{1}{T} \int_0^T X(s) \, ds. \tag{8.5}$$

In plain English, the above theorem says that

<u>P</u>oisson <u>A</u>rrivals <u>S</u>ee <u>T</u>ime <u>A</u>verages.

This theorem is known as PASTA (with capital letters) and has immediate applications in simulations and queuing theory. (On the other hand, pasta, with lowercase letters, has applications in other culinary sciences.)

There is an *anti-PASTA* property as well.[1]

---

[1] Not to be confused with antipasto or any food of Italian origin.

Theorem 8.3 *If an arrival process sees time averages, it is Poisson.*

8.1.4   Alternative Definitions of the Poisson Process

The following definitions of the Poisson process with parameter $\lambda > 0$ were also proposed in the past.

1. Infinitesimal definition

Consider a random process $\{N(t),\ t \in [0, \infty)\}$. Let $I(t, s) \triangleq N(t) - N(s)$ be the increment of the process in $[s, t)$. The process is called a Poisson random process if the following properties are satisfied:
1. $N(0) = 0$ with probability 1.
2. $\{N(t)\}$ is a process with IID increments.
3. The distribution of $I(t, s)$ does not depend on $N(s)$ or s.
4. $P[I(t, s) = 1] = \lambda(t - s) + o(t - s)$, as $s \to t$.
5. $P[I(t, s) > 1] = o(t - s)$, as $s \to t$.

2. Limit of binomials

Consider a sequence of IID Bernoulli random variables, with parameter $0 < p < 1$. We know that the sum process

$$S_n = \sum_{i=1}^{n} X_i$$

is a binomial random variable. Moreover, the process $\{S_n\}$ has IID increments.

We can associate a continuous-time random process $\{N(t),\ t \in [0, \infty)\}$ with $\{S_n\}$ as follows. For a fixed $t > 0$, subdivide the time interval $[0, t)$ into n subintervals of equal length. Then X_i can be given the interpretation of counting "events" occurring in the ith such subinterval. Consider next finer and finer subdivisions of $[0, t)$ such that as $n \to \infty$, $np \to \lambda t$, where $\lambda > 0$ is a parameter. The limiting process is then a Poisson process.

8.1.5   Variations of the Poisson Process

1. Shot noise

Let $\{A_n,\ n = 1, 2, \ldots\}$ represent as usual the arrival instant sequence associated with the Poisson process. Define a new random process $\{X(t), t \in [0, \infty)\}$ as follows:

$$X(t) = \sum_{i=1}^{\infty} \delta(t - A_i).$$

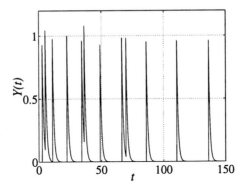

Figure 8.7 Filtered Poisson input.

Here $X(t)$ is simply a convenient mathematical representation of the arrival sequence, in continuous time. And $X(t)$ is very useful when viewed as the input to a filter with impulse response $h(t)$.[2]

When $X(t)$ is the input to a linear, time-invariant system with impulse response $h(t)$, the output is another random process $Y(t)$, which we can represent as

$$Y(t) = \sum_{i=1}^{\infty} h(t - A_i)$$

[see Figure 8.7, where $h(t)$ is chosen as an exponentially decaying function, i.e., a low-pass filter]. This process has been primarily investigated in studies of photoelectric devices. In such devices, electrons can "hit" photodetectors in arbitrary time instants A_i (they are modeled as a train of delta functions). The resulting electric current can be then represented as another train of functions, such as the ones in Figure 8.7. Hence the name **shot noise**.

2. Interrupted Poisson process

The **interrupted Poisson process** (IPP) was studied extensively in the late 1980s and is widely used in studies of high-speed network performance. It can be viewed as follows: Consider two *states*, labeled *on* and *off*, as Figure 8.8 depicts.

While the "system" is in the on state, arrivals can occur according to a Poisson process with rate λ. While the system is in the off state, arrivals cannot occur (i.e., they occur with rate 0). The system switches between the two states as follows: while in the on state, it remains in that state for an amount

---

[2] We say more about random processes as input to filters in Chapter 9.

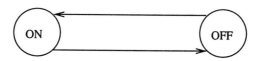

Figure 8.8   IPP.

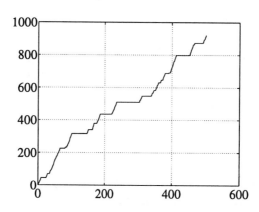

Figure 8.9   Number of arrivals, typical IPP sample path.

of time that is an exponentially distributed random variable with parameter σ_{on}. While in the off state, it remains in that state for an amount of time that is an exponentially distributed random variable with parameter σ_{off}. (We explain in Chapter 10 that the transitions between states are governed by a continuous-time Markov Chain.) The arrival process is then called IPP for obvious reasons.

The IPP can capture "burstiness," as the typical sample path in Figure 8.9 depicts. Comparing Figure 8.9 to Figure 8.13, we see why burstiness is captured by the IPP and not the Poisson process. The flat regions in the sample path are mainly due to the system's staying in the off state.

The pmf of $N(t)$, the number of arrivals or events in the time interval $[0, t)$, is not known in explicit form. However, Z transforms of $N(t)$ are known [21], through which the mean $EN(t)$ and variance $\text{var}(N(t))$ can be calculated. The random variable T, representing interarrival times, is more interesting in this process. The notion of burstiness is captured by the parameter c_T^2, the *coefficient of variation*, defined as follows:

$$c_T^2 \triangleq \frac{ET^2}{(ET)^2}.$$

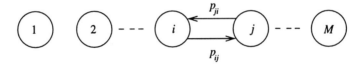

Figure 8.10 MMPP.

The higher the value of this coefficient, the burstier the process. For the Poisson process, we can easily calculate that $c_T^2 = 1$. Typical values of this coefficient for human speech processes are between 15 and 20, indicating, as expected, the burstiness of the "talk-spurt-silence" human voice process. For the IPP, the coefficient of variation is given by [21]

$$c_T^2 = 1 + \frac{2\sigma_{\text{on}}\lambda}{(\sigma_{\text{on}} + \sigma_{\text{off}})^2}.$$

3. Markov modulated Poisson process

The Markov modulated Poisson process (MMPP) was also used extensively in the late 1980s as a tool in high-speed network performance studies [28]. It is a generalization of the IPP, which captures both burstiness and correlation properties, that the Poisson process does not have. It can be viewed as follows: Consider M states, labeled $1, 2, \ldots, M$, as Figure 8.10 depicts.

While the system is in state i, arrivals can occur according to a Poisson process with rate $\lambda_i \geq 0$. The system stays in state i for an amount of time that is an exponentially distributed random variable with parameter σ_i. It then switches instantaneously to state j with probability p_{ij}. Figure 8.11 depicts a typical sample path of an MMPP. Notice that the burstiness effect is present there; the different slopes in the curve are produced by the different rates λ_i.

The pmf of $N(t)$, the number of arrivals or events in $[0, t)$, is only known as a Z transform [21]. Let T denote the time interval between two arrivals. For the special case of two states (that is, $M = 2$), the coefficient of variation for the MMPP is given by [21]

$$c_T^2 = 1 + \frac{2\sigma_1\sigma_2(\lambda_1 - \lambda_2)^2}{(\sigma_1 + \sigma_2)^2(\lambda_1\lambda_2 + \lambda_2\sigma_1 + \lambda_1\sigma_2)}.$$

Both the IPP and the MMPP can be better described and studied via a larger class of random processes, namely, Markov Chains. We present the MMPP model afresh in Chapter 10.

4. Compound Poisson process

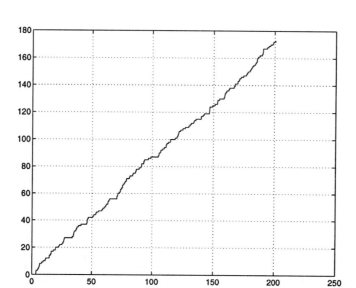

Figure 8.11   A typical MMPP sample path.

Let $\{N(t)\}$ be a Poisson process with parameter $\lambda > 0$. Let $\{Y_i\}_{i=1}^{\infty}$ be a sequence of IID, integer-valued random variables, independent of the process $\{N(t)\}$, with pmf $f(k)$. The random process $\{X(t)\}$, defined by

$$X(t) = \sum_{i=0}^{N(t)} Y_i,$$

is called a **compound Poisson process**. (In the above definition, Y_0 can be any random variable independent of $\{N(t)\}$ and the sequence $\{Y_i\}$.)

This process has been used as a model of systems in physics and telecommunications. It can model, for example, the emission of charged particles, where $N(t)$ represents the number of emitted particles and Y_i represents a property of the ith particle, such as charge, ionization level, or number of secondary emitted particles. In a telecommunication environment, $N(t)$ may represent the number of arriving packets in a buffer, while Y_i may represent its size in bits, transmission time, etc.

The compound Poisson process is a random sum process, so its properties can be derived from Theorem 6.1. From its definition we can see that it has independent, stationary increments. As we see in Chapter 10, it is a special case of a Markov Chain; in anticipation of that material, the behavior of the process can be characterized completely if the conditional probability $P[X(t) =$

$j|X(s) = i]$, $t \geq s$, can be determined. We have

$$
\begin{aligned}
P[X(t) = j|X(s) = i] &= P[X(t) - X(s) = j - i|X(s) = i] \\
&= P[X(t) - X(s) = j - i],
\end{aligned}
$$

where the last equality follows from the independent increment property. The random variable $X(t) - X(s)$ is a random sum by itself. Let $A \triangleq P[X(t) - X(s) = j - i]$. Conditioning on the number of arrivals in $[t, s)$, and using the total probability theorem for expected values (see Problem 3.194), we get

$$
\begin{aligned}
A &= E\{P[X(t) - X(s) = j - i|N(t) - N(s)]\} \\
&= \sum_{k=0}^{\infty} P[X(t) - X(s) = j - i|N(t) - N(s) = k]\frac{[\lambda(t - s)]^k}{k!}e^{-\lambda(t-s)}.
\end{aligned}
$$

Now, when we condition on the event $\{N(t) - N(s) = k\}$, the random variable $X(t) - X(s)$ is equal to the sum of exactly k random variables from the sequence $\{Y_i\}$. Therefore, $P[X(t) - X(s) = j - i|N(t) - N(s) = k]$ is equal to the k-fold convolution of the known pmf $f(\cdot)$. To summarize, the conditional probability $P[X(t) = j|X(s) = i]$ can then be found, at least in principle.

5. Binomial process in discrete time

The Poisson process is a continuous-time process. One wonders whether there is a discrete-time process counterpart that enjoys similar properties. The answer is yes, and the counterpart in discrete time is the binomial random process, defined as follows:

Definition: The random process $\{G_n, n = 0, 1, 2, \ldots\}$ is a called a **binomial random process** if the following conditions are met:
1. $G_0 = 0$ with probability 1.
2. The sequence $\{G_n\}_{n=0}^{\infty}$ is an IID increment sequence.
3. The random variable G_n, $n = 1, 2, \ldots$, is binomially distributed, i.e.,

$$
P[G_n = k] = \binom{n}{k}p^k(1 - p)^{n-k}, \quad k = 0, 1, \ldots, n,
$$

where $0 < p < 1$ is a fixed parameter.

In Problem 8.23 we identify several properties of this process in greater detail.

8.1.6 Ergodicity and Rates

The *rate* of arrivals at a given computer node, the *throughput* of a communication line, and the *speed* of a given software algorithm are all instances of performance measures that are defined as rates, i.e., as some quantity (packets, bits, instructions) per unit of time (usually seconds). In practice, we define such performance measures[3] as follows. We *count* the number of units $U(0, T)$ in an interval of time $[0, T)$; then the rate in the interval $[0, T)$ is defined as

$$r(T) \triangleq \frac{U(0, T)}{T}. \tag{8.6}$$

Intuitively, we expect that for large T, the rate $r(T)$ converges to a constant r,

$$r(T) \to r, \tag{8.7}$$

which we call the rate of arrivals, speed of the line, throughput of the protocol, etc.

A moment's reflection will show you that for any T, $r(T)$ in Equation 8.6 is a random variable since $U(0, T)$ is a random variable. So the convergence in 8.7 must be carefully interpreted. *Almost-sure* convergence to a constant is what we actually want; in the *ftp* scenario, for example, this means that the value of r we calculate in a single experiment run does not depend on the experiment (i.e., the day we perform the experiment). In simulations, almost-sure convergence is again what we need, since otherwise the results cannot be trusted. We formally show next that this is indeed the convergence in 8.7. The random process $N(t)$ can be more general than Poisson.

Let $\{S_i\}_{i=1}^{\infty}$ be the sequence of interarrival times associated with $N(t)$. In general, it is a sequence of IID positive random variables with mean $ES_i \triangleq ES = m = 1/\lambda > 0$. When S_i is not exponentially distributed, the collection of random variables $N(t)$ in Equation 8.8 is called in general a **counting process**.

Let $A_n \triangleq \sum_{i=1}^{n} S_i$, $A_0 \triangleq 0$. For any fixed $t > 0$, we can express the random variable $N(t)$ as

$$N(t) = \sup_{n} \{n : A_n \leq t\}. \tag{8.8}$$

Theorem 8.4 *Let $N(t)$ be a counting process associated with the IID sequence $\{S_i\}_{i=1}^{\infty}$. Then with probability 1,*

$$\lim_{n \to \infty} \frac{N(t)}{t} = \frac{1}{ES} = \lambda.$$

Proof: Consider a fixed $t > 0$. Now $A_{N(t)}$ is a random sum that represents the time of the most recent arrival that occurred before time t. (If no arrival occurred before t, the value of the sum is 0.) We have therefore

$$A_{N(t)} \leq t < A_{N(t)+1} \tag{8.9}$$

---

[3] We did this in the *ftp* example in Section 1.9.

with probability 1. Divide[4] now all three numbers in Inequality 8.9 by $N(t)$. We have

$$\frac{A_{N(t)}}{N(t)} \le \frac{t}{N(t)} < \frac{A_{N(t)+1}}{N(t)}. \tag{8.10}$$

Consider now the term $A_{N(t)}/N(t)$. We have

$$\frac{A_{N(t)}}{N(t)} = \frac{1}{N(t)} \sum_{i=1}^{N(t)} S_i. \tag{8.11}$$

As $t \to \infty$, we can easily see from Equation 8.8 that $N(t) \to \infty$ with probability 1 as well, since $A_n \to \infty$. Therefore the sum in Equation 8.11 contains an infinite number of IID random variables; from the Strong Law of Large Numbers (Theorem 6.5), then, as $t \to \infty$,

$$\frac{A_{N(t)}}{N(t)} \to ES = m$$

with probability 1. Consider next the term $A_{N(t)+1}/N(t)$. Since

$$\frac{A_{N(t)+1}}{N(t)} = \frac{A_{N(t)+1}}{N(t)+1} \frac{N(t)+1}{N(t)}, \tag{8.12}$$

we conclude that

$$\frac{A_{N(t)+1}}{N(t)} \to ES = m$$

with probability 1, as $t \to \infty$, since the first fraction in Equation 8.12 approaches ES and the second approaches 1. □

8.1.7   Operations on the Poisson Process

1. Sums of independent Poisson processes

Consider $M < \infty$ independent Poisson processes $\{N_i(t)\}$, defined on some sample space S, each with parameter λ_i. Define the sum process $\{N(t)\}$ on S as follows:

$$N(t, \zeta) = \sum_{i=1}^{M} N_i(t, \zeta), \quad t > 0, \ \zeta \in S.$$

The process $\{N(t)\}$ is widely used in communication networks; it can model the combined telephone call traffic to a switch, the combined packet arrival

---

[4] We can assume without loss of generality that $N(t) > 0$. Otherwise, Inequality 8.10 will hold trivially.

process from all the incoming links at a router, the output of a multiplexer, etc.

From our discussion in Section 4.11.8, it is clear that the process $\{N(t)\}$ is a Poisson process with parameter $\lambda = \sum_{i=1}^{M} \lambda_i$. The assumption of independence is indispensable; sums of dependent processes may not be Poisson (see Problem 8.5).

Example 8.2. Consider two Poisson processes $N_1(t)$ and $N_2(t)$ such that $N_1(t) = N_2(t)$ for all t. Then the process $N(t) = N_1(t) + N_2(t) = 2N_1(t)$ cannot be Poisson since

$$P[N(t) = 1] = P[N_1(t) = 0.5] = 0.$$

Can you derive another proof of dependence?

2. Bernoulli splitting

Consider the following operation on the sample paths of a Poisson process. At the ith arrival instant A_i, we flip a coin that shows heads with probability p, $0 < p < 1$. Flips are independent. If the coin shows tails, we "erase" the arrival; if it shows heads, we let the arrival proceed. Mathematically, we construct a new random process $\tilde{N}(t)$ as follows:

$$\tilde{N}(t) = \sum_{i=1}^{N(t)} I_i,$$

where the sequence $\{I_i\}$ is an IID sequence of Bernoulli random variables with parameter p. Pictorially, a typical sample path of the new process may look like that in Figure 8.12. We say that the process $\{\tilde{N}(t)\}$ is obtained from $\{N(t)\}$ through *Bernoulli splitting*. Processes obtained through Bernoulli splitting are also called *thinned* processes, for obvious reasons.

We show now that the thinned process is also Poisson, and we determine its parameter. Observe that $\tilde{N}(0) = 0$ and that $\tilde{N}(t)$ is an IID increment process. The IID property comes from the fact that the increment $\tilde{N}(t_2) - \tilde{N}(t_1)$ depends solely on the increment $N(t_2) - N(t_1)$ and the Bernoulli sequence. Thus the first two properties of the definition of a Poisson process are satisfied. The pmf of the single random variable $\tilde{N}(t)$ is easily computed through the sequence of interarrival times $\tilde{S}_n$. Using the total probability theorem (Theorem 2.6), we have

$$\begin{aligned}
P[\tilde{S}_1 > t] &= P[\text{all arrivals in } [0,t) \text{ erased}\}] \\
&= \sum_{k=0}^{\infty} P[\{N(t) = k\} \cap \{k \text{ erasures}\}]
\end{aligned}$$

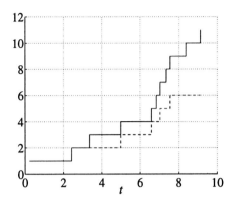

Figure 8.12   A "thinned" Poisson process.

$$= \sum_{k=0}^{\infty} P[N(t) = k] P[k \text{ erasures}]$$

$$= e^{-\lambda t} \sum_{k=0}^{\infty} \frac{(\lambda t)^k}{k!} (1-p)^k$$

$$= e^{-\lambda t} \sum_{k=0}^{\infty} \frac{[(1-p)\lambda t]^k}{k!} = e^{-\lambda p t},$$

and thus the new interarrival time is exponentially distributed with parameter λp. Independence of the random variables $\tilde{S}_k$ is easily established, along the lines of Theorem 8.1. Therefore the random process $\tilde{N}(t)$ is Poisson, with parameter λp.

Can you calculate $P[\tilde{N}(t) = k]$ directly, interpreting $\tilde{N}(t)$ as, say, a compound Poisson process?

8.1.8   Generation of Values of a Poisson Process

How do we generate sample paths for the Poisson process? Since it is an IID increment process, all we need to do is to generate variates for a single random variable. We have a number of choices here: We can generate exponential variates that represent the interarrival times, or we can generate Poisson variates that represent the number of arrivals in $[0, t)$. We discuss the first choice here, and we leave the second as an exercise.

Suppose that we have generated k exponentially distributed variates $\{S_l\}_1^k$, using any of the methods in Section 3.13. Then the sample path for the Poisson

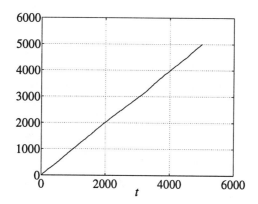

Figure 8.13   Typical Poisson sample path, based on 5000 arrivals.

process is simply given by

$$N(t) = \sum_{i=1}^{k} u\left(t - \sum_{l=1}^{i} S_l\right),$$

where $u(x)$ denotes the step function. A typical sample path with a few arrivals will look like Figure 8.1. You can interpret that graph as a microscopic view of the arrival process. A typical sample path with 5000 arrivals looks like Figure 8.13. This can be viewed as the macroscopic view of the arrival process. Notice that the sample path looks like an almost-straight line, the slope of which is of course λ!

8.2   GAUSSIAN RANDOM PROCESS

8.2.1   Definition of a Gaussian Process

Let $I = [0, \infty)$ or $I = (-\infty, \infty)$ be an index set. Consider a continuous-time random process $\{X(t), t \in I\}$ that takes values in R. For any fixed integer $n > 0$, let $t_1, \ldots, t_n \in I$ be a sequence of given time indices. For ease of notation, let $X_i \overset{\Delta}{=} X(t_i)$; define $\mathbf{X} \overset{\Delta}{=} (X_1\ X_2\ \cdots\ X_n)$, $\mathbf{x} \overset{\Delta}{=} (x_1\ x_2\ \cdots\ x_n)$. Let $\mathbf{m} \overset{\Delta}{=} (EX_1\ EX_2\ \cdots\ EX_n)$ and $C \overset{\Delta}{=} \text{cov}(X_1, \ldots, X_n)$.

Definition: The random process $\{X(t),\ t \in I\}$ is called a **Gaussian random process** if for any $n > 0$, the n-dimensional jpdf of the random variables

$X_1, \ldots, X_n$ is given by

$$f_{X_1 X_2 \cdots X_n}(x_1, x_2, \ldots, x_n) = \frac{1}{(2\pi)^{\frac{n}{2}} |C|^{\frac{1}{2}}} e^{\{-\frac{1}{2}(\mathbf{x}-\mathbf{m})^T C^{-1}(\mathbf{x}-\mathbf{m})\}}.$$

Note that the random variables $X_1, \ldots, X_n$ need not be independent; however, they are individually Gaussian-distributed.

From our discussions in Chapter 7, we can easily see the following:

Theorem 8.5 *If the Gaussian random process $X(t)$ is WSS, it is strictly stationary.*

8.2.2   Examples of Gaussian Processes

We present next two special cases of Gaussian random processes. We see more cases (and in particular *white-noise models*) in Chapter 9, after frequency domain representations of random processes are introduced.

1. The Wiener process

The Wiener[5] process is one of the most widely studied models of random processes. A large number of mathematicians have worked on it; Wiener formally introduced it in 1921.

Definition: The **Wiener process** with parameter $\sigma > 0$ is a Gaussian process on $I = [0, \infty)$, with stationary and independent increments, mean $m_X(t) = 0$, and autocovariance function

$$C_X(t_1, t_2) = \sigma^2 \min(t_1, t_2), \quad t_1, t_2 \in I.$$

Therefore, its one-dimensional density is given by

$$f_{X(t)}(x) = \frac{1}{\sqrt{2\pi t}\,\sigma} e^{-\frac{x^2}{2t\sigma^2}}, \quad t \in I, \tag{8.13}$$

and for any $t_1, \ldots, t_n \in I$, $x_1, \ldots, x_n \in R$, the n-dimensional density is

$$\begin{aligned}
f_{X_1 \cdots X_n}(x_1, \ldots, x_n) &= f_{X_1}(x_1) f_{X_2 - X_1}(x_2 - x_1) \cdots f_{X_n - X_{n-1}}(x_n - x_{n-1}) \\
&= \frac{e^{-\frac{1}{2}\left(\frac{x_1^2}{\sigma^2 t_1} + \frac{(x_2 - x_1)^2}{\sigma^2(t_2 - t_1)} + \cdots + \frac{(x_n - x_{n-1})^2}{\sigma^2(t_n - t_{n-1})}\right)}}{\sqrt{(2\pi\sigma^2)^n t_1(t_2 - t_1) \cdots (t_n - t_{n-1})}}.
\end{aligned} \tag{8.14}$$

Notice that as $t \to \infty$, the variance of the random variable $X(t)$ grows to ∞, while its mean stays at zero. Intuitively, we expect the sample paths of the Wiener process to oscillate "wildly" as $t \to \infty$. Figure 8.14 shows a typical sample path of this process.

---

[5] After Norbert Wiener, 1894–1964, a great contemporary U.S. mathematician.

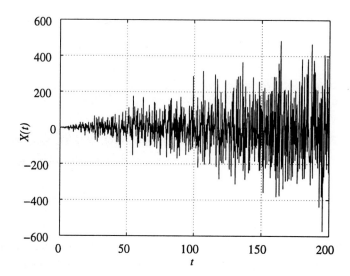

Figure 8.14   Wiener process sample path.

The process has been used primarily to describe *Brownian motion*, i.e., the motion of particles in a fluid under the influence of other particles.[6] It is also connected to another widely known process, namely, the random walk process we introduced in Section 7.5.1.

It is easy to generate sample paths of the Wiener process since it enjoys independent, stationary increments. Here is a simple algorithm:

1. Let $\{t_k\}_{k=1}^n$ be the time instants of interest.

2. Generate zero-mean, unit-variance Gaussian variates $\{G_k\}_{k=1}^n$.

3. Output $X(t_k) = \sigma\sqrt{t_k}\,G_k$.

A number of other models can be derived from the Wiener process, via appropriate transformations.

---

[6] This motion was first observed in 1827, by R. Brown, a botanist. It immediately attracted the attention of a lot of physicists, who tried to explain it. In 1906, a young physicist by the name of Albert Einstein published his doctoral dissertation on Brownian motion, explaining correctly its probabilistic behavior. He produced Equation 8.13.

2. Ornstein-Uhlenbeck process

A Gaussian random process $\{X(t),\ t \geq 0\}$ is called an **Ornstein-Uhlenbeck process** with parameters $a > 0$ and $\sigma > 0$ if

$$EX(t) = 0, \quad \text{cov}(X(t_1), X(t_2)) = \frac{\sigma^2}{2a}\left(e^{-a|t_1 - t_2|} - e^{-a(t_1 + t_2)}\right).$$

It can be shown that the Ornstein-Uhlenbeck process is related to the Wiener process; as we see in Chapter 9, it is the output of a first-order filter with impulse response e^{-at} and input a Wiener process with parameter σ.

8.2.3 Karhunen-Loève Expansion of the Wiener Process

Consider an interval $[0, T]$ and a Wiener process with covariance function

$$C_X(t_1, t_2) = \sigma^2 \min(t_1, t_2).$$

From Equation 7.35, for any $t \in [0, T]$, the eigenfunctions must satisfy the relationship

$$\int_0^T \sigma^2 \min(s, t)\, \phi(s)\, ds = \lambda \phi(t),$$

where λ is the eigenvalue. Now

$$\lambda \phi(t) = \sigma^2 \int_0^t s\phi(s)\, ds + \sigma^2 t \int_t^T \phi(s)\, ds.$$

We can solve the above integral equation by differentiating twice. We have

$$\lambda \frac{d^2\phi(t)}{dt^2} + \sigma^2 \phi(t) = 0, \tag{8.15}$$

with initial conditions

$$\phi(0) = \left.\frac{d\phi(t)}{dt}\right|_{t=T} = 0.$$

The solution to Equation 8.15 is known to be

$$\phi_n(t) = \sqrt{\frac{2}{T}} \sin \frac{(2n-1)\pi t}{2T}, \quad n = 1, 2, \ldots,$$

and the desired expansion is given by

$$X(t) = \sqrt{\frac{2}{T}} \sum_{n=1}^{\infty} X_n \sin \frac{(2n-1)\pi t}{2T}, \quad t \in [0, T],$$

where the random variables $\{X_n\}$ are uncorrelated Gaussian with

$$EX_n = 0, \quad EX_n^2 = \frac{\sigma^2 T^2}{(n - \frac{1}{2})^2 \pi^2}.$$

8.2.4   Relationship to Other Processes

We saw in Chapter 6 that the Gaussian random variable is related to normalized sums of IID random variables (see Theorem 6.6). The Gaussian random process, and in particular the Wiener process, plays a similar, important role in the theory of random processes. It is the counterpart of the Gaussian random variable in a version of a central limit theorem for random processes.

Consider a sequence of IID random variables $\{Y_i\}_{i=1}^{\infty}$; let $S_m = \sum_{i=1}^{m} Y_i$, for $m > 0$ and $S_0 = 0$. Define the sequence of random processes $\{X_n(t)\}_{n=1}^{\infty}$ as follows: For a fixed $n \geq 1$, and $t > 0$, let

$$X_n(t) = \frac{1}{\sqrt{n}} S_{[nt]}, \qquad (8.16)$$

where the notation $[a]$ represents the integer part of the positive real number a. We state the following result without proof.

Theorem 8.6 *Consider a sequence of independent, identically distributed random variables $\{Y_i\}_{i=1}^{\infty}$, with common mean $EY = 0$ and finite variance $\sigma^2 > 0$. Then the processes $X_n(t)$ defined in Equation 8.16 converge in distribution to $X(t)$, a Wiener process with parameter σ.*

This fundamental result has been extended for sequences of random processes that do not have the IID structure we assumed above; see [19] for more details.

8.2.5   Processing the Gaussian Process

1. Sums of independent Gaussian processes

Consider $M < \infty$ independent Gaussian processes $\{X_i(t), t \in I\}$ defined on some sample space S with parameters $\mathbf{m_i}$ and $\mathbf{C_i}$. Define

$$X(t, \zeta) = \sum_{i=1}^{M} X_i(t, \zeta), \quad \forall \zeta \in S,\ t \in I.$$

Then we can easily see that the random process $\{X(t)\}$ is Gaussian, with parameters

$$\mathbf{m} = \sum_{i=1}^{M} \mathbf{m_i}, \quad \mathbf{C} = \sum_{i=1}^{M} \mathbf{C_i}.$$

2. Linear transforms of Gaussian processes

In Chapter 9 we encounter a number of operations on the random variables of a process $\{X(t)\}$, mainly in the form of linear transformations (e.g., filtering, prediction) that can be described functionally by

$$Y(t) = \int h(t-s)X(s)\, ds.$$

From our discussion in Theorem 4.17, it is expected that when $X(t)$ is Gaussian, the resultant process $Y(t)$ is also Gaussian. Of course, the difficulty comes from the fact that in an integral like the above, the number of Gaussian random variables involved is infinite. The random process case, even though plausible, requires more sophisticated machinery; thus we only state the result here, without proof. See [23] for a rigorous proof.

Lemma 8.3 *Consider a Gaussian random process $X(t)$. Then the process $Y(t)$ defined by*

$$Y(t) = \int h(t-s)X(s)\, ds$$

is also Gaussian.

This fact explains the wide applicability of the Gaussian random process model.

8.2.6   Generation of Values of a Gaussian Process

How do we generate sample paths for the Gaussian process? In principle, for the general case the algorithm in Section 4.12 could be used. In practice, the amount of computational work required in this approach is prohibitive.

For the Wiener process, things are much easier since it is an independent increment process. However, since the one-dimensional distributions in 8.14 depend on time, the amount of work is not again trivial.

We see in Chapter 9 that things are a bit better with white noise, the most widely used variation of a Gaussian process.

8.3   SUMMARY OF MAIN POINTS

- The **Poisson process** is a continuous-time random process with IID increments that follow the Poisson pmf.

- The Poisson process is the main **model** for counting occurrences of events (e.g., arrivals in computer networks), due to its mathematical tractability.

- In a Poisson process, **interarrival times** are exponential; **times until the nth arrival** are Erlang.

- Variations of the Poisson process include **shot noise**, the **compound Poisson process**, the **IPP**, and the **MMPP**.

- The **Gaussian process** is a continuous-time random process, with or without IID increments that follow the Gaussian pdf.

- The **Gaussian process** is the limit of IID sequences (in the spirit of Theorem 8.6); it is the main **model** for noise in communication channels because linear transformations of a Gaussian process result again in a Gaussian process.

- Special cases of the Gaussian process include white noise, the Wiener process, and the Ornstein-Uhlenbeck process.

- Generating variates of the Poisson and Gaussian processes is not difficult.

8.4   CHECKLIST OF IMPORTANT TOOLS

- The PASTA property

- Summation and thinning of a Poisson process

- Linear transformations of a Gaussian process

8.5   PROBLEMS

Poisson Process; Simple Skills

8.1 Prove that the three definitions of the Poisson process are equivalent.

8.2 Consider a Poisson process $\{N(t)\}$. What is the value of the probability
(a) $P[N(t) = \infty$, for some $t < \infty]$? Why?
(b) $\lim P[N(t) = k]$, as $t \to \infty$, for any fixed k?

8.3 Consider a Poisson process $\{N(t)\}$. Find the joint pdf of the M arrival instants, given that exactly M arrivals have occurred in the interval $[0, T)$, $T > 0$.

8.4 Consider a Poisson process $\{N(t)\}$. Find the pdf of the nth arrival, given that $K \geq n$ arrivals have occurred in the interval $[0, T)$, $T > 0$.

8.5 Show by counterexample that the sum of two dependent Poisson processes *may not* be Poisson.

8.6 Is the Poisson process stationary?

8.7 Show that the Poisson process is mean square continuous.

8.8 Consider a Poisson process $\{N(t)\}$. For a fixed $T > 0$, define the new process $\{Y(t)\}$ by

$$Y(t) = \frac{N(t+T) - N(t)}{T}.$$

(a) Calculate the autocovariance of $Y(t)$.
(b) Use this result to determine whether the derivative of the Poisson process exists.

8.9 Consider a Poisson process $\{N(t)\}$ with parameter λ. Define the new process

$$Y(t) = \cos(2\pi f_0 t + \pi N(t)), \quad t \geq 0,$$

where f_0 is a given positive constant.
(a) Plot some sample paths of $\{Y(t)\}$.
(b) Calculate $f_{Y(t)}(y)$, $EY(t)$, and $\text{var}(Y(t))$.
(c) Calculate the joint pdf $f_{Y(t_1)Y(t_2)}(x, y)$.
(d) Calculate the autocovariance of $\{Y(t)\}$.
(e) Is $\{Y(t)\}$ stationary? WSS? Cyclostationary?
(f) Does $\{Y(t)\}$ have an integral process? A derivative process?
(g) Is $\{Y(t)\}$ mean square continuous?

Poisson Process; *Advanced Skills*

8.10 Evaluate the probabilities

$$P[N(t) = j | N(t+s) = k], \quad P[N(t) = j, N(u) = i | N(t+s) = k],$$

where $s > 0$ and $0 < u < t$.

8.11 Consider two independent Poisson processes $\{N_1(t)\}$ and $\{N_2(t)\}$ with parameters λ_1 and λ_2. Define a new random process as

$$Y(t) = N_1(t) - N_2(t), \quad t \geq 0.$$

(a) Evaluate $EY(t)$ and $\text{var}(Y(t))$.
(b) Show that

$$\sum_{k=-\infty}^{\infty} P[Y(t) = k]z^k = e^{-(\lambda_1 + \lambda_2)t} e^{\lambda_1 zt + \lambda_2 tz^{-1}}.$$

8.12 Consider a Poisson process $\{N(t)\}$. Let A be a positive random variable independent of the process. Denote the number of arrivals in the interval $[t, t + A)$ by $N_A(t)$.

(a) Find the probability generating function of the random variable $N_A(t)$.
(b) Find the pmf of $N_A(t)$ when A is exponential.

8.13 Consider two independent Poisson processes $\{N_1(t)\}$ and $\{N_2(t)\}$ with parameters λ_1 and λ_2. Let A_1, A_2 denote two successive arrival instants of the $N_1(t)$ process. Let $B \overset{\triangle}{=} N_2(A_2) - N_2(A_1)$ be the number of arrivals in the $N_2(t)$ process. Show that

$$P[B = n] = \frac{\lambda_1}{\lambda_1 + \lambda_2} \left(\frac{\lambda_2}{\lambda_1 + \lambda_2} \right)^n, \quad n = 0, 1, \ldots.$$

8.14 Consider two independent Poisson processes $\{N_1(t)\}$ and $\{N_2(t)\}$ with parameters λ_1 and λ_2. Let $N_1(s) = n_1$ and $N_2(s) = n_2$, where $s > 0$ is a given constant. Let $n > \max\{n_1, n_2\}$ be a given integer. Calculate the probability $P[N_1(t) = n, N_2(t) < n]$ for $t > s$.

8.15 Consider the sample space $S = [0, 1]$ with a uniform probability assignment on it. Define the trivial random process

$$X(t, \zeta) = te^{-t}, \quad t \in [0, \infty), \quad \zeta \in S.$$

Consider the sequence

$$Y_k = X(A_k), \quad k = 0, 1, \ldots,$$

where A_k is a sequence of Poisson arrival instants. We can view the sequence $\{Y_k\}$ as samples of the $\{X(t)\}$ process, taken in a Poisson fashion. Can you evaluate the following?

$$\lim_{n \to \infty} \frac{1}{n} \sum_{k=1}^{n} Y_k$$

8.16 Consider the case of Bernoulli splitting a Poisson process $\{N(t)\}$, with parameter λ. Let $\{N_1(t)\}$ [or $\{N_2(t)\}$] denote the process that results when the outcome of the Bernoulli trial is a success (failure).
(a) Find the conditional joint pmf of $N_1(t)$ and $N_2(t)$, given $N(t) = n$.
(b) Show that $\{N_1(t)\}$ and $\{N_2(t)\}$ are independent Poisson processes.

8.17 Consider a Poisson process $\{N(t)\}$ with parameter λ. Define the new process $\{Y(t)\}$ via $Y(t) = N(t) - tN(1)$.
(a) Are the sample paths of $\{Y(t)\}$ nondecreasing?
(b) Find $EY(t)$ and $\text{var}(Y(t))$.
(c) Calculate the autocovariance of $Y(t)$.

8.18 Let $\{N(t)\}$ be a Poisson process with parameter λ. For a fixed $t > 0$, let

$$Y_t = t - A_{N(t)}, \quad Z_t = A_{N(t)+1} - t,$$

where A_k denotes the kth arrival instant.
(a) Determine the range of the random variables Y_t, Z_t.
(b) Find the joint pdf of the random variables Y_t, Z_t.
(c) Is the random variable $A_{N(t)+1} - A_{N(t)}$ exponentially distributed?

8.19 Let $N(t)$ be a Poisson process with parameter λ. Show that

$$[N(t) - \lambda t]/(\sqrt{\lambda t}) \to X$$

in distribution, where X is a Gaussian random variable with zero mean and unit variance.

8.20 Let $\{N(t)\}$ be a Poisson process. Show that $P[\lim_{s \to t} N(s) = N(t)] = 1$.

8.21 Consider the shot noise process. Evaluate the mean and variance when the filter has impulse response $h(t) = e^{-at}u(t)$ for given $a > 0$.

8.22 Consider a Poisson process $\{N(t)\}$ with parameter λ and arrival instants $\{A_n\}$. Suppose that $\{B_n\}$ is an IID sequence independent of $\{N(t)\}$. For a given filter $h(t)$, define the new process $\{X(t)\}$ by

$$X(t) = \sum_{n=1}^{\infty} B_n h(t - A_n).$$

(a) Find $EX(t)$ and $R_X(t, s)$.
(b) Repeat part (a) for the special case $h(t) = u(t)$, the usual step function.

8.23 Consider a binomial process $\{G_n\}$.
(a) Prove the memoryless property.
(b) Show that sums of binomial processes are also binomial.
(c) Show that thinning of a binomial process results in binomial processes.

Poisson Process; Modeling

8.24 Suppose that telephone calls arrive at a telephone switch according to a Poisson process at a rate of 1000 calls per minute. The switch must be taken out of operation every Sunday morning, for 10 min, for preventive maintenance.
(a) Calculate the probability of calls being blocked during that time.
(b) If the telephone company wants to make the probability of calls being blocked during maintenance less than 0.001, how long can the maintenance period be?

8.25 Suppose that packets arrive at a computer network node according to a Poisson process at a rate λ. The probability of a packet's being received correctly is p.

(a) Find the pmf of the number of packets received correctly in the time interval $[0, t)$.

(b) Find the pmf of the number of packets received incorrectly in the time interval $[0, t)$.

8.26 Suppose that packets arrive at a router (with a huge memory) according to a Poisson process at a rate λ. Packet i requires X_i ms of processing, at the router's microprocessor, before it can be sent out. The output link has a speed of K kbytes/s. Suppose that the sequence $\{X_i\}$ is a sequence of IID random variables. In answering the following questions, identify any missing or extraneous information.

(a) How fast should the microprocessor be if we want the throughput of the router to be at least 1000 packets per second? 50,000 packets per second?

(b) Suppose that the speed of the microprocessor is S millions of instructions per second, and the packet arrival rate is r packets per second. The routing protocol takes on average O instructions per packet. What is the largest possible value of O for which the router can sustain a throughput of r packets per second?

8.27 Suppose that packets arrive at a router (with an infinitely fast microprocessor) according to a Poisson process at a rate λ. Packet i requires X_i bytes of storage. The output link has a speed of K kbytes/s. Suppose that the sequence $\{X_i\}$ is a sequence of IID random variables. The size of the router's memory is B Mbytes.

(a) Simulate the operation of the router and calculate the probability that the router will drop a packet due to storage overflow, in a 1-h period of operation.

(b) How big should the memory be if that probability must be kept below 0.000001?

8.28 Suppose that packets arrive at a router according to a Poisson process at a rate λ. The packet size is fixed and equal to 1500 bytes. The number of bit errors in packet i, denoted by X_i, is a binomial random variable with parameters $n = 12,000$ and $p = 0.0001$.

(a) Find the probability that a packet is received error-free.

(b) Long files are routinely divided into 1500-byte packets. Find the probability that a 1-Mbyte file is received error-free.

(c) Find the probability that the router will receive error-free packets for 3 consecutive h, either analytically or by simulation.

8.29 A router accepts packets from two input links. Suppose that the packet arrival process at link i is a Poisson process with a rate λ_i, $i = 1, 2$. The two arrival processes are independent of each other.

(a) Find the probability that a packet arrives at the router in a 3-s interval.

(b) Suppose that due to a transient interruption, the router is taken out for x

ms. Find the probability that the router loses a packet.

(c) Find the probability that the lost packet arrived on link 1.

(d) Find the pmf of the number of arrivals in the interval $[0, t)$.

8.30 Can you generalize Problem 8.29 for the case of $N > 2$ input links?

8.31 Suppose that file read requests arrive at a file server according to a Poisson process $\{R(t)\}$ with rate λ_R. The ith read request results in a random number of disk accesses X_i; we can model the sequence $\{X_i\}$ as an IID sequence independent of the process $\{R(t)\}$. Let $D(t)$ denote the total number of disk accesses in the interval $[0, t)$.

(a) Find $P[D(t) = j | R(t) = i]$.

(b) Find $P[D(t) = j]$.

(c) If possible, obtain data from your local system and validate or disprove this model.

8.32 Two communicating applications are exchanging files. Files are divided into two packets, before transmission. Suppose that the packet arrival process at the receiver is a Poisson process $\{N(t)\}$, with rate λ. Before the receiver forwards the files to the application, it must wait for both packets to arrive. Upon reception of the second packet, the file is immediately forwarded. Define D_i, the reassembly delay for file i, as the time interval between the arrivals of the first and second packets.

(a) Find the pdf of D_i, $i \geq 1$.

(b) Is $\{D_i\}$ an IID sequence?

(c) Just for fun, simulate this system and try to answer questions (a) and (b) experimentally.

Wiener Process; Simple Skills

8.33 Consider a Wiener process $\{X(t)\}$. Define the *rectified* Wiener process $\{Y(t)\}$ by $Y(t) = X^2(t)$.

(a) Evaluate $EY(t), \mathrm{var}(Y(t)), EX(t)Y(t)$.

(b) Find $f_{Y(t)}(y)$.

(c) Find $f_{Y(t)Y(s)}(y_1, y_2)$.

8.34 Consider a Wiener process $\{X(t)\}$. Define the *shifted* Wiener process $\{Y(t)\}$ by $Y(t) = X(t) + mt$, where m is a real constant.

(a) Evaluate $EY(t), \mathrm{var}(Y(t)), EX(t)Y(t)$.

(b) Find $f_{Y(t)}(y)$.

(c) Find $f_{Y(t)Y(s)}(y_1, y_2)$.

8.35 (a) Show that the Wiener process does not have a derivative.

(b) Show that is has an integral.

(c) Find the mean and autocorrelation of its integral.

Wiener Process; Advanced Skills

8.36 Consider a Wiener process $\{X(t)\}$. Define the process $\{Y(t)\}$ by

$$Y(t) = \frac{X(t+\epsilon) - X(t)}{\epsilon},$$

where $\epsilon > 0$ is a given constant.
(a) Show that $\{Y(t)\}$ is a stationary Gaussian random process.
(b) Show that

$$R_Y(\tau) = \begin{cases} \frac{\sigma^2}{\epsilon}\left(1 - \frac{|\tau|}{\epsilon}\right), & |\tau| < \epsilon, \\ 0, & |\tau| \geq \epsilon. \end{cases}$$

c) What is the limit of $R_Y(\tau)$ as $\epsilon \to 0$?

8.37 Consider a Wiener process $\{X(t)\}$. Define the process $\{Y(t)\}$ by $Y(t) = tX(t^{-1})$. Find $R_Y(\tau)$.

8.38 Consider a Wiener process $\{X(t)\}$. Define the process $\{Y(t)\}$ by $Y(t) = X(t) - tX(1)$. Find $EY(t)$ and $R_Y(t, s)$.

8.39 Consider a Wiener process $\{X(t)\}$ with parameter σ^2. Define the process $\{Y(t)\}$ by $Y(t) = e^{-at}X(e^{2at})$, where $a > 0$ is a given constant.
(a) Show that $\{Y(t)\}$ is a stationary Gaussian random process.
(b) Show that $R_Y(\tau) = \sigma^2 e^{-a|\tau|}$.

8.40 Consider a Wiener process $\{X(t)\}$. Let $Y(t) = c^{-1}X(c^2 t)$ where $c > 0$ is a given constant. Show that $\{Y(t)\}$ is also a Wiener process and find its parameter.

8.41 Consider a Wiener process $\{X(t)\}$. Show that the pdf of $X(t)$, given $X(s) = a$ and $X(u) = b$, where $s < t < u$, is Gaussian, with mean and variance

$$m = a + \frac{b-a}{u-s}(t-s), \quad \sigma^2 = \frac{(u-t)(t-s)}{u-s}.$$

8.42 Consider a Wiener process $\{X(t)\}$. Let $Y = \int_0^1 X(t)\, dt$. Find $\text{cov}(Y, X(t))$.

8.43 Find the Karhunen-Loève expansion of the random process of Problem 8.38 over the interval $[0, 1]$. For simplicity let $\sigma^2 = 1$.

Gaussian Process; Simple Skills

8.44 Consider two jointly Gaussian random processes $\{X(t)\}$ and $\{Y(t)\}$. Under what conditions, if any, are the notions of independence, uncorrelatedness, and orthogonality equivalent?

8.45 Consider a Gaussian random process $\{X(t)\}$ with $EX(t) = 0$ and $R_X(\tau) = 4e^{-|\tau|}$.
(a) Find the jpdf of $X(t)$ and $X(s)$.
(b) Find the pdf of $X(t)$.

8.46 Consider a Gaussian random process $\{X(t)\}$. Define the new process $\{Y(t)\}$ by $Y(t) = X(t) - X(t + d)$, where $d > 0$ is a given real constant.
(a) Find the pdf of $Y(t)$.
(b) Find $EY(t), \mathrm{var}(Y(t))$.
(c) Find the joint pdf $f_{Y(t)Y(s)}(x, y)$.
(d) Find $R_Y(t, s)$.

8.47 Consider a Gaussian random process $\{X(t)\}$. Define the new process $\{Y(t)\}$ by $Y(t) = X^2(t)$. Find $EY(t), \mathrm{var}(Y(t))$, and $R_Y(t, s)$.

Gaussian Process; Advanced Skills

8.48 Consider a zero-mean Gaussian random process $\{X(t)\}$. A sampler samples the process at time instants $t_k = k$, $k = 1, 2, \ldots$, producing a sequence of random variables $\{X_k\}$. Then a *decimator* drops every other sample, forming the sequence
$$Y_n = X_{2n}, \quad n = 1, 2, \ldots.$$
(a) Calculate $f_{Y_n}(y)$, EY_n, and $\mathrm{var}(Y_n)$.
(b) Calculate the joint pdf $f_{Y_n Y_m}(x, y)$.
(c) Calculate the autocovariance of Y_n.
(d) Is $\{Y_n\}$ stationary? WSS? Cyclostationary?

8.49 Consider two independent, zero-mean Gaussian random processes $\{X(t)\}$ and $\{Y(t)\}$ with autocorrelation functions
$$R_X(\tau) = \sigma_X^2 \rho_X^{|\tau|}, \quad R_Y(\tau) = \sigma_Y^2 \rho_Y^{|\tau|},$$
where $\sigma_X^2, \rho_X, \sigma_Y^2, \rho_Y$ are given constants. A sampler samples these random processes at time instants $t_k = k$, $k = 1, 2, \ldots$, producing two sequences of random variables $\{X_k\}$ and $\{Y_k\}$. Then a multiplexer produces a third sequence of random variables $\{Z_k\}$, by interleaving the two samples as follows:
$$X_1, \ X_2, \ Y_1, \ Y_2, \ X_3, \ X_4, \ Y_3, \ Y_4, \ X_5, \ X_6, \ \ldots.$$

(a) Write Z_k as a function of $\{X_k\}$ and $\{Y_k\}$.
(b) Calculate the pdf, mean, and variance of Z_k.
(c) Calculate the joint pdf of Z_k, Z_l.
(d) Calculate the autocovariance of the $\{Z_k\}$ process.
(e) Is $\{Z_k\}$ stationary? WSS? Cyclostationary?

Projects

8.50 Consider the simulated simple buffer project in the Web site.
(a) Plot sample paths of the arrival process, transmission time process, and departure process for $t = 100$ and $t = 5000$ s.
(b) Generate histograms of the interarrival and transmission times. Are they close to exponential pdf's? What are μ, the transmission rate, and λ, the arrival rate?
(c) Plot a histogram of the interdeparture times. What density can you fit? What is the average of that density? Anything expected or unusual?
(d) Plot a histogram for the waiting and total delay time. Comment as in part (c).
(e) Plot a histogram of the idle and busy period duration. Comment.
(f) Let $U(t)$ denote the system utilization at time t; in other words,

$$U(t) = \begin{cases} 1, & \text{if system busy at time } t, \\ 0, & \text{if system idle at time } t. \end{cases}$$

Is $U(t)$ a random process? Why? Plot $U(t)$ for $t \in [0, 30]$.
(g) For $t \in [0, 5000]$, plot the time-average quantity

$$\bar{U}(t) \triangleq \frac{1}{t} \int_0^t U(s)\, ds.$$

(h) Observe the limiting behavior of $\bar{U}(t)$ as $t \to \infty$. Compare it to the ratio λ/μ.
(i) Suppose that after $t \approx 1000$ s, the system reaches steady state. Plot a histogram for the random variable Q, the total number of packets in the system at steady state. What distribution can you fit?
(j) Sample the random process $U(t)$ in a Poisson fashion, at various rates λ. Verify that PASTA holds.

8.51 In the Web site, and in the same project directory, you will also find a C simulator of this buffer. Compile and run the file main.c. Enter 10 to 15 arrival and transmission rate parameter values that give you $\rho = \lambda/\mu$ values in the range $[0.2, 0.95]$.

(a) Plot the average total delay ET versus ρ.

(b) Explain intuitively your results.

Computer Problems

8.52 Plot sample paths of the IPP for various values of its parameters. Determine the rate of the process experimentally. Observe how burstiness changes as a function of the process parameters.

8.53 Plot sample paths of the interarrival times associated with the Poisson process. Determine the rate of the process experimentally, using these interarrival times.

8.54 Generate M variates of the Poisson process. Derive and plot experimental histograms of the following random variables:

(a) Interarrival time,

(b) Time until the nth arrival, $n = 1, 2, 3, 10$.

(c) One conditional arrival time in $[0, 1]$.

(d) Two conditional arrival times in $[0, 1]$.

(e) Conditional arrival time, given that an arrival did not occur in 2 s. Hence verify the memoryless property defined in Problem 3.118.

(f) Number of arrivals in $[0, 4]$.

8.55 Can you verify experimentally whether the PASTA property holds, using the Dow Jones data?

8.56 Calculate experimentally the rate of an MMPP with given parameters.

8.57 Thin a Poisson process. Calculate means, variances, and histograms related to the thinned processes.

8.58 Compound a Poisson process. Calculate means, variances, and histograms related to the compound process.

8.59 Plot sample paths of the Wiener process. Calculate means, variances, and two-dimensional histograms of the random variables $X(t), Y(t)$ in Problem 8.33.

8.60 Simulate the Ornstein-Uhlenbeck process. Plot a few sample paths. Calculate means, variances, and two-dimensional histograms for the random variables $X(t), X(t+1)$.

8.61 Verify the CLT theorem, Theorem 8.6, using exponential variates.

PROCESSING OF RANDOM PROCESSES

9.1 INTRODUCTION

Random processes or, more accurately, sample paths of random processes, are very often inputs to a system that "transforms" them into another time function. For example, the FM signal in Example 7.4 is the input to a communication channel (the air-space between the station antenna and the antenna of your radio). The system (the channel in this case) transforms the typical sample path by adding noise into it, attenuating it, etc.

As another example, consider a file server connected to a set of client workstations through an Ethernet. Consider the (discrete-time) random process $\{T_i\}$, where T_i represents the time instant at which the ith file is sent to a requesting station. This random process is the input to the "network system," which transforms it into another sequence $\{D_i\}$, where D_i represents the time instant at which the ith file is received at the requesting station.

From the above examples, we see that the general processing problem can be stated as follows (see Figure 9.1): We are given a random process $\{X(t)\}$, in continuous or discrete time, and a system $g(\cdot)$. We are interested in characterizing the "probabilistic behavior" of the output process $Y(t)$.[1]

---

[1] From our discussion in the previous chapter, characterizing the n-dimensional pdf's would suffice.

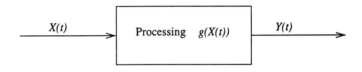

$X(t)$ → Processing $g(X(t))$ → $Y(t)$

Figure 9.1 The general processing problem.

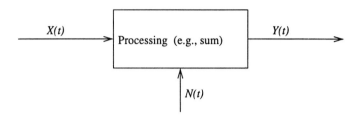

Figure 9.2   Multiple inputs, one output.

Of course, the immediate theoretical question is the following: Given a "system" $g(\cdot)$ and an input random process $X(t)$, is the output $Y(t)$ a random process?[2] As in the case of a single random variable and a random vector, the answer here is also yes if the function $g(\cdot)$ satisfies certain mild conditions; a rigorous discussion, however, is beyond the scope of this book.

We can immediately classify some special cases of the general problem. The alert student will recognize them as almost identical to the transformations presented in Chapters 3 and 4.

1. Single random process as input, single random process as output. Figure 9.1 depicts this case. We have two special subcases, depending on the function $g(\cdot)$:

a. $g(\cdot)$ linear. We treat this case in great detail in Section 9.3. Questions relating the input to the output of a deterministic linear system with deterministic inputs are easily answered in the *frequency domain*; we use this convenient domain in the realm of random processes as well. So we must define a frequency domain representation of a random process. We do that in Section 9.2, where we introduce the fundamental concept of the *power spectral density* of a random process.

b. $g(\cdot)$ nonlinear. We do not discuss this case here since only limited, specialized results have been produced so far.

2. Multiple random processes as input, single random process as output. Figure 9.2 depicts this case, for two inputs. The special case of signal plus noise is treated in Sections 9.4.6 and 9.5.

3. Multiple random processes as input, multiple random processes as output. We do not cover this case here.

How do we characterize the probabilistic behavior of the output random process? Does it inherit any of the characteristics of the input process? For example, if the input process is stationary, is the output stationary as well? What is the mean of the output process? Its covariance? We address such questions

---

[2] This may sound like a trivial question, but recall our discussion in Section 3.8.

in Section 9.3; the foundation for this development, namely, the notion of the power spectral density, is introduced in Section 9.2.

In the past, filtering noise effects out of measurements was the main application of the theory of random processes. So the focus was on frequency domain processing, with *spectral factorization* being one of the major achievements of the theory (see Section 9.4.3). Nowadays, much processing is done in specialized hardware chips or computers. When the processing system is digital, the characteristics of the output random process are best described in the time domain. We use this approach in Section 9.5, where we describe the Kalman filter, one of the most widely used filters in applications.

In order to apply the results of the theory, we need to evaluate parameters of the random process, such as mean values, autocorrelations, and power spectral densities, from measurements of the random process. In Section 9.6 we discuss how the power spectral density can be estimated. For the estimation of the autocorrelation function $R_X(\tau)$, see Problem 7.73. For an estimate of the mean m_X, we can apply the SLLN theory in Chapter 6 or ergodicity in Chapter 7.

9.2 POWER SPECTRAL DENSITY FUNCTION

9.2.1 How to Define It

In the deterministic case, the *Fourier transform* of a signal $s(t)$ represents the frequencies present in the signal, in the sense that $s(t) = \cos(2\pi f_0 t)$ contains just one frequency (f_0), while

$$s(t) = \cos(2\pi f_0 t) + \cos\left(2\pi f_1 t - \frac{\pi}{2}\right)$$

contains two (f_0 and f_1), and $s(t) = u(t-T) - u(t)$ contains an infinite number of frequencies [all in the interval $[0, \infty)$]. The Fourier transforms of these three signals are shown in Figure 9.3, for $f_0 = 1$ and $f_1 = 3$.

Since for a fixed $\zeta \in S$ the sample path $X(t, \zeta)$ is a deterministic time signal, the intuitive approach to determining a frequency domain representation for a random process is to define Fourier transforms for each sample path. Consider, therefore, the function of the frequency parameter $f \in (-\infty, \infty)$:

$$\bar{X}(f, \zeta) \triangleq \int_{-\infty}^{\infty} X(t, \zeta) e^{-j2\pi ft} \, dt. \tag{9.1}$$

There are a few points worth mentioning about the function in Equation 9.1. From the theoretical point of view, the immediate question is of course, Is the

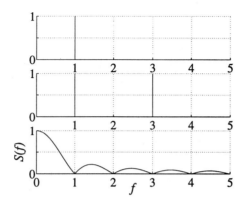

Figure 9.3   The frequency spectra of three signals.

integral well defined for all ζ? In what sense is equality in Equation 9.1 to be interpreted? Mean square? Almost surely? From the practical point of view, two issues arise: First, we *must know* the dependence of the sample paths on time, in order to evaluate Equation 9.1. Therefore, this definition cannot be applied to most of the random processes, not even Poisson and Gaussian, since the functions $X(t,\zeta)$ are not known! Even if we do know the sample path behavior, another problem arises, as the following example demonstrates.

Example 9.1. Suppose that the random process has only two sample paths, say, $X(t,\zeta_1) = \cos(100t)$ and $X(t,\zeta_2) = \cos(1000t)$. Suppose now that $P[\zeta_1] = 0.999999$ and $P[\zeta_2] = 0.000001$. Since probabilities of sample paths do not appear in Equation 9.1, the Fourier transform of this process will simply contain the two frequencies 100 and 1000, without any indication of their relative probabilities.

It is apparent that we must fix those problems if we are going to have a frequency representation that's both computable and useful. We see next how we can avoid all problems, via the autocorrelation function. We limit our discussion in the sequel to WSS processes, for reasons of simplicity.

From Equation 9.1 and the standard definition of the energy of a signal in the frequency band $[f, f + df)$, we see that the random variable

$$\bar{X}_T(f) \triangleq \frac{1}{2T} \left[\int_{-T}^{T} X(t)e^{-j2\pi ft}\, dt \right] \cdot \left[\int_{-T}^{T} X(s)e^{j2\pi fs}\, ds \right]$$

is an estimate of the energy of the signal $X(t, \zeta)$ in the band $[f, f + df)$. The expected value of that energy is then given by

$$E\bar{X}_T(f) = \int_{-T}^{T} \left(1 - \frac{|\tau|}{2T}\right) R_X(\tau)e^{-j2\pi f\tau}\, d\tau.$$

So, as $T \to \infty$, it is reasonable to expect that

$$E\bar{X}_T(f) \to \int_{-\infty}^{\infty} R_X(\tau)e^{-j2\pi f\tau}\, d\tau \qquad (9.2)$$

and to define the power spectral density as the value of that limit. A rigorous proof of Equation 9.2 was given by nobody less than Albert Einstein in the early 1900s and subsequently by Wiener and Khinchine in the 1930s. We summarize the above discussion in the following definitions (which are actually deep theorems, known as the Wiener-Khinchine-Einstein theorems).

Definition: The **power spectral density (psd)** of the WSS, continuous-time process $\{X(t)\}$ is defined as the Fourier transform of its autocorrelation function $R_X(\tau)$:

$$S_X(f) \triangleq \int_{-\infty}^{\infty} R_X(\tau)e^{-j2\pi f\tau}\, d\tau. \qquad (9.3)$$

Definition: The power spectral density of the WSS, discrete-time process $\{X_k\}$ is defined as the Fourier transform of its autocorrelation function $R_X(k)$:

$$S_X(f) \triangleq \sum_{k=-\infty}^{\infty} R_X(k)e^{-j2\pi kf}. \qquad (9.4)$$

The Fourier transform of the cross-correlation function $R_{XY}(\tau)$ of two jointly WSS random processes will be useful in Section 9.3.

Definition: The **cross-power spectral density** of the two jointly WSS, continuous-time processes $\{X(t), Y(t)\}$ is defined as the Fourier transform of the cross-correlation function $R_{XY}(\tau)$:

$$S_{XY}(f) \triangleq \int_{-\infty}^{\infty} R_{XY}(\tau)e^{-j2\pi f\tau}\, d\tau. \qquad (9.5)$$

Definition: The cross-power spectral density of the jointly WSS, discrete-time processes $\{X_k, Y_k\}$ is defined as the Fourier transform of the cross-correlation function $R_{XY}(k)$:

$$S_{XY}(f) \triangleq \sum_{k=-\infty}^{\infty} R_{XY}(k)e^{-j2\pi kf}. \qquad (9.6)$$

To summarize, for the vast majority of random processes, the autocorrelation function $R_X(\tau)$ is known, instead of the actual sample paths as functions of time. Moreover, the relative probabilities of individual sample paths are taken into account, in a definition like 9.3. Thus Definitions 9.3–9.6 are both computable and useful in practice. We see how we can calculate the psd in practice in Section 9.6. Since its introduction by A. Khinchine, in 1938, in his seminal work on WSS processes [39], much work has been done on efficient calculation of the psd.

9.2.2   Properties of PSD

Before we discuss how to apply the concept of psd, let's see some of its properties. Only continuous-time processes are considered here since results for the discrete-time case are entirely similar.

1. $S_X(f)$ is a real function of f

Since the autocorrelation function of a WSS process is an even function of τ (see Theorem 7.2), we get

$$
\begin{aligned}
S_X(f) &= \int_{-\infty}^{\infty} R_X(\tau)e^{-j2\pi f\tau}\, d\tau = \int_{-\infty}^{\infty} R_X(\tau)[\cos(2\pi f\tau) - j\sin(2\pi f\tau)]\, d\tau \\
&= \int_{-\infty}^{\infty} R_X(\tau)\cos(2\pi f\tau)\, d\tau
\end{aligned}
\tag{9.7}
$$

since $\int_{-\infty}^{\infty} R_X(\tau)\sin(2\pi f\tau)\, d\tau = 0$.

2. $S_X(f)$ is an even function of f

This follows directly from Equation 9.7 since $\cos(2\pi f\tau)$ is an even function of f.

3. $S_X(f)$ is a nonnegative function of f

This proof is presented a bit later, when ideal filters are discussed in Section 9.3.4.

4. $S_X(f)$ uniquely determines $R_X(\tau)$

This follows directly from the uniqueness of the Fourier transform and the inverse Fourier transform relationship:

$$
R_X(\tau) = \int_{-\infty}^{\infty} S_X(f)e^{j2\pi f\tau}\, df.
\tag{9.8}
$$

5. Energy of $X(t)$

From Relationship 9.8, and for $\tau = 0$, we get

$$R_X(0) \stackrel{\triangle}{=} EX^2(t) = \int_{-\infty}^{\infty} S_X(f)\, df.$$

No properties similar to those five can be shown for the cross-power spectral density $S_{XY}(f)$.

Suppose we are given a function $S(f)$; under what conditions can it be the power spectral density of a WSS random process $\{X(t)\}$? The following theorem provides the answer for the continuous-time case. Its proof is omitted (see [23]).

Theorem 9.1 *The function $S(f)$, $f \in R$, is the power spectral density of a real-valued, WSS, continuous-time random process $\{X(t)\}$ if and only if $S(f)$ is a nonnegative, real, even function of f.*

9.2.3 Examples

Let's see some examples of psd functions now.

Example 9.2. *Random Sinusoid.* Consider the random process with sample paths given by

$$X(t, \zeta) = A\sin(2\pi f_0 t + \zeta), \quad t \in R, \tag{9.9}$$

where A, f_0 are constants and ζ is uniformly distributed in $[0, 2\pi)$. We can easily check that $\{X(t)\}$ is WSS, with

$$R_X(\tau) = \frac{A^2}{2}\cos(2\pi f_0 \tau).$$

From any table of Fourier transforms, we can see that

$$S_X(f) = \frac{A^2}{4}\delta(f - f_0) + \frac{A^2}{4}\delta(f + f_0).$$

Figure 9.4 depicts this psd. Properties 1, 2, and 3 are easily confirmed; the energy of the signal is also easily found as

$$\int_{-\infty}^{\infty} S_X(f)\, df = \frac{A^2}{4} + \frac{A^2}{4} = \frac{A^2}{2}.$$

Example 9.3. *Delayed Versions of Random Processes.* Consider a WSS random process $\{X(t)\}$, with a known pdf $S_X(f)$. Consider $\{Y(t)\}$, the delayed version of $\{X(t)\}$, defined by $Y(t) = X(t - d)$, where d is a given deterministic constant. We have

$$R_Y(t, s) \stackrel{\triangle}{=} EY(t)Y(s) = EX(t - d)X(s - d) = R_X(t - s),$$

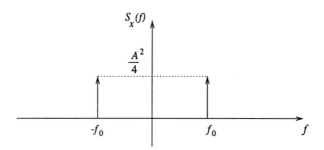

Figure 9.4   The psd for a sinusoid random process.

and thus $\{Y(t)\}$ is WSS, with $S_Y(f) = S_X(f)$. Therefore, equality of psd functions or autocorrelation functions of two random processes does not imply that their sample paths will be equal as well. Consider now the cross-correlation function. We have

$$R_{YX}(t, s) \triangleq EY(t)X(s) = EX(t - d)X(s) = R_X(t - d - s),$$

and $\{X(t), Y(t)\}$ are indeed jointly WSS. We can easily see that the cross-power spectral density is equal to

$$S_{YX}(f) = S_X(f)e^{-j2\pi f d},$$

and thus the cross-power spectral density is not necessarily a real-valued function.

Example 9.4.   *White Noise.* Let $W > 0$ be a given real constant. Consider the function $S(f)$, defined as

$$S(f) = \begin{cases} c, & f \in [-W, W], \\ 0, & \text{otherwise,} \end{cases} \tag{9.10}$$

and shown in Figure 9.5(a). From Theorem 9.1, $S(f)$ is a valid psd; therefore, there exists a continuous-time WSS random process $\{X(t)\}$ with $S_X(f) = S(f)$.

The autocorrelation function for this random process is easily found from the inverse Fourier transform of Equation 9.10 [see Figure 9.5(b)]:

$$R_X(\tau) = \frac{\sin(2\pi W \tau)}{\pi \tau}.$$

Definition: The random process $\{X(t)\}$ with the psd of Equation 9.10 is called **band-limited white noise**, with a band $[-W, W]$.

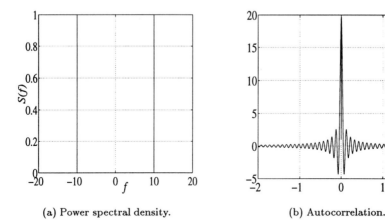

(a) Power spectral density.   (b) Autocorrelation.

Figure 9.5   Band-limited white noise.

Notice that certain random variables in the process $\{X(t)\}$ are uncorrelated since $R_X(\tau) = 0$ for some values of τ (which ones?). This process is very frequently used in noise studies in telecommunication systems. When the band W becomes larger and larger, the correlation decreases. In the limit as $W \to \infty$, we see that

$$R_X(\tau) = \delta(\tau),$$

and therefore, any two random variables $X(t_1)$ and $X(t_2)$ (with $t_1 \neq t_2$) become uncorrelated.

Definition: The random process $\{X(t)\}$ with psd

$$S_X(f) = \sigma^2, \quad f \in (-\infty, \infty),$$

or, equivalently, autocorrelation function

$$R_X(\tau) = \sigma^2 \delta(\tau), \quad \tau \in (-\infty, \infty),$$

is called **white noise** with parameter σ^2. Figure 9.6 displays $S_X(f)$ and $R_X(\tau)$, with $\sigma^2 = c > 0$. When the random variables $X(t)$ are Gaussian, the model is called **white Gaussian noise**. This is the most widely studied and applied model for noise in electric circuits and communication systems. There are a number of reasons for that popularity; mathematical tractability is the single most important, as we witness in subsequent sections of this chapter. Kolmogorov and Karhunen [37] are credited with studying the properties of white noise as input to a linear system (in discrete and continuous time, respectively).

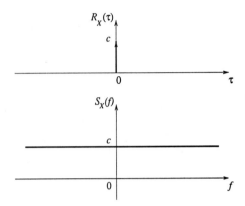

Figure 9.6   White noise.

It is easy to generate sample paths of white Gaussian noise. Since for any two time instants t_1, t_2 the random variables $X(t_1), X(t_2)$ are uncorrelated and hence independent, here is a simple algorithm for generating values of this process:

1. Let $\{t_k\}_{k=1}^n$ denote the time instants of interest.

2. Generate zero-mean, unit-variance Gaussian variates $\{G_k\}_{k=1}^n$.

3. Output $X(t_k) = \sigma G_k$.

9.3   RESPONSE OF LINEAR SYSTEMS TO RANDOM PROCESSES

9.3.1   Review of Deterministic Linear System Theory

Consider a continuous-time system with a deterministic signal $x(t)$ as input and another deterministic signal $y(t)$ as output. The system is *linear and time-invariant*, with impulse response $h(t)$, if the input and output signals are related as

$$y(t) = h(t) \star x(t) \triangleq \int_{-\infty}^{\infty} h(s)x(t-s)\,ds = \int_{-\infty}^{\infty} h(t-s)x(s)\,ds,$$

where $\star$ denotes convolution. The impulse response $h(t)$ is the output of the system when the input is the $\delta(t)$ function. The system is *causal* if $h(t) = 0$

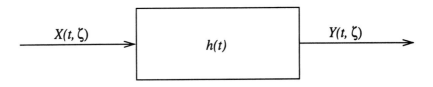

Figure 9.7   Linear system with a random process as input.

when $t < 0$. The Fourier transform of the function $h(t)$, denoted by $H(f)$, is called the *transfer function* of the system.

A discrete-time system with a deterministic signal x_n as input and another deterministic signal y_n as output is linear and time-invariant, with impulse response h_n, if the input and output are related as

$$y_n = h_n \star x_n \triangleq \sum_{k=-\infty}^{\infty} h_k x_{n-k} = \sum_{k=-\infty}^{\infty} h_{n-k} x_k.$$

The impulse response h_n is the output of the system when the input is the discrete-time δ_n signal, defined as

$$\delta_n = \begin{cases} 1, & n = 0, \\ 0, & n \neq 0. \end{cases}$$

The system is causal if $h_n = 0$ when $n < 0$. The Fourier transform of the function h_n, denoted by $H(f)$, is called the *transfer function* of the system. For a more detailed description of linear systems, see the classic book by Oppenheim and Willsky [54].

We investigate next the properties of a linear system when its input is a sample path of a random process. We focus primarily on continuous-time systems, and we state the corresponding results for discrete time ones without proof.

9.3.2   Linear Systems with Random Processes as Input

Consider a linear, time-invariant system, with impulse response $h(t)$ and transfer function $H(f)$. In the remainder of this section, let $I = (-\infty, \infty)$. Consider a random process $\{X(t), t \in I\}$, the sample paths of which are inputs to the linear system, as shown in Figure 9.7.

For every fixed outcome $\zeta \in S$, the deterministic signal

$$Y(t,\zeta) \triangleq \int_{-\infty}^{\infty} h(s) X(t-s, \ \zeta) \, ds \tag{9.11}$$

is the output of the linear system. Is the collection $\{Y(t),\ t \in I\}$ a random process? The following theorem, which we state without proof, provides the answer, under very mild assumptions on the linear system (see [23] for a proof):

Theorem 9.2 *Consider a linear, time-invariant system, with impulse response $h(t)$. Let the random process $\{X(t)\}$, with autocovariance $C_X(s, u)$, be the input to this system. Then if*

$$\int_{-\infty}^{\infty} |h(t)|^2\, dt < \infty, \quad \int_{-\infty}^{\infty} \int_{-\infty}^{\infty} h(t-s)C_X(s, u)h(t-u)\, ds\, du < \infty,$$

the output $\{Y(t)\}$ is a random process. Moreover, the random variable $Y(t)$ in Equation 9.11 is equal to the random variable $\int_{-\infty}^{\infty} h(s)X(t-s)\, ds$ in the mean square sense.

Having established that Equation 9.11 does define a random process, we can now proceed to answer the natural question: What are the n-dimensional cdf's or pdf's of the output random process? As we will see, a complete answer to that question is possible only when the input process is Gaussian.[3] Otherwise, we have to settle for partial information. This discussion should also eliminate any hopes for completely characterizing the output of a nonlinear system; it should also explain why we do not study nonlinear systems in this chapter.

9.3.3   Probabilistic Behavior of Output

We start by evaluating the mean of the output $Y(t)$. We have

$$EY(t) = E\left[\int_{-\infty}^{\infty} h(s)X(t-s)\, ds\right] = \int_{-\infty}^{\infty} h(s)EX(t-s)\, ds, \qquad (9.12)$$

assuming that we can interchange the expectation and integral operators. Equation 9.12 holds true for a nonstationary random process $\{X(t)\}$. It is, however, only of theoretical value since evaluating the integral in 9.12 for general $h(t)$ and $EX(t)$ is rather difficult. When we restrict our attention to a WSS random process $\{X(t)\}$, as we do very often in the sequel, we can obtain a sharper result:

Theorem 9.3 *Consider a linear, time-invariant system, with impulse response $h(t)$ and with a WSS random process $\{X(t),\ t \in I\}$ as input. Then the output process $\{Y(t),\ t \in I\}$ is also WSS. Moreover, the processes $\{X(t), Y(t)\}$ are jointly WSS. Further,*

$$S_Y(f) = |H(f)|^2 S_X(f), \qquad (9.13)$$

---

[3]Which attests to the special consideration given to this process and to its prominent role in modeling.

$$S_{XY}(f) = H^*(f)S_X(f), \quad S_{YX}(f) = H(f)S_X(f).$$

Proof: Let m_X, $R_X(\tau)$ be the mean and autocorrelation, respectively, of $X(t)$. From Equation 9.12 we have that

$$EY(t) = \int_{-\infty}^{\infty} h(s)m_X \, ds = m_X \int_{-\infty}^{\infty} h(s) \, ds = m_X H(0) \overset{\Delta}{=} m_Y,$$

which shows that the output process has a constant mean. [It is implicitly assumed here that $|H(0)| < \infty$, that is, the system is stable.] Moreover,

$$
\begin{aligned}
R_Y(t,s) \overset{\Delta}{=} \ & EY(t)Y(s) = E\left[\int_{-\infty}^{\infty} h(v)X(t-v)\,dv \cdot \int_{-\infty}^{\infty} h(u)X(s-u)\,du\right] \\
= \ & \int_{-\infty}^{\infty}\int_{-\infty}^{\infty} h(v)h(u)E[X(t-v)X(s-u)]\,dv\,du \\
= \ & \int_{-\infty}^{\infty}\int_{-\infty}^{\infty} h(v)h(u)R_X(t-s-v+u)\,dv\,du. \qquad (9.14)
\end{aligned}
$$

Since the last equality in Equation 9.14 depends only on the time difference $\tau \overset{\Delta}{=} t - s$, we conclude that $R_Y(t,s) = R_Y(\tau)$ and thus the output process is WSS. Equation 9.13 now follows directly as the Fourier transform of Equation 9.14.

It remains to show that the two random processes are jointly WSS. To that end, consider the cross-correlation $R_{YX}(t,s)$. Let again $\tau = t - s$. We have

$$
\begin{aligned}
R_{YX}(t,s) \overset{\Delta}{=} \ & EY(t)X(s) = E\left[\int_{-\infty}^{\infty} h(v)X(t-v)\,dv \cdot X(s)\right] \\
= \ & \int_{-\infty}^{\infty} h(v)EX(t-v)X(s)\,dv = \int_{-\infty}^{\infty} h(v)R_X(t-s-v)\,dv \\
= \ & \int_{-\infty}^{\infty} h(v)R_X(\tau-v)\,dv = h(\tau) * R_X(\tau), \qquad (9.15)
\end{aligned}
$$

which shows that the two processes are jointly WSS. The form of $S_{YX}(f)$ follows easily from the last equality in Equation 9.15 and the convolution property of Fourier transforms. The form of $S_{XY}(f)$ follows immediately from the fact that $R_X(\tau)$ is an even function of τ and $R_{YX}(\tau) = R_{XY}(-\tau)$. $\qquad \square$

It should be expected that, in general, the input and output random processes are dependent. A characterization of even the joint density $f_{X(t)Y(t)}(x,y)$

is not, in general, possible; we should take comfort in proving the above theorem, however limited it may sound. Finally, note that the theorem applies to the discrete-time case, even though we stated it for continuous-time random processes only.

9.3.4   Examples

Example 9.5. *First-Order Filter.* Consider the random process in Equation 9.9 and the linear system with $h(t) = e^{-at}u(t)$. This is a first-order filter, typically implemented as an RC circuit. We evaluate the probabilistic behavior of the output process $Y(t)$. Since $m_X = 0$, we have

$$EY(t) = m_X H(0) = 0.$$

Now, from Theorem 9.3, we have

$$
\begin{aligned}
S_Y(f) &= |H(f)|^2 S_X(f) = \frac{1}{a^2 + 4\pi^2 f^2} \frac{A^2}{2} [\delta(f - f_0) + \delta(f + f_0)] \\
&= \frac{A^2}{2} \frac{1}{a^2 + 4\pi^2 f_0^2} [\delta(f - f_0) + \delta(f + f_0)].
\end{aligned}
$$

Therefore,

$$
\begin{aligned}
R_Y(\tau) &= \frac{A^2}{2} \frac{1}{a^2 + 4\pi^2 f_0^2} \delta(\tau), \\
S_{XY}(f) &= H^*(f) S_X(f) = \frac{1}{a - j2\pi f} \frac{A^2}{4} [\delta(f - f_0) + \delta(f + f_0)], \\
S_{YX}(f) &= H(f) S_X(f) = \frac{1}{a + j2\pi f} \frac{A^2}{4} [\delta(f - f_0) + \delta(f + f_0)].
\end{aligned}
$$

Example 9.6. *White Noise as Input to a Filter.* Consider a general filter with impulse response $h(t)$ and a white noise random process $\{X(t)\}$, with parameter $\sigma^2 = 1$ as its input. From Theorem 9.3, the psd of the filter output is given by

$$S_Y(f) = |H(f)|^2 S_X(f) = |H(f)|^2.$$

The above equation gives us an algorithm to generate variates for a WSS process with a desired psd $S_Y(f)$ or autocorrelation function $R_Y(\tau)$. We outline next an algorithm (shown for a discrete-time random process).

WSS process generation:

1. Calculate $H(f) = \sqrt{S_Y(f)}$ and $h(t)$.

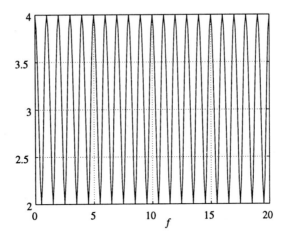

Figure 9.8 Autoregressive process.

2. **Generate** a sequence $\{X_n\}$ of variates of the white noise process.

3. **Compute** the convolution $Y_n = X_n \star h(t)$.

Example 9.7. *ARMA Process.* The **autoregressive, moving-average** (ARMA) process was defined in Section 7.5.2 via the recursion

$$Y_n = \sum_{k=1}^{M} a_k Y_{n-k} + \sum_{l=0}^{L} b_l X_{n-l}, \quad n = 1, 2, \ldots. \qquad (9.16)$$

(Note that the notation in Equation 7.22 is slightly different). The transfer function of the linear system in Equation 9.16 is given by

$$H(f) = \frac{\sum_{l=0}^{L} b_l e^{-j 2\pi l f}}{1 - \sum_{k=1}^{M} a_k e^{-j 2\pi k f}}.$$

When the sequence $\{X_n\}$ is white noise with parameter 1, the psd of the ARMA process is given by

$$S_Y(f) = |H(f)|^2.$$

Figure 9.8 depicts the special case of $L = 1, b_0 = 1, b_1 = 3$, and all $a_k = 0$.

The ARMA process is widely used in signal processing, both deterministic and random. We see in Section 9.5 a very widely used ARMA model for the Kalman filter.

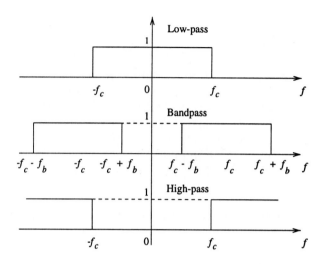

Figure 9.9   Transfer functions of ideal filters.

Example 9.8. *Ideal Filters.* The ideal low-pass, bandpass, and high-pass filters are defined via their transfer functions as follows:

$$H_{\text{low}}(f) \;=\; \begin{cases} 1, & |f| \leq f_c, \\ 0, & \text{otherwise.} \end{cases}$$

$$H_{\text{band}}(f) \;=\; \begin{cases} 1, & |f - f_c| \leq f_b, \\ 0, & \text{otherwise.} \end{cases}$$

$$H_{\text{high}}(f) \;=\; \begin{cases} 1, & |f| > f_c, \\ 0, & \text{otherwise.} \end{cases}$$

Here f_c, f_b are frequency parameters of the filters; their meaning should be clear from Figure 9.9.

Consider an ideal bandpass filter with an arbitrary process $\{X(t)\}$ as its input. The power spectral density of the output is

$$S_Y(f) = \begin{cases} S_X(f), & |f - f_c| \leq f_b, \\ 0, & \text{otherwise.} \end{cases}$$

Now,

$$EY^2(t) = R_Y(0) = \int_{f_c-f_b}^{f_c+f_b} S_X(f)\, df.$$

Since $EY^2(t) \geq 0$, the integral is nonnegative; since the frequencies f_b, f_c are arbitrary, we deduce that $S_X(f) \geq 0$, proving Property 3 in Section 9.2.2.

It is well known from deterministic linear system theory that an ideal filter will cut off all frequencies outside its band. Is the same true when a random process becomes its input? We show at the end of the next section that this is true, not with probability 1 but in the mean square sense.

Example 9.9. *Amplitude Modulation and Demodulation.* Amplitude modulation (AM) is one way of transmitting an information signal (such as speech and music) through a transmission medium (typically the airwaves). Let $X(t)$ denote the information signal; we assume that $X(t)$ is a WSS random process, with a psd function that is "low-pass" in nature since most speech and music signals have frequencies no higher than a few kilohertz. In AM, $X(t)$ is multiplied by the carrier signal $\cos(2\pi f_c t + \Phi)$, where f_c is the carrier frequency and Φ is a random variable describing the phase. We assume that the phase is independent of the information signal and has a uniform distribution in the interval $[0, 2\pi)$. Let $Y(t) = X(t)\cos(2\pi f_c t + \Phi)$. We calculate the statistics of the random process $\{Y(t)\}$ next.

We can easily see that $EY(t) = EX(t) \cdot E\cos(2\pi f_c t + \Phi) = 0$; moreover,

$$
\begin{aligned}
R_Y(\tau) &= E[X(t+\tau)X(t)\cos(2\pi f_c(t+\tau) + \Phi)\cos(2\pi f_c t + \Phi)] \\
&= EX(t+\tau)X(t) \cdot E\cos(2\pi f_c(t+\tau) + \Phi)\cos(2\pi f_c t + \Phi) \\
&= R_X(\tau)E[0.5\cos(2\pi f_c \tau) + 0.5\cos(2\pi f_c(2t+\tau) + 2\Phi)] \\
&= 0.5R_X(\tau)\cos(2\pi f_c \tau).
\end{aligned}
$$

Therefore, $\{Y(t)\}$ is also a WSS random process; its psd is then given by

$$
S_Y(f) = 0.25S_X(f + f_c) + 0.25S_X(f - f_c).
$$

If the carrier frequency f_c is much bigger than the highest frequency in the low-pass information signal, then the AM signal $Y(t)$ is a bandpass signal. We can demodulate it at a receiver by multiplying it again with the carrier signal. Let $Z(t) = 2Y(t)\cos(2\pi f_c t + \Theta)$, where the coefficient 2 represents a gain and Θ is again some random phase, independent of $Y(t)$. Now,

$$
S_Z(f) = 0.5S_Y(f + f_c) + 0.5S_Y(f - f_c),
$$

and finally

$$
S_Z(f) = 0.5[S_X(f + 2f_c) + S_X(f - 2f_c)] + S_X(f).
$$

Passing the signal $Z(t)$ through a low-pass filter with a cutoff frequency less than $2f_c$ will restore the original signal.

9.4   OPTIMAL LINEAR ESTIMATION

In this section, we develop several techniques for processing random processes in both the time and frequency domains. The theory we present here is a direct extension of the theory in Chapter 5 and in particular of the *orthogonality principle*. The basic difference between Chapter 5 and the case we study here is that now we have (in general) an infinite number of measurements to process. This difference introduces a lot of mathematical subtleties; the general problem was first formulated in the context of discrete-time random processes by Kolmogorov in the 1930s; Wiener formulated and solved the problem for the continuous-time case in the 1940s.

9.4.1   Problem Definition

Consider an experiment in which the measurements are functions of time. Let I denote the time interval over which we take measurements. What we have then is a collection of measurements $\{X(s),\ s \in I\}$. We are interested in estimating the values of another random process $Y(t)$, for $t \in I'$, given the measurements of the random process $X(t)$. It is *not* necessary that $X(t) = Y(t)$ or $I' = I$.

Example 9.10. Consider a network node. We can view the buffer contents as a sample path $X(t, \zeta)$ of the random process $\{X(t)\}$, where $X(t)$ is a random variable defining the number of packets in the node memory at time t. Such measurements are routinely taken; we may be interested then in estimating performance measures such as throughput, losses, and delays through the node.

Two features distinguish this problem from the one in Chapter 5. First, the number of measurements is infinite (potentially uncountably infinite); second, the number of random variables we have to estimate may be more than 1 (also potentially uncountably infinite).

In Chapter 5 implementation of nonlinear estimators proved quite difficult; therefore, we focus on **optimal linear** estimators only. Then we can define the problem of **optimal linear estimation of random processes** as follows:

Given the set of measurements $\{X(s), s \in I\}$, derive the best linear estimator of $\{Y(t), t \in I'\}$.

Of course, we have to specify the criterion for deciding the best estimator. Let $\hat{Y}(t)$ denote the estimator of $Y(t)$. Let $e(t) = Y(t) - \hat{Y}(t)$, $\forall t \in I'$, denote the estimation error. Based on our discussions in Section 5.1, we focus only on **mean square error** criteria of the form

$$Ee^2(t) = E[Y(t) - \hat{Y}(t)]^2, \quad \forall t \in I'.$$

From now on, unless otherwise noted, we assume that the random processes $\{Y(t), X(t)\}$ are jointly WSS, with zero means, and known autocorrelations $R_Y(\tau), R_X(\tau)$ and cross-correlations $R_{XY}(\tau), R_{YX}(\tau)$.

For fixed $t \in I'$ and $s \in I$, let $h(t - s)$ denote the coefficient of $X(s)$ in the linear estimator. [The reason for labeling the coefficient of $X(s)$ as $h(t - s)$ will become apparent shortly.] The most general form of a linear estimator is a weighted sum of the measurements

$$\hat{Y}(t) = \sum_{s \in I} h(t - s)X(s), \quad \forall t \in I',$$

when a countable set of measurements is involved or

$$\hat{Y}(t) = \int_I h(t - s)X(s)\, ds, \quad \forall t \in I',$$

when an uncountable set of measurements is involved. Of course, these equalities hold true in the mean square sense.

We determine the best estimators in three special cases:

1. **Filtering.** (Also known as *causal filtering*.) In essence, we are using "past and present" measurements $X(s)$ in order to estimate the "present" value of $Y(t)$. Here common choices are $I = (-\infty, t]$ or $I = (a, t]$ (in both cases an infinite number of measurements are involved) and $I' = \{t\}$.

2. **Prediction.** Here we are using "past" measurements $X(s)$ in order to estimate a "future" value of the random variable $Y(t)$. It is very usual (even though not necessary, as we see in Section 9.4.6) to have $Y(t) = X(t)$, that is, we are using measurements to predict the process itself. Here common choices are

$$I = (-\infty, t] \quad \text{and} \quad I' = \{a\}$$

or

$$I = \{t, t+1, \ldots, a - 1\} \quad \text{and} \quad I' = \{a\},$$

where $a > t$.

3. **Smoothing.** Here we are using "past, present, and future" measurements in order to estimate a value of $Y(t)$. Common choices are

$$
\begin{aligned}
I &= (-\infty, \infty) \quad \text{and} \quad I' = \{t\}, \\
I &= (-\infty, a] \quad \text{and} \quad I' = \{t\}, \quad a > t.
\end{aligned}
$$

All our results are derived as simple applications of the orthogonality principle developed in Chapter 5. For convenience, we restate this principle in the next section, using the new notation for random processes.

9.4.2   Orthogonality Principle Revisited

As we saw in Chapter 5, the error $e(t)$ and each individual measurement are orthogonal, in the sense that

$$E[e(t) \cdot X(s)] = 0, \quad \forall t \in I', \quad \forall s \in I. \tag{9.17}$$

Equation 9.17 specifies a (potentially infinite) number of equations, that the optimum estimator must satisfy, one for each pair of t, s. From Equation 9.17 we get

$$E[e(t) \cdot X(s)] = E\{[Y(t) - \hat{Y}(t)] \cdot X(s)\} = 0,$$

and thus

$$EY(t)X(s) = E\hat{Y}(t)X(s), \quad \forall t \in I', \quad \forall s \in I. \tag{9.18}$$

Equation 9.18 will form the basis for developing the best estimators in the special cases we consider next. Note that Equation 9.18 holds true for I countable or uncountable.

9.4.3   Causal Filtering

Pioneering work on this problem was performed by Norbert Wiener (continuous time) and A. N. Kolmogorov (discrete time), in the 1930s and 1940s. We can state the problem as follows. Consider a set of measurements $\{X(s), \ s \in I\}$, where $I = (-\infty, t]$.[4] For any given value of $t \in R$, we wish to obtain the best estimate of the random variable $Y(t)$, based on the set $\{X(s), s \in I\}$. The main result, stated here for the continuous-time case, is given by the following theorem.

Theorem 9.4 *Let $\{X(s), Y(s)\}$ be zero-mean, jointly WSS random processes. Let*

$$\hat{Y}(t) \triangleq \int_{-\infty}^{t} h(t - s)X(s)\, ds \tag{9.19}$$

be the estimate of $Y(t)$. The best causal linear filter $h(s)$ is given as the solution to the equation

$$R_{YX}(\tau) = \int_{-\infty}^{\tau} h(\tau - s)R_X(s)\, ds. \tag{9.20}$$

The minimum mean square error $e_{\min}$ is equal to

$$e_{\min} = R_Y(0) - \int_{-\infty}^{t} h(t - s)R_{YX}(t - s)\, ds. \tag{9.21}$$

---

[4] The case $I = (a, t]$ is similar and left as an exercise. Moreover, in discrete time, we have $I = \{-\infty, \ldots, t - 1, t\}$.

Proof: Let $\tau = t - s$. From Equations 9.18 and 9.19 we have

$$
\begin{aligned}
R_{YX}(\tau) &= E\hat{Y}(t)X(s) = EX(s)\int_{-\infty}^{t} h(t-v)X(v)\,dv \\
&= \int_{-\infty}^{\tau} h(\tau - v)R_X(v)\,dv
\end{aligned}
$$

and Equation 9.20 is proved. For Equation 9.21, observe that

$$
\begin{aligned}
Ee^2(t) &= Ee(t)[Y(t) - \hat{Y}(t)] = Ee(t)Y(t) - Ee(t)\hat{Y}(t) \\
&= Ee(t)Y(t) - Ee(t)\int_{-\infty}^{t} h(t-s)X(s)\,ds \\
&= Ee(t)Y(t) - \int_{-\infty}^{t} h(t-s)E[e(t)X(s)]\,ds = Ee(t)Y(t)
\end{aligned}
$$

since the measurements and the error are orthogonal. Therefore,

$$
Ee^2(t) = Ee(t)Y(t) = E[Y(t)-\hat{Y}(t)]Y(t) = R_Y(0) - \int_{-\infty}^{t} h(t-s)R_{YX}(t-s)\,ds,
$$

and Equation 9.21 is proved. $\qquad\qquad\qquad\qquad\qquad\qquad\qquad\qquad$ $\square$

The same result can be also expressed in the frequency domain. The proof of the following theorem is left as an exercise.

Theorem 9.5 *Let $\{X(s), Y(s)\}$ be zero-mean, jointly WSS random processes. The best causal linear filter for the random variable $Y(t)$, given the set of measurements $\{X(s), s \in I\}$, is given by*

$$
H(f) = \frac{S'_{YX}(f)}{S_X(f)},
$$

where

$$
S'_{YX}(f) \triangleq \int_{0}^{\infty} R_{YX}(\tau)e^{-j2\pi f\tau}\,d\tau. \tag{9.22}
$$

Remark: Note that in the above equation, $S'_{YX}(f)$ is *not* the cross-power spectral density function since the integral does not extend from $-\infty$ to ∞.

Example 9.11. *Filtering White Noise.* Suppose that $R_X(\tau) = \delta(\tau)$. Moreover, assume that the cross-correlation $R_{YX}(\tau)$ is given. Then, from Equation 9.20 we have

$$
R_{YX}(\tau) = \int_{-\infty}^{\tau} h(\tau - s)R_X(s)\,ds = \int_{-\infty}^{\tau} h(\tau - s)\delta(s)\,ds = h(\tau),
$$

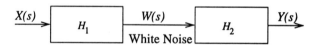

Figure 9.10   Whitening filter.

and thus the best filter is given by

$$h(\tau) = \begin{cases} R_{YX}(\tau), & \tau \geq 0, \\ 0, & \tau < 0. \end{cases}$$

Example 9.12. *Whitening Filters.* In general, determining the best causal filter through direct calculation of Equation 9.22 is difficult. A different approach, which capitalizes on the solution of the previous example, involves determining the best filter in two steps, as Figure 9.10 depicts.

The basic idea is to convert the measurement process $X(s)$ to white noise $W(s)$ with parameter 1. Then the best filter $H_2(f)$ can be determined from the previous example. The filter $H_1(f)$ is called a *whitening filter* for obvious reasons. It can be determined as follows:

From Theorem 9.3 we have

$$1 = |H_1(f)|^2 S_X(f) = |H_1(f)|^2 |A_X(f)|^2,$$

assuming for the moment that we can write

$$S_X(f) = |A_X(f)|^2 = A_X(f) A_X^*(f).$$

Then

$$H_1(f) = \frac{1}{A_X(f)}$$

is the required whitening filter. Of course, the implicit requirement is that this filter [and $A_X(f)$] be causal. It can be shown that under the so-called Paley-Wiener[5] condition

$$\int_{-\infty}^{\infty} \frac{|\log(S_X(f))|}{1+f^2} \, df < \infty,$$

$S_X(f)$ can be factored out as $|A_X(f)|^2$, with both $A_X(f)$ and $1/A_X(f)$ being causal filters [66]. To summarize, here is the whole procedure (known as *spectral factorization*) in algorithmic form:

1. **Factor** the given $S_X(f)$ to obtain the whitening filter

---

[5] After Raymond E. A. C. Paley, 1907–1933, a U.S. mathematician, and N. Wiener.

$$H_1(f) = 1/A_X(f).$$

2. **Obtain** $S_{YW}(f) = S_{YX}(f)/A_X^*(f)$; obtain $R_{YW}(\tau)$.

3. **Set** $H_2(f) = \int_0^\infty R_{YW}(\tau)e^{-j2\pi f\tau}\,d\tau.$

4. **Set** $H(f) = H_1(f)H_2(f).$

Let's see an example next.

Example 9.13. Suppose that

$$S_{YX}(f) = \frac{16}{1+f^2}, \quad S_X(f) = \frac{16}{1+f^2} + 9.$$

We can factor $S_X(f)$ as follows:

$$S_X(f) = |A_X(f)|^2 = \frac{|5+3jf|^2}{|1+jf|^2} = \frac{5+3jf}{1+jf}\frac{5-3jf}{1-jf},$$

and thus the whitening filter is

$$H_1(f) = \frac{1}{A_X(f)} = \frac{1+jf}{5+3jf}.$$

We have now

$$S_{YW}(f) = \frac{16}{1+f^2}\frac{1-jf}{5-3jf} = \frac{2}{1+jf} + \frac{6}{5-3jf}$$

The causal part is $\frac{2}{1+jf}$, and thus

$$H_2(f) = \frac{2}{1+jf}.$$

Finally,

$$H(f) = H_1(f)H_2(f) = \frac{1+jf}{5+3jf}\frac{2}{1+jf} = \frac{2}{5+3jf}.$$

9.4.4 Prediction

Consider a set of measurements $\{X(s),\ s \in I\}$. For simplicity, consider the case where the future of the process $X(t)$ itself must be predicted, and therefore take $Y(t) \triangleq X(t)$. Furthermore, assume that the number of measurements n is finite; therefore, let $I = \{t-n, t-n+1, \ldots, t-1\}$.[6]

---

[6] The case of an infinite number of measurements is similar and left as an exercise. Observe that the problem of finite measurements is identical to that of Section 5.3; only the notation is different.

Theorem 9.6 *Let $\{X(s)\}$ be a zero-mean WSS random process. Let*

$$\hat{Y}(t) \overset{\triangle}{=} \sum_{s \in I} h(t - s) X(s) \tag{9.23}$$

be the estimate of $Y(t) \overset{\triangle}{=} X(t)$. The best linear predictor satisfies the equation

$$R_X(\tau) = \sum_{s \in I} h(s) R_X(\tau - s), \quad \tau \in \{1, \ldots, n\}. \tag{9.24}$$

The minimum mean square error $e_{\min}$ is equal to

$$e_{\min} = R_X(0) - \sum_{s \in I} h(s) R_X(s).$$

Proof: Let $\tau = t - s$. From Equations 9.18 and 9.23 we have

$$\begin{aligned}
R_{YX}(\tau) &= R_X(\tau) = E\hat{Y}(t)X(s) = \sum_{v \in I} h(t - v) E X(v) X(s) \\
&= \sum_{v \in I} h(t - v) R_X(v - s) = \sum_{s \in I} h(s) R_X(\tau - s).
\end{aligned}$$

(In the last equality, we used the change of variables $s = t - v$.) Following an argument similar to that used for calculating the errors in filtering, we have

$$e_{\min} = E[X(t) - \hat{Y}(t)]X(t) = R_X(0) - \sum_{s \in I} h(s) R_X(s).$$

$\square$

From Equation 9.24 we see that the coefficients of the best predictor can be found by solving a linear system of n equations into the n unknown parameters $h(1), \ldots, h(n)$. Rewriting Equation 9.24 in matrix form, we get the *Yule-Walker equations* (observe the special structure of the rows or columns of this matrix):

$$\begin{bmatrix} R_X(0) & R_X(1) & \cdots & R_X(n-1) \\ R_X(1) & R_X(0) & \cdots & R_X(n-2) \\ \cdots & \cdots & \cdots & \cdots \\ R_X(n-1) & R_X(n-2) & \cdots & R_X(0) \end{bmatrix} \begin{bmatrix} h(1) \\ h(2) \\ \vdots \\ h(n) \end{bmatrix} = \begin{bmatrix} R_X(1) \\ R_X(2) \\ \vdots \\ R_X(n) \end{bmatrix}.$$

Of course, solving this system for large values of n by any traditional technique (e.g., Gauss-Seidel-like methods) is out of the question. Fortunately, we can take advantage of the special structure of the above matrix to solve it *recursively in n.* The following algorithm, due to Levinson, allows us to evaluate the coefficients of the best filter recursively. Its proof can be found in [47].

Denote the mean square error in the lth iteration of the algorithm as E_l. Denote the values of the predictor coefficients, calculated during iteration l, as $h_{l,j}$, $j = 1, \ldots, l-1$. The algorithm is as follows:

Initialize $E_0 = R_X(0)$.

For $l = 1$ to n, do

  Calculate: $\quad c_l = \dfrac{R_X(l) - \sum_{k=1}^{l-1} h_{k,l-1} R_X(l-k)}{E_{l-1}}$

  Set: $\quad h_{l,l} = c_l, h_{l,j} = h_{l-1,j} - c_l h_{l-j,l-1}, j = 1, \ldots, l-1.$

  Calculate: $\quad E_l = (1 - c_l^2) E_{l-1}.$

end for

The coefficients $h(j)$, of the best predictor, are found at the nth iteration of the algorithm:

$$h(j) = h_{j,n}.$$

9.4.5 Smoothing

In this case, we have a set of measurements $\{X(s), \ s \in I\}$ and wish to estimate the value of the random variable $Y(t) = X(t)$ for $t \in I$. Typical cases are $I = (-\infty, \infty)$ and $t < \infty$ or $I = (-\infty, T]$ and $t < T$. The name *smoothing* comes from a typical application where we are processing "future" measurements in order to remove high-frequency noise and thus "smooth out" a low-frequency signal. Sometimes the name *off-line smoothing* is also used, to suggest that the measurements must be stored and processed off-line. In the rest of this section, we consider the case $I = (-\infty, \infty)$ only, the other case being treated easily in a similar fashion.

Let the smoothing estimator be

$$\hat{Y}(t) = \sum_{s=-\infty}^{\infty} h(t-s)X(s), \quad t \in I,$$

when a countable set of measurements is involved or

$$\hat{Y}(t) = \int_{-\infty}^{\infty} h(t-s)X(s)\,ds, \quad t \in I,$$

when an uncountable set of measurements is involved.

Theorem 9.7 *Let* $\{X(s)\}$ *be a zero-mean WSS random process. The best smoothing filter for the random variable* $Y(t) = X(t)$, *given the set of measurements* $\{X(s), s \in (-\infty, \infty)\}$, *is given by*

$$H(f) = \frac{S_{XY}(f)}{S_X(f)}. \tag{9.25}$$

Proof: From the orthogonality Equation 9.18, we get

$$R_{XY}(t) = \sum_{s=-\infty}^{\infty} h(s) R_X(t-s)$$

in the discrete-time case or

$$R_{XY}(\tau) = \int_{-\infty}^{\infty} h(s) R_X(\tau - s)\, ds$$

in the continuous-time case. Equation 9.25 follows now directly from the convolution property of the Fourier transform.   □

9.4.6   Case of Signal Plus Noise

The material in this section is a special but very important case, with significant applications in telecommunications and analysis of measurements, so we present it here in some detail. Consider the case in which a "desired" signal $Y(t)$ (in general a random process) is contaminated with additive noise. This case is very common in a lot of measurement scenarios in which noise is added due to imperfections in the measurement tools or procedures. It is also very common in communication channels, as Figure 9.11 suggests. Typically, a stream of bits (the first signal in Figure 9.11) is modulated (via a Manchester encoding algorithm for this example) and sent over the channel. Electromagnetic noise (commonly known as EMI, or electromagnetic interference) can be added to the electric waveforms traveling through a channel. Then the signal at the receiver may look like the one in the second graph. The function of the receiver is to process the received signal (via a matched filter in this example) and "recover" the original bit stream (last signal in the figure), hopefully with no errors (as is the case in this example).

In general, then, for any time s, the measurement $X(s)$ and the signal $Y(s)$ we have to estimate are related by

$$X(s) = Y(s) + N(s), \quad s \in I, \tag{9.26}$$

where $N(s)$, the *noise*, is another random process.

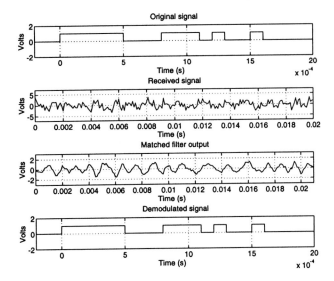

Figure 9.11 Additive noise.

We wish to filter, predict, and smooth the signal $Y(t)$, $t \in I'$, based on the set of noisy measurements $X(s)$, $s \in I$. Our hope is, of course, to remove or at least reduce the effects of $N(s)$. [Why don't we simply subtract the noise value $N(s)$ from $X(s)$?]

We assume again that the signal and noise processes are jointly WSS, with zero means and known or computable autocorrelation functions $R_Y(\tau)$, $R_N(\tau)$. It is common and reasonable to assume that the signal and noise processes are uncorrelated, so we further assume that $R_{YN}(\tau) = 0$.

From Equation 9.26 and the assumptions we made about $Y(s)$ and $N(s)$, we have easily

$$EX(s) = 0, \ s \in I, \quad R_X(\tau) = R_Y(\tau) + R_N(\tau), \ \tau \in R.$$

Then

$$
\begin{aligned}
R_{XY}(\tau) \ &\triangleq \ EX(v)Y(v - \tau) = E\{[Y(v) + N(v)]Y(v - \tau)\} \\
&= \ EY(v)Y(v - \tau) + EN(v)Y(v - \tau) = EY(v)Y(v - \tau) \\
&= \ R_Y(\tau).
\end{aligned}
\tag{9.27}
$$

The basis of our analysis is, of course, the orthogonality principle 9.18:

$$E\hat{Y}(t)X(s) = EY(t)X(s), \quad \forall t \in I', \ \forall s \in I.$$

From Equation 9.27, the orthogonality principle for the signal plus noise can be specialized to

$$E\hat{Y}(t)X(s) = R_Y(t-s). \tag{9.28}$$

We develop next the formulas for the best estimator, predictor, and smoothing filter, based on Equation 9.28.

1. Filtering

Let $I = (-\infty, t]$. The best filter is given of course by

$$\hat{Y}(t) = \int_{-\infty}^{t} h(s)X(t-s)\,ds$$

and the coefficients $h(s)$ solve (see Equation 9.20):

$$R_Y(\tau) = \int_{-\infty}^{\tau} h(s)[R_Y(\tau-s) + R_N(\tau-s)]\,ds. \tag{9.29}$$

The optimum filter in Equation 9.29 is called in this case the Wiener filter.

2. Smoothing

Let $I = (-\infty, \infty)$. The best smoothing filter is given by (see Equation 9.25)

$$H(f) = \frac{S_{XY}(f)}{S_X(f)} = \frac{S_Y(f)}{S_Y(f) + S_N(f)}.$$

3. Prediction

Let $I = \{t-1, t-2, \ldots, t-n\}$. We wish to predict the value of $Y(t)$, based on the n measurements $\{X(s),\ s \in I\}$. The best predictor filter is given by

$$R_Y(\tau) = \sum_{s \in I} h(s)[R_Y(\tau-s) + R_N(\tau-s)]$$

Example 9.14. *Ideal Filters.* Consider the case where the psd of the signal $Y(t)$ and the noise $N(t)$ have no common frequencies, as Figure 9.12 depicts. Then filtering $X(t) = Y(t) + N(t)$ with the ideal low-pass filter of Figure 9.9, with a cutoff frequency $f_s \le f_c \le f_n$, should remove the noise completely and give back the signal $Y(t)$, as Figure 9.13 suggests. Let $W(t)$ denote the output of the low-pass filter when its input is the signal $X(t)$. We calculate the psd of the random process $W(t)$ as follows:

$$S_W(f) = |H_{\text{low}}(f)|^2 S_Y(f) + |H_{\text{low}}(f)|^2 S_N(f) = S_Y(f).$$

So the output has the same psd as the input $Y(t)$. But this does not imply that $W(t) = Y(t)$ with probability 1, as we saw in Example 9.3. What we really have

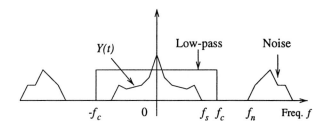

Figure 9.12 Signal and noise without common frequencies.

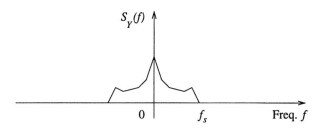

Figure 9.13 Ideal low-pass filtering.

is that $W(t) = Y(t)$ in the mean square sense. Indeed, let $V(t) \overset{\triangle}{=} W(t) - Y(t)$. Then

$$R_V(\tau) = R_W(\tau) - R_{WY}(\tau) - R_{YW}(\tau) + R_Y(\tau),$$

and thus

$$
\begin{aligned}
S_V(f) &= S_W(f) - S_{WY}(f) - S_{YW}(f) + S_Y(f) \\
&= S_Y(f) - H_{\text{low}}(f)S_Y(f) - H_{\text{low}}^*(f)S_Y(f) + S_Y(f) = 0.
\end{aligned}
$$

Therefore, $R_V(\tau) = 0$ for all τ; since

$$E[W(t) - Y(t)]^2 = R_V(0) = 0,$$

we conclude that $W(t) = Y(t)$ in the mean square sense.

9.5   KALMAN FILTER

This filter has been one of the crowning achievements of the theory of random processes; it has been very widely implemented, with applications ranging from airplane navigation to geological data processing to alcohol distilleries.

The filter was invented in 1960 by R. E. Kalman, a contemporary Hungarian mathematician [33, 34]. A tremendous number of variations and extensions have followed since. As we will see, the filter is essentially a *recursive* solution of System 9.18, in *the time domain*.

Perhaps the single most important reason for the success of the filter was the *timing* of its invention. Before the 1960s, the Wiener filter and the frequency-domain-based technique of spectral factorization (see Section 9.4.3) were the preferred way of processing measurements. The Wiener filter was essentially a nonrecursive solution to System 9.18. However, the introduction of computers and digital hardware, in the 1950s and 1960s opened up new possibilities for recursive solutions *directly* in the time domain.

The Kalman filter, in its general form, applies to multidimensional signals, in continuous or discrete time, and correlated noises. We present first the simplest possible case and outline some proofs. The vector case is presented at the end of this section, without proof. For a more complete presentation of the filter, and a more thorough investigation of the principle of innovations, which is central to the theoretical development of the filter, refer to the original papers [33, 34] or [17].

9.5.1   Discrete-Time, One-Dimensional Case

Consider a discrete-time random process $\{Y_n, n = 0, 1, \ldots\}$ described by the equation

$$Y_n = \phi_{n-1} Y_{n-1} + W_{n-1}, \quad n = 1, 2, \ldots, \tag{9.30}$$

where ϕ_{n-1} is a constant (possibly dependent on n) and Y_0 is a random variable with a zero mean and variance σ_0. (This process is a special case of the ARMA process in Sections 7.5.2 and 9.3, with $M = 1$ and $L = 0$.) The random process $\{W_n, n = 0, 1, 2, \ldots\}$ is a sequence of random variables with

$$EW_n = 0, \quad EW_n W_k = Q_n \cdot \delta_{nk}, \quad n, k = 0, 1, 2, \ldots, \tag{9.31}$$

where δ_{nk} is 1 for $n = k$ and 0 otherwise. In other words, the sequence is uncorrelated, and $\text{var}(W_n) = Q_n$. Since Y_0 and W_n have zero mean, the random variable Y_n has a zero mean also.

The random process $\{X_n, n = 0, 1, 2, \ldots\}$ is related to the random process $\{Y_n\}$ as follows:

$$X_n = c_n Y_n + N_n, \quad n = 0, 1, 2, \ldots, \tag{9.32}$$

where c_n is another constant, possibly dependent on n. The random process $\{N_n, n = 0, 1, 2, \ldots\}$ is a sequence of random variables with

$$EN_n = 0, \quad EN_n N_k = R_n \cdot \delta_{nk}, \quad n, k = 0, 1, 2, \ldots.$$

In other words, the sequence is uncorrelated, and $\text{var}(N_n) = R_n$. Since Y_n and N_n have zero mean, the random variable X_n has a zero mean also. A final assumption we make is that the random variable Y_0, the sequence $\{N_n\}$, and the sequence $\{W_m\}$ are uncorrelated. Then we can easily calculate that

$$R_X(n) = c_n^2 R_Y(n) + R_n.$$

Equations 9.30 and 9.32 have a natural interpretation as the signal and measurement equations. Equation 9.30 represents a desirable "signal" Y_n (possibly random, if $\sigma_0 \neq 0, Q_n \neq 0$); Equation 9.32 represents the "noisy" measurements of this signal. The parameter c_n is an amplification constant, while the sequence $\{N_n\}$ represents the measurement noise.

Remark 1: The signal $\{Y_n, n = 1, 2, \ldots\}$ is not necessarily WSS, as was the case with the Wiener filter. Hence the Kalman filter has wider applicability. The measurement process $\{X_n\}$ is not necessarily WSS either, even when $\{Y_n\}$ is.

Remark 2: If the initial condition Y_0 and the sequence $\{W_n\}$ are Gaussian, then the signal process is also Gaussian since it is a linear transformation of Y_0 and $\{W_n\}$. If the noise process is Gaussian, then the measurement process is Gaussian as well.

Remark 3: The variance of a random variable V can be thought of as a measure of power or energy when V represents a (random) voltage across a unit resistor. From that interpretation, it can be easily seen that the variances σ_0, Q_n, and R_n determine the well-known signal-to-noise ratio.

The problem we address in this section is the following: Suppose we take n measurements $X_0, \ldots, X_{n-1}$; what is the best linear estimate (predictor) of the signal Y_n, *in recursive form?*

The recursiveness requirement on the estimator is what makes the results of this section unique; otherwise, the best predictor is given by the results in Section 9.4.4. So let

$$\hat{Y}_n \triangleq \sum_{k=1}^{n} h_k(n) X_{n-k} \tag{9.33}$$

and

$$\hat{Y}_{n+1} \triangleq \sum_{k=1}^{n+1} h_k(n+1) X_{n+1-k} \tag{9.34}$$

denote the estimates of Y_n and Y_{n+1}. We use the notation $h_k(n)$ to denote the fact that the best filter will be a function of time n.

Our goal is to express the filter $h_k(n+1)$, $k = 1, \ldots, n+1$, in terms of $h_k(n)$, $k = 1, \ldots, n$. We do that through a series of lemmas. To that end, observe from Equations 9.33 and 9.34 that the same measurement X_l, for all $l < n$, is multiplied by $h_l(n)$ and $h_l(n+1)$, respectively. Using the orthogonality principle results in Section 9.4.2, once for each of the two filters, we have

$$R_{YX}(n,m) \;=\; \sum_{k=1}^{n} h_k(n) R_X(n-k,m), \quad m = 0,1,\ldots,n-1, \quad (9.35)$$

$$R_{YX}(n+1,m) \;=\; \sum_{k=1}^{n+1} h_k(n+1) R_X(n+1-k,m), \quad m = 0,1,\ldots,n.$$

$$(9.36)$$

We can express the left-hand sides of Equations 9.35 and 9.36 in a recursive form as follows.

Lemma 9.1 *For all* $n \geq 1$,

$$R_{YX}(n+1,m) = \phi_n R_{YX}(n,m), \quad m = 0,1,\ldots,n-1.$$

Proof: From its definition,

$$
\begin{aligned}
R_{YX}(n+1,m) \;&\overset{\triangle}{=}\; E[Y_{n+1} X_m] = E[(\phi_n Y_n + W_n) X_m] \\
&=\; E[\phi_n Y_n X_m] + E[W_n X_m] = \phi_n R_{YX}(n,m),
\end{aligned}
$$

since the random variables W_n and X_m are uncorrelated for all n and $m < n$. $\square$

Lemma 9.2 *For all* $n \geq 1$,

$$R_{YX}(n,m) = \frac{R_X(n,m)}{c_n}, \quad m = 0,1,\ldots,n-1.$$

Proof: From its definition,

$$
\begin{aligned}
R_{YX}(n,m) \;&\overset{\triangle}{=}\; E[Y_n X_m] = E\left[\frac{X_n - N_n}{c_n} X_m\right] = \frac{1}{c_n} E X_n X_m - \frac{1}{c_n} E N_n X_m \\
&=\; \frac{1}{c_n} E X_n X_m = \frac{R_X(n,m)}{c_n}.
\end{aligned}
$$

The first equality follows from the definition of the measurement X_n; the third equality follows from the fact that the noise random variable N_n and the measurement X_m for any $m < n$ are not correlated. (Can you justify why?) $\square$

From Lemmas 9.1 and 9.2, we have the desired recursive formula:

Lemma 9.3 *For all* $n \geq 1$,

$$h_{k+1}(n+1) = h_k(n)\left[\phi_n - h_1(n+1)c_n\right], \quad k = 1, \ldots, n+1. \tag{9.37}$$

Proof: Consider an $m < n$. From Lemma 9.1 and Equation 9.36, we have

$$R_{YX}(n+1, m) = \phi_n R_{YX}(n, m) = \sum_{k=1}^{n+1} h_k(n+1)R_X(n+1-k, m)$$

$$= h_1(n+1)R_X(n, m) + \sum_{k=2}^{n+1} h_k(n+1)R_X(n+1-k, m).$$

We can use now Lemma 9.2 to rewrite the above expression as

$$\left[\phi_n - h_1(n+1)c_n\right]R_{YX}(n, m) = \sum_{k=2}^{n+1} h_k(n+1)R_X(n+1-k, m)$$

$$= \sum_{l=1}^{n} h_{l+1}(n+1)R_X(n-k, m).$$

From the above expression and Equation 9.35, we can immediately identify the coefficients $h_k(n)$ as

$$h_k(n) = \frac{h_{k+1}(n+1)}{\phi_n - h_1(n+1)c_n}.$$

QED. □

From Equations 9.37, 9.33, and 9.34, we can easily relate the two predictors as

$$\hat{Y}_{n+1} = \phi_n \hat{Y}_n + h_1(n+1)(X_n - c_n \hat{Y}_n). \tag{9.38}$$

This equation has an intuitive and pleasing interpretation as follows: The filter processes the measurements X_n *one at a time, instead of the entire sequence at once.*[7] The "impact" of the "entire past" from time 0 to time n is summarized in the previous estimate $\hat{Y}_n$. The relative impacts of the "new" measurement and the past are weighed by the coefficients ϕ_n and $h_1(n+1)$, which we still have to determine! Traditionally, the coefficient $h_1(n+1)$ has been called the *gain* of the filter and denoted as $K(n)$. We follow this convention from now on.

Deriving a recursive formula for the gain coefficient $K(n)$ is a bit lengthy. So, here we go. Let's rewrite Equation 9.38 as

$$\hat{Y}_{n+1} = \phi_n \hat{Y}_n + K(n)I_n, \tag{9.39}$$

---

[7] This is the major difference between this filter and the best estimators in Section 5.3.

where I_n, the *innovation sequence*, is defined as $I_n = X_n - c_n \hat{Y}_n$. Define now the error sequence

$$e_n = Y_n - \hat{Y}_n, \quad n = 1, 2, \ldots,$$

with initial condition $e_0 \stackrel{\triangle}{=} 0$. We can express the error sequence in a recursive form:

$$\begin{aligned} e_{n+1} \quad \stackrel{\triangle}{=} \quad & Y_{n+1} - \hat{Y}_{n+1} = (\phi_n Y_n + W_n) - [K(n)I_n + \phi_n \hat{Y}_n] \\ = \quad & [\phi_n - K(n)c_n]e_n + W_n - K(n)N_n, \quad n = 1, 2, \ldots \end{aligned}$$

Since $Ee_n W_n = Ee_n N_n = EW_n N_n = 0$ (check that), the variance of the estimation error is given by

$$\begin{aligned} Ee_{n+1}^2 \quad = \quad & [\phi_n - K(n)c_n]^2 Ee_n^2 + EW_n^2 + K^2(n)EN_n^2 \\ = \quad & [\phi_n - K(n)c_n]^2 Ee_n^2 + Q_n + K^2(n)R_n. \end{aligned} \qquad (9.40)$$

The best $K(n)$ is of course the one that minimizes Ee_{n+1}^2; differentiating Equation 9.40, we get finally

$$K(n) = \frac{\phi_n c_n Ee_n^2}{R_n + c_n^2 Ee_n^2}. \qquad (9.41)$$

Equations 9.41, 9.40, and 9.39 define now an algorithm for calculating the best predictor in a recursive form. We summarize the algorithm as follows:

Kalman Filter

   Initialize: $\hat{Y}_0 = 0; Ee_0^2 = \sigma_0$.
   For $n = 0, 1, 2, \ldots$ **do**
      Set $K(n) = \frac{\phi_n c_n Ee_n^2}{R_n + c_n^2 Ee_n^2}$.
      Set $Ee_{n+1}^2 = [\phi_n - K(n)c_n]^2 Ee_n^2 + Q_n + K^2(n)R_n$.
      Output $\hat{Y}_{n+1} = K(n)[X_n - c_n \hat{Y}_n] + \phi_n \hat{Y}_n$.
   end

Let's see now an example of the filter.

Example 9.15.   Consider the system

$$\begin{aligned} Y_n \quad &= \quad 0.9 Y_{n-1} + W_n, \quad n = 1, 2, \ldots, \\ X_n \quad &= \quad 2Y_n + N_n, \quad\quad\,\, n = 1, 2, \ldots, \end{aligned}$$

where $Q_n = 1, \sigma_0 = 1$, and $R_n = 0.1$. Figure 9.14(a) shows the signal (solid line) and its estimate (dotted line); Figure 9.14(b) shows the errors, while Figure 9.15 shows how fast the filter gains converge.

   We present next the multidimensional case, without any proof (see the original papers [33, 34]).

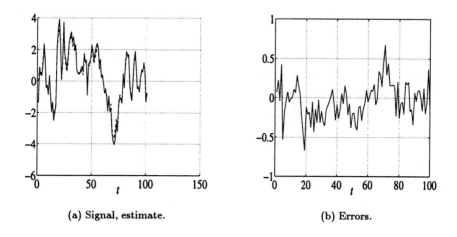

(a) Signal, estimate.

(b) Errors.

Figure 9.14   Kalman filter performance.

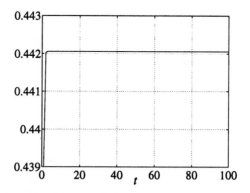

Figure 9.15   The filter gains in Example 9.15.

9.5.2   Discrete-Time, Multidimensional Case

Let Y_n and W_n be l-dimensional vectors, X_n and N_n be j-dimensional vectors, and Φ_n and C_n be $l \times l$ and $j \times l$ dimensional matrices. The system evolution in the multidimensional case is given by

$$Y_n = \Phi_n Y_{n-1} + W_n, \quad n = 1, 2, \ldots,$$

and the measurement equation is given by

$$X_n = C_n Y_n + N_n, \quad n = 1, 2, \ldots.$$

We assume that the noise process N_n, the process W_n, and the initial value Y_0 are independent of one another. And N_n is a white Gaussian noise with covariance matrix R_n. Also, W_n is Gaussian with covariance matrix Q_n. The Kalman filter equations are given by

$$\hat{Y}_{n+1} = \Phi_n \hat{Y}_n + K(n)[X_n - C_n \hat{Y}_n],$$

where $K(n)$ can be computed as

$$
\begin{aligned}
K(n) &= \Phi_n E e_n^2 C_n^T (R_n + C_n E e_n^2 C_n^T)^{-1} \\
E e_{n+1}^2 &= \Phi_n E e_n^2 \Phi_n^T + Q_n.
\end{aligned}
$$

9.6   PERIODOGRAMS

We saw that the psd function plays a central role in the analysis and processing of random processes through linear systems. So how do we measure the psd in practice?

A large body of literature exists on the subject; here we develop only some basic ideas and point out some of the problems that arise in practice. For a more detailed description, see [55]. We saw in Section 9.3 that the estimates

$$\bar{X}_T(f) = \frac{1}{2T} \int_{-T}^{T} X(t) e^{-j2\pi ft}\, dt \cdot \int_{-T}^{T} X(s) e^{j2\pi fs}\, ds \qquad (9.42)$$

(in continuous time) and

$$\bar{X}_n(f) = \frac{1}{n} \left| \sum_{k=0}^{n-1} X_k e^{-j2\pi fk} \right|^2$$

(in discrete time) can be used to approximate $S_X(f)$. We study now the properties of these estimates in greater detail. We present and prove these properties in a series of lemmas.

Recall that for any fixed T or n (and thus for any fixed number of measurements) the estimate is a random variable. The first lemma evaluates the mean of this random variable.

Lemma 9.4 *For a fixed* $f \in (-\infty, \infty)$ *and any* $n < \infty$ *or* $T < \infty$,

$$E\bar{X}_n(f) = \sum_{k=-(n-1)}^{n-1} \left(1 - \frac{|k|}{n}\right) R_X(k) e^{-j2\pi fk},$$

$$E\bar{X}_T(f) = \int_{-T}^{T} \left(1 - \frac{|s|}{2T}\right) R_X(s) e^{-j2\pi fs} \, ds.$$

Proof: We prove the discrete-time case only since the continuous one is a direct analog. From the definition of the estimate, we have

$$E\bar{X}_n(f) = \frac{1}{n} E \left(\sum_{k=0}^{n-1} X_k e^{-j2\pi fk}\right) \left(\sum_{m=0}^{n-1} X_m e^{-j2\pi fm}\right)^*$$

$$= \frac{1}{n} \sum_{k=0}^{n-1} \sum_{m=0}^{n-1} E(X_k X_m) e^{-j2\pi f(k-m)}$$

$$= \frac{1}{n} \sum_{k=0}^{n-1} \sum_{m=0}^{n-1} R_X(k-m) e^{-j2\pi f(k-m)}.$$

Using the same trick for evaluating the double summation, as we did with the proof of Theorem 7.10, we can rewrite the last expression as

$$E\bar{X}_n(f) = \frac{1}{n} \sum_{k=-(n-1)}^{n-1} \left(1 - \frac{|k|}{n}\right) R_X(k) e^{-j2\pi fk},$$

completing the proof. □

The following lemma evaluates the variance of the estimate. Note that to determine the variance, knowledge of only $R_X(s)$ does not suffice. *Fourth*-order moments of the process must be known. The estimate in the next lemma is based on Gaussian approximation assumptions. Its proof is rather complicated; consult [32] for a detailed analysis.

Lemma 9.5 *For a fixed* $f \in (-\infty, \infty)$ *and any* $n < \infty$ *or* $T < \infty$,

$$\text{var}(\bar{X}_n(f)) = [S_X(f)]^2 \left[1 + \left(\frac{\sin(2\pi fn)}{n \sin(2\pi f)}\right)^2\right].$$

The next two lemmas describe how the estimate converges as we process more and more measurements.

Lemma 9.6 *For any fixed $f \in (-\infty, \infty)$,*

$$\lim_{n \to \infty} E\bar{X}_n(f) = S_X(f), \quad \lim_{T \to \infty} E\bar{X}_T(f) = S_X(f),$$

in the mean square sense.

Proof: The result follows from Lemma 9.4, by letting $n \to \infty$ and $T \to \infty$. $\square$

However, and this may be a bit surprising at first, we can show the following:

Lemma 9.7 *For any fixed $f \in (-\infty, \infty)$,*

$$\lim_{n \to \infty} E\bar{X}_n(f) \neq S_X(f), \quad \lim_{T \to \infty} E\bar{X}_T(f) \neq S_X(f),$$

in the almost-sure sense.

Lemma 9.7 could be expected since the random variables in the sequence $\{\bar{X}_T(f)\}$ (or $\{\bar{X}_n(f)\}$) are not independent. The practical consequence of this lemma is that we cannot process our measurements $X(t)$ as Equation 9.42 suggests. If we want to have almost-sure convergence, we must somehow create independent measurements. In the next section we explore this idea further.

9.6.1   Smoothing of Periodograms

The basic idea is to generate IID measurements; then the SLLN will guarantee convergence to the mean value $S_X(f)$, in the almost-sure sense. The brute-force way of doing so is to repeat the experiment N times; during the ith repetition, we create a sequence of measurements $\{X_i(k), \ k = 0, 1, \ldots, n-1\}$. Then the estimates in Equation 9.42 can be rewritten as

$$\bar{X}_n^i(f) = \frac{1}{n} \left| \sum_{k=0}^{n-1} X_i(k) e^{-j2\pi f k} \right|^2, \quad i = 1, \ldots, N,$$

and the overall estimate can be written as

$$\bar{X}(f, N) = \frac{1}{N} \sum_{i=1}^{N} \bar{X}_n^i(f).$$

We can evaluate the mean and variance of the estimate, as the following lemma suggests.

Lemma 9.8 *For a fixed $f \in (-\infty, \infty)$ and any $n < \infty$,*

$$E\bar{X}(f, N) = \sum_{k=-(n-1)}^{n-1} \left(1 - \frac{|k|}{n}\right) R_X(k) e^{-j2\pi fk},$$

$$\mathrm{var}(\bar{X}(f, N)) \approx \frac{S_X(f)}{N},$$

$$\lim_{N \to \infty} E\bar{X}(f, N) = S_X(f), \quad \text{almost surely.}$$

Proof: The first two relations follow immediately from the theory of sums of IID random variables in Chapter 6, namely, Theorem 6.1. The third one follows from the Strong Law of Large Numbers. $\square$

Repeating the same experiment N times can be very time-consuming, as we discussed in Section 6.6.1. A faster procedure is to use only one experiment and to produce the N segments of measurements $X_i(k)$ from this single run. Of course, it is not absolutely guaranteed that the random variables in different segments are IID. The choice of the data size N and the segment size must often be made on a trial-and-error basis.

9.7 SUMMARY OF MAIN POINTS

- Random processes can be processed in the time or frequency domain. General results can only be obtained when they are **linearly processed**.

- The autocorrelation or **power spectral density** (psd) functions play an important role in linear processing of WSS processes. **Periodograms** are estimates of the psd function.

- **(Causal) filtering, smoothing**, and **prediction** of random processes are three common MMSE processing problems.

- The **Levinson** and **Kalman** filters are recursive algorithms that are easily implementable on a computer or specialized hardware. **Wiener filtering** is used in the frequency domain.

9.8 CHECKLIST OF IMPORTANT TOOLS

- Linear processing of a random process, Theorem 9.3

- Causal filtering

- Smoothing

- Prediction

- Kalman filter

- Levinson's algorithm

- Periodograms

9.9   PROBLEMS

Power Spectral Density Function; Simple Skills

9.1 Determine whether any of the functions in Figure 9.16 represents the psd of a random process $\{X(t)\}$.

9.2 Let $s(x)$, $p(x, A)$ denote the functions in Figure 9.17. Find the psd that corresponds to the following autocorrelation functions $R_X(\tau)$:

$$s(\tau), \quad p(\tau, 10), \quad s(\tau)|\tau|, \quad e^{-|\tau|}\left[4\cos(|\tau|) + 3\sin(|\tau|)\right].$$

9.3 Let again $s(x)$, $p(x, A)$ denote the functions in Figure 9.17. Find the autocorrelation function that corresponds to the psd:

$$S_X(f) = 10s(5f), \quad S_X(f) = 3p(\tau, 10) + 2p(\tau, 3).$$

9.4 Determine whether any of the following functions can be the autocorrelation function of a random process $\{X(t)\}$:
(a) $R(\tau) = ce^{a\tau^{2b}}$, $a, b, c \in R$.
(b) $R(\tau) = a + b\cos(c\tau + d)$, $a, b, c, d \in R$.
(c) If you answered yes to part (a) or (b), calculate the psd function.

9.5 Determine whether any of the following functions can be the psd of a random process $\{X(t)\}$:

$$\frac{1}{f^2}, \quad \frac{1}{100 + f^2}, \quad \frac{f^2}{1 + f^8}, \quad \frac{\cos(5f)}{1 + f^4}, \quad \frac{1}{1 - f^2}, \quad \frac{1}{1 + e^{-f^2}}, \quad \frac{e^{-|f|}}{(1 + f)^4}.$$

9.6 Let $\{X(t)\}$ be a WSS random process. Define the new process $\{Y(t)\}$ via $Y(t) = AX(t) + B$, where A, B are constants. Determine the psd of $Y(t)$.

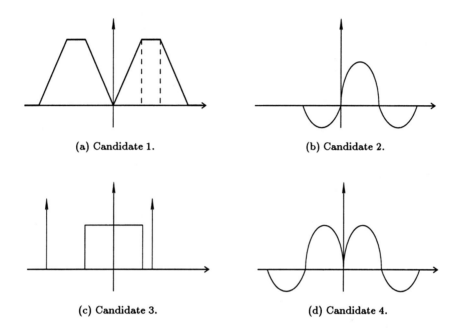

(a) Candidate 1.　　　　　　　　(b) Candidate 2.

(c) Candidate 3.　　　　　　　　(d) Candidate 4.

Figure 9.16 Candidate psd functions.

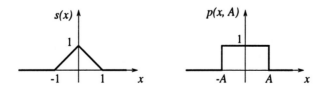

Figure 9.17 The functions $s(x)$, $p(x, A)$.

9.7 Let $\{X(t)\}$ have psd $S_X(f)$. Define the new process $\{Y(t)\}$ via $Y(t) = X(t+1) - X(d)$, where d is a given constant. Find the psd $S_Y(f)$.

9.8 Consider the sequences $\{X_n\}$, $\{Z_n\}$ in Problem 7.24.
(a) Find the cross-correlation function of the sequences $\{X_n\}$, $\{Z_n\}$.
(b) Find the cross-power spectral density of Z_n, X_n.

9.9 Consider an IID sequence $\{X_n\}$ of Bernoulli random variables with parameter p and an IID sequence $\{Y_n\}$ of Bernoulli random variables with parameter ϵ, independent of $\{X_n\}$. Consider the sequence $\{Z_n\}$ defined via $Z_n = X_n \oplus Y_n$, where $\oplus$ denotes modulo-2 addition.

Suppose that X_n represents a bit transmitted over a communication channel with noise; then $Y_n = 0$ means that the bit was received without errors, while $Y_n = 1$ indicates a bit reversal, i.e., a transmission error.
(a) Find EZ_n and $R_Z(n,k)$.
(b) Is the sequence $\{Z_n\}$ WSS? If yes, what is its power spectral density?

9.10 Consider two independent sequences $\{X_n\}$ and $\{Y_n\}$. A multiplexer produces the output

$$Z_n = \begin{cases} X_{\frac{n}{2}}, & n = 2, 4, \ldots, \\ Y_{\frac{n+1}{2}}, & n = 1, 3, \ldots \end{cases}$$

(that is, $Z_1 = Y_1, Z_2 = X_1, Z_3 = Y_2, Z_4 = X_2$, etc.).
(a) Is that system linear or nonlinear?
(b) Suppose that $\{X_n\}$ and $\{Y_n\}$ are jointly stationary. Is $\{Z_n\}$ stationary?
(c) Suppose that $\{X_n\}$ and $\{Y_n\}$ are jointly WSS. Is $\{Z_n\}$ WSS? If yes, find its psd.
(d) Suppose that $\{X_n\}$ and $\{Y_n\}$ are jointly cyclostationary. Is $\{Z_n\}$ cyclostationary?

9.11 Let $\{X(t)\}$ and $\{Y(t)\}$ be two jointly stationary, Gaussian random processes, with zero mean and equal autocovariance functions. Suppose further that the two processes are uncorrelated, that is, $EX(t)Y(s) = 0$. Define a new random process as follows: Fix a frequency f_0 and let

$$Z(t) = X(t)\cos(2\pi f_0 t) + Y(t)\sin(2\pi f_0 t).$$

(a) Is that system linear or nonlinear?
(b) Calculate $f_{Z(t)}(z)$ and $f_{Z(t)Z(s)}(z_1, z_2)$.
(c) Calculate $EZ(t)$ and $R_Z(t,s)$.

9.12 Let $\{X_n\}_{-\infty}^{\infty}$ be an IID sequence of Bernoulli random variables with parameter p.
(a) Let $Y_n = (-1)^{X_n}$. Calculate $EY_n, R_Y(j,k)$, and $S_Y(f)$. Is that system linear or nonlinear?

(b) Let $Y_n = \sum_{k=0}^{n} 2^{-k} X_k$. Calculate $EY_n, R_Y(j, k)$. Is that system linear or nonlinear?

Power Spectral Density Function; Simple Skills

9.13 Let Y be a random variable with characteristic function $\Phi_Y(\omega)$ and Z be a uniform random variable with parameters $-\pi, \pi$, independent of Y. Define the random process $\{X(t)\}$ via $X(t) = A\cos(2\pi Y t + Z)$, where A is a given constant.
(a) Find the psd $S_X(f)$.
(b) Use this result to explain how you can construct a random process with a given $S_X(f)$.

Linear Systems; Simple Skills

9.14 Consider an IID Gaussian sequence $\{X_n\}_{-\infty}^{\infty}$, with mean m and variance σ^2. The sequence is fed into a linear system with impulse response $h(k) = 1/k, k > 0, h(0) = 1$. Let Y_n denote the output of this system.
(a) Find, if possible, EY_n, $R_{XY}(i, j)$, and $R_Y(i, j)$.
(b) Is the output stationary? WSS?

9.15 Repeat Problem 9.14 when $h(k) = 1/k^2$, $k > 0$, $h(0) = 1$.

9.16 Prove Theorem 9.5.

9.17 Consider two independent white noise sequences $\{X_n\}$ and $\{Y_n\}$ with variances σ_X^2 and σ_Y^2. Consider the system $Z_n = 3X_n + Y_{n-3}$. Calculate the autocovariance of the output Z_n.

9.18 A random process $\{X(t)\}$ with autocorrelation $R_X(\tau) = s(\tau)$ (see Figure 9.17) is passed through a linear filter with impulse response $h(t) = e^{-t}$. Let $Y(t)$ denote the output of this filter. Find $EY(t)$, $R_Y(\tau)$, and $S_Y(f)$.

9.19 A random process $\{X(t)\}$ with psd $S_X(f) = p(f, 1)$ (see Figure 9.17) is passed through a linear filter with impulse response $h(t)$. Let $Y(t)$ denote the output of this filter. Is there a proper filter for achieving $R_Y(\tau) = s(\tau)$?

9.20 The WSS random process $\{X(t)\}$ with $R_X(\tau) = 5e^{-3|\tau|}$ is the input to a filter with impulse response $h(t) = 10e^{-2t}u(t)$. Evaluate $EY(t)$, $\text{var}(Y(t))$, $R_Y(\tau)$, and $S_Y(f)$.

9.21 A random process $\{X(t)\}$ with psd $S_X(f) = (1 + \pi^2 f^2)^{-1}$ is input to a filter with transfer function $H(f) = (1 + j2\pi f)^{-1}$. Find the autocorrelation and spectral density of the output process.

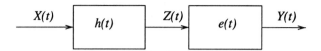

Figure 9.18   Two cascaded filters.

9.22 Let $\{X_n\}_{-\infty}^{\infty}$ be an IID sequence of exponential random variables with parameter λ. Let $\{Y_n\}_{-\infty}^{\infty}$ be another IID sequence of exponential random variables with parameter μ, independent of $\{X_n\}$. Define the new sequence $\{Z_n\}_{-\infty}^{\infty}$ as a "noisy" ARMA process, of the form

$$Z_n = X_n + X_{n-1} + X_{n-2} + Y_n,$$

where Y_n is considered to be the noise. Calculate EZ_n and $R_Z(n,k)$.

9.23 Let $\{X_n\}$ be an IID process; define $Y_n = X_n - X_{n-1}$.
(a) Is $\{Y_n\}$ an IID process?
(b) Does it have stationary and independent increments?
(c) Find EY_n and $R_Y(j,k)$.
(d) Repeat questions (a)–(c) when $Y_n = aX_n + bX_{n-1}$, where a,b are given constants.

9.24 Let $\{X(t)\}, \{Y(t)\}$ be the input and output processes of a linear, time-invariant system $h(t)$. Define a new random process $\{Z(t)\}$ via

$$Z(t) = X(t)Y(t+T),$$

where $T > 0$ is a given constant. Evaluate $EZ(t)$, in terms of $h(t)$ and the statistical parameters of the processes $\{X(t)\}, \{Y(t)\}$.

9.25 Consider the two *cascaded* filters $h(t), e(t)$ shown in Figure 9.18. Evaluate $EY(t)$ and $R_Y(t,s)$.

9.26 Consider the cascaded filters shown in Figure 9.18 with a WSS random process $\{X(t)\}$ as input.
(a) Calculate the psd's $S_Y(f)$ and $S_Z(f)$.
(b) Calculate $S_{XY}(f)$, $S_{XZ}(f)$, and $S_{ZY}(f)$.
(c) Evaluate your answers when the input is zero-mean, white Gaussian noise and $h(t) = e(t) = e^{-t}u(t)$.

9.27 Consider the two *parallel* filters $h(t)$ and $e(t)$ in Figure 9.19. Evaluate the cross-correlation $R_{YZ}(t,s)$. Under what conditions on the input $X(t)$ are the two outputs jointly WSS? Cyclostationary?

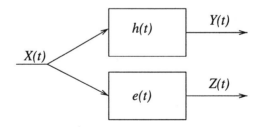

Figure 9.19 Two parallel filters.

9.28 Let $\{X_n\}$ be an IID process; define the *moving-average* process $\{Y_n\}$ via

$$Y_n = \frac{1}{n} \sum_{k=1}^{n} X_{n-k}.$$

Calculate EY_n, $R_Y(j,k)$, and $R_{XY}(j,k)$.

9.29 Let $\{X(t)\}$ be a WSS random process. Define the new process $\{Y(t)\}$ via $Y(t) = AdX(t)/dt$, where A is a constant. Determine the psd of $Y(t)$.

9.30 Let $\{X(t)\}$ be a WSS random process. Define the new processes

$$Y(t) = X(t)\cos(2\pi f_1 t + \Theta), \quad Z(t) = X(t)\cos(2\pi f_2 t + \Theta),$$

where f_1, f_2 are positive constants and Θ is a uniform random variable in $[0, 2\pi)$, independent of the random process $\{X(t)\}$.
(a) Determine conditions, if any, that make the random processes $\{Y(t)\}$ and $\{Z(t)\}$ jointly WSS. Evaluate their psd functions.
(b) Evaluate, if possible, the cross-correlation function $R_{YZ}(t,s)$ and the cross-power spectral density.

9.31 The WSS random process $\{X(t)\}$ with autocorrelation

$$R_X(\tau) = \begin{cases} 1 - |\tau|, & |\tau| \leq 1, \\ 0, & \text{otherwise} \end{cases}$$

is the input to an integrator with parameter $T > 1$:

$$Y(t) = \frac{1}{T} \int_{t-T}^{t} X(s)\, ds.$$

(a) Verify that this is a linear time-invariant system.
(b) Evaluate the mean $EY(t)$ and variance $\text{var}(Y(t))$.
(c) Evaluate the autocorrelation $R_Y(\tau)$ and psd $S_Y(f)$.

(d) Evaluate $R_{XY}(t, s)$ and $S_{XY}(f)$.

(e) What happens when $T < 1$?

9.32 Let $\{X_n\}$ be a white noise process. Define the ARMA process $\{Y_n\}$ via $Y_n = X_n + aX_{n-1} + bX_{n-2}$, where a, b are given constants. Evaluate $R_Y(k)$ and $S_Y(f)$.

9.33 Let $\{X_n\}$ be a white noise process. Define the ARMA process $\{Y_n\}$ via $Y_n = aY_{n-1} + bY_{n-2} + X_n$, where a, b are given constants. Evaluate $R_Y(k)$ and $S_Y(f)$.

Linear Systems; Advanced Skills

9.34 Consider the two *parallel* filters $h(t)$ and $e(t)$ shown in Figure 9.19. Suppose that their transfer functions coincide over the frequency range $[-f_0, f_0]$. Suppose that the input to the system is a random process $\{X(t)\}$ with psd that is zero outside the band $[-f_0, f_0]$. Show that the outputs of the two filters are identical in the mean square sense.

9.35 Two zero-mean, jointly WSS, Gaussian processes $\{X(t)\}$ and $\{Y(t)\}$ are inputs to two filters with impulse responses $h_1(t)$ and $h_2(t)$. Show that if $|H_1(f)| \cdot |H_2(f)| = 0$ (i.e., the transfer functions do not overlap), the two outputs are independent processes.

9.36 Consider a random process $\{X(t)\}$, with psd that is zero outside the band $[-f_0, f_0]$. Show that

$$\frac{R_X(0) - R_X(\tau)}{R_X(0)} \le \frac{(2\pi f_0 \tau)^2}{2};$$

this bound gives us an estimate of how "slowly" a low-pass process can change.

9.37 Let $\{X_n\}$ be a sequence of IID random variables. Consider the periodic pulse shown in Figure 9.20. Define the random process $\{X(t)\}$ via

$$X(t) = \sum_{k=-\infty}^{\infty} X_k p(t - kT).$$

The process $\{X(t)\}$ is an example of a pulse amplitude-modulated system.

(a) Plot a few sample paths.

(b) Is $\{X(t)\}$ stationary? WSS?

(c) Let now Θ be a uniform random variable. Define a new random process $\{Y(t)\}$ via $Y(t) = X(t + \Theta)$. Is $\{Y(t)\}$ stationary? WSS?

9.38 Consider a WSS random process $\{X(t)\}$. Define the process $\{Z(t)\}$ via $Z(t) = A \cos(2\pi f_0 t + X(t))$, where A, f_0 are positive constants. This is an

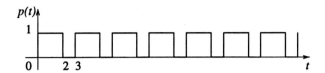

Figure 9.20 A periodic pulse, $T = 3$.

example of a *phase-shift key* system.
(a) Is $\{Z(t)\}$ stationary? WSS?
Let $V(t) = Z(t + \Theta)$, where Θ is a random variable uniformly distributed in $(0, 1/f_0]$.
(b) Find the cross-correlation $R_{ZV}(t, s)$.
(c) Are the two processes $\{V(t)\}, \{Z(t)\}$ jointly WSS?

9.39 The zero-mean Gaussian WSS random process $\{X(t)\}$ with autocorrelation function $R_X(\tau)$ is the input to a filter with impulse response

$$h(t) = \begin{cases} \frac{1}{T}, & 0 < t < T, \\ 0, & \text{otherwise,} \end{cases}$$

where $T > 0$ is a given constant. Show that the variance of the output is bounded by

$$\text{var}(Y(t)) \leq \frac{1}{T} \int_{-\infty}^{\infty} |R_X^2(s) + R_X(T + s)R_X(T - s)| \, ds.$$

9.40 Consider a zero-mean Gaussian WSS random process $\{X(t)\}$ with autocorrelation function $R_X(\tau)$. Let $Y(t) = X(t)X(t - T)$, where $T > 0$ is a fixed constant. Show that

$$R_Y(\tau) = R_X^2(T) + R_X^2(\tau) + R_X(\tau + T)R_X(\tau - T).$$

(*Hint:* Use the result of Problem 4.174.)

9.41 A zero mean WSS Gaussian random process $\{X(t)\}$ with $R_X(\tau) = e^{-2|\tau|}$ is the input to two filters with impulse responses

$$h_1(t) = 4e^{-6t}u(t), \quad h_2(t) = te^{-t}u(t).$$

(a) Evaluate the joint pdf of the two outputs $Y_1(t), Y_2(t)$.
(b) Evaluate $R_{Y_1Y_2}(\tau)$ and the cross-power spectral density $S_{Y_1Y_2}(f)$.

9.42 Consider a discrete-time zero-mean WSS process $\{X_n, n = 0, \pm 1, \pm 2, \ldots\}$. Let

$$Y_n = \frac{X_{n+1} + X_n + X_{n-1}}{3}.$$

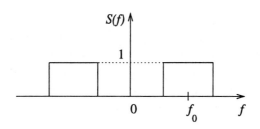

Figure 9.21   Band-limited noise psd $S(f)$.

(a) Find the psd $S_Y(f)$.

(b) Explain why the process $\{Y_n\}$ is called a "smoothed version" of $\{X_n\}$.

9.43 Consider a random process $\{N(t)\}$, with the psd $S(f)$ shown in Figure 9.21. Let

$$X(t) = A\cos(2\pi f_0 t + \Theta) + N(t),$$

where A, Θ are random variables and f_0 is a given constant. Show that we can write

$$X(t) = A_t \cos(2\pi f_0 t + \Theta_t);$$

determine the appropiate A_t, Θ_t.

Minimum Mean Square Estimation; Simple Skills

9.44 Show how the estimation problem formulation will change if $Y(t)$ and $X(t)$ have nonzero means.

9.45 Formulate the prediction problem for infinite measurements.

9.46 Let $X_n = Y_n + N_n$, where $R_Y(k) = 0.5^{|k|}$ and $\{N_n\}$ is white noise with parameter 1. Find the optimum filter for estimating Y_n given the measurement X_{n-1}, and evaluate the minimum mean square error.

9.47 Let X_n, Y_n, N_n be as in Problem 9.46. Find the optimum filter for estimating Y_n given M measurements $X_{n-1}, \ldots, X_{n-M}$. Evaluate the minimum mean square error.

9.48 Let X_n, Y_n, N_n be as in Problem 9.46. Find the optimum filter for estimating Y_n given X_1 and X_n only.

9.49 Consider a discrete-time Wiener process $\{X_n\}$. Calculate the MMSE estimate of X_n given the measurements $X_1, \ldots, X_{n-1}$.

9.50 Consider a Poisson process $\{N(t)\}$ and a given positive increasing sequence of time indices $\{t_n\}$. Define the *sampled* process $\{X_n\}$ via $X_n = N(t_n)$. Calculate the MMSE estimate of X_n given the measurements $X_1, \ldots, X_{n-1}$.

9.51 Let $\{X(t)\}$ be a WSS random process. We wish to form an estimate of the random variable $X(t)$, based on three measurements of the process at fixed points $t_1, t_2,$ and t_3, as follows:

$$\hat{X}(t) = a_1 X(t_1) + a_2 X(t_2) + a_3 X(t_3).$$

(a) Find the optimal values of the coefficients a_1, a_2, a_3.
(b) Evaluate the minimum mean square error.

9.52 Let $X(t) = Z(t) + N(t)$, where the random processes $\{Z(t)\}$ and $\{N(t)\}$ are independent, zero-mean, jointly WSS, with

$$R_Z(\tau) = e^{-|\tau|} \cos(|\tau|), \quad R_N(\tau) = e^{-4|\tau|}.$$

We want to predict $Z(t)$ in terms of $X(t)$. Find the equations that the best predictor must satisfy.

9.53 Let $\{X_n\}$ be an ARMA process. Use Levinson's algorithm to find the best predictor of X_n in terms of X_{n-1} and X_{n-2}.

9.54 Let $\{X(t),\ t \geq 0\}$ be a zero-mean WSS process with a given autocorrelation function $R_X(\tau)$. Let $T > 0$ be a fixed constant. We wish to estimate

$$Z = \frac{1}{T} \int_0^T X(t)\, dt$$

via $Y = h_0 X(0) + h_1 X(T)$, where h_0 and h_1 are the unknown filter parameters. Show that for the best estimate we must have $h_0 = h_1$.

Smoothing

9.55 Evaluate the minimum mean square error for the smoothing filter in Section 9.4.5.

9.56 Evaluate the smoothing filter in Section 9.4.5 and its error when $\{Y(t)\}$ is bandlimited white noise and $\{N(t)\}$ is white noise.

9.57 Evaluate the smoothing filter in Section 9.4.5 and its error when $\{Y(t)\}$ is an ARMA process and $\{N(t)\}$ is white noise.

9.58 Let X_n, Y_n, N_n be as in Problem 9.46. Find the optimum filter for estimating Y_n given $2M + 1$ measurements $X_{n-M}, \ldots, X_{n+M}$. Evaluate the minimum mean square error.

9.59 Consider the processes in Problem 9.52. We want to smooth $Z(t)$ in terms of $X(t)$. Find the best smoothing filter.

9.60 Let $X(t) = Z(t) + N(t)$, where

$$S_Z(f) = \frac{1}{(1 + 4\pi^2 f^2)^2}, \quad S_N(f) = \frac{1}{1 + 4\pi^2 f^2}.$$

Find the best smoothing filter.

9.61 Repeat Problem 9.60 when

$$S_Z(f) = \frac{1}{1 + 4\pi^2 f^2}, \quad S_N(f) = \frac{1}{4 + 16\pi^4 f^4}.$$

9.62 Consider a random process $\{X(t)\}$ with psd that is zero outside the band $[-f_0, f_0]$. Does it satisfy the Paley-Wiener condition?

Advanced Skills

9.63 Suppose that Y_1, Y_2 are two optimal estimates of the random variable Y. Show that $Y_1 = Y_2$ almost surely.

9.64 Let $\{X(t)\}$ be a zero-mean WSS random process with continuous auto-correlation function $R_X(\tau)$.
(a) Find the optimal predictor of $X(t + s)$ in terms of $X(s)$.
(b) Show that if the optimal predictor of $X(t+s)$ in terms of $X(s)$ is the same as the optimal predictor in terms of $X(s)$ and $X(0)$, then $R_X(\tau) = \sigma^2 e^{-a|\tau|}$, and vice versa.

9.65 Let $\{X(t)\}$ be a WSS random process with $R_X(\tau) = e^{-2|\tau|}$. Two estimators are used for $X(t)$:

$$\hat{X}_1(t) = aX(0), \quad \hat{X}_2(t) = bX(t - 10).$$

Find the best values for a, b. Is $\hat{X}_2(t)$ better than $\hat{X}_1(t)$ for all t?

9.66 Let $\{N_n\}$ be a sequence of IID Gaussian random variables with zero mean and unit variance. Suppose that $X_n = Y + N_n$, $n = 1, 2, \ldots, M$, is a sequence of M noisy measurements of an unknown, zero-mean, Gaussian random variable Y, with variance σ^2, that is independent of $\{N_n\}$. Let $\hat{Y}_n = E(Y | X_1, \ldots, X_n)$ be a MMSE estimate of Y. Define the random variable ξ_n as

$$\xi_n = X_n - E(X_n | X_1, \ldots, X_{n-1}).$$

(a) Find $E\xi_n$, $E\xi_n\xi_l$.
(b) Show that $\hat{Y}_n = \hat{Y}_{n-1} + k_n\xi_n$. Determine k_n.

9.67 We wish to measure the resistance R of a resistor by supplying a current I and measuring the voltage V across the resistor. We repeat the experiment M times, each time supplying a current $I_n = n$, $n = 1, 2, \ldots, M$. The measurements are noisy, so $V_n = nR + N_n$, where $\{N_n\}$ is a sequence of IID Gaussian random variables with zero mean and unit variance. Let

$$\hat{R} \triangleq \sum_{n=1}^{M} h_n V_n$$

be an estimator of R. Determine the best MMSE estimator.

9.68 Consider a WSS random process $\{X(t)\}$, with autocorrelation function $R_X(\tau) = e^{-|\tau|} \cos(2\pi f_0 |\tau|)$, where $f_0 \gg 1$.
(a) Find the spectral factorization of $S_X(f)$.
(b) Determine the whitening filter.
(c) Suppose that $X(t)$ is the input to a filter with transfer function $|H(f)|^2 = |2\pi f|$. Find the autocorrelation of the output process.

9.69 Consider the periodogram estimate in Equation 9.42. Show that

$$\mathrm{var}(\bar{X}_T(f)) \leq \frac{2}{T} \int_0^T R_X(s) \, ds.$$

Computer Problems

9.70 Consider a random process $\{X(t)\}$ and the clipping transformation in Figure 3.40. Evaluate experimentally the pdf of $Y(t)$ and its autocorrelation. For a WSS $\{X(t)\}$, evaluate the psd also.

9.71 Plot sample paths of the Wiener filter for various values of its parameters.

9.72 Plot sample paths of outputs of ideal filters. Evaluate experimentally the psd of the outputs; thus "verify" that no frequencies outside the ideal band are passed.

9.73 Evaluate and plot the output of a predictor when the input is an ARMA process.

9.74 Apply the Kalman filter to an input of your choice. Plot the output, errors, and gain values. Observe how the estimation errors behave as a function of the measurement noise variance.

9.75 Evaluate experimentally the periodogram for a process with given $R_X(\tau)$.

10

MARKOV CHAINS

The random process models we developed in Chapters 7 and 9 are essentially the IID increment and the WSS process (in continuous and discrete time). We have been able to obtain a complete characterization (the n-dimensional cdf's) and a lot of results for the IID model. We obtained a lot of results for the WSS model as well, even though we were not able to provide its complete characterization.

One drawback of the IID model is that it cannot capture any system dynamics since independence is at the heart of it. WSS models are not a great help either because they are based on partial information only (mean and autocovariance function) and thus they do not provide information on distribution functions. On the other hand, a generic random process, with a "full-fledged" cdf can fully capture the dynamics of an arbitrary system, its time dependencies, etc. However, very limited results, if any, could be stated for such an all-encompassing model.

In 1906, A. A. Markov[1] put forth a theory that now[2] bears his name, which aims at capturing dependencies between the random variables of a random process [52]. The **Markov random process** model bridges the gap between an IID model and a general random process. It essentially allows for the simplest possible form of dependencies, namely, the dynamics of a first-order system. As a result of this restriction, many results can be stated about the model. No extensive theory has been developed for second- or higher-order systems (see Problems 10.25 and 10.29). The Markov process model has seen dramatic theoretical developments and practical applications in its less than 100-year history, mainly because of its mathematical tractability and the very good approximations it provides for realistic systems.

---

[1] A Russian mathematician and a student of Chebyshev's.

[2] The Markov processes were named as such in 1934, by Khinchine, as a tribute to Markov's work.

Markov applied this theory (for the discrete-time and finite state space case) in a study of dependencies in linguistics. In just 1 year, P. and T. Ehrenfest used this theory in a problem regarding the macroscopic (i.e., on average) behavior of gas molecules [7]; this way, they were able to settle an apparent conflict between Boltzmann's kinetic theory of gases and thermodynamics, regarding molecule states. In a series of articles [42, 43, 44] in the 1930s, Kolmogorov extended Markov's theory into continuous time and countable space, provided most of the basic definitions and basic equations, and set forth the foundation for a rich theory, which constitutes perhaps the single most important tool that probability theory has to offer for studying system dynamics. Other scientists who made fundamental contributions include Chung [12], Doob [16], and Dynkin [18].

In Section 10.1 we provide the basic definitions and some examples. In the next section we introduce discrete-time Markov Chains, the class of Markov Chains with the richest theory. In Section 10.3 we present conditions and results for the steady-state behavior of a discrete-time Markov Chain. In Section 10.4 we introduce the notion of drifts and present a few results that connect it to the steady-state behavior. In Section 10.5 we discuss continuous-time Markov Chains. We present an application of the theory to a local-area network in some detail in Section 10.6. Finally, we discuss how to generate values of a Markov Chain in Section 10.7.

10.1   DEFINITION AND CLASSIFICATION

Consider a random process $\{X(t),\ t \in I\}$ taking values in a set E. In this chapter, we call E the *state space*. Intuitively, the random process is a Markov process if "its future is independent of the past, when the present is known," in the same sense as for the deterministic system described by

$$x(n + 1) = Ax(n) + Bu(n),$$

when we know $x(n)$ (the "present"), knowledge of $x(n-1), x(n-10)$, etc. (the "past") is not necessary to determine $x(n + 1)$ (the "future"). More precisely, fix an integer $n > 1$ and a sequence of indices $t_0 < t_1 < \cdots < t_{n+1}$ in I.

Definition: The random process $\{X(t),\ t \in I\}$ is a **Markov process** if

$$P[X(t_{n+1}) \leq j | X(t_n) = i, X(t_{n-1}) = i_{n-1}, \ldots, X(t_1) = i_1, X(t_0) = i_0]$$
$$= P[X(t_{n+1}) \leq j | X(t_n) = i]$$

$$(10.1)$$

for any choice of $n, \{t_k\} \in I$, and $i, j, \{i_k\} \in E$.[3]

The definition can be interpreted as follows: if t_n represents the present of the random process, the conditioning event in Equation 10.1 expresses its past and present; the left-hand side is the conditional cdf of the "future" random variable $X(t_{n+1})$. For a Markov process, then, this cdf depends only on the present. In other words, the present "summarizes" the entire history of the process.

Note that Equation 10.1 does not imply that the random variable $X(t_{n+1})$ is independent of any of the random variables $X(t_{n-1}), \ldots, X(t_0)$. What is stipulated is simply *conditional independence, given the value of* $X(t_n)$.

Remark: The definition can be stated in a number of different but equivalent ways involving the pmf or pdf or events other than $\{X(t_{n+1}) \leq j\}$. When the set E is discrete, an equivalent definition in terms of the pmf is based on

$$P[X(t_{n+1}) = j | X(t_n) = i, X(t_{n-1}) = i_{n-1}, \ldots, X(t_1) = i_1, X(t_0) = i_0]$$
$$= P[X(t_{n+1}) = j | X(t_n) = i]$$
(10.2)

for any choice of $n, \{t_k\} \in I$, $i, j, \{i_k\} \in E$. As we will see, these probabilities, called the **transition probabilities**, will play an important role. When the jpdf's exist, another definition can be based on

$$f_{X(t_{n+1})|X(t_n)X(t_{n-1})\cdots X(t_1)X(t_0)}(x|y, i_{n-1}, \ldots, i_1, i_0) = f_{X(t_{n+1})|X(t_n)}(x|y).$$

10.1.1   Classification

There are three major classifications of Markov processes, based on the nature of the state space E, the index set I, and the conditional probabilities $P[X(t_{n+1}) \leq j | X(t_n) = i]$.

When the state space E is a discrete set, the process is called a **Markov Chain (MC)**. The name *chain* is a reminder of the defining property, as Figure 10.1 suggests. Traditionally, the term **Markov process**, which we have used for the generic process, is reserved for the case where the state space is continuous. A Markov Chain or process can be **discrete-time** or **continuous-time**, according to whether the index set I is discrete or continuous. The chain or process is **homogeneous** if the conditional probabilities $P[X(t_{n+1}) \leq j | X(t_n) = i]$ depend on the difference $t_{n+1} - t_n$ only. It is **nonhomogeneous** if these probabilities depend on both t_{n+1} and t_n. Finally, a Markov Chain can be further

---

[3] Provided of course that the conditioning events have nonzero probability; see Examples 10.6 and 10.7 for some subtleties that arise in the contrary case. When the random variables are continuously valued, the definition should be interpreted in a limiting sense, as we did in Section 4.6.2.

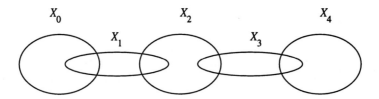

Figure 10.1   The dependencies in a Markov process.

classified as **finite** or **infinite**, based on whether E contains a finite or an infinite number of elements.

Powerful results have been developed for both Markov Chains and Markov processes. However, the mathematical sophistication for Markov processes is beyond the scope of this book. (The interested reader may consult [24].) For this reason, we focus on Markov Chains exclusively; we present results for both types of chains, discrete- and continuous-time, emphasizing mainly homogeneous chains.

10.1.2   Multidimensional pmf

We will see now that Definition 10.2 suffices to characterize the probabilistic behavior of the random process completely.

Theorem 10.1 *Let $p_0(i)$ denote the pmf of the random variable $X(t_0)$. The multidimensional pmf of the Markov Chain $\{X(t)\}$ is given by*

$$P[X(t_{n+1}) = i_{n+1}, X(t_n) = i_n, \ldots, X(t_0) = i_0]$$
$$= p_0(i_0) \cdot \prod_{k=0}^{n} P[X(t_{k+1}) = i_{k+1} | X(t_k) = i_k].$$
$$(10.3)$$

Proof: Using the definition of conditional probabilities, we have

$$P[X(t_{n+1}) = i_{n+1}, X(t_n) = i_n, \ldots, X(t_0) = i_0]$$
$$= P[X(t_{n+1}) = i_{n+1} | X(t_n) = i_n, \ldots, X(t_0) = i_0] \cdot$$
$$P[X(t_n) = i_n, \ldots, X(t_0) = i_0] \stackrel{\triangle}{=} A.$$

We can use now the defining property of the Markov Chain to simplify the conditioning event. We have

$$A = P[X(t_{n+1}) = i_{n+1} | X(t_n) = i_n] \cdot P[X(t_n) = i_n, \ldots, X(t_0) = i_0]. \quad (10.4)$$

Finally, we can apply this procedure recursively to the last probability in Equation 10.4, to get

$$
\begin{aligned}
A &= P[X(t_{n+1}) = i_{n+1}|X(t_n) = i_n] \cdot P[X(t_n) = i_n|X(t_{n-1}) = i_{n-1}] \\
&\quad \cdots P[X(t_1) = i_1|X(t_0) = i_0] \cdot P[X(t_0) = i_0].
\end{aligned}
$$

QED. □

Therefore, an initial pmf and the conditional probabilities of Definition 10.2 suffice to characterize the random process completely. Observe that we were not able to derive a similar result for a generic random process; moreover, make sure you understand how Equation 10.3 simplifies for the IID model.

Given a set of conditional probabilities that satisfy Definition 10.2, is it guaranteed that there exists a Markov random process with those as its conditional pmf's? Theorem 10.1 and Kolmogorov's Theorem 7.1 should provide an answer in the affirmative. Having set aside this theoretical question, we turn now to some examples.

10.1.3 Examples

Example 10.1. *IID Processes.* Consider a sequence of integer-valued, independent, identically distributed random variables $\{X(t_i)\}_{i=1}^{\infty}$. Is it a Markov Chain?

From Definition 10.2, the conditional probability on the left-hand side is

$$
\begin{aligned}
P[X(t_{n+1}) = j|X(t_n) = i, X(t_{n-1}) = i_{n-1}, \dots, X(t_1) = i_1, X(t_0) = i_0] \\
= P[X(t_{n+1}) = j],
\end{aligned}
$$

while the conditional probability on the right-hand side is

$$
P[X(t_{n+1}) = j|X(t_n) = i] = P[X(t_{n+1}) = j].
$$

Thus the sequence is a Markov Chain. We can repeat this argument for a sequence of real-valued, IID random variables, using Definition 10.1. As a consequence, the white noise process in Section 9.2.3 is a Markov process.

Example 10.2. *Independent Increment Processes.* Consider a sequence of integer-valued, IID random variables $\{X(t_i)\}_{i=1}^{\infty}$. Let now $S(t_n) = \sum_{i=1}^{n} X(t_i)$ be the sum process associated with the sequence $\{X(t_i)\}_{i=1}^{\infty}$. Is $\{S(t_i)\}_{i=1}^{\infty}$ a Markov Chain?

From Definition 10.2 again, we get

$$
\begin{aligned}
P[S(t_{n+1}) = j|S(t_n) = i, S(t_{n-1}) = i_{n-1}, \dots, S(t_0) = i_0] \\
= P[S(t_{n+1}) - S(t_n) = j - i|S(t_n) = i, S(t_{n-1}) = i_{n-1}, \dots, S(t_0) = i_0] \\
= P[X(t_{n+1}) = j - i], \quad\quad\quad\quad (10.5)
\end{aligned}
$$

and similarly,

$$P[S(t_{n+1}) = j|S(t_n) = i] = P[X(t_{n+1}) = j - i].$$

Therefore the sequence $\{S(t_i)\}_{i=1}^{\infty}$ is a Markov Chain. As a consequence, the Poisson process in Chapter 8 is a Markov Chain.

Example 10.3. *ARMA Process.* Consider the ARMA process in Section 9.3.4, defined (with a slight change in notation) as

$$X_n = \sum_{k=1}^{M} a_k X_{n-k} + \sum_{l=0}^{L} b_l Y_{n-l}. \tag{10.6}$$

Under what conditions on $\{a_k\}, \{b_l\}, \{Y_m\}, L$ and M is it a Markov process?

Obviously, if $M > 1$, the random variable X_n *does* depend on X_{n-2} and possibly more variables from the sequence $\{X_k\}$. Therefore, for the ARMA process to be Markov, it is necessary that $M = 1$. It should be clear that a_1, being a simple scaling factor, does not change the Markovian nature of the process, so a_1 can be arbitrary. So the real question is, What conditions on $\{Y_m\}$ and L are sufficient to make $\{X_k\}$ a Markov process?

Consider the case $L = 0$ and $\{Y_m\}$ an IID sequence. Then from Equation 10.6, we get

$$P[X_{n+1} = j|X_n = i, X_{n-1} = i_{n-1}, \ldots, X_0 = i_0]$$
$$= P[X_{n+1} - a_1 X_n = j - a_1 \cdot i|X_n = i, X_{n-1} = i_{n-1}, \ldots, X_0 = i_0]$$
$$= P[Y_{n+1} = \frac{j - a_1 i}{b_0}],$$

and similarly,

$$P[X_{n+1} = j|X_n = i] = P\left[Y_{n+1} = \frac{j - a_1 i}{b_0}\right].$$

Thus the sequence is a Markov Chain.

Consider now the case where $L > 0$. It is reasonable to expect that the sequence $\{X_k\}$ loses its Markovian property since the same random variable Y_n will affect now *both* X_n and X_{n+1}. If, for example, Y_n assumes a "huge" value, which makes X_n large, we expect that X_{n+1} might also be large.

To make those ideas a bit more precise mathematically, consider the even simpler case of $a_k = 0, b_0 = b_1 = 1$, and $Y_n \in \{0, 1\}$. We have then

$$X_n = Y_n + Y_{n-1}.$$

Consider the probability $P[X_3 = 1|X_2 = 1, X_1 = 2]$. Suppose that $Y_0 = 1$. The conditioning event is equivalent to the event $\{Y_1 = 1$ and $Y_2 = 0\}$ as we can easily see. Then

$$P[X_3 = 1 | X_2 = 1, X_1 = 2] = P[Y_3 = 1],$$

while $P[X_3 = 1 | X_2 = 1] \neq P[Y_3 = 1]$ necessarily.

To summarize this discussion, the ARMA process is not in general a Markov process, except for special cases.

Example 10.4. *The Process* $X_{n+1} = f_n(X_n) + W_{n+1}$. Let $\{W_i\}_{i=0}^{\infty}$ be an IID sequence. Let $I = \{0, 1, \ldots\}$. Let $X_0 = W_0$. Define a new random process by the nonlinear recursion

$$X_{n+1} = f_n(X_n) + W_{n+1}, \quad n = 0, 1, \ldots, \tag{10.7}$$

where $f_n(\cdot)$ is a nonlinear function. Is the sequence $\{X_i\}_{i=1}^{\infty}$ a Markov process?

We can check again the definition directly; however, since we know of the functional dependence between X_{n+1} and X_n, we can supply another argument as well. From Equation 10.7, we see that X_{n+1} depends on two random variables only; so if we show that W_{n+1} does not depend on $X_{n-1}, \ldots, X_0$, the sequence will be a Markov process.

Now, X_0 depends on W_0 only. From Equation 10.7, we see that X_1 depends on W_0 and W_1 only, and in general, X_l, for $l \leq n$, depends on $W_0, \ldots, W_l$ only. Therefore, W_{n+1} is independent of $X_{n-1}, \ldots, X_0$, and the sequence $\{X_i\}_{i=1}^{\infty}$ is a Markov process.

As a special case, the process $\{Y_i\}_{i=1}^{\infty}$ in Equation 9.30 associated with the Kalman filter in Section 9.5 is a Markov process.

Example 10.5. *Student GPA.* Consider the sequence of random variables $\{G_i\}_{i=1}^{8}$, where G_i describes a student's grade in semester i.[4] Define

$$X_k = \frac{1}{k} \sum_{i=1}^{k} G_i, \quad k = 1, \ldots, 8, \tag{10.8}$$

as the student's overall GPA in semester k. Is $\{X_i\}_{i=1}^{8}$ a MC?

We can rewrite Equation 10.8 as

$$X_k = \frac{k-1}{k} X_{k-1} + \frac{1}{k} G_k,$$

from which we see that $\{X_k\}$ is a special case of the process in Example 10.4. However, are the $\{G_i\}$ IID? We can easily argue that they are not, so the GPA sequence is not a Markov Chain. (If you did well in the last two semesters, chances are high you will do well in the future, too.)

---

[4]The implicit assumption here is that the student is an undergraduate; moreover, we optimistically assume that the student will take eight semesters to graduate.

Example 10.6. *Deterministic Functions of Time.* Consider an integer-valued deterministic function of time $f(t)$ and an arbitrary sample space S. Define the random process $\{X(t)\}$ as follows:

$$X(t, \zeta) = f(t), \quad \forall \zeta \in S;$$

in other words, all sample paths are deterministic. Is $\{X(t)\}$ a Markov process? We have

$$P[X(t_{n+1}) = j | X(t_n) = i] = \begin{cases} 1, & i = f(t_n), \ j = f(t_{n+1}), \\ 0, & \text{otherwise.} \end{cases}$$

On the other hand, the probability $P[X(t_{n+1}) = j | X(t_n) = i, X(t_{n-1}) = k]$ is not defined for $k \neq f(t_{n-1})$, since the conditioning event has zero probability. Therefore the process is not Markovian, since the probabilities in Definition 10.2 are not equal.

Example 10.7. *A Random Process with Sinusoidal Sample Paths.* The trouble we had in the previous example cannot necessarily be attributed to the deterministic nature of the sample paths. Consider the sinusoid random process in Chapter 7:

$$X(t, \zeta) = \zeta \sin(2\pi t), \quad \zeta \in [0, 1], \quad t \in I,$$

where $I = [0, \infty)$ and ζ is uniformly distributed in $[0, 1]$. Can you argue whether the process $\{X(t), \ t \in I\}$ is Markov?

Example 10.8. *Buffer Contents in Network Nodes.* Consider the buffer example in Section 1.9. Define the integer-valued, continuous-time random process $\{N(t), \ t \in [0, \infty)\}$ as the number of packets being stored in the buffer's memory and waiting for transmission at time t. Let $\{A_k, k = 1, 2, \ldots\}$ be the instant the kth packet arrives at the node buffer; let $\tilde{S}_k$ be the size of this packet, in bits. The transmission time S_k for this packet is equal to $\tilde{S}_k / C$ seconds, where C is the line capacity (in bits per second). And $N(t)$ will change value at time instants when a packet arrives or completes transmission. Denote the sequence of time instants when a change occurs as $\{t_l, \ l = 1, 2, \ldots\}$. For example, in Figure 10.2 we have

$$t_1 = A_1, \quad t_2 = t_1 + A_2, \quad t_3 = t_1 + S_1, \quad t_4 = t_3 + S_2,$$

and so on. Consider the sequence of random variables

$$N(t_{k+1}) = N(t_k) + \xi_k,$$

where

$$\xi_k \triangleq \begin{cases} +1, & k\text{th change is due to an arrival,} \\ -1, & k\text{th change is due to a transmission completion.} \end{cases}$$

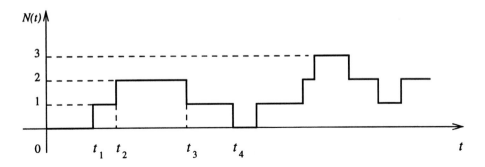

Figure 10.2 Buffer-size sample path.

Therefore, the sequence is a sum process. Under what conditions is the sequence $\{N(t_k),\ k = 1, 2, \ldots\}$ a Markov Chain? Clearly, $\{\xi_k,\ k = 1, 2, \ldots\}$ being IID is a sufficient condition. As a matter of fact, independence only is sufficient. All we need to show is that the random variable ξ_k depends only on $N(t_k)$ [and not, say, on $N(t_{k-1})$ or $N(t_{k+1})$ etc.].

Let's take a closer look at this sequence. The random variable ξ_k depends on the arrival process, the packet size sequence, and the buffer contents at time t_k. Suppose that the arrival process is Poisson. Suppose that packet size is an IID exponential sequence. Consider the definition of ξ_k. The arrival that constitutes the kth change is independent of all arrivals that determined $N(t_k)$, since the arrival process has independent increments. Whether the kth event is a transmission completion depends on two random variables. First, if $N(t_k) = 0$, the event cannot be a transmission completion. Second, given that $N(t_k) > 0$, $\xi_k = -1$ will depend on a packet size S_n. Since the packet size sequence is assumed IID, we conclude that ξ_k can only depend on $N(t_k)$, and thus the sequence $\{N(t_k),\ k = 1, 2, \ldots\}$ is a Markov Chain.[5]

10.2   DISCRETE-TIME MARKOV CHAINS

Recall from our classification in Section 10.1 that for a discrete-time Markov Chain (DTMC) both the state space E and the index set I are discrete. Since any discrete set can be mapped onto the set of integers (see Problem 10.1), the elements of the state space can be relabeled as integers. This relabeling gives us significant savings in notation. From now on, therefore, we assume, without

---

[5] The time instants $\{t_k\}$ that define the random variables $N(t)$ are random themselves and determined by the MC. What we have used here is the **strong Markov property**. For more details on this property and its ramifications, see reference [38].

loss of generality, that $E = \{\ldots, -2, -1, 0, 1, 2, \ldots\}$; occasionally, we assume that $E = \{0, 1, 2, \ldots\}$ as well. Since the index set is discrete as well, we can further assume that $I = \{0, 1, 2, \ldots\}$, that is, $t_i = i$, $i = 0, 1, \ldots$. We can then write X_n for the random variable $X(t_n)$. With these simplifications, we can restate the definition of a DTMC as follows:

Definition: The random process $\{X_n, \ n = 0, 1, \ldots\}$ is a **discrete-time Markov Chain** if

$$P[X_{n+1} = j | X_n = i, X_{n-1} = i_{n-1}, \ldots, X_0 = i_0] = P[X_{n+1} = j | X_n = i]$$
$$(10.9)$$

for any choice of n and $i, j, \{i_k\} \in E$. Let's define a matrix $P(n)$, the ijth element of which is given by

$$P_{ij}(n) \triangleq P[X_{n+1} = j | X_n = i].$$

The argument n denotes that the Markov Chain may be nonhomogeneous.[6] When the chain is homogeneous, it is customary to drop the argument n and denote the matrix as P.

Definition: The matrix $P(n) = \{P_{ij}(n)\}$ is called the **one-step transition matrix** (or simply the **transition matrix**) of the DTMC.

$$P(n) = \begin{bmatrix} \cdots & \cdots & & \cdots & & \cdots & \cdots & & \cdots \\ \cdots & P_{00}(n) & P_{01}(n) & \cdots & P_{0l}(n) & \cdots \\ \cdots & P_{10}(n) & P_{11}(n) & \cdots & P_{1l}(n) & \cdots \\ \cdots & \cdots & & \cdots & & \cdots & \cdots \\ \cdots & P_{l0}(n) & P_{l1}(n) & \cdots & P_{ll}(n) & \cdots \\ \cdots & \cdots & & \cdots & & \cdots & \cdots \end{bmatrix}.$$

The exact dimensionality of the matrix depends on the cardinality of the state space E. If E contains an infinite number of elements, then the matrix will be infinite-dimensional. We look at some specific examples in later sections.

As we saw, the conditional probabilities $P[X_{n+1} = j | X_n = i]$ in Equation 10.9 provide a complete characterization of the DTMC. Therefore, the transition matrix $P(n)$ is all we have to provide in order to completely characterize the DTMC.[7]

An equivalent but graphical way of displaying the contents of the transition matrix is a **state transition diagram**. In such a diagram, the elements of E

---

[6] It is unfortunate that we use the symbol P both for denoting the generic probability assignment in our probability model $(S, \mathcal{Q}, P)$ *and* as the matrix P. Hopefully, no confusion will arise, and which is which will be clear from the context.

[7] To fully appreciate the notational advantage of dealing with integers instead of an arbitrary E, try to identify what would replace $P_{ij}(n)$ in the general case.

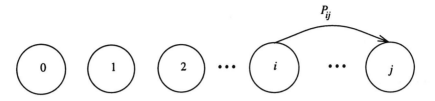

Figure 10.3 The state transition diagram.

are depicted as circles, and the element $P_{ij}(n)$ of the transition matrix $P(n)$ is depicted as a label on the directed arc connecting state i to state j. Figure 10.3 depicts this idea.

10.2.1 Various Probabilities Related to a DTMC

There are a number of probabilities related to a DTMC, which we are interested in calculating. Then all can, in principle, be evaluated from the pmf in Equation 10.3. We derive here explicit formulas in terms of the defining transition matrix elements, for the following probabilities:

1. The joint pmf $P[X_n = i_n, X_{n-1} = i_{n-1}, \ldots, X_1 = i_1, X_0 = i_0]$

2. The marginal pmf $P[X_n = j]$

3. The conditional pmf $P[X_n = j | X_0 = i]$, also known as the n-step transition probability

4. The limiting pmf $\lim_{n \to \infty} P[X_n = j]$

5. The limiting pmf $\lim_{n \to \infty} P[X_n = j | X_0 = i]$

1. $P[X_n = i_n, X_{n-1} = i_{n-1}, \ldots, X_1 = i_1, X_0 = i_0]$

This is the fundamental probability, from which everything else follows. We derived it already in Equation 10.3. We restate it now in terms of the transition matrix elements.

$$P[X_n = i_n, X_{n-1} = i_{n-1}, \ldots, X_1 = i_1, X_0 = i_0]$$
$$= p_0(i_0) \cdot \prod_{k=0}^{n-1} P[X_{k+1} = i_{k+1} | X_k = i_k] = p_0(i_0) \cdot \prod_{k=0}^{n-1} P_{i_k i_{k+1}}(k).$$
$$(10.10)$$

When the MC is homogeneous, Equation 10.10 simplifies slightly. We state its new form as a lemma.

Lemma 10.1 *For a homogeneous MC,*

$$P[X_n = i_n, X_{n-1} = i_{n-1}, \ldots, X_1 = i_1, X_0 = i_0] = p_0(i_0) \cdot \prod_{k=0}^{n} P_{i_k i_{k+1}}.$$

2. Marginal pmf $P[X_n = j]$

This probability shows us how the initial probability assignment $P[X_0 = i_0]$ has evolved after n time steps or transitions. It expresses, and is useful in studies of, the "transient behavior" of the process $\{X_n\}$. We can express it as follows:

Theorem 10.2 *For any $n \geq 1$, we have*

$$P[X_n = j]$$
$$= \sum_{i_{n-1}} \sum_{i_{n-2}} \cdots \sum_{i_0} P[X_0 = i_0] P_{i_{n-1}j}(n-1) P_{i_{n-2}i_{n-1}}(n-2) \cdots P_{i_0 i_1}(0).$$

$$(10.11)$$

Proof: Of course, $P[X_n = j]$ is the marginal pmf of the multidimensional pmf $P[X_n = i_n, X_{n-1} = i_{n-1}, \ldots, X_1 = i_1, X_0 = i_0]$. Therefore, we can express it as follows:

$$P[X_n = j] = \sum_{i_{n-1}} \sum_{i_{n-2}} \cdots \sum_{i_0} P[X_n = j, X_{n-1} = i_{n-1}, \ldots, X_0 = i_0].$$

The result now follows from Equation 10.10.   □

Needless to say, the computational effort required to calculate $P[X_n = j]$, as n gets larger and larger, may be prohibitive. A recursive expression is also possible:

Theorem 10.3 *For any $n \geq 1$, we have*

$$P[X_n = j] = \sum_{i \in E} P_{ij}(n-1) P[X_{n-1} = i].   (10.12)$$

Proof: From the total probability theorem (see Equation 2.10) we have

$$P[X_n = j] = \sum_{i \in E} P[X_n = j | X_{n-1} = i] \cdot P[X_{n-1} = i],$$

and Equation 10.12 follows.   □

Can you say whether Equation 10.12 is faster to compute than Equation 10.11?

When the chain is homogeneous, the above two theorems can be specialized further. Let P^n denote the nth power of the matrix P, that is, $P^n = P \cdot P \cdots P$ (product of n matrices). Let $p(n) \triangleq (\ldots, P[X_n = j], \ldots)$ be a row vector that contains all the probabilities $P[X_n = j]$.

Theorem 10.4 *Consider a homogeneous MC. For any $n \geq 1$, we have*

$$p(n) = p(0) \cdot P^n.$$

Proof: We can interpret Equation 10.12 as a multiplication of the jth column of the matrix $P(n-1)$ $(= P$ in this case) and the vector $p(n-1)$. The result follows by induction on n. $\qquad\square$

3. Conditional pmf $P[X_n = j | X_0 = i]$

This probability shows us how the initial state $X_0 = i$ affects the state evolution after n time steps or transitions. It is another measure of the transient behavior of the process $\{X_n\}$.

Theorem 10.5 *For any $n \geq 1$, we have*

$$P[X_n = j | X_0 = i] = \sum_{i_{n-1}} \sum_{i_{n-2}} \cdots \sum_{i_1} P_{i_{n-1}j}(n-1) P_{i_{n-2}i_{n-1}}(n-2) \cdots P_{ii_1}(0).$$

Proof: The theorem follows from Equation 10.11, upon conditioning on the event $\{X_0 = i\}$. Observe that from the Markov property, this conditioning does not affect any probability $P_{kl}(m)$, $m > 0$. Moreover,

$$P[X_0 = i_0 | X_0 = i] = \begin{cases} 1, & i = i_0, \\ 0, & i \neq i_0. \end{cases}$$

QED. $\qquad\square$

There is a recursive version of Theorem 10.5 as well:

Theorem 10.6 *For any $n \geq 1$, we have*

$$P[X_n = j | X_0 = i] = \sum_{l \in E} P_{lj}(n-1) P[X_{n-1} = l | X_0 = i].$$

Proof: The theorem follows from Equation 10.12, upon conditioning on the event $\{X_0 = i\}$. Observe that from the Markov property, this conditioning does not affect the transition probabilities on the right-hand side of Equation 10.12. $\qquad\square$

does not affect the transition probabilities on the right-hand side of Equation 10.12. □

Another form for the probability $P[X_n = j|X_0 = i]$ is possible when we condition on the event $\{X_r = l\}$, for some arbitrary $l \in E$ and $0 < r < n$. To simplify our notation a bit, we consider only the homogeneous case here and leave the general case as an exercise.

For an arbitrary $m \in I$, let's define[8] $P_{ij}^n \triangleq P[X_{n+m} = j|X_m = i]$. Then we have the following:

Theorem 10.7 *For any $n \geq 1$,*

$$P_{ij}^n = \sum_{l \in E} P_{il}^{n_1} P_{lj}^{n_2}, \tag{10.13}$$

where the positive integers n_1, n_2 satisfy $n_1 + n_2 = n$.

Proof: Fix an integer n_1 such that $n > n_1 > 0$. Consider the event $\{X_{n_1} = l\}$. From the total probability theorem, we have

$$
\begin{aligned}
P_{ij}^n &= P[X_n = j|X_0 = i] \\
&= \sum_{l \in E} P[X_n = j|X_0 = i, X_{n_1} = l] \cdot P[X_{n_1} = l|X_0 = i] \\
&= \sum_{l \in E} P[X_n = j|X_{n_1} = l] \cdot P[X_{n_1} = l|X_0 = i].
\end{aligned}
$$

Equation 10.13 now follows easily from the new notation we introduced.   □

Equation 10.13 is one form of the **Chapman-Kolmogorov equations,** specialized here for the case of a DTMC. We see another form of those equations in Section 10.5.5.

4. Limiting pmf $\lim_{n \to \infty} P[X_n = j]$

This probability expresses our intuitive notion of the "steady state" or "equilibrium" of a dynamic system.

Suppose that the MC is homogeneous, with transition matrix P. Assume for the moment that the limit exists for all $j \in E$, and let $\pi_j \triangleq \lim_{n \to \infty} P[X_n = j]$. We derive conditions under which this limit exists in the next section. Let π be a row vector with probability π_j as its jth element.

Theorem 10.8 *The limiting pmf satisfies the system of linear equations*

$$\pi = \pi P.$$

---

[8]Here n is a superscript, *not* a power symbol.

Proof: From Equation 10.12 we have

$$P[X_n = j] = \sum_{i \in E} P_{ij} P[X_{n-1} = i];$$

therefore

$$\lim_{n \to \infty} P[X_n = j] = \lim_{n \to \infty} \sum_{i \in E} P_{ij} P[X_{n-1} = i].$$

Assuming that we can interchange the sum and limit operators, we get

$$\pi_j = \sum_{i \in E} P_{ij} \lim_{n \to \infty} P[X_{n-1} = i],$$

and the result follows. □

5. *Limiting pmf* $\lim_{n \to \infty} P[X_n = j | X_0 = i]$

This probability is another measure of the steady-state behavior of the DTMC. Intuitively, if a system reaches steady state, the effects of any initial conditions should "die out." We expect then that $\lim_{n \to \infty} P[X_n = j | X_0 = i]$ is independent of i. For a proof of this fact, see [38].

Theorem 10.9 *The limiting conditional pmf satisfies the equation*

$$\lim_{n \to \infty} P[X_n = j | X_0 = i] = \pi_j.$$

We address the two issues of existence of limits and of interchanging the limit and sum operators, in a more rigorous fashion, in the following section; there we derive conditions for a homogeneous MC to reach steady state.

10.3 STEADY STATE OF MARKOV CHAINS

To better appreciate the subject of a random process approaching steady state, consider the following examples.

Example 10.9. *Steady-State Behavior.* Consider a Markov Chain with two states $\{0, 1\}$, transition matrix

$$P = \begin{bmatrix} 0.5 & 0.5 \\ 0.5 & 0.5 \end{bmatrix}, \tag{10.14}$$

and $I = \{0, 1, 2, \ldots\}$. From Theorem 10.3, for all n and $j = 0, 1$,

$$\begin{aligned} P[X_n = j] &= 0.5 P[X_{n-1} = 1] + 0.5 P[X_{n-1} = 0] \\ &= 0.5 \cdot (P[X_{n-1} = 1] + P[X_{n-1} = 0]) = 0.5. \end{aligned}$$

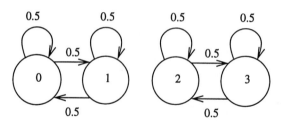

Figure 10.4   Initial-condition effects.

Therefore, regardless of the initial probability distribution $P[X_0 = j]$, the probability $P[X_n = j]$ does not change for $n > 0$. This chain exhibits steady-state behavior in the sense that $\lim_{n \to \infty} P[X_n = j]$ exists.

Example 10.10. *Oscillatory Behavior.* Consider now the same MC, but with transition matrix

$$P = \begin{bmatrix} 0 & 1 \\ 1 & 0 \end{bmatrix}.$$ (10.15)

Consider the conditional probability $P[X_n = 1 | X_0 = 1]$. It is intuitively clear that X_n can only be in state 1 when n is even. Therefore,

$$P[X_n = 1 | X_0 = 1] = \begin{cases} 1, & n \text{ even,} \\ 0, & n \text{ odd.} \end{cases}$$

Since this probability oscillates, the limit $\lim_{n \to \infty} P[X_n = j | X_0 = 1]$ cannot exist. This chain does not reach steady state.

Example 10.11. *Initial Condition Effects Never Die Out.* Consider a four-state MC with transition matrix

$$P = \begin{bmatrix} 0.5 & 0.5 & 0 & 0 \\ 0.5 & 0.5 & 0 & 0 \\ 0 & 0 & 0.5 & 0.5 \\ 0 & 0 & 0.5 & 0.5 \end{bmatrix}$$

(see Figure 10.4).

It is clear that when $X_0 = 3$, $P[X_n = 1 | X_0 = 3] = 0$, while when $X_0 = 0$, $P[X_n = 1 | X_0 = 0] = 0.5$. Therefore, the initial condition *does* affect the limiting probabilities of this MC.

Example 10.12. *Divergence.* Consider two independent Markov Chains $\{X_n\}$ and $\{Y_n\}$ for which steady state exists. Let π_j^X, π_j^Y denote the steady-state probabilities for the two MCs. Define a new Markov Chain Z_n (check that Z_n is indeed a MC) as follows:

$$Z_n = \begin{cases} X_n, & 2^{2k} \le n \le 2^{2k+1} - 1, & \text{for some } k \ge 0, \\ Y_n, & 2^{2k+1} \le n \le 2^{2k+2} - 1, & \text{for some } k \ge 0. \end{cases}$$

To get a feel for this construction, spell out Z_1 through Z_{64}.

Consider the limit $\lim_{n \to \infty} P[Z_n = j]$ along the subsequence $n_m = 2^{2m+1} - 1$, $m = 0, 1, \ldots$. From the definition of Z_n, we have

$$\lim_{m \to \infty} P[Z_{n_m} = j] = \lim_{m \to \infty} P[X_{n_m} = j] = \pi_j^X .$$

Consider now the limit $\lim_{n \to \infty} P[Z_n = j]$ along a different subsequence $l_m = 2^{2m+2} - 1$, $m = 0, 1, \ldots$. From the definition of Z_n, we have

$$\lim_{m \to \infty} P[Z_{l_m} = j] = \lim_{m \to \infty} P[Y_{l_m} = j] = \pi_j^Y .$$

Therefore, the sequence $\lim_{n \to \infty} P[Z_n = j]$ cannot converge, since it approaches two different limits along two subsequences.

10.3.1 State Classification

It is apparent that for steady-state results we need to rule out behavior such as the one described in Examples 10.10–10.12. The following state classification will help us to do so. Recall that we have restricted our attention to a homogeneous MC. Also, recall that for such a chain $P_{ij}^n \triangleq P[X_{n+k} = j | X_k = i]$, for all k.

Definition: State j is **accessible** from state i if $P_{ij}^n > 0$ for at least one $n > 0$.

Intuitively, since $P_{ij}^n \geq 0$ for *all* n, this definition asserts that it is possible to reach state j in a *finite* number of steps, provided that we start at time 0 in state i. It should be clear from the definition that, if for a given MC, there is a state j which is not accessible from state i, then this MC cannot reach steady state.

Example 10.13. The MC in Figure 10.4 is an example of a MC that cannot reach steady state.

Definition: State j **communicates with** state i if both states are accessible from each other.

We write $i \longleftrightarrow j$ to denote that state j communicates with state i. The relation $\longleftrightarrow$ is an equivalence relation (see Problem 10.47); therefore it partitions the state space E into equivalence classes, called *communicating classes*.

Definition: The MC is **irreducible** if it contains only one communicating class. Otherwise, it is called **reducible**.

It is apparent that a reducible MC fails to reach steady state.

Example 10.14. We can easily check that the MC in Figure 10.4 is reducible since it contains two communicating classes.

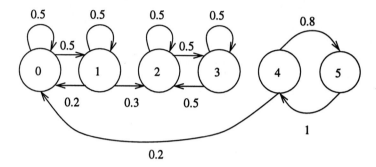

Figure 10.5   A reducible MC.

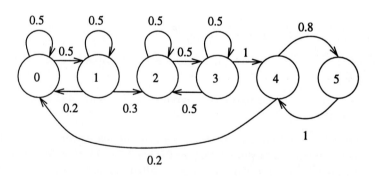

Figure 10.6   An irreducible MC.

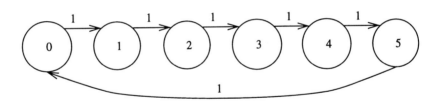

Figure 10.7 A periodic MC.

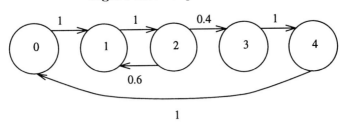

Figure 10.8 An aperiodic MC.

Example 10.15. Figure 10.5 depicts a reducible MC.

Example 10.16. Figure 10.6 depicts an irreducible MC.

Consider the probability P_{jj}^n, that is, the probability that the MC will "return to state j" after n steps. Let

$$I_j \triangleq \{n \in I :\ P_{jj}^n > 0\}$$

be the set of times that the MC can return to state j, given that it started at state j at time $n = 0$.

Definition: The **period** $d(j)$ of state j is the greatest common divisor of the elements of set I_j.

If it happens that $P_{jj}^n = 0$, for all n, then $I_j = \emptyset$, and we define the period of state j as 0.

Definition: The MC is **aperiodic** if all states have period equal to 1. The MC is **periodic** if all states have period greater than 1.

Example 10.17. The MC in Equation 10.15 is not aperiodic; thus in general steady state is not guaranteed for such a MC.

Example 10.18. Figure 10.7 depicts a periodic MC.

Example 10.19. Figure 10.8 depicts an aperiodic MC.

From the above definitions, it is apparent that irreducibility and aperiodicity are necessary conditions for existence of steady state.

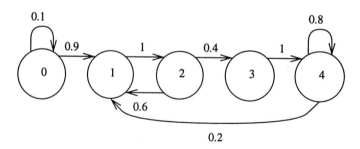

Figure 10.9   Transient and recurrent states.

1. Criteria for irreducibility and aperiodicity

How do we establish irreducibility and aperiodicity of a given MC?

Irreducibility must be established on a state-by-state basis since we must show that all states belong to the same communicating class. It is best checked by *inspection*, via a state transition diagram. The following lemma (which we state without proof, see [38]) enables us to examine the period of a single state only, when checking for aperiodicity. It thus reduces the amount of effort we have to expend in order to establish this property:

Lemma 10.2 *All states in a communicating class have the same period.*

Definition: State j is **recurrent** if the MC returns to it infinitely often (or with probability 1). Otherwise, the state is **transient**.

The chain is called recurrent (or transient) if all states are recurrent (transient). Intuitively, a transient state will have zero probability in steady state, since the event $\{X_n = j\}$ will occur only a finite number of times.

Example 10.20.   State 0 of the chain in Figure 10.9 is transient. What can you say about the other states?

Are all states of the MC of Equations 10.14 and 10.15 recurrent? The answer is given by the criterion we present next.

2. Criterion for recurrency

From Theorem 6.13 the event

$$A \triangleq \{X_n = j \text{ occurs infinitely often}\}$$

has probability $P[A] = 1$ if

$$\sum_{n=1}^{\infty} P_{jj}^n = \infty$$

and $P[A] = 0$ if

$$\sum_{n=1}^{\infty} P_{jj}^n < \infty.$$

Therefore, we can use the following theorem as a criterion to establish the recurrent or transient nature of a state.

Theorem 10.10 *State j is recurrent if*

$$\sum_{n=1}^{\infty} P_{jj}^n = \infty,$$

and state j is transient if

$$\sum_{n=1}^{\infty} P_{jj}^n < \infty.$$

Example 10.21. With the aid of this theorem, we can see that both states of the MC in Equation 10.14 are recurrent.

The next theorem simplifies our job of establishing recurrency for an entire chain.

Theorem 10.11 *All states in a communicating class are either recurrent or transient.*

Proof: Consider two communicating states i and j, and suppose that state j is recurrent. Since $i \longleftrightarrow j$, there exist finite integers n and m such that

$$P_{ij}^n > 0, \quad P_{ji}^m > 0.$$

From Theorem 10.7, for any integer $l > 0$,

$$P_{ii}^{n+l+m} \geq P_{ij}^n \cdot P_{jj}^l \cdot P_{ji}^m$$

(verify that). Therefore,

$$\sum_{l=m+n+1}^{\infty} P_{ii}^l \geq P_{ij}^n P_{ji}^m \sum_{l=1}^{\infty} P_{jj}^l. \tag{10.16}$$

We can now easily see from Theorem 10.10 that if state j is recurrent, so is state i, since the sum on the left-hand side of Inequality 10.16 must be equal to ∞. Suppose now that state i is transient, and thus the sum on the left-hand side of 10.16 is finite. Then the sum $\sum_{l=1}^{\infty} P_{jj}^l$ must be also finite, and state j is also transient. $\qquad\square$

10.3.2   Limiting Probabilities

We are ready now to state the fundamental result regarding steady-state behavior of a DTMC. Its rigorous proof is a bit lengthy and is beyond the scope of this book. Consult [38] for a detailed proof.

Theorem 10.12 *If the DTMC is irreducible and aperiodic, then the limit*

$$\pi_j \overset{\triangle}{=} \lim_{n \to \infty} P[X_n = j]$$

exists for all $j \in E$. Moreover, for all $i \in E$,

$$\lim_{n \to \infty} P[X_n = j | X_0 = i] = \pi_j. \tag{10.17}$$

Remark: The above theorem does not preclude the possibility that $\pi_j = 0$. The next theorem sharpens the result of Theorem 10.12. Its proof can also be found in [38].

Theorem 10.13 *If the DTMC is irreducible, aperiodic, and positive recurrent, then the system*

$$\pi_j = \sum_{i \in E} P_{ij} \pi_i, \quad \sum_{i \in E} \pi_i = 1 \tag{10.18}$$

has a unique solution with $\pi_j > 0$, $\forall j \in E$.

When the chain is periodic, we have seen with a counterexample that the limit $\lim_{n \to \infty} P[X_n = j]$ may not exist. However, a limit along a suitable subsequence exists, as the following theorem asserts.

Theorem 10.14 *Consider an irreducible, periodic, and positive recurrent DTMC with period d. Then*

$$\lim_{n \to \infty} \frac{1}{d} P[X_{nd} = j | X_0 = j] = \pi_j,$$

where $\{\pi_j\}$ is the solution to the system

$$\pi_j = \sum_{i \in E} P_{ij} \pi_i, \quad \sum_{i \in E} \pi_i = 1.$$

The quantity π_j can be loosely interpreted as the steady-state probability of state j, even though it is not so. As we see in the next section, it can be better interpreted as the "fraction of time the MC spends in state j."

10.3.3 Ergodicity of MC

In Chapter 7, we saw that a random process is ergodic if time averages of sample paths approach statistical means, as time becomes arbitrarily large. We see now that an irreducible, aperiodic, and positive recurrent MC is ergodic, in the sense that the limit

$$\lim_{n \to \infty} \frac{1}{n} \sum_{i=1}^{n} X_i$$

is equal to a constant value (with probability 1). Intuitively, for any finite $M > 0$, the contribution of the first M random variables in this sum is negligible since the sum $\sum_{i=1}^{M} X_i$ is divided by the arbitrarily large n. Now, for M sufficiently large, the pmf of the random variables involved in the sum

$$\sum_{i=M+1}^{n} X_i \tag{10.19}$$

should not be "very different" from the limiting pmf π_j. Let X_∞ denote a random variable whose pmf is equal to the steady-state pmf $\{\pi_j\}$. The X_i in the sum of 10.19 are therefore "almost" identically distributed to X_∞. What is surprising is that even though we know that the X_i are *not independent*, a result similar to the Strong Law of Large Numbers holds for a MC as well!

In order to make these intuitive statements more precise, we need to define two new concepts first, namely, the *sojourn* and *return times*.

Definition: The **kth sojourn time** in state j, denoted by O_k^j, $k = 1, 2, \ldots$, is the amount of time the MC spends in state j at its kth return to state j.

Example 10.22. Suppose that at time $n = 12$, the MC entered state j for the third time. Then if

$$X_{12} = j, \quad X_{13} = j, \quad X_{14} = j, \quad X_{15} = j, \quad X_{16} = j, \quad X_{17} = j + 100,$$

we have $O_3^j = 4$.

Definition: The **kth return time** to state j, denoted by R_k^j, $k = 1, 2, \ldots$, is the amount of time that elapses between the $k - 1$st and kth returns to state j.

Example 10.23. Suppose that at time $n = 12$ the MC returned to state j for the third time. Then if

$$X_{12} = j, \quad X_{13} = j+1, \quad X_{14} = j+20, \quad X_{15} = j-3, \quad X_{16} = j+100, \quad X_{17} = j,$$

we have $R_4^j = 5$.

Example 10.24. Based on the sample path shown in Figure 10.10, we have

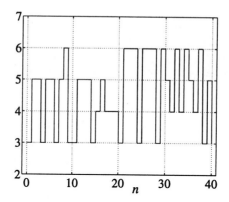

Figure 10.10   A typical sample path.

$$R_1^4 = 3, \quad R_2^4 = 7, \quad R_3^4 = 4, \quad R_4^4 = 7,$$
$$O_1^3 = 0, \quad O_2^1 = 2, \quad O_3^3 = 2, \quad O_4^3 = 0.$$

We can characterize the pmf of these variables, as the next two lemmas specify. To avoid trivialities, we assume that the MC is irreducible and recurrent.

Lemma 10.3 *For a fixed $j \in E$, the sequence $\{O_k^j, \ k = 1, 2, \ldots\}$ is an IID sequence of geometric random variables with parameter P_{jj}.*

Proof: Fix an integer $k > 0$, and suppose that for some $n < \infty$, the MC enters state j for the kth time at time n; therefore, $X_n = j$. Then for any $l = 1, 2, \ldots,$ we have

$$P[O_k^j = l] = P[X_{n+l+1} \neq j, X_{n+l} = j, X_{n+l-1} = j, \ldots, X_{n+1} = j | X_n = j].$$

It is important here to realize that $P[O_k^j = l]$, by the definition of the random variable O_k^j, is a conditional probability. Then from the Markov property and the chain rule for probabilities (see Problem 2.90), we can easily get

$$P[O_k^j = l] = P[X_{n+l+1} \neq j | X_{n+l} = j] P[X_{n+l} = j | X_{n+l-1} = j]$$
$$\cdots P[X_{n+1} = j | X_n = j],$$

and since the MC is assumed homogeneous,

$$P[O_k^j = l] = (1 - P_{jj}) P_{jj}^l.$$

This shows that the random variable O_k^j has a geometric pmf, with parameter P_{jj}. Observe that since the transition probabilities do not depend on time,

$P[O_k^j = l]$ does not depend on k either. Therefore, the random variables $\{O_k^j, k = 1, 2, \ldots\}$ are identically distributed.

To see why they are independent as well, consider two indices k and k', with $k > k'$. For fixed l, l', consider the conditional probability

$$p \triangleq P[O_k^j = l | O_{k'}^j = l'].\tag{10.20}$$

Let's define

$$A \triangleq \{X_{n+l+1} \neq j, X_{n+l} = j, \ldots, X_{n+1} = j\},$$

$$B \triangleq \{X_{n'+m+1} \neq j, X_{n'+m} = j, \ldots, X_{m+1} = j\},$$

with $n > m + n'$. We can rewrite the probability in Equation 10.20 as

$$p = P[A | X_n = j, B, X_m = j].$$

From the Markovian property, we have

$$p = P[A | X_n = j] = P[O_k^j = l],$$

and thus the two random variables are independent. We can show in a similar fashion (but with a tremendous burden in notation) that any N random variables from the collection $\{O_k^j, k = 1, 2, \ldots\}$ are independent. The important point is that given that $X_n = j$, we can ignore conditioning on any other random variable X_k, with $k < n$. □

In the above proof, we assumed that there are an infinite number of random variables in the sequence $\{O_k^j\}$. The recurrence of the MC guarantees that any recurrent state j will be visited infinitely often. If a state is not recurrent, then the sequence contains a finite number of random variables only.

It turns out that the return times are also IID; however, their pmf does not have a simple characterization.

Lemma 10.4 *For a fixed $j \in E$, the sequence $\{R_k^j, k = 1, 2, \ldots\}$ is an IID sequence of random variables.*

Proof: The proof is quite similar to that of Lemma 10.3. Fix an integer $k > 0$, and let $n < \infty$ be the time the MC returns to state j for the $k - 1$st time ($n < \infty$ is possible, since the MC is recurrent). Then, for any $l = 1, 2, \ldots$, we have

$$P[R_k^j = l] = P[X_{n+l} = j, X_{n+l-1} \neq j, \ldots, X_{n+1} \neq j | X_n = j].\tag{10.21}$$

The right-hand side of Equation 10.21 can be expressed in terms of conditional probabilities only. Indeed, we can write the joint probability as

$$P[R_k^j = l] =$$
$$\sum_{i_{n+l-1} \neq j} \cdots \sum_{i_{n+1} \neq j} P[X_{n+l} = j, X_{n+l-1} = i_{n+l-1}, \ldots, X_{n+1} = i_{n+1} | X_n = j],$$

$$(10.22)$$

and each term in Equation 10.22 can now be expressed in terms of the transition matrix elements P_{ij}. Since these elements are not time-dependent, the sequence $\{R_k^j, \ k = 1, 2, \ldots\}$ is identically distributed. Independence can be shown in a manner similar to that used to show independence of O_k^j and $O_{k'}^j$, so we do not repeat it here.   □

We are now ready to state the main result of this section. Its rigorous proof is again outside the scope of this book; the interested reader can consult [7, 38].

Theorem 10.15 *Consider an irreducible, aperiodic, and positive recurrent MC. Then the MC is ergodic in the almost-sure sense, i.e.,*

$$\lim_{n \to \infty} \frac{1}{n} \sum_{k=1}^{n} X_i = EX_\infty \triangleq \sum_{j \in E} j\pi_j, \quad \text{almost surely.}$$

10.3.4   Examples

Example 10.25. *IID Random Processes.* Consider a sequence of independent, identically distributed random variables $\{X_i\}_{i=1}^{\infty}$. Suppose that the random variables can take any integer value, positive or negative. Let

$$p_j \triangleq P[X_k = j] > 0, \ k = 1, 2, \ldots, \ j \in \{\ldots, -1, 0, 1, \ldots\}.$$

We can easily check that the chain is homogeneous since the probabilities p_j do not depend on k. The state transition diagram for this chain is shown in Figure 10.11. By inspection of this diagram, we can easily see that the chain is irreducible since every state is accessible from every other state in one step. (The definition of accessibility applies with $n = 1$.) Since $p_0 > 0$, a self-loop exists (in every state), so the chain is also aperiodic. Even if $p_0 = 0$, the chain will be aperiodic if

$$p_1 > 0, \ p_2 > 0, \ p_{-1} > 0.$$

This is the case because for any state i, the set I_i in the definition of the state period would contain the numbers 2 and 3 (check that), whose greatest common divisor is 1.

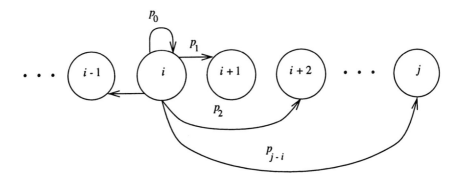

Figure 10.11 An IID process MC.

Recurrency is very easy to check since the probability

$$P_{jj}^n = P[X_n = j | X_0 = j] = P[X_n = j] = p_j > 0$$

in the sum of Theorem 10.10 is independent of n. Therefore,

$$\sum_{n=1}^{\infty} P_{jj}^n = \sum_{n=1}^{\infty} p_j = \infty,$$

and trivially the chain is recurrent.

Example 10.26. *Sums of IID Random Variables.* Let $S_n = \sum_{i=1}^{n} X_i$ be the sum of an IID sequence $\{X_i\}_{i=1}^{\infty}$ of nonnegative, integer-valued random variables. Let

$$p_j \triangleq P[X_k = j] > 0, \quad k = 1, 2, \ldots, \quad j \in \{0, 1, \ldots\}.$$

We can easily check that the sequence $\{S_n\}_{n=1}^{\infty}$ is a homogeneous chain (see Equation 10.5). The state transition diagram for this chain is shown in Figure 10.12. We can easily see from this diagram that there are an infinite number of communicating classes, each containing only one state. Thus the chain is reducible. Since

$$P_{ii}^1 = p_0 > 0, \quad \forall i \in E,$$

the period of all states is 1. It is easy to check recurrency for this chain since for any $l > 0$ and $j \in E$, we have $P_{jj}^l = p_0^l$. Therefore,

$$\sum_{l=1}^{\infty} P_{jj}^l = \sum_{l=1}^{\infty} p_0^l = \frac{p_0}{1 - p_0} < \infty,$$

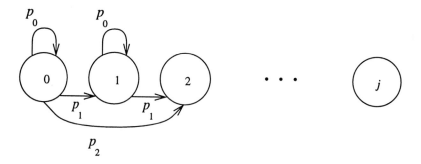

Figure 10.12   Sums of IID random variables.

and thus all states are transient.

Example 10.27. *Poisson Process.* The Poisson process is a special case of the sum process in Example 10.26 with

$$P_{ij} = p_j = e^{-\lambda}\frac{\lambda^j}{j!}, \quad \forall i, j \in \{0, 1, \ldots\}. \tag{10.23}$$

Therefore it is transient. Its increments form an IID process, with the same $P_{ij} = p_j$ of Equation 10.23. We can easily see that the increment process is recurrent.

Example 10.28. *Random Walks.* Consider the random walk process introduced in Section 7.5.1. A typical sample path of this process is shown again in Figure 10.13. The state transition diagram is shown in Figure 10.14.

   We study the recurrence properties of this process in detail since it is a very fundamental model in a lot of studies. Let $I = \{0, 1, \ldots\}$. Consider an IID sequence $\{X_i\}_{i=0}^{\infty}$ of ± 1-valued random variables, where

$$p \triangleq P[X_i = +1], \quad \forall i \in \{0, 1, \ldots\},$$
$$q \triangleq P[X_i = -1], \quad \forall i \in \{0, 1, \ldots\}.$$

Assume that $p + q = 1$. The cases $p = 0$, $p = 1$ are trivial; from now on, we assume that both $p > 0$ and $q > 0$.

   From the state transition diagram in Figure 10.14, we can see that the chain is irreducible since all states can communicate with each other. To see this more formally, we can use Theorem 10.7 to bound P_{ij}^n as follows: Suppose that $j > i$, and let $j - i = k$. Then for $n \geq k$, we have

$$P_{ij}^n = \sum_{l \in E} P_{il}^{n_1} P_{lj}^{n_2} \geq p^k.$$

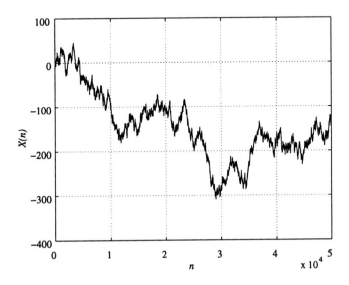

Figure 10.13 A typical sample path of a random walk.

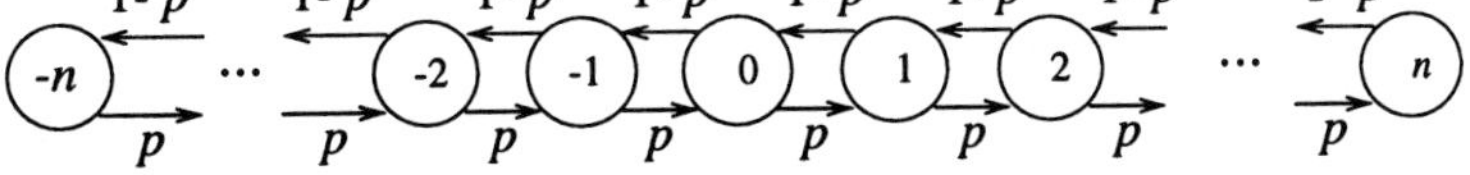

Figure 10.14 State transition diagram of a random walk.

Since $p > 0$, any state higher than i is accessible from state i. Similarly, suppose that $i > j$; let $i - j = m$. Then for $n \geq m$, we have

$$P_{ij}^n = \sum_{l \in E} P_{il}^{n_1} P_{lj}^{n_2} \geq q^m.$$

Since $q > 0$, any state lower than i is accessible from state i. Therefore, since $p > 0$ and $q > 0$, the chain is irreducible.

Aperiodicity is a bit more difficult to check since there are no self-transitions. We must calculate P_{ii}^n instead. We have

$$P_{ii}^1 = 0, \quad P_{ii}^2 = \sum_{l \in E} P_{il}^{n_1} P_{lj}^{n_2} \geq pq > 0, \quad P_{ii}^3 = \sum_{l \in E} P_{il}^2 P_{lj}^1 = 0,$$

and a pattern should appear now:

$$P_{ii}^n \begin{cases} = 0, & n \text{ odd}, \\ > 0, & n \text{ even}. \end{cases}$$

Therefore,

$$I_i = \{2, 4, 6, 8, \ldots\}$$

and thus easily $d(i) = 2$. Therefore the chain is periodic with period 2. Since the chain is periodic, it cannot reach steady state. However, from Theorem 10.14, the limit

$$\pi_j \overset{\triangle}{=} 0.5 \lim_{n \to \infty} P[X_{2n} = j | X_0 = j]$$

exists and is given by the solution to the system

$$\pi_j = \sum_{i \in E} P_{ij} \pi_i, \quad \sum_{i \in E} \pi_i = 1.$$

Let's check now whether the states are recurrent or transient. Consider state 0. We can easily see that

$$P_{00}^{2n+1} = 0, \quad n = 0, 1, \ldots;$$

moreover, as we ask you to show in Problem 10.98,

$$P_{00}^{2n} = \binom{2n}{n} p^n q^n = \frac{(2n)!}{n!n!} p^n q^n. \tag{10.24}$$

To check whether the sum in Theorem 10.10 diverges, we can use an asymptotic expansion of the probability in Equation 10.24, using Stirling's approximation formula for $n!$ (see Equation B.2):

$$n! \approx \sqrt{2\pi} n^{n+0.5} e^{-n}.$$

After some elementary algebra, we obtain the approximation

$$P_{00}^{2n} \approx \frac{(4pq)^n}{\sqrt{\pi n}}$$

We can easily verify that $4pq \leq 1$, with equality if and only if $p = q = 0.5$. Therefore, $\sum_{n=1}^{\infty} P_{jj}^n = \infty$ if and only if $p = q = 0.5$, and the chain is null recurrent whenever $p \neq q$.

10.4   DRIFTS AND ERGODICITY

Consider a DTMC $\{X_n, n = 0, 1, \ldots\}$, with $E = \{0, 1, 2, \ldots\}$. Define the **drift** at state i as

$$D_i \overset{\Delta}{=} E[X_{n+1} - X_n | X_n = i].$$

Intuitively, the drift is a measure of the "average change" of the MC. We expect then that a positive drift in state i means that, "on average," the next state after we visit i is "larger" than i, while a negative drift should imply that the next state is "lower." We make those ideas more precise shortly. Observe that

$$D_i = E[X_{n+1} | X_n = i] - E[X_n | X_n = i] = \sum_{j \in E} j P_{ij} - i.$$

The main result is Theorem 10.16, known as **Pakes' lemma**. It provides a sufficient condition for positive recurrence, based on the transition matrix information only, without the need to calculate any further probabilities, as were needed in Theorem 10.10. In that sense, it is more widely applicable. For its proof, we need the following lemma, which provides a sufficient condition for positive recurrence, that also uses information about the transition matrix only.

Lemma 10.5 *Consider an irreducible, aperiodic DTMC with $E = \{0, 1, 2, \ldots\}$ and transition probabilities P_{ij}. If there exists a positive integer m such that the inequalities*

$$\sum_{j=0}^{\infty} P_{ij} y_j \leq y_i - 1, \quad \forall i \geq m,$$

have a nonnegative solution $\{y_j\}$, with

$$s_i \overset{\Delta}{=} \sum_{j=0}^{\infty} P_{ij} y_j < \infty, \quad \forall i \leq m - 1, \tag{10.25}$$

then the chain is positive recurrent.

Proof: It suffices to prove that $\pi_j > 0$ for one state j. Since the chain is irreducible and aperiodic, the limit $\pi_j \stackrel{\triangle}{=} \lim_{n\to\infty} P[X_n = j | X_0 = i]$ in Equation 10.17 exists for all $j \in E$.

Fix a $j \in E$. Define the collection of sequences $\{y_i^n,\ n = 1, 2, \ldots\}$ for $i \leq j$ as follows: Set $y_i^1 = y_i$, and compute

$$y_i^{n+1} = \sum_{j=0}^{\infty} P_{ij}^n y_j.$$

We have

$$
\begin{aligned}
y_i^{n+1} &= \sum_{j=0}^{\infty} \sum_{l=0}^{\infty} P_{ij}^{n-1} P_{jl} y_l \\
&\leq \sum_{j=0}^{m-1} P_{ij}^{n-1} s_j + \sum_{j=m}^{\infty} P_{ij}^{n-1}(y_j - 1) \\
&\leq \sum_{j=0}^{m-1}(1 + s_j) P_{ij}^{n-1} + y_i^n - 1.
\end{aligned}
\qquad (10.26)
$$

Since y_i^1 is finite, we conclude that the entire sequence $\{y_i^n\}$ is finite, for all i and n. From Inequality 10.26 we get

$$y_i^{n+1} \leq \sum_{j=0}^{m-1}(1 + s_j) \sum_{l=1}^{n-1} P_{ij}^l + y_i^1 - (n-1),$$

so

$$\frac{1}{n} y_i^{n+1} \leq \sum_{j=0}^{m-1}(1 + s_j) \frac{1}{n} \sum_{l=1}^{n-1} P_{ij}^l + \frac{y_i^1}{n} - \frac{n-1}{n}.$$

Letting $n \to \infty$ and rearranging terms, we get

$$\sum_{j=0}^{m-1}(1 + s_j)\pi_j \geq 1,$$

which shows that at least one (and thus all) π_j must be strictly positive. $\square$

We can now state and prove the following:

Theorem 10.16 *Pakes' lemma. Consider an irreducible, aperiodic DTMC, with state space $E = \{0, 1, 2, \ldots\}$ and drifts $\{|D_i| < \infty\}$. Suppose that there exist an integer $m \geq 0$ and a positive real number $d > 0$ such that*

$$D_i \leq -d, \quad \forall i > m.$$

Then the chain is positive recurrent.

Proof: Consider the sequence $\{y_i, \ i = 0, 1, 2, \ldots\}$, defined as $y_i = i/d$. For all $i > m$, we have from the definition of drift D_i,

$$\sum_{j=0}^{\infty} P_{ij} y_j = \sum_{j=0}^{\infty} P_{ij} \frac{j}{d} = \frac{D_i + i}{d} \leq y_i - 1.$$

Since the drifts are finite, Inequality 10.25 is also satisfied, and from Lemma 10.5 the MC is positive recurrent. □

We state for completeness a converse to this theorem, which provides a sufficient condition for transience. It is also based on transition matrix information only. Its proof can be found in [35].

Theorem 10.17 *Consider an irreducible, aperiodic DTMC, with state space $E = \{0, 1, 2, \ldots\}$ and drifts $\{|D_i| < \infty\}$. Suppose that there exist integers $m \geq 0$ and $k > 0$ such that*

$$D_i > 0, \quad \forall i > m \tag{10.27}$$

and

$$P_{ij} = 0, \quad \forall i, j \in E, \quad \text{such that} \ \ 0 \leq j \leq i - k. \tag{10.28}$$

Then the chain is null recurrent.

The basic idea behind Condition 10.28 is that transitions to states "around 0" are not possible.

10.4.1 Examples

Example 10.29. *Positive Recurrent MC.* Consider an irreducible, aperiodic DTMC, with a finite number M of states, that is recurrent. The chain must be positive recurrent since at least one state j must have $\pi_j > 0$, otherwise, $\sum_{j=1}^{M} \pi_j \neq 1$. Can you apply Pakes' lemma to this chain? Why or why not?

Example 10.30. Consider now the MC in Figure 10.15. The transitions out of states $i < M$ are arbitrary, subject to the condition

$$|D_i| < \infty, \quad i < M.$$

For $i \geq M$, the chain shows some regularity, and therefore we can easily calculate the drifts:

$$D_i = \sum_{j \in E} j P_{ij} - i = \frac{1}{3}(i - 1) + \frac{1}{4}(i + 1) + \frac{5}{12} i - i = -\frac{1}{12}.$$

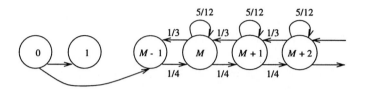

Figure 10.15   A positive recurrent MC.

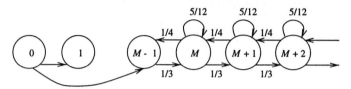

Figure 10.16   A null recurrent MC.

Therefore, Theorem 10.16 holds true with $d = 1/12$, $m = M$, and the MC is positive recurrent.

Example 10.?1. *Null Recurrent MC.* Consider the MC shown in Figure 10.16. For all $i > M$, we have

$$D_i = \frac{1}{4}(i-1) + \frac{1}{3}(i+1) - i = \frac{1}{12},$$

and thus Inequality 10.27 is satisfied, with $m = M$. Equation 10.28 is satisfied, with $k = 2$, as we can easily check. Therefore the chain is null recurrent.

Example 10.?2. *Random Walk with a Reflecting Barrier at Zero.* Let's investigate now the recurrency of the random walk in Section 7.5.1, where $a = \infty$ (i.e., no absorbing barrier). The drift in any state $i > 0$ is

$$D_i = \sum_{j \in E} jP_{ij} - i = (i+1)p + (i-1)q - i = p - q.$$

Therefore, from Pakes' lemma, if $p < q$, or, equivalently, if $q > 0.5$, the random walk is ergodic. From Theorem 10.17, we see that if $p > q$, the chain is null recurrent, since for $m = 1$ we have

$$D_i > 0, \quad \forall i > m,$$

and for $k = 2$ we also have

$$P_{ij} = 0, \quad \forall i, j \in E, \quad \text{such that} \quad 0 \leq j \leq i - k.$$

10.5 CONTINUOUS-TIME MARKOV CHAINS

10.5.1 Definition and the Transition Matrix

For a continuous-time Markov Chain (CTMC) the state space E is discrete, and the index set I is continuous. As we saw with the DTMC, we can assume without loss of generality that E is the set of integers or a proper subset of it. A convenient choice for I is the set $[0, \infty)$.

Definition: The random process $\{X(t), \ t \in I\}$ is a **continuous-time Markov Chain** if

$$P[X(t_{n+1}) = j | X(t_n) = i, X(t_{n-1}) = i_{n-1}, \ldots, X(t_1) = i_1, X(t_0) = i_0]$$
$$= P[X(t_{n+1}) = j | X(t_n) = i]$$

for any choice of $n, \{t_k\} \in I$, and $i, j, \{i_k\} \in E$.

Let $P_{ij}(t_n, t_{n+1}) \triangleq P[X(t_{n+1}) = j | X(t_n) = i]$. The matrix $P(t_n, t_{n+1})$ with elements $P_{ij}(t_n, t_{n+1})$ is called the **transition matrix** of the CTMC. From now on, we restrict our attention to **homogeneous** CTMCs only, for which the transition matrix will depend on the time difference $t \triangleq t_{n+1} - t_n$ only. So the defining matrix is denoted as

$$P(t) = \begin{bmatrix} \cdots & \cdots & \cdots & \cdots & \cdots & \cdots \\ \cdots & P_{00}(t) & P_{01}(t) & \cdots & P_{0l}(t) & \cdots \\ \cdots & P_{10}(t) & P_{11}(t) & \cdots & P_{1l}(t) & \cdots \\ \cdots & \cdots & \cdots & \cdots & \cdots & \cdots \\ \cdots & P_{l0}(t) & P_{l1}(t) & \cdots & P_{ll}(t) & \cdots \\ \cdots & \cdots & \cdots & \cdots & \cdots & \cdots \end{bmatrix}.$$

This matrix has two properties:

1. $P(0) = I_d$, where I_d is the identity matrix. To see why this is the case, notice that $P_{ij}(0) = \delta_{ij}$.

2. $\sum_{j \in E} P_{ij}(t) = 1$, for all $i \in E$ and $t \in I$.

10.5.2 Probabilities Related to a CTMC

As in the discrete-time case (see Section 10.2.1), the following five probabilities

1. The n-dimensional pmf $P[X(t_n) = i_n, \ldots, X(t_0) = i_0]$

2. The marginal pmf $P[X(t_n) = j]$

3. The conditional pmf $P[X(t_n) = j | X(t_0) = i]$

4. The limiting pmf $\lim_{n \to \infty} P[X(t_n) = j]$

5. The limiting pmf $\lim_{n \to \infty} P[X(t_n) = j | X(t_0) = i]$

are of interest in a CTMC as well. (In the last two probabilities, it is implicitly assumed that $t_n \to \infty$ as $n \to \infty$.) We can develop relationships for these probabilities in terms of the matrix $P(t)$, as we did in Section 10.2.1. Except for the obvious changes in notation, these relationships are very similar to the ones developed in Section 10.2.1. Therefore, we do not present them again here.

Instead, we focus on the marginal pmf $P[X(t_n) = j]$ and its limiting form $\lim_{n \to \infty} P[X(t_n) = j]$. We develop next some new relationships, in terms of a new matrix.

10.5.3   Transition Rate Matrix Q

Even though the transition matrix $P(t)$ is sufficient to completely characterize the probabilistic behavior of the CTMC, another matrix—the transition *rate* matrix—is more convenient to work with in practice. In a sense, the rate matrix describes the behavior of the CTMC for very small intervals of time, as opposed to the transition matrix $P(t)$, which describes the behavior of the CTMC in arbitrary intervals of length t.

In order to derive the relationship between the two matrices, we need to examine the concept of state sojourn times a bit more closely.

1. State sojourn times

As in the discrete-time case, the **kth sojourn time** O_k^j in state j is defined as the amount of time the MC spends in state j, at its kth return to state j.

Arguing as in Lemma 10.3, we can see that the random variables $\{O_k^j\}$ are IID; we identify the common pdf next.

Lemma 10.6 *For a fixed $j \in E$, the sequence $\{O_k^j, \ k = 1, 2, \ldots\}$ is an IID sequence of exponential random variables.*

Proof: Identical distribution and independence can be shown in a manner entirely similar to that for the discrete-time case. Therefore we omit the proof here.

Consider an arbitrary sojourn time O_k^j. In order to show exponentiality, consider the conditional probability $P[O_k^j > t + \tau | O_k^j > \tau]$. This is the probability that the MC will remain in state j for an additional t units of time, given that it is already in state j for τ units. Since the chain is homogeneous, we may assume without loss of generality that at time 0, $X(0) = j$.

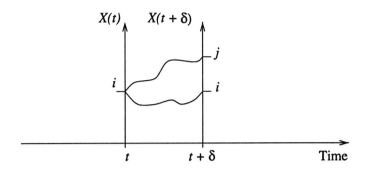

Figure 10.17 Transition rates.

For any $u > 0$, the events $\{O_k^j > u\}$ and $\{X(s) = j,\ 0 \le s \le u\}$ are equivalent. Therefore,

$$P[O_k^j > t + \tau | O_k^j > \tau] = P[\{X(s) = j, 0 \le s \le t + \tau\} | \{X(s) = j, 0 \le s \le \tau\}].$$

From the Markovian property, we have

$$P[O_k^j > t + \tau | O_k^j > \tau] = P[\{X(s) = j, 0 \le s \le t + \tau\} | \{X(\tau) = j\}] = P[O_k^j > t].$$

We recognize this as the *memorylessness* property, which uniquely determines the exponential distribution. □

Remark: We have not identified yet the parameter of this distribution. In general, it will depend on the state j. We see in the next section how to determine this parameter. For the time being, we denote this parameter as ν_j.

2. Transition rate matrix

Let's examine closely now the *rate* at which state changes take place. Suppose that at time t the MC is in state i, as Figure 10.17 depicts.

We evaluate the probabilities $P_{ij}(\delta)$ next, for "small" δ, and investigate the ratios $P_{ij}(\delta)/\delta$ as $\delta \to 0$. These limiting ratios should naturally provide information about the *rate* at which the probabilities change, as a function of time.

Consider the case $j = i$ first. Since the sojourn times are IID, let O^i denote a generic random variable from that sequence. We have

$$P_{ii}(\delta) \triangleq P[X(t + \delta) = i | X(t) = i] \approx P[O^i > \delta] + o(\delta).$$

This approximate equality can be justified as follows: Since successive sojourn times are IID exponential, the number of visits to state i in an interval of length

t is Poisson. Therefore, for small δ, two or more visits will have probability $o(\delta)$.

We may expand $P[O^i > \delta]$ by using its Taylor series expansion for small δ as follows:

$$P[O^i > \delta] = 1 - \nu_i \cdot \delta + o(\delta),$$

and therefore,

$$P_{ii}(\delta) \approx 1 - \nu_i \cdot \delta + o(\delta). \tag{10.29}$$

Consider now the case $j \neq i$. When the MC leaves state i, it will enter another state j with some probability $\tilde{q}_{ij}$. Since the probability of leaving state i is $1 - P_{ii}(\delta)$, we have

$$P_{ij}(\delta) = [1 - P_{ii}(\delta) + o(\delta)]\tilde{q}_{ij}$$

and thus from Equation 10.29

$$P_{ij}(\delta) = \nu_i \cdot \delta\tilde{q}_{ij} + o(\delta) = r_{ij}\delta + o(\delta),$$

whereby we have defined $r_{ij} = \nu_i \cdot \tilde{q}_{ij}$, for simplicity of notation. The ratio limits as $\delta \to 0$ are now easy to obtain from the above two equations.

$$q_{ij} \stackrel{\triangle}{=} \lim_{\delta \to 0} \frac{P_{ij}(\delta)}{\delta}, \quad i \neq j, \tag{10.30}$$

$$q_{jj} \stackrel{\triangle}{=} \lim_{\delta \to 0} \frac{P_{jj}(\delta) - 1}{\delta}. \tag{10.31}$$

Notice that q_{jj} cannot be positive.

We are ready to define now the rate matrix:

Definition: The matrix Q whose ijth element is equal to the rate q_{ij} is called the **transition rate matrix** of the CTMC.

$$Q = \begin{bmatrix} \cdots & \cdots & \cdots & \cdots & \cdots & \cdots \\ \cdots & q_{00} & q_{01} & \cdots & q_{0l} & \cdots \\ \cdots & q_{10} & q_{11} & \cdots & q_{1l} & \cdots \\ \cdots & \cdots & \cdots & \cdots & \cdots & \cdots \\ \cdots & q_{l0} & q_{l1} & \cdots & q_{ll} & \cdots \\ \cdots & \cdots & \cdots & \cdots & \cdots & \cdots \end{bmatrix}.$$

Remark 1: Matrix Q is *not* a probability matrix; therefore, its elements are not numbers in $[0, 1]$ necessarily.

Remark 2: From the definition of q_{ij}, we must have $\sum_{j \in E} Q_{ij} = 0$. [Compare this property to $\sum_{j \in E} P_{ij}(t) = 1$.]

How can we calculate the transition rate matrix in practice? Usually, we can only *postulate* about it, as the examples of the next section show. Then, as we have seen with the Poisson process in Section 8.1.6, we can *estimate* the rates through a sequence of IID measurements.

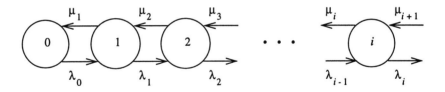

Figure 10.18 The general birth-death process.

10.5.4 Examples

Example 10.33. *Birth-Death Process.* Consider a process with a state space of the set of nonnegative integers. Consider a sequence of given parameters $\lambda_i \geq 0, \mu_i \geq 0$, with $\mu_0 \triangleq 0$. The transition rate matrix of the birth-death process is given by

$$q_{ij} = \begin{cases} \lambda_i, & j = i+1, \\ \mu_i, & j = i-1, \\ -(\lambda_i + \mu_i), & j = i, \\ 0, & \text{otherwise.} \end{cases} \tag{10.32}$$

In other words,

$$Q = \begin{bmatrix} -\lambda_0 & \lambda_0 & 0 & 0 & 0 & \cdots \\ \mu_1 & -(\lambda_1 + \mu_1) & \lambda_1 & 0 & 0 & \cdots \\ 0 & \mu_2 & -(\lambda_2 + \mu_2) & \lambda_2 & 0 & \cdots \\ \cdots & \cdots & \cdots & \cdots & \cdots & \cdots \end{bmatrix}.$$

A graphical way to represent the elements of Q is to put all nondiagonal elements in a state transition diagram, as we did in the discrete-time case. The state diagram for this process is depicted in Figure 10.18.

Example 10.34. *Yule Process.* The Yule[9] process is a special case of a birth-death process where $\mu_i = 0$ and $\lambda_i = \lambda \cdot i$. The transition rate matrix is given by

$$Q = \begin{bmatrix} 0 & 0 & 0 & 0 & \cdots \\ 0 & \lambda & -\lambda & 0 & \cdots \\ 0 & 0 & 2\lambda & -2\lambda & \cdots \\ \cdots & \cdots & \cdots & \cdots & \cdots \end{bmatrix}.$$

The state diagram is depicted in Figure 10.19.

Example 10.35. *Markov Modulated Poisson Process.* We briefly saw the Markov modulated Poisson process (MMPP) in Chapter 8. A more formal definition of this process is the following.

---
[9] After G. Y. Yule, 1871–1951, an English mathematician.

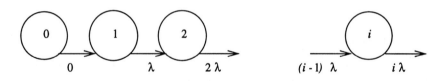

Figure 10.19   The Yule process.

Consider a CTMC with finite state space $E = \{1, 2, \ldots, M\}$ and transition rate matrix Q, defined as

$$Q = \begin{bmatrix} -\sigma_1 & \sigma_{12} & \cdots & \sigma_{1M} \\ \sigma_{21} & -\sigma_2 & \cdots & \sigma_{2M} \\ \cdots & \cdots & \cdots & \cdots \\ \sigma_{M1} & \sigma_{M2} & \cdots & -\sigma_M \end{bmatrix},$$

where

$$\sigma_i = \sum_{\substack{j=1 \\ j \neq i}}^{M} \sigma_{ij},$$

and σ_{ij} is the transition rate from state i to state j. While the chain is in state i, arrivals occur according to a Poisson process with rate $\lambda_i \geq 0$. The MMPP is fully characterized by the transition rate matrix Q and the M rates $\lambda_1, \ldots, \lambda_M$ [28].

Let A_n denote the instant of the nth arrival. Let $S_n = A_{n+1} - A_n$ denote the interarrival time between the $(n + 1)$st and nth arrivals. There are three quantities of interest about the MMPP:

(a) The arrival rate, which we can define as

$$\lambda \triangleq \lim_{n \to \infty} \frac{n}{A_n},$$

(b) The coefficient of variation, which we can define as

$$c^2 \triangleq \lim_{n \to \infty} \frac{ES_n^2}{(ES_n)^2},$$

(c) The correlation r between successive interarrival times, which we can define as

$$r = \lim_{n \to \infty} \frac{E\{[S_{n-1} - E(S_{n-1})](S_n - ES_n)\}}{\operatorname{var}(S_n)}$$

For the special case $M = 2$, the correlation can be found in explicit form:

$$r = \frac{\lambda_1 \lambda_2 (\lambda_1 - \lambda_2)^2 \sigma_1 \sigma_2}{c^2 (\sigma_1 + \sigma_2)^2 (\lambda_1 \lambda_2 + \lambda_2 \sigma_1 + \lambda_1 \sigma_2)^2}.$$

10.5.5 Marginal pmf $P[X(t) = j]$

We are ready now to calculate the marginal pmf $P[X(t) = j]$ in terms of the matrix Q. For simplicity, let $p_j(t)$ denote the probability $P[X(t) = j]$.

Theorem 10.18 *Consider a homogeneous CTMC, with rate matrix Q. For any $t \in I$, we have*

$$\frac{dp_j(t)}{dt} = \sum_{i \in E} Q_{ij} p_i(t), \quad \forall j \in E. \tag{10.33}$$

Proof: For any $\delta > 0$,

$$p_j(t + \delta) \overset{\triangle}{=} P[X(t + \delta) = j] = \sum_{i \in E} P_{ij}(\delta) p_i(t)$$

$$= \sum_{i \neq j} P_{ij}(\delta) p_i(t) + P_{jj}(\delta) p_j(t). \tag{10.34}$$

Therefore, adding and subtracting equal terms in Equation 10.34, we get

$$p_j(t + \delta) - p_j(t) = \sum_{i \neq j} P_{ij}(\delta) p_i(t) + [P_{jj}(\delta) - 1] p_j(t).$$

Divide now both sides by δ; Equation 10.33 follows from the definition of Q_{ij} in Equations 10.30 and 10.31, as $\delta \to 0$. □

Remark: The linear system of differential equations in Equation 10.33 is another form of the **Chapman-Kolmogorov equations** we already saw in discrete time (see Equation 10.13).

Let's see some examples now.

Example 10.36. *Birth-Death Process.* Consider the birth-death process introduced in the previous section, with transition rates given in Equation 10.32. The system of differential equations in Equation 10.33 can be written as

$$\frac{dp_j(t)}{dt} = -(\lambda_j + \mu_j) p_j(t) + \lambda_{j-1} p_{j-1}(t) + \mu_{j+1} p_{j+1}(t) \tag{10.35}$$

for $j > 0$; for the boundary case $j = 0$, we have

$$\frac{dp_0(t)}{dt} = -\lambda_0 p_0(t) + \mu_1 p_1(t). \tag{10.36}$$

There are an infinite number of differential equations to solve, in order to determine the marginal probabilities $p_j(t) \overset{\triangle}{=} P[X(t) = j]$. We see in the next

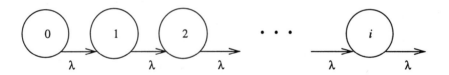

Figure 10.20   State diagram, Poisson process.

section a recursive technique that exploits the structure of matrix Q. Another, more general technique for solving systems of equations with an infinite number of unknowns is, of course, Z transforms. For more details on how to use these transforms, see [41].

Example 10.37. *Poisson Process.* When $\lambda_i = \lambda > 0$ and $\mu_i = 0$, we have a process with the state diagram depicted in Figure 10.20. Equations 10.35 and 10.36 specialize for this case to

$$\frac{dp_j(t)}{dt} = -\lambda p_j(t) + \lambda p_{j-1}(t), \qquad (10.37)$$

$$\frac{dp_0(t)}{dt} = -\lambda p_0(t). \qquad (10.38)$$

It is easy to derive a solution to the above system recursively. The solution to Equation 10.38 is simply

$$p_0(t) = e^{-\lambda t},$$

as we can easily verify by substitution. Then from Equation 10.37 for $j = 1$,

$$\frac{dp_1(t)}{dt} = -\lambda p_1(t) + \lambda p_0(t),$$

which is a simple first-order differential equation. Its solution is

$$p_1(t) = \lambda t e^{-\lambda t}.$$

We can now get the solution for the general case by induction:

$$p_j(t) = e^{-\lambda t} \frac{(\lambda t)^j}{j!}.$$

We recognize this as the Poisson process in Chapter 8.

10.5.6   Limiting pmf $\lim_{t \to \infty} P[X(t) = j]$

Assume for the moment that the limit exists for all $j \in E$. As usual, let $\pi_j \triangleq \lim_{t \to \infty} P[X(t) = j]$, and let π be a row vector whose jth element is equal to π_j. Then $\lim_{t \to \infty} dP[X(t) = j]/dt = 0$ and thus

$$\sum_{i \in E} Q_{ij} \pi_i = 0. \tag{10.39}$$

In matrix form,

$$\pi \cdot Q = 0,$$

where 0 is a row vector with the number of elements equal to the number of states. Rearranging terms in Equation 10.39, we get

$$\sum_{i \neq j} Q_{ij} \pi_i + Q_{jj} \pi_j = 0.$$

Since $Q_{jj} = -\sum_{l \neq j} Q_{jl}$, we have

$$\sum_{i \neq j} Q_{ij} \pi_i = \pi_j \sum_{l \neq j} Q_{jl}.$$

We can summarize this discussion in the following:

Theorem 10.19 *Consider a homogeneous CTMC, with rate matrix Q. The steady-state pmf is given by the solution to the system*

$$\pi_j \sum_{l \neq j} Q_{jl} = \sum_{i \neq j} Q_{ij} \pi_i, \quad \sum_{i \in E} \pi_i = 1. \tag{10.40}$$

Traditionally, the system of equations in 10.40 is referred to as the **global balance equations** (GBEs). It is the analog of System 10.18 for the discrete-time case.

When does System 10.40 have a nonzero solution? We saw in Section 10.3.2 that the properties of aperiodicity, irreducibility, and positive recurrency guarantee a nonzero solution in the discrete-time case. There are no direct, analogous conditions for the continuous-time case. The concept of irreducibility can be applied to a CTMC, but there is no obvious way to extend the concept of aperiodicity.

Let's see some examples next.

Example 10.38. *Birth-Death Process.* Consider the general birth-death process, with $\lambda_i > 0$, $\mu_i > 0$. The GBEs are given by

$$\lambda_0 p_0 = \mu_1 p_1, \quad j = 0, \tag{10.41}$$

$$(\lambda_j + \mu_j) p_j = \lambda_{j-1} p_{j-1} + \mu_{j+1} p_{j+1}, \quad j > 0. \tag{10.42}$$

Rearranging Equation 10.42, we have

$$\lambda_j p_j - \mu_{j+1} p_{j+1} = \lambda_{j-1} p_{j-1} - \mu_j p_j, \quad j > 0.$$

Substituting Equation 10.41 into Equation 10.42 and using $j = 1$, we get

$$\lambda_1 p_1 = \mu_2 p_2,$$

and inductively then $\lambda_j p_j = \mu_{j+1} p_{j+1}$. Let

$$\rho_j \triangleq \frac{\lambda_j}{\mu_{j+1}}, \quad j = 1, 2, \ldots;$$

then for all $j > 0$, we can write

$$p_j = p_0 \prod_{k=1}^{j} \rho_k.$$

The unknown probability p_0 can now be calculated from the normalizing condition $\sum_{j=0}^{\infty} p_j = 1$:

$$p_0 = \frac{1}{\sum_{j=0}^{\infty} \prod_{k=1}^{j} \rho_k}.$$

Therefore, the solution for the limiting probabilities is given by

$$p_j = \frac{\prod_{k=1}^{j} \rho_k}{\sum_{j=0}^{\infty} \prod_{k=1}^{j} \rho_k}.$$

For a nonzero solution, $p_0 > 0$, and thus

$$\sum_{j=0}^{\infty} \prod_{k=1}^{j} \rho_k < \infty.$$

Example 10.29. *Condition Arrival Rate < Departure Rate.* Consider a birth-death process with constant birthrates and death rates, equal to λ and μ, respectively. In that case, the condition $p_0 > 0$ becomes equivalent to

$$\sum_{j=0}^{\infty} \left(\frac{\lambda}{\mu}\right)^j = \frac{\mu}{\mu - \lambda} < \infty,$$

or, equivalently, the system will reach steady state if and only if $\lambda < \mu$. In other words, for equilibrium to be reached, the birthrate must be strictly less than the death rate. In queuing theory, births correspond to arrivals and deaths to departures, hence the name of the condition. It turns out that this stability condition holds true for very general systems; a large class of such systems is queuing systems with IID arrival processes and IID service time processes.

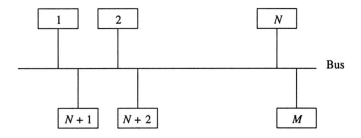

Figure 10.21 A simplified Ethernet LAN.

10.6 APPLICATION TO ETHERNET LANS

10.6.1 Physical System

Ethernet local-area networks (LANs) are simple, inexpensive ways to interconnect computers, printers, file servers, etc., together over small distances. Consider a simplified model of the Ethernet LAN that interconnects M "stations," as shown in Figure 10.21.

The protocol of communication over an Ethernet is quite simple. Station i "listens" to the transmissions on the bus. Whenever a packet is destined to station i, station i copies it off the bus. Packets not destined to station i are not copied. Whenever station i has a packet to transmit, it does so if and only if it detects that the bus is idle. Packets have all equal length (typically around 1500 bytes). When a station starts a transmission while another transmission is in progress, a *collision* occurs. Colliding packets must be retransmitted. The retransmission protocol is a simple randomization; in other words, if a station experienced a collision, it must wait for a random amount of time (drawn from some distribution) and then retransmit the packet. The basic question we want to address here regards the *stability* of this protocol. In particular, we want to investigate the throughput of this protocol as the number of stations M becomes very large. To do so, we need to make some assumptions regarding the stochastic behavior of this system.

10.6.2 Stochastic Model

Let M denote the number of stations sending packets over the Ethernet bus. Since it will not affect our throughput considerations, assume for simplicity that all stations have identical stochastic behavior. In particular, suppose that the packet arrival process at each station is Poisson, independent of all other packet arrival processes; its rate is denoted by λ/M, so the total system rate is λ. The stations are assumed to have no buffers.

We assume that the system operates in a synchronous manner: Transmissions of packets will last one time unit (a slot), and all stations that have decided to transmit must do so at the beginning of such a slot. All stations whose packets have experienced a collision are said to be *backlogged*. Backlogged stations must retransmit the collided packets. To simplify our analysis, we assume that a backlogged station will attempt retransmissions, until successful transmission, in a series of IID Bernoulli trials, with parameter p. In other words, if during a slot a decision is made not to retransmit, during the following slot another Bernoulli trial is made, independent of previous ones. Moreover, the stations are assumed to have no buffers, and thus any new packets arriving at a backlogged station will be lost. This assumption will simplify our analysis; even though it is not realistic, it does not diminish our results since we are only interested in (maximum) throughput, not delays.

Let X_k denote the number of backlogged stations at the beginning of slot k, $k \geq 1$. Based on the above assumptions, the sequence of random variables $\{X_k\}$ is a discrete-time MC with state space $E = \{0, 1, \ldots, M\}$. We evaluate the transition probabilities of this chain and investigate (through drifts) its behavior when $M \to \infty$.

Suppose that $X_k = i$, that is, at the beginning of slot k, exactly i stations were backlogged. And $M - i$ stations can accept new arrivals and therefore transmit; since arrivals to a station are Poisson with rate λ/M, the probability a station will transmit is $p_a = 1 - e^{-\lambda/M}$. On the other hand, i stations can attempt a retransmission, each with probability p. Let $P_{nb}(l, i)$ denote the probability that l nonbacklogged stations transmit given that $X_k = i$; let $P_b(l, i)$ be the probability that l backlogged stations transmit given that $X_k = i$. It is easy to see that

$$P_{nb}(l, i) = \binom{M - i}{l}(1 - p_a)^{M-i-l}p_a^l, \tag{10.43}$$

$$P_b(l, i) = \binom{i}{l}(1 - p)^{i-l}p^l. \tag{10.44}$$

Let P_{ij} denote the usual transition probability from state i to state j. We have

$$P_{ij} = \begin{cases} P_{nb}(0, i)P_b(1, i), & j = i - 1, \\ P_{nb}(l, i)P_b(0, i) + P_{nb}(0, i)[1 - P_b(1, i)], & j = i, \\ P_{nb}(1, i)[1 - P_b(0, i)], & j = i + 1, \\ P_{nb}(j - i, i), & j = i + 2, \ldots, M. \end{cases} \tag{10.45}$$

We explain only the case $j = i + 1$, the others being similar. For a transition to a state with one more backlogged station, exactly one nonbacklogged station must transmit, and at least one backlogged station must also transmit (resulting in a collision). Figure 10.22 depicts this MC.

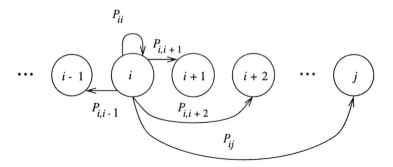

Figure 10.22 The Markov Chain model.

We can easily calculate the drifts for this MC; let D_i denote the drift in state i. From Equation 10.45 we have

$$D_i = (M - i)p_a - P_{nb}(1, i)P_b(0, i) - P_{nb}(0, i)P_b(1, i). \qquad (10.46)$$

If we define $P_{\text{dep}}(i) = P_{nb}(1, i)P_b(0, i) + P_{nb}(0, i)P_b(1, i)$, Equation 10.46 has the interpretation that the drift is equal to the arrival rate $(M - i)p_a$ minus the departure rate $P_{\text{dep}}(i)$. For small values of p, p_a, we can simplify the expression for $P_{\text{dep}}(i)$, using the approximation $(1 - a)^b \approx e^{-ab}$, in Equations 10.43 and 10.44. After some elementary algebra,

$$P_{\text{dep}}(i) \approx R(i)e^{-R(i)},$$

where we define $R(i) = (M - i)p_a + ip$. Therefore, from Equation 10.46, we have

$$D_i = (M - i)p_a - R(i)e^{-R(i)}.$$

In Figure 10.23, we plot the two components of the drift as a function of i (for $M = 100, p_a = 0.0035, p = 0.05$). For convenience, we plot the functions as if i were a continuous variable. Notice that the arrival and departure curves intersect at three points, for $i = 7$, 57, and 81, approximately; therefore, the drifts of the Markov Chain are positive in the intervals $[0, 6]$ and $[58, 80]$ and negative in the intervals $[8, 56]$ and $[82, 100]$.

Suppose now that the number of stations in the network increases while the parameter p_a decreases such that Mp_a, the total load on the network, remains constant. Observe now from Figure 10.23 that as $M \to \infty$, the arrival rate curve approaches a horizontal line, which intersects the departure rate curve only at two points. After the second intersection point, $D_i > 0$ for all states, and from Pakes' lemma the MC is not ergodic. The number of backlogged stations that are involved in retransmissions of previously collided packets grows without

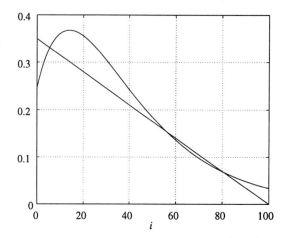

Figure 10.23 The drift as a function of state.

bounds. Therefore, this simple model predicts that the protocol is inherently unstable; this is also observed in practice.

10.7   GENERATION OF VALUES OF A MARKOV CHAIN

10.7.1   Discrete-Time MC

Consider a homogeneous DTMC $\{X_n\}$, described by the transition matrix P. How do we generate sample paths of $\{X_n\}$? There are two issues involved here:

- Only steady-state results are of interest.

- Transient results are of interest as well.

In the first case, we are essentially interested in generating values for a single random variable X_∞, with pmf $\{\pi_j, \ j \in E\}$, that describes the steady-state behavior of the MC. Since only a single, one-dimensional pmf is involved, the algorithms in Section 3.13 will suffice. Of course, the problem here is solution of the system $\pi = \pi P$.

In the second case, we must generate variates for the entire sequence of random variables $X_0, X_1, \ldots, X_n$. Intuitively we can achieve that as follows: We can generate a variate x_0 for X_0, using the given pmf $P[X_0 = i_0]$. Then, using the pmf $P[X_1 = j | X_0 = x_0]$ (i.e., using the x_0 row of the transition matrix P), we can generate a variate x_1 for the random variable X_1. The two variates x_0, x_1 correspond to the two-dimensional pmf $P[X_1 = i_1, X_0 = i_0]$. The general step for arbitrary n should now be obvious. More precisely, this algorithm can be summarized as follows:

0. Let $k = 0$.

1. Generate x_k, a variate distributed according to $P[X_k = i_0]$.

2. For $k = 1$ to n do

　　Set $i = x_k$.

　　Set $\tilde{p}_j = P_{ij}, \ j \in E$.

　　Generate a variate distributed according to $\tilde{p}_j$.

end do

The above algorithm involves generation of a variate invoking a potentially different pmf at each step. This is time-consuming. One can gain a significant shortening in execution time when the MC can be expressed in the form

$$X_{n+1} = f(X_n) + W_n, \tag{10.47}$$

with the sequence $\{W_n\}$ IID. Sums of IID random variables are a special case, so white noise and the Poisson process are special cases.

In this case, we can take advantage of Relationship 10.47 to avoid step 2 of the algorithm completely. We can instead generate a sequence of variates $\{w_i, i = 1, \ldots, n\}$, using the single pmf of the sequence $\{W_n\}$, and generate variates x_k recursively as

$$x_{k+1} = f(x_k) + w_k, \quad k = 0, \ldots, n - 1.$$

10.7.2　Continuous-Time MC

The easiest way to generate sample paths of a CTMC takes advantage of Lemma 10.6. Since sojourn times in state j are IID random variables with an exponential distribution, all we need is a DTMC and a generator for IID exponential random variables. Therefore, an algorithm for generating sample paths of a CTMC can be stated as follows:

0. Let $k = 0$.

1. Generate x_k, a variate distributed according to $P[X_k \leq x]$.

2. For $k = 1$ to n do

　　Set $i = x_k$.

　　Set $\tilde{p}_j = P_{ij}, \ j \in E$.

　　Generate a sojourn time variate distributed according to

an exponential distribution with parameter ν_i.

Generate a next-state variate distributed according to $\tilde{p}_j$.

end do

10.8   SUMMARY OF MAIN POINTS

- A **Markov Chain** captures the dynamic behavior of a first-order system.

- A Markov Chain is mathematically tractable. All we need to specify is the **transition probability matrix** P (in discrete time) or the **transition rate matrix** Q (in continuous time).

- A Markov Chain sequence $\{X_n\}$ converges to a random variable X_∞ with probability 1 if the chain is **ergodic**. The pmf of X_∞ is the solution to the linear system $\pi = \pi \cdot P$ or $\pi Q = 0$.

- The **global balance equations** enable us to calculate the steady-state probabilities π.

- The **drifts** of a Markov Chain provide us with a simple means to check ergodicity.

- Generating variates of a Markov Chain is fairly easy.

10.9   CHECKLIST OF IMPORTANT TOOLS

- Global balance equations

- Drifts

10.10   PROBLEMS

In the following problems, if the state space of a MC is not explicitly specified, the set of all integers is assumed.

Definition and Classification; Simple Skills

10.1 Show how a finite set can be mapped onto the set of all integers. Show how the set $S = \{0, 1, 2, \ldots\} \times \{0, 1, 2, \ldots\}$ can be mapped onto the set of all integers. Generalize to an arbitrary discrete set.

10.2 Consider a MC with state space $E = \{-3, -1, 1\}$, known transition probabilities P_{ij}, and known $P[X_1 = l]$. Compute $P[X_4 = -1, X_2 = 1]$.

10.3 Consider a MC with two states, labeled 0 and 1, known transition probability matrix P, and known initial pmf $P[X_0 = i]$. Find

$$P[X_1 = 1|X_0 = 0, X_2 = 1], \quad P[X_0 = X_1], \quad P[X_1 \neq X_2].$$

10.4 Consider a nonstationary Markov Chain $\{X_n\}$ with states $1, 2, 3$ and

$$P[X_{n+1} = j|X_n = i] = \frac{n}{n+1}\delta_{ij} + \frac{1}{2(n+1)}(1 - \delta_{ij}).$$

Here δ_{ij} is the usual Kronecker symbol; that is, $\delta_{ij} = 1$ if $i = j$ and $\delta_{ij} = 0$ if $i \neq j$. Compute $P[X_3 = 1, X_2 = 1, X_1 = 1]$, when
(a) $P[X_0 = 0] = 1$,
(b) $P[X_0 = 1] = 1$.

10.5 Consider a Markov Chain $\{X_n\}$. Since $P[X_3 = j|X_2 = i, X_1 = k]$ does not depend on k, can we say that X_3 is independent of X_1? Give a proof, example, or counterexample.

10.6 Consider a Markov Chain $\{X_n\}$. Show that the random variables X_k, X_{k+m} are in general dependent, for all k, m.

10.7 Consider a Markov Chain $\{X_n\}$. Is the following true?

$$P[X_{n+1} = j|X_{n-2} = i, X_{n-10} = k] = P[X_{n+1} = j|X_{n-2} = i].$$

10.8 Consider a Markov Chain $\{X_n\}$ with known transition probabilities P_{ij}. Compute $P[X_n = i|X_{n+1} = j]$ and $P[X_n = i|X_{n+2} = k]$.

10.9 Consider a Markov Chain $\{X_n\}$ and an integer $m > 0$. Is the following true?

$$A \overset{\Delta}{=} P[X_{n+m} = j_{n+m}, \ldots, X_{n+1} = j_{n+1}|X_n = j_n]$$

$$= \prod_{i=1}^{m} P[X_{n+i} = j_{n+i}|X_{n+i-1} = j_{n+i-1}].$$

10.10 Let $A_n, \ldots, A_0$ be subsets of the state space E. Show that the definition of a MC implies that

$$P[X_{n+1} = j|X_n \in A_n, \ldots, X_0 \in A_0] = P[X_{n+1} = j|X_n \in A_n].$$

10.11 Show that $P[X_0 = i|X_1 = j, X_2 = i_2, \ldots, X_n = i_n] = P[X_0 = i|X_1 = j]$.

10.12 Calculate the conditional moment $E[X_{n+1}^3 | X_n]$ in terms of the transition probability matrix.

10.13 Write explicitly the first 64 random variables from the sequence $\{Z_n\}$ in Example 10.12. Is the MC homogeneous?

Markov Chains and Previous Concepts

10.14 Is a homogeneous MC a WSS process? Is Gaussian white noise a Markov process?

10.15 Is a MC an independent-increments random process? Does a homogeneous MC have stationary increments?

10.16 Consider a finite Markov Chain $\{X_n\}$ with $EX_n = 0$. Let $M_n = \frac{1}{n}\sum_{i=1}^n X_i$. Under what conditions does M_n satisfy the result of the WLLN?

10.17 Consider the sample space $S = [0, 1]$. Can you define a sequence of three random variables on S, such that they form (a) a stationary MC, (b) a nonstationary MC, and (c) a Poisson sequence?

Concepts

10.18 Consider a sequence of independent tosses of a coin with probability of heads p. Let X_n be the total number of tosses which showed heads among the first n tosses.
(a) Check that $\{X_n\}$ is a MC. What is the state space?
(b) Determine the transition probability matrix. Is the chain stationary?
(c) Repeat parts (a) and (b) when X_n is defined as the total number of tosses which showed heads minus the total number of tosses which showed tails (again among the first n tosses).

10.19 Give an example of a MC from your everyday life. Give an example of a random process from your everyday life that is not a MC.

10.20 Give an example of two sequences $\{X_n\}$, $\{Y_n\}$, which are not MC but whose difference is. Give an example in which their difference is not a MC.

10.21 Consider a Markov Chain $\{X_n\}$. Define a new sequence $Y_n = X_{2n}$. In other words, we ignore all odd-numbered random variables in the original sequence. Is $\{Y_n\}$ a MC?

10.22 Consider a Markov Chain $\{X_n\}$.
(a) Let $Y_n = X_{n+10}$. Is the sequence $\{Y_n\}$ also a MC?
(b) Define now Y_n by "randomly" choosing a subset of the random variables X_n.

Is the sequence $\{Y_n\}$ also a MC? Define carefully how you interpret *randomly* in the previous statement.

10.23 Consider two independent Markov Chains $\{X_n\}$, $\{Y_n\}$ with common transition probability matrix P.
(a) Let $T = \inf\{n : X_n = Y_n\}$. Show that T is a stopping time.
(b) Let

$$Z_n = \begin{cases} X_n, & n > T, \\ Y_n, & n \leq T. \end{cases}$$

Show that $\{Z_n\}$ is a MC.

10.24 Consider an integer-valued, IID sequence $\{X_n\}$; define

$$\overline{M}_n = \max\{X_1, \ldots, X_n\}, \quad \underline{M}_n = \min\{X_1, \ldots, X_n\}.$$

Is $\overline{M}_n$ a MC? Is $\underline{M}_n$ a MC?

10.25 *Sequences with Memory.* Consider a sequence of integer-valued random variables $\{X_n\}_{n=1}^{\infty}$, with the property that for all i, j, k in the state space

$$
\begin{aligned}
P_{ijk} &\triangleq P[X_{n+1} = j | X_n = i, X_{n-1} = k] \\
&= P[X_{n+1} = j | X_n = i, X_{n-1} = k, \ldots, X_1 = i_1].
\end{aligned}
$$

(a) Can you derive the joint pmf $P[X_n = i, X_{n-1} = k, \ldots, X_1 = l]$? What information do you need for that?
(b) Comment on how difficult this is, compared to the case where $\{X_n\}_{n=1}^{\infty}$ is a MC. If working with the general case is too difficult for you, you can consider the special cases of $n = 3$ and $n = 4$ only.
(c) Is this sequence stationary?

Extensions

10.26 Extend the definition of a MC for a *vector* random variable.

10.27 Two independent random variables X_0 and X_1 have cdf's F_1 and F_2.
(a) Is the random process $\{X_n\}$, defined via $X_{n+1} = X_n + X_{n-1}$, for $n > 0$, a MC?
(b) If you answer no, define a (column) vector $Y_n = (X_n \ X_{n-1})'$. Does the sequence $\{Y_n\}$, which depends on $\{X_n\}$, have the Markov property?

10.28 Consider the pure autoregressive model $\{X_n\}$ in Section 7.5.2, with $M > 1$, $L = 0$. We saw that $\{X_n\}$ is not a MC. Find a vector transformation of X_n that is a MC.

10.29 Consider a sequence $\{X_n\}$ with discrete state space that satisfies the property

$$P[X_{n+1} = j | X_n = i_n, \ldots, X_0 = i_0]$$
$$= P[X_{n+1} = j | X_n = i_n, \ldots, X_{n-r+1} = i_{n-r+1}]$$

for some integer $r > 0$. Such a sequence is called rth-order Markov. Find a vector transformation to make this a MC.

10.30 Consider a Markov Chain $\{X_n\}$. Is the vector (X_n, X_n) a MC? If yes, find the transition probability matrix.

10.31 Consider a Markov Chain $\{X_n\}$. Is the vector $(X_n, X_{n+1} - X_n)$ a MC? If yes, calculate the transition probability matrix.

10.32 Consider two independent Markov Chains $\{X_n\}$, $\{Y_n\}$ defined on the same state space. Is (X_n, Y_n) a MC? If yes, find the transition probability matrix.

Markov Chains and Modeling

10.33 Let X_n, $n = 1, 2, 3, 4$, denote the score of the home team at the end of the nth quarter in an NFL game. (If you do not know what NFL stands for, you are in serious trouble.)
(a) Is the sequence $\{X_n, \ n = 1, 2, 3, 4\}$ IID?
(b) Is it a MC?

10.34 Consider an NBA game. Define two sequences of random variables as follows:

$$X_n = i, \text{ if home team scores } i \text{ points on } n\text{th possession,}$$
$$Y_n = i, \text{ if guest team scores } i \text{ points on } n\text{th possession.}$$

Here $i = 0, 1, 2, 3$, so no four-point plays are considered.
(a) Is $\{X_n\}$ a MC? Explain in detail and state your assumptions clearly.
(b) Define $Z_n = X_n - Y_n$. Is $\{Z_n\}$ a MC? Why?

10.35 Consider the all-time NBA scoring record. Can it be described by a MC?

10.36 Consider the sport of tennis. Suppose you want to model a *game* as a MC. Specify X_n, the state space, and transition diagram. Discuss in detail any assumptions you make (e.g., about player competence levels).

10.37 Consider the Dow Jones example in Section 1.9. Let $n = 1, 2, 3, 4, 5$ denote the nth business day of the week. Let X_n, $n = 1, 2, 3, 4, 5$, denote the closing value of the index during the nth business day of the week. Discuss whether $\{X_n\}$ is a MC. Provide as many arguments as possible.

10.38 Consider the closing values of the Dow Jones example again. Let now the index n denote the business days in a whole year.
(a) Discuss whether $\{X_n\}$ is a MC. Provide as many arguments as possible.
(b) Describe how exactly you would use the available data to support your claims.

10.39 The following model is useful in studying extinction problems in physics and genetics alike. Let X_n denote the population during generation n. Initially, let $X_0 = 1$ be the population (e.g., of photons, protons, α particles,) of the first generation. The population of the second generation, denoted by X_1, is the offspring of all X_0 members of the first generation. More formally, let $\xi_{k,l}$ denote the number of offspring of member l in generation k. Then

$$X_{n+1} = \sum_{i=1}^{X_n} \xi_{n,i}.$$

Consider a (double) sequence $\{\xi_{k,l}\}$ of IID, nonnegative random variables. Show that $\{X_n\}$ is a MC, and evaluate its transition probability matrix. Is the chain stationary?

10.40 *Tough!* Suppose that humankind will exist forever. Do you think your family's name will survive? State your assumptions clearly. No formal proof is necessary.

10.41 Consider a group of N politicians. They must vote on a bill until a consensus is reached, either for or against it. Let X_n denote the number of politicians in favor of the bill at the nth voting. Assume that the politicians vote independently (an unrealistic assumption?) and randomly (kind of realistic?).
(a) Show that $\{X_n\}$ is a MC.
(b) Calculate the transition probability matrix P.

10.42 A communication link has an available bandwidth of M frequency units (slots). The link carries traffic submitted by two types of users; users of both types require one frequency slot for service. Users of type 1 cannot be queued; i.e., upon arrival they are either given a bandwidth slot, if there is any available, or they are rejected. Users of type 2 can be queued.

Assume that users of type i, $i = 1, 2$, arrive according to a Poisson process at rate λ_i. Their transmission times are independent, exponential random variables, with rate μ_i. All random variables are assumed independent of each other.

Let t_k denote the time instant of the kth "event," an event being an arrival or departure. Let N_{ik} denote the total number of users of type i in the system at time t_k.

(a) Explain why the vector (N_{1k}, N_{2k}) is a discrete-time MC.

(b) Specify the state space.

(c) For $M = 3$, draw the transitions in and out of states $(1, 1)$ and $(2, 2)$.

10.43 A computer chip works without problems for a random period of time that is distributed according to a positive, integer-valued random variable L. Upon failure, the chip is replaced with an identical, new one. At time n, let X_n denote the age of the chip in service. Is the sequence $\{X_n\}$ a MC?

10.44 Consider a machine with two parts that fail independently of each other. The parts are also repaired independently. Define X_n as the number of working parts during, say, day n. The probability that a part fails during a day is p_f. The probability that a part is repaired is p_r, and it takes 1 day to repair a failed part.

(a) Is $\{X_n\}$ a MC? What is the range for n? What is the range for X_n?

(b) If you answered yes in part (a), determine the transition probabilities.

10.45 Consider the pseudorandom number generator in Equation 3.89.

(a) Show that its output is a MC.

(b) Specify the state space and calculate its transition probability matrix.

DTMC; Simple Skills

10.46 Prove Theorem 10.13 for the nonhomogeneous case.

10.47 Show that the $i \longleftrightarrow j$ relation is an equivalence relation.

10.48 Consider a finite state DTMC with state space E and transition probability matrix P. The matrix P satisfies the property

$$P_{ii}^{2m} \geq P_{ij}^{2m} \tag{10.48}$$

for all integers $m > 1$ and all $i, j \in E$.

(a) Interpret this property intuitively.

(b) Prove that if P is a symmetric matrix and its rows are permutations of each other, then P satisfies Inequality 10.48.

10.49 Consider a DTMC $\{X_n\}$ with transition probability matrix

$$P = \begin{bmatrix} 0 & 1 & 0 & 0 \\ p_1 & p_2 & p_3 & p_4 \\ 1 & 0 & 0 & 0 \\ 0 & 0 & 0 & 1 \end{bmatrix}.$$

For what values (if any) of the parameters p_i, $i = 1, 2, 3, 4$, is this chain irreducible and aperiodic?

10.50 A DTMC on states $\{0, 1, 2, 3, 4\}$ has transition probability matrix

$$P = \begin{bmatrix} 1 & 0 & 0 & 0 & 0 \\ 0 & 0.5 & 0.5 & 0 & 0 \\ 0 & 0.1 & 0.9 & 0 & 0 \\ 0.1 & 0.3 & 0 & 0.2 & 0.4 \\ 0.3 & 0 & 0.1 & 0.2 & 0.4 \end{bmatrix}.$$

(a) Draw the state transition diagram.
(b) For each state, determine the set of accessible states.
(c) Determine the communicating classes.
(d) Find the period of each state.

10.51 Consider a DTMC $\{X_n\}$ with states 1, 2, 3, 4 and transition matrix

$$P = \begin{bmatrix} 1 & 0 & 0 & 0 \\ 1 - \frac{p}{2} & \frac{p}{4} & \frac{p}{4} & 0 \\ 0 & \frac{p}{4} & 1 - \frac{p}{2} & \frac{p}{4} \\ 0 & 0 & 0 & 1 \end{bmatrix}.$$

Here $0 < p < 1$ is a given parameter.
(a) Which states are accessible from state 1? From state 2?
(b) How many communicating classes are there?
(c) Find the period of each state.
(d) Classify each state as transient or recurrent.

10.52 Consider a DTMC with states $\{0, 1, 2, \ldots\}$ and transition probabilities given by

$$P_{ij} = \begin{cases} p, & j = i + 3, \\ 1 - p, & j = 0, \\ 0, & \text{otherwise.} \end{cases}$$

Is the chain homogeneous? Aperiodic? Irreducible?

10.53 Draw the state transition diagram for a DTMC $\{X_n\}$ with the following properties:
(a) $E = \{1, 2, \ldots, 10\}$,
(b) $d(2) = d(4) = d(6) = d(8) = d(10) = 2$,
(c) $d(1) = d(3) = d(5) = d(7) = d(9) = 3$,
(d) The chain is reducible.

10.54 Consider the MC in Problem 10.39. Is the MC aperiodic? Irreducible?

10.55 Repeat Problem 10.54 for the MC in Problem 10.41.

10.56 Repeat Problem 10.54 for the MC in Problem 10.45.

10.57 Show that if for some $k > 0$, the matrix P^k has identical rows, then P^{k+1} has identical rows as well.

10.58 Consider an irreducible MC with the property that $P^2 = P$.
(a) Show that the chain is aperiodic.
(b) Show that $P_{ij} = P_{jj}$, $\forall i, j \in E$.

10.59 Can you construct an irreducible DTMC with two periods? Explain.

10.60 Can you construct a DTMC with an infinite number of periods? If yes, give an example. If no, provide a proof.

10.61 Consider a DTMC with transition probability matrix

$$P = \begin{bmatrix} 1-p & p \\ q & 1-q \end{bmatrix}.$$

Use induction on n to prove that

$$P^n = \frac{1}{p+q} \begin{bmatrix} q & p \\ q & p \end{bmatrix} + \frac{(1-q-p)^n}{p+q} \begin{bmatrix} p & -p \\ -q & q \end{bmatrix}.$$

10.62 Consider a DTMC with transition probability matrix

$$P = \begin{bmatrix} 0 & 1 & 0 \\ 1-p & 0 & p \\ 0 & 1 & 0 \end{bmatrix}.$$

(a) Show that $P^4 = P^2$.
(b) Use part (a) to find P^n, $n > 0$.

10.63 Consider a finite, irreducible DTMC with N states.
(a) Show that the linear system

$$x_i = \sum_{k=1}^{N} P_{ik} x_k, \quad i = 1, \ldots, N,$$

has a solution $x_i = c$, $i = 1, \ldots, N$.
(b) Is the condition of irreducibility necessary?
(c) Is the finiteness condition necessary?

10.64 Show that a convex combination of steady-state distributions (on the same state space) is also a steady-state distribution.

10.65 Consider a DTMC with steady-state distribution $\{\pi_i\}$. Show that if for all i, we have that $P_{ij} = cP_{ik}$, for some states j, k and a constant $c > 0$, then $\pi_j = c\pi_k$.

10.66 Consider an irreducible, aperiodic DTMC with transition probabilities

$$P_{ij} = \begin{cases} p, & i = j+1, \\ q, & i = j-1, \\ 0, & \text{otherwise.} \end{cases}$$

Suppose that $P[X_0 = i] = \pi_i > 0$, where $\{\pi_i\}$ is the steady-state distribution. Show that $P[X_0 = i | X_1 = j] = P_{ji}$.

10.67 Define a DTMC with states that are all transient. In other words, specify the state space E and the transition probability matrix.

10.68 Consider the machine repair DTMC in Problem 10.44. Determine its steady-state distribution.

DTMC; Advanced Skills

10.69 Consider a homogeneous, finite, aperiodic, and irreducible MC. Show that for all $i, j \in E$,

$$P_{ij}^n \neq 0$$

for all sufficiently large n.

10.70 Consider a finite, irreducible DTMC. Prove that the chain is aperiodic if and only if for some $n > 0$, $P_{ij}^n > 0$, $\forall i, j \in E$.

10.71 Consider an ergodic, finite Markov Chain $\{X_n\}$ with transition probability matrix P. Let $A = \lim_{n \to \infty} P^n$. Show that
(a) the rows of A are equal,
(b) $\sum_j A_{ij} = 1$, for all i,
(c) $A_{ij} > 0$, for all i.

10.72 Consider a DTMC on $E = \{0, 1, 2, \dots, \}$ with transition probability matrix

$$P = \begin{bmatrix} p_0 & p_1 & p_2 & p_3 & \cdots \\ p_0 & p_1 & p_2 & p_3 & \cdots \\ 0 & p_0 & p_1 & p_2 & \cdots \\ 0 & 0 & p_0 & p_1 & \cdots \\ \cdots & \cdots & \cdots & \cdots & \cdots \end{bmatrix}.$$

The constants p_i are nonzero for all i. Classify the states of this chain.

10.73 Consider a finite DTMC, with N states. Suppose that state j is accessible from i, that is, $P[X_n = j | X_0 = i] > 0$, where n is the smallest such integer. Let $i_1, i_2, \dots, i_{n-1}$ be the sequence of states through which state j is accessible from state i.
(a) Show that $P_{ii_1} P_{i_1 i_2} P_{i_2 i_3} \cdots P_{i_{n-2} i_{n-1}} P_{i_{n-1} j} > 0$.

(b) Show that $i_1, i_2, \ldots, i_{n-1}$ are all distinct states.
(c) Show that $n < N$.

10.74 Consider a finite DTMC, with $N + 1$ states labeled $0, 1, \ldots, N$. The transition probabilities are given by

$$P_{ij} = \begin{cases} 1 - \frac{i}{N}, & j = i+1, i = 1, \ldots, N-1, \\ \frac{i}{N}, & j = i-1, i = 1, \ldots, N-1, \\ 1, & i = 0, j = 1, \\ 1, & i = N, j = N-1, \\ 0, & \text{otherwise.} \end{cases}$$

This chain is known as the *Ehrenfest model*.
(a) Is the chain irreducible?
(b) What is the period of the chain?
(c) Suppose that $P[X_0 = k] = \binom{N}{k} 2^{-N}$ (a binomial pmf with parameters $n = N$ and $p = 0.5$). Find $P[X_1 = k]$ and $P[X_2 = k]$.
(d) Find P^2 and P^3 when $N = 3$.
(e) Suppose that the chain is used to model the contents of a memory buffer. Give an intuitive interpretation of the transition probabilities.

10.75 Consider a DTMC that can take only nonnegative values. Suppose that its transition probability matrix has the property

$$\sum_{j \in E} j \cdot P_{ij} = a \cdot i + b, \quad \forall i \in E,$$

where a, b are real constants.
(a) Show that $EX_{n+1} = aEX_n + b$.
(b) Using part (a), show that if $a \neq 1$, then

$$EX_{n+1} = \frac{b}{1-a} + a^{n+1}\left(EX_0 - \frac{b}{1-a}\right).$$

10.76 Consider the Ehrenfest Markov Chain $\{X_n\}$ in Problem 10.74.
(a) Show that $\{X_n\}$ satisfies the assumptions in Problem 10.75.
(b) Calculate $E(X_n|X_0)$.

10.77 Consider a finite DTMC with $E = \{0, 1, \ldots, N\}$ whose transition probability matrix has the property that

$$\sum_{j \in E} j P_{ij} = i, \quad \forall i \in E.$$

Assume, moreover, that $P_{00} = P_{NN} = 1$.
(a) Show that state 0 is accessible from states $1, 2, \ldots, N-1$.
(b) Show that states $1, 2, \ldots, N-1$ are transient.

10.78 Consider a DTMC with transition probability matrix that has the property $P_{ij} = f(i)$, $\forall i, j \in E$, that is, all entries in a row are constant.
(a) Is the chain aperiodic?
(b) Show that the chain is ergodic.
(c) Show that $\pi_i = f(i)$.

10.79 The MC in Problem 10.78 has identical rows. Can we have a MC with identical columns?

10.80 Consider a finite DTMC with N states and $P_{ij} > 0$, for all $i, j \in E$. Let $p = \min\{P_{ij}\}$. Show that for all integers $n > 0$,

$$|P_{ij}^n - \pi_j| \le (1 - Np)^n, \quad \forall i, j \in E.$$

10.81 Let $\{Y_n\}$ be an IID sequence of nonnegative, continuously valued random variables. Define the series of *record times* as follows: Set $X_1 = 1$, and for $n > 1$, recursively set

$$X_n = \inf_k \{k \ge X_{n-1} + 1 : Y_k \ge \max\{Y_1, \ldots, Y_{k-1}\}\}.$$

Show that $\{X_n\}$ is a MC. Find the transition probability matrix.

10.82 Consider a finite DTMC with the following property: The set of states can be partitioned into two subsets A_1, A_2 such that the chain jumps from a state $k \in A_i$ to a state in $l \in A_j$ with equal probability (i.e., $P_{lk} = 1/|A_j|$, where $|A_j|$ denotes the cardinality of A_j).
(a) Determine a suitable transition probability matrix.
(b) Show that the chain has period 2.
(c) Show that the chain is irreducible.
(d) Find an expression for P^n.
(e) Determine the limiting distribution.

10.83 Consider an irreducible and aperiodic Markov Chain $\{X_n\}$.
(a) Show that $Y_n = (X_{n+1}, X_n)$ is an irreducible and aperiodic MC.
(b) Calculate the steady-state distribution of Y_n in terms of that of X_n.

Modeling

10.84 Consider an IID, ± 1-valued sequence $\{Y_i\}$, with $P[Y_i = +1] = p$.
(a) Let $X_n = Y_n Y_{n+1}$. Is $\{X_n\}$ a MC? If yes, find P_{ij} and determine whether it is aperiodic and irreducible.
(b) Let $X_n = Y_n Y_{n-1}$. Is $\{X_n\}$ a MC? If yes, find P_{ij} and determine whether it is aperiodic and irreducible.

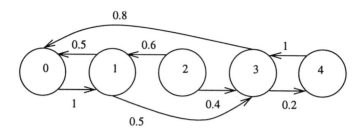

Figure 10.24   A finite MC.

10.85 Consider a sequence of IID random variables $\{X_i\}$, each taking values in the set $A = \{0, 1, 2, \ldots\}$. Suppose that $P[X_i = k] > 0$, for all k. Consider their sample mean

$$M_n = \frac{1}{n} \sum_{i=1}^{n} X_i.$$

(a) Specify the range of values that the sequence $\{M_n\}$ can take. Does the process have continuous or discrete space?
(b) Verify that the sequence $\{M_n\}$ is a Markov Chain.
(c) Is it homogeneous?
(d) Derive the transition probability matrix.
(e) Is $\{M_n\}$ irreducible? Aperiodic?

10.86 Would any of your answers to Problem 10.85 change if $A = \{1, 2, \ldots\}$? If $A = \{\ldots, -1, 0, 1, 2, \ldots\}$?

10.87 Let $\{\xi_n\}_0^\infty$ be a sequence of random variables, each taking values in the set $\{0, 1, 2, \ldots\}$, with $P[\xi_n = j] = (1 - p)p^j$. Let $X_0 = 0$. Define a sequence of random variables $\{X_n\}_1^\infty$ via $X_{n+1} = a_n X_n + b_n \xi_n$, where a_n, b_n are nonnegative, nonrandom integers.
(a) Is $\{X_n\}$ a DTMC?
(b) Find, if possible, conditions (on $\{\xi_n\}, a_n, b_n,$) that make $\{X_n\}$ a MC with stationary transition probabilities.
(c) Is $\{X_n\}$ irreducible? Aperiodic?

Drifts

10.88 Consider the MC in Figure 10.24. Calculate the drifts for all states.

10.89 Calculate the drifts for the chains in Problems 10.74 and 10.50.

10.90 Can you construct a DTMC with D_i, the drift in state i, equal to:
(a) 0.5, all states; (b) $3 - i$, all states; (c) $\frac{i}{2}$, all states?

10.91 Consider an irreducible, aperiodic DTMC with state space $E = \{0, 1, 2 \ldots\}$. The chain has the property

$$D_i \triangleq E(X_{n+1} - X_n | X_n = i) = \begin{cases} 1, & i = \text{even}, \\ -i, & i = \text{odd}. \end{cases}$$

Can you determine whether π_∞ is 0 or 1? If no, why?

10.92 Can you construct an ergodic, nonstationary MC?

CTMC; Simple Skills

10.93 Consider a CTMC with three states 1, 2, 3 and

$$Q = \begin{bmatrix} -1 & 1 & 0 \\ 1 & -2 & 1 \\ 1 & 0 & -1 \end{bmatrix}.$$

Write the equations for $P[X(t) = i], i = 1, 2, 3$. If you can, find $P[X(t) = i]$.

10.94 A CTMC has two states, labeled 0 and 1. Its parameters are $q_{00} = -\lambda_0$ and $q_{11} = -\lambda_1$, respectively. Calculate $P_{00}(t)$.

10.95 Consider the Wiener process $\{X(t)\}$ in Chapter 8.
(a) Verify that it is a continuous-time Markov process.
(b) Is it continuous or discrete space?
(c) Is it homogeneous?

10.96 Is the process $\{Y(t)\}$ in Problem 8.33 a Markov process?

10.97 Is the process $\{Y(t)\}$ in Problem 8.34 a Markov process?

CTMC; Advanced Skills

10.98 Consider a birth-death process with parameters $\lambda_i = 0$, $\mu_i = i\mu$, where $\mu > 0$ is a given constant (pure death process). Let $X(0) = i$.
(a) Write the Chapman-Kolmogorov equations.
(b) Find $p_i(t)$.
(c) Find $EX(t), \text{var}(X(t))$.

10.99 Consider a birth-death process with parameters $\lambda_i = \lambda i$, $\mu_i = \mu i$, where $\lambda > 0$, $\mu > 0$ are given constants.
(a) Let $m_i(t) \triangleq E[X(t)|X(0) = i]$. Show that $m_i(t) = ie^{(\lambda - \mu)t}$.
(b) Let $v_i(t) \triangleq E[X^2(t)|X(0) = i]$. Show that

$$\frac{dv_i(t)}{dt} = 2(\lambda - \mu)v_i(t) + (\lambda + \mu)m_i(t).$$

(c) Show that

$$v_i(t) = \begin{cases} ie^{2(\lambda-\mu)t}[i + \frac{\lambda+\mu}{\lambda-\mu}(1 - e^{-(\lambda-\mu)t})], & \lambda \neq \mu, \\ i(i + 2\lambda t), & \lambda = \mu. \end{cases}$$

10.100 Consider a birth-death process with linear death rates. What should the birthrates be, for the process to be positive recurrent?

10.101 Determine the recurrency properties of the birth-death CTMC with

$$q_{ij} = \begin{cases} \frac{i+2}{2(i+1)}, & j = i + 1, \\ \frac{i}{2(i+1)}, & j = i - 1, \\ 0, & \text{otherwise.} \end{cases}$$

Modeling

10.102 Consider the following simplistic model of voice sources. A source always alternates between periods of speech and silence. Suppose that speech periods are independent random variables, exponentially distributed with parameter λ_1; similarly, silence periods are independent random variables, exponentially distributed with parameter λ_2. Speech and silence periods are independent of each other.
(a) Model the source activity as a CTMC. Determine the rate matrix.
(b) Consider now N independent voice sources, the outputs of which are multiplexed. Let $X(t)$ denote the number of sources that are not silent at time t. Find the rate matrix for the CTMC $\{X(t)\}$.
(c) Write the global balance equations and solve them.

10.103 Consider a full duplex connection between two computer network nodes. The two links fail independently. The times between failures are IID, exponentially distributed random variables, with parameter $p > 0$. The times to repair either link are also IID, exponentially distributed random variables, with parameter $q > 0$. At time 0, both links are operational. Calculate the probability that both links will be operational at time $t > 0$.

10.104 A telephone trunk has a capacity to handle M telephone calls simultaneously (i.e., the trunk has M channels). Calls arrive at the trunk according to a Poisson process at rate $\lambda > 0$. The call durations form an IID sequence of exponential random variables, with parameter μ.
(a) Find the probability that in steady state, k channels are busy, $0 \leq k \leq M$.
(b) Find the probability that an arriving call will be blocked.

10.105 Consider a telephone switch with a capacity so large it can be practically thought of as infinite. Suppose that telephone calls arrive at the switch

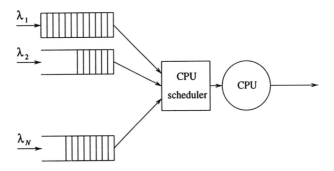

Figure 10.25 The queuing model.

according to a Poisson process at rate λ. Call durations form an IID sequence, with rate μ. Let $X(t)$ denote the number of calls in progress at time $t > 0$. Under what conditions on the pdf of the call durations is $X(t)$ a birth-death process?

10.106 Consider a population $X(t)$ that produces subsequent generations as follows: A member of the present generation waits a random amount of time T from its birth and then produces two offspring with probability p and dies or dies without any offspring with probability $1 - p$. Time T is exponentially distributed with parameter λ. Can $X(t)$ be modeled as a CTMC?

Projects

10.107 Consider the following CPU scheduling problem. (User) programs submitted for execution are partitioned into N classes. The rate of submission for class i is λ_i programs per second. The execution time of a class i program is a random variable with mean EX_i and variance σ_i^2. We may assume that the sequences of program interarrival and execution times form independent, IID processes. A queuing model for this system is shown in Figure 10.25.

The queues have infinite capacity, so no program submission is blocked. For simplicity we may assume that an executing program cannot be preempted (we can rectify that later).

The objective of the scheduling algorithm is the following. We are interested in keeping the average response time for programs of class i, call it ER_i, below a class-dependent, given threshold g_i. In math lingo, we want to devise a CPU scheduling algorithm to solve the following problem:

$$ER_i \leq g_i, \qquad i = 1, \ldots, N.$$

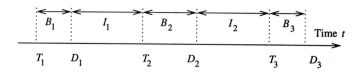

Figure 10.26   Busy and idle periods.

Here is an idea that uses dynamic priorities to solve this problem. Let T_n denote a program arrival instant such that the program finds the system empty for the nth time (see Figure 10.26). Let D_n denote a departure instant such that a finished program leaves the system empty for the nth time. And B_n is called the nth busy cycle; I_n is the nth idle period.

We wish to set up class priorities only at the beginning of each busy cycle. The priorities will be kept constant throughout the busy cycle. Of course, they may change from busy cycle to busy cycle. From now on, we consider the special case of two program classes only, for simplicity of presentation.

Consider the following quantity:

$$P_i(T_n) = \frac{1}{\lambda_i} \frac{1}{g_i} \frac{1}{T_n} \int_0^{T_n} \eta_i(t)\, dt, \quad i = 1, 2.$$

Here $\eta_i(t)$ is the queue size (i.e., the number of programs in the system memory) including the program currently being executed, if any, of class i at time t. The integral (divided by T_n) is the time-average queue size; if the system is ergodic and if n is very large, its value should be equal to the average queue size in steady state. Then, from Little's law, the ratio

$$\frac{1}{\lambda_i} \frac{1}{T_n} \int_0^{T_n} \eta_i(t)\, dt$$

should converge to ER_i.

We propose to use $P_1(T_n)$ and $P_2(T_n)$ as the priority indices since (at least for large time values) they represent estimates of ER_i/g_i. The scheduler can be now described as follows:

At time T_n compute $P_i(T_n)$:

If $P_1(T_n) > P_2(T_n)$, give higher priority to class 1.

If $P_1(T_n) < P_2(T_n)$, give higher priority to class 2.

If $P_1(T_n) = P_2(T_n)$, give higher priority to class 2 (arbitrary).

The rationale for such an algorithm is the following: By giving it higher priority, such an algorithm always helps the class with the highest current delay. It is dynamic (it needs to measure queue sizes), and thus we expect it to be "adaptive" (whatever that means.) Do the following:

(a) Express $P_i(T_{n+1})$ recursively, in terms of $P_i(T_n)$.

(b) Is $\{P_i(T_n)\}_{n=1}^{\infty}$ a MC? Explain in any detail you like.

(c) Is $\{P_1(T_n) - P_2(T_n)\}_{n=1}^{\infty}$ a MC? Explain.

(d) If you answered no to (b) and (c), can you identify a MC related to this system? Discuss qualitatively the drifts of this chain.

(e) Consider the $(P_1(T_n), P_2(T_n))$ plane. The two priority indices determine a point in this plane (for any given n). As n varies, this point moves around the plane. Observe that the priority indices are random variables since they depend on the queue size. Assume for the time being that the priority indices converge to some constant values as n approaches ∞. Can you identify the region of the plane where the point should lie, in steady state?

(f) Describe the response of this scheduling algorithm to an overload.

(g) If you found that response unsatisfactory, suggest ways to improve it.

(h) Simulate the system for various λ, EX_i, σ_i^2 values. Plot the behavior of the priority indices as a function of time.

(i) Simulate a few scheduling algorithms of your own choice. Calculate P_1 and P_2, the steady-state values of the priority indices. Plot these values in the $P_1 P_2$ plane for all the scheduling algorithms you have simulated. Do the points fall on a straight line? Can you explain?

Computer Problems

10.108 Plot a few sample paths of the Ehrenfest MC in Problem 10.74.

10.109 Evaluate and plot the pmf $P[X_n = k]$ of the Ehrenfest MC in Problem 10.74. Assume that $P[X_0 = i] = 1$ for some $i \neq 0, N$.

10.110 Calculate and plot $\{\pi_k\}$, the stationary distribution for the Ehrenfest MC in Problem 10.74.

10.111 Evaluate numerically how fast the pmf $P[X_n = k]$ converges to π_k for the Ehrenfest MC in Problem 10.74. Vary the parameter N.

10.112 Calculate a histogram of sojourn times for the Ehrenfest MC in Problem 10.74.

10.113 Define a nontrivial MC that exhibits oscillatory behavior.
(a) Plot a few sample paths.
(b) Plot the pmf $P[X_n = k]$ for various n.

10.114 Define a nontrivial MC that exhibits divergent behavior.
(a) Plot a few sample paths.
(b) Plot the pmf $P[X_n = k]$ for various n.

10.115 Simulate and plot sample paths of the Yule process. Calculate experimentally the pdf of the time till $X(t) = 10$, given that $X(0) = 1$.

10.116 Simulate and plot sample paths of a birth-death process. Evaluate the steady-state distribution experimentally.

10.117 Consider a random walk on the set $\{0, 1, 2, \ldots, M\}$ with an absorbing barrier at state M and reflecting barrier at state 0. Simulate the walk, and calculate experimentally the pmf of the time until absoprtion. Assume that $X_0 = 0$.

10.118 Consider a random walk on the set $\{0, 1, 2, \ldots, M\}$ with reflecting barriers at states 0 and M. Simulate the walk, and calculate experimentally the long-term proportion of time the walk spends in state i, $i = 0, 1, 2, \ldots, M$. Assume that $X_0 = 0$.

11

CASE STUDY: A BUS-BASED SWITCH ARCHITECTURE

In this chapter, we analyze in detail a realistic communication network problem. The goal is twofold. First, we see how the *transition* can be made, from the description of a fairly complex physical system to the parameters of a simplified, mathematical, stochastic model. Second, we see how to select among the specific models of the theory of probability and random processes.

The system under study is a computer network switch that is based on a bus architecture. The switch is used to interconnect fairly high-speed communication lines, namely, T1, T3, and OC3c links. A T1 line has a raw capacity of approximately 1.5×10^6 bits per second (1.5 Mb/s); a T3 link has a raw capacity of about 45 Mb/s, while the OC3c link, based on ATM technology, has a capacity of 155 Mb/s.

The analysis of this switch is done by simulation. The simulator was written in C. The architecture is simulated accurately, so the only difference between the simulation and an actual experiment is due solely to the characteristics of the traffic patterns generated.

In Section 11.1 we present the switch architecture and briefly describe the basic rules of its operation. In Section 11.2 we describe in detail how the mathematical model of the switch can be obtained. In Section 11.3 we present the system model. In Section 11.4 we introduce the criteria by which the performance of the switch can be judged. In Section 11.5 we describe in detail the input parameters that affect this performance. In Section 11.6 we analyze the results of the simulation. Finally, a brief description of the simulator is given in Section 11.7.

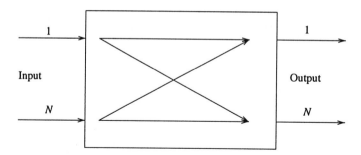

Figure 11.1   An $N \times N$ switch.

11.1   SWITCH ARCHITECTURE AND OPERATION

Computer network switches are devices that interconnect *input links* to *output links*, as Figure 11.1 depicts. Usually, but not necessarily, the number of input links is equal to the number of output links, in which case we talk about an $N \times N$ switch. The special hardware used to interface between the switch and a communication line is called an *adapter*.

The basic operation of a switch is to *receive* a packet from the input link, *store* it in a buffer until a decision is made about which output link should be selected, and finally *transmit* the packet over the outgoing link.

The interconnection between the input and output links (in other words, the switch *architecture*) can be implemented in a number of ways. Two of the most basic and common architectures are the *bus* and *crossbar* switch (see Figure 11.2).

There are advantages and disadvantages to both; mainly, a bus-based architecture is less expensive. A cross-bar architecture is faster since it can achieve parallelism—two or more packets can be transferred from the input to the output part simultaneously. The architecture we consider in this study is a bus-based one, shown in greater detail in Figure 11.3(a).

A *master bus* (MB) interconnects a number N of adapters. The adapters can support T1, T3, or OC3c links. In our study, a T1 adapter can support up to four output and four input T1 links. A T3 adapter can support up to two input and two output links. An OC3c adapter can support one input and one output link only.

Regardless of its type, each adapter can support three priority classes of traffic. The three classes are labeled HP, MP, and LP (high, medium, and low priority). For simplicity of buffer management, the buffer space available

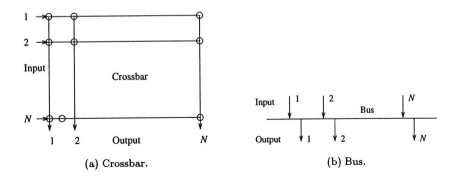

Figure 11.2 Switch architectures.

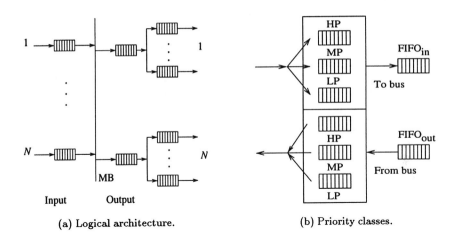

Figure 11.3 Bus-based switch architecture.

at each adapter is *partitioned* between the three traffic classes (i.e., no buffer sharing), as Figure 11.3(b) depicts.

From the logical configuration point of view, the adapter can be partitioned into two parts, namely, input and output [see Figure 11.3(b)]. At the input part, packets arriving from the input links are routed to the corresponding class buffer. An *internal bus* transfers these packets to the $FIFO_{in}$ queue, which acts as an interface to the MB. The transfer of packets from the three input buffers follows a *nonpreemptive, exhaustive priority* rule. In other words, before a packet from a lower-priority buffer is transferred via the internal bus (or, as we say, before it is served), all higher-priority buffers must be emptied; but once a lower-priority packet is on the bus, a higher-priority packet must wait until the transfer is complete.

Once the packets are in the $FIFO_{in}$ queue, however, they are served (by the MB) in a *first-in first-out* (FIFO) order. This is done to simplify the protocol of the MB operation since implementing a FIFO scheme requires a few pointers only. The MB interconnects the various adapters. Its operation follows a *gated, round-robin* service. The MB protocol can be briefly described as follows:

The MB operates in *MB cycles*. At the beginning of such a cycle, the ith FIFO queue records N_i, the number of packets present in the $FIFO_{in}$ buffer at that time. Only N_i packets are transferred during this cycle. Packet arrivals during this cycle are served at the next cycle (gated service). Therefore, during the given cycle, a total of $\sum_i N_i$ packets are transferred by the MB, from the $FIFO_{in}$ queues to the corresponding $FIFO_{out}$ queues of the adapters. The identification of the destination adapter for a given packet is stored in a field in the packet header; the output queue controllers examine this field and determine whether the packet is copied or not.

Once a packet is deposited into the $FIFO_{out}$ queue of an adapter, the *internal output bus* of the adapter routes the packet to the corresponding priority output buffer [see Figure 11.3(a)]. Packets are transferred by the internal bus in a FIFO order. Once again, the output buffer controllers determine the priority of a packet and decide whether to copy a packet or not. The total output buffer space is also partitioned among the three priority classes. Finally, from the output buffers, the packets are transferred to the output link(s). This transfer follows again a priority mechanism, preemptive or not. In T1 adapters, we assume that this mechanism is preemptive. In T3 and OC3c adapters, the mechanism is nonpreemptive.

11.2   TRANSITION TO A STOCHASTIC MODEL

What is random here? There are a variety of sources that introduce randomness in the operation of this physical system. Let's look at the broader picture of a network, part of which is our switch S (see Figure 11.4). The users of this network will generate traffic, part of which will go through switch S. The amount of traffic through S is a random process, for five basic reasons:

1. The *number* of users in the network is random.
2. The *volume* of packets a given user generates is random.

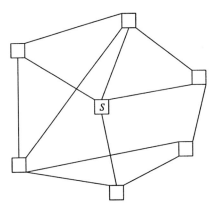

Figure 11.4   The network picture.

3. The exact *time* a user will generate a packet is random.
4. The *address* of the user receiving these packets is random.
5. The *size* of the packets is random.

Therefore, the analysis of such a system can only be done via the theory we developed in this book. The question is, then, How do we build a *mathematical or simulation model* out of the technical descriptions given in Section 11.1 and the information about the nature of the randomness given above? It is the primary goal of this case study to outline this transition from a realistic system to a model. Therefore, we proceed through the steps of this transition in detail.

11.3   SYSTEM MODEL

The system operates in continuous time, so it is natural to think of continuous-time random processes to describe it. There are, however, a number of events taking place in more or less discrete-time instants. Indeed, packet transmission times are multiples of some basic time unit (e.g., a 1-bit transmission time). Packet arrival times are also determined, at least in part, by transmission times since packets will arrive at the switch S through a T1, T3, or OC3c link. One might argue that the actual start time of transmission can occur at any real-valued instant; however, as far as operation and the performance measures of the system (to be introduced shortly) are concerned, we can think of those instants as discrete valued as well.

Based on the above considerations, we decided to model the system as a discrete-time one; therefore, its behavior can be captured by a *sequence of random variables*, as opposed to a *random process*. Moreover, all arrival processes can be modeled by sequences of random variables. The advantage of the discrete- over continuous-time choice is simplicity in generating sample paths of all those processes.

The next question is, What kind of a model should we adopt for the arrivals of packets over a link? This model should be able to capture the randomness caused by factors 1 through 3 in the list of the previous section. We chose to model the arrival process at each link as an interrupted Bernoulli process (IBP). This is very similar to the IPP in Section 8.1.5, except that it operates in discrete time: Given that at the beginning of a time slot the process is in the on state (off state) at the end of a time slot the process remains in the on state (off state) with probability p_{on} (p_{off}). The only difference is that while the system is in the on state, arrivals will occur with probability 1 (as opposed to rate λ). Such a model offers two advantages: It is simple and can capture burstiness very well, as does its IPP counterpart. We have assumed for simplicity that the three priority classes contribute equally to the overall link load.

There is a system requirement to direct all OC3c traffic to T1 output links. (Thus, for example, we can study the effects of mismatching speeds of communication links.) Based on that requirement, the fourth cause of randomness is modeled as follows: An incoming packet is routed to an output link proportionately to the capacity of that link. Thus the routing was uniform across all output links. One way to implement this uniform destination address selection is via a sequence of IID random variables since addresses of different packets do not have any correlation among them. The IID modeling assumption will be satisfactory if the sequence of packets on an input link is thoroughly multiplexed between a large number of users, making successive addresses independent. If there is enough reason to believe that there is correlation between addresses, a different model that can capture correlation should be used. For example, a Markov modulated Bernoulli process (similar to MMPP, but in discrete time) could be a good choice.

Finally, based on system requirements again, the model for the packet size, the fifth cause of randomness, is the following: The highest priority class contributes packets of a constant size equal to 280 bytes. The other two classes have packets of variable length, which is chosen uniformly in the region $[12, 2048]$ bytes. This also results in representing the packet sizes with an IID sequence of discrete uniform random variables.

The system model of an input adapter is shown in Figure 11.5. Arrivals are IBP; a Bernoulli-like splitting (see Section 8.1.7) of the input process is used to generate input to the three priority buffers. The internal adapter bus (IAB) empties these buffers into the $FIFO_{in}$ queue, in a nonpreemptive priority fashion. The output adapter operates in a similar fashion. The $FIFO_{out}$ queue receives packets from the master bus. The internal adapter bus puts the packets

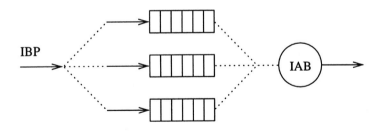

Figure 11.5   The input adapter model.

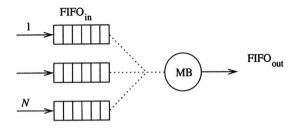

Figure 11.6   The master bus model.

in the corresponding priority queues. Then the packets exit the system on one of the output links. The master bus (MB in Figure 11.6) serves all the $FIFO_{in}$ queues, in a gated round-robin fashion, with no overhead spent on empty queues.

Having identified the models, next we select their parameters. We already saw that, due to system requirements, the packet size parameters were fixed. We describe the rest in Section 11.5. Before we do that, let's identify the performance measures.

11.4   PERFORMANCE MEASURES

In general, performance measures can be defined for steady-state or transient operation, since this is a dynamic system. The performance measures chosen for this study reflect only steady-state operation (for simplicity). They are the following:

1. Probability of packet loss per link
2. Total probability of packet loss per priority class
3. Average packet delay per priority class
4. Average packet delay per link

How are these measures determined from the simulation (or in actual operation of the system)? Let's discuss in detail the probability of packet loss per link; the reasoning for the rest is similar.

Let's define the following random variables, for $i = 1, 2, \ldots$ (losses are confined to the link in discussion only):

$$L_i \triangleq \begin{cases} 1, & \text{packet } i \text{ is lost,} \\ 0, & \text{packet } i \text{ is not lost.} \end{cases}$$

It is natural to define (and measure) P_l, the probability of packet loss on the link in discussion, as

$$P_l \triangleq \lim_{n \to \infty} \frac{1}{n} \sum_{i=1}^{n} L_i. \tag{11.1}$$

In a sense, we are observing one sample path of the system only; then, appealing to ergodicity, we calculate a statistical average (P_l) as a time average (through Equation 11.1). Since we have assumed that all sequences of random variables in the system are IID, it is not difficult to see that the sequence $\{L_i\}$ will have the ergodicity property. Indeed, let's denote by TL_k the cumulative packet losses during busy cycle k, on the link in discussion. Then we can rewrite the sum in Equation 11.1 as a sum over busy cycles, as follows:

$$P_l \triangleq \lim_{n \to \infty} \frac{1}{n} \sum_{k=1}^{n} TL_k. \tag{11.2}$$

We can appeal to the Strong Law of Large Numbers to justify Equation 11.2 since the sequence $\{TL_k\}$ is an IID sequence. Finally, based on our discussions in Section 6.6.1, simulating a small number of busy cycles should suffice to approximate the limit in Equation 11.2.

The following system parameters have a major effect on these performance measures.

1. Input packet rate. Since we chose an IID sequence, we can see from our discussions in Section 8.1.6 that the inverse of the average interarrival time can be used as the rate of the input packet process.

2. Input burstiness. We see in the next section how the burstiness can be described as a function of the model parameters for our IBP.

3. Buffer size (per priority class).

4. Priority allocation (in all buses).

5. Bus speeds.

These parameters may be fixed throughout the operation of the system. For example, once a bus speed is chosen, it remains unchanged. Priority allocations may be changed throughout the life of the switch, in order to adapt to changing traffic profiles, for example. However, such changes are in practice very rare and may happen only a few times, due to hardware, cost, and complexity considerations. Therefore, we opted to fix the priority allocation (in all buses) and bus speed in our model. We chose to fix the input packet rate so that an overall utilization of 85 percent was achieved in most of our experiments. Utilizations as low as 65 percent were also studied. Therefore, the only *free* variables in this study are input burstiness and the buffer partitioning.

11.5   INPUT DATA DESCRIPTION

The following parameters were used as input data to the simulation model. As mentioned above, some were given fixed values throughout the experiments.

11.5.1   Fixed Data Description

The master bus has a speed of 1.6 Gb/s. It operates on a 40-nanosecond (ns) clock and can transfer 64 bits in parallel. The input and output adapter buses are much slower; they can carry 32 bits in parallel. For T1 adapters, the speed is about 272 Mb/s and the clock cycle 118 ns. For T3 adapters, the numbers are 370 Mb/s and 87 ns; for OC3c adapters, 450 Mb/s and 71 ns. All buses introduce some overhead per packet transfer; for this study, we fixed it at 5 bytes per transfer. In all adapters, the capacity of the $FIFO_{in}$ and $FIFO_{out}$ queues was chosen equal to 16 kbytes.

11.5.2   Variable Data Description

1. C^2 coefficient of variation

The coefficient of variation C^2 is a widely used, simple measure of the burstiness of a traffic source. It is given by

$$C^2 = \frac{EX^2}{(EX)^2},$$

where X denotes the interarrival time between successive packets. We chose three values for this parameter, in an effort to cut down the number of possible combinations: $C^2 = 5, 10, 20$. Purely Poisson traffic has a C^2 value equal to 1, while digitized voice traffic has a coefficient of variation of about 20. No consensus has been reached in the scientific community on the coefficient value that represents data traffic. In general, the higher this value, the more bursty the traffic.

that represents data traffic. In general, the higher this value, the more bursty the traffic.

The coefficient of variation C^2 and the utilization ρ of a link can be determined from the $p_{\text{on}}, p_{\text{off}}$ values of the IBP input model. The formulas for the conversion are

$$C^2 = \frac{(p_{\text{on}} + p_{\text{off}})(1 - p_{\text{on}})}{(2 - p_{\text{on}} - p_{\text{off}})^2},$$

$$\rho = \frac{1 - p_{\text{off}}}{2 - p_{\text{on}} - p_{\text{off}}}.$$

2. Buffer size

Initially, we ran our experiments for various C^2 values and fixed buffer allocations of 64 kbytes for the two lower-priority classes. The allocation for the highest-priority class was 16 kbytes for T1 adapters and 16 kbytes for T3 and OC3c adapters. We noticed that the buffer utilization for this class was low, due to the priority bus scheduling scheme. We also observed that the middle-priority class was experiencing some loss, in some experiments, while the lowest-priority class was experiencing significant loss. Then we decided to vary the buffer allocations for the two lower-priority classes, in an effort to determine minimal buffer allocations for which the loss would be negligible for all classes.

11.6   DISCUSSION OF RESULTS

The main conclusion we were able to derive from this performance study was the following. The lowest-priority class experienced a considerable loss and delay. Figure 11.7 depicts the probability of loss of typical runs for a given value of coefficient of variation, varying utilization, and varying buffer size for LP traffic. The loss becomes worse as input burstiness increases. Most of this loss is incurred at the output part of the adapter. Figure 11.8 depicts average delay results (in milliseconds for the LP class, in seconds for the MP class) for the same experiments.

This behavior is primarily due to the *head-of-line* (HOL) blocking phenomenon, which we describe briefly as follows. Since buffer space at the output part of the adapter is assigned according to priority classes, sometimes the HOL packets are destined for the same output link(s). Therefore, one or more output links may be effectively blocked, even though packets in positions other than the HOL may be available. This phenomenon effectively reduces the capacity of the output links. It causes a drop in the output link utilization and increases delay and loss. Its impact is more profound on the lowest-priority class. Its impact is of course eliminated when an adapter supports only one output link.

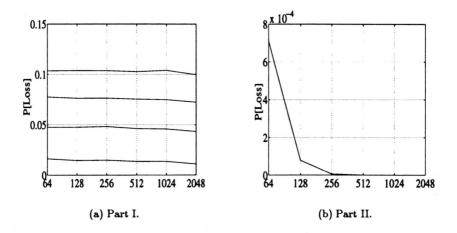

(a) Part I.

(b) Part II.

Figure 11.7 Loss performance.

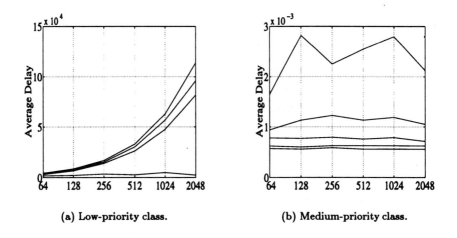

(a) Low-priority class.

(b) Medium-priority class.

Figure 11.8 Average delay performance.

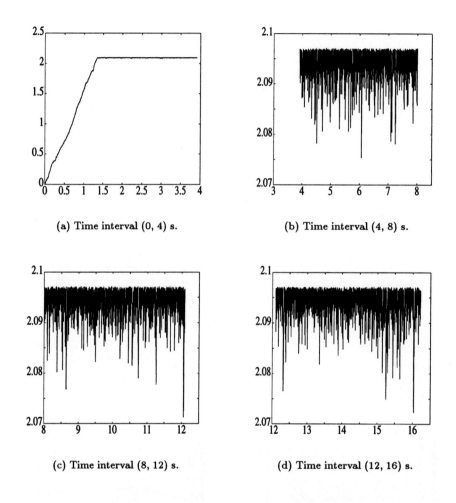

(a) Time interval (0, 4) s.

(b) Time interval (4, 8) s.

(c) Time interval (8, 12) s.

(d) Time interval (12, 16) s.

Figure 11.9   Buffer contents, in megabytes.

A more complex buffer management scheme is needed to avoid the effects of this phenomenon, which appears to be the main reason for the losses incurred. Indeed, buffer size did not have a significant effect on loss in the cases where more than one output link was supported per adapter. Figures 11.9 and 11.10 depict typical sample paths of the buffer contents in units of megabytes and thousands of packets, for an actual operation of about 16 s.

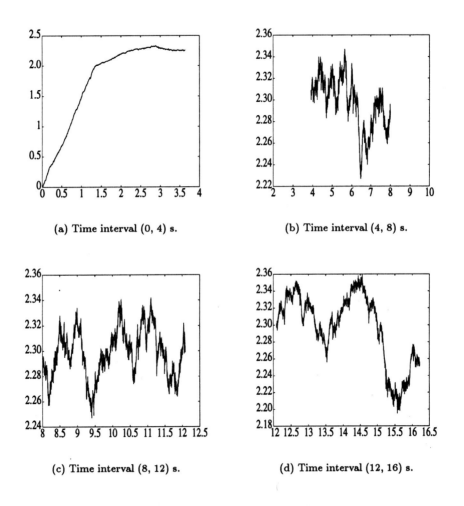

(a) Time interval (0, 4) s.

(b) Time interval (4, 8) s.

(c) Time interval (8, 12) s.

(d) Time interval (12, 16) s.

Figure 11.10   Buffer contents, in thousands of packets.

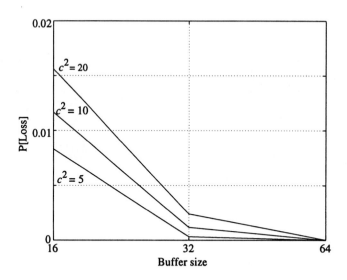

Figure 11.11   Loss for 85 percent utilization.

However, buffer size had an effect when only one link was supported. Figure 11.11 shows loss measures for 85 percent utilization and one link per adapter. The buffer sizes used were the smallest ones for which nonzero loss was observed. At lower utilizations (and for the same buffer sizes) we did not observe any losses.

Figure 11.12 depicts a typical sample path of the LP traffic buffer contents, at the output part of a T3 adapter, in units of kilobytes. The capacity of the buffer is 64 kbytes, utilization is 85 percent, and the coefficient of variation of the input traffic is 5. The adapter supports multiple links in this experiment. Typical sample paths were observed in all other adapters and experiment setups.

Figure 11.13 depicts the same sample paths (in units of megabytes) for buffer capacity of 2048 kbytes.

The main result derived from those graphs is the HOL blocking effect. Regardless of the buffer capacity, the buffer contents increase to capacity and fluctuate stochastically around that value, even though (at least theoretically) 15 percent of the time the buffer should be empty.

Higher-priority traffic monopolizes a bus for relatively long periods. This trend is more evident when input traffic is burstier. Lower-priority traffic is given then a larger chance to accumulate and start experiencing delays (and ultimately losses).

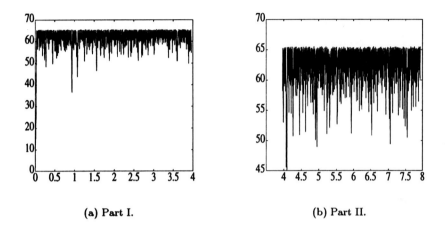

(a) Part I. (b) Part II.

Figure 11.12   LP traffic buffer contents, 64-kbyte buffer.

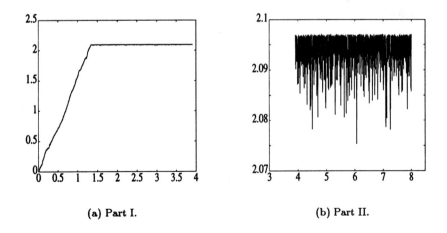

(a) Part I. (b) Part II.

Figure 11.13   LP traffic, 2048-kbyte buffer.

There are two ways to improve performance for the lowest-priority class: (1) implement a more flexible priority mechanism and (2) increase the buffer allocation for this class. We opted for the second approach since it is easier to implement. Buffer increase can be effected by buffer-sharing rules. Alternatively, we can retain the partitioning scheme and simply increase the allocated buffer space. We followed this approach in this study.

11.7   DESCRIPTION OF THE SIMULATOR

The simulator was written in C; it compiles under Unix and AIX. All our experiments were run on a DEC 3100 machine at North Carolina State University. The main parts of the simulator are described below.

The gated algorithm for the MB protocol is coded in procedure *Bus-algo*. The operation of the internal input bus is coded in procedure *IntBus-in*. The operation of the internal output bus is coded in procedure *IntBus-out*. Provision is taken to follow a preemptive or nonpreemptive policy, given the adapter type.

The traffic specifications and processing are handled in routines *next-arrival*, *traffic-in*, and *traffic-out*. The collection *tools.c* contains some general-purpose queue manipulation routines, such as *insert* (insertion of a packet to the end of a queue), *remv* (removal of a packet from the front of a queue), and *switch* (transfer of a packet from the front of one queue to the end of another).

The collection *event.c* contains event manipulation procedures, such as *remove-event* (remove a preempted packet departure from a list of future events), *delete-event* (remove an event from the front of a list), and *insert-event* (insert a future event in the proper position in a list of future events).

The main program, *main.c*, starts with parameter input and system variable initialization. It essentially implements an event-driven simulation and calculates the performance measures of interest by using the method of batches. The main loop $while(t_{global} < t_{stop})$ implements the event processing cycle. Time advances from event to event and appropriate routines take actions based on the event that has just occurred (i.e., external packet arrival, internal bus transfer, MB transfer, or packet departure). Performance statistics are collected inside the *traffic-out* routine (delay) and the *insert* routine (loss). The main program ends with an output of the performance results.

A

SET THEORY PRIMER

In this appendix we review several set-theoretic concepts we need in Chapters 2, 3, and 4. For a good introduction to set theory, see [20, 27].

A.1 SETS AND SUBSETS

A.1.1 Definitions and Examples

Definition: A **set** is a collection of **elements**.

Example A.1. The integers 1, 2, 10 are elements of the set of all positive integers. The set of all months contains 12 elements.

The elements of a set are unique. We write $a \in A$ to denote that object a is an element of set A. We write $a \notin A$ to denote that object a is *not* an element of set A.

There are three ways to specify a set: via element enumeration, through a relationship that its elements satisfy, and the graphical method. Some of the sets we use very frequently in this book are given in the next three examples, expressed via element enumeration and/or through a relationship that their elements satisfy. And R is the set of real numbers.

Example A.2. The set of all positive integers N is

$$N = \{1, 2, 3, 4, 5, \ldots\}.$$

Example A.3. The familiar two-dimensional plane R^2 is

$$R^2 = \{(x, y) : x \in R, \ y \in R\}.$$

The notation here reads as follows: R^2 is the set of all ordered pairs (x, y) such that x is a real number and y is a real number. The following set is the

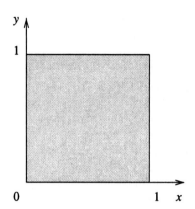

Figure A.1   The set of all points in the unit square.

Euclidean n-dimensional space.

$$R^n = \{(x_1, \ldots, x_n) : x_1 \in R, \ldots, x_n \in R\}.$$

Example A.4. The line segment $[0, 1]$ and the unit square are, respectively,

$$[0,1] = \{x \in R : 0 \le x \le 1\},$$

$$[0,1]^2 = \{(x,y) \in R^2 : 0 \le x \le 1 \text{ and } 0 \le y \le 1\}.$$

A graphical representation of the set $[0,1]^2$ is given in Figure A.1.

In describing a set, very often an English description is also used, as an equivalent form of enumeration, such as

$$A = \{\text{the entire population in the United States}\};$$

such a notation is handy when the set contains a large number of elements. When the exact nature of the set is not important, a **Venn diagram**[1] is customarily used to represent a set. In such a diagram, a two-dimensional, closed curve, usually a circle or rectangle, represents a set; the generic element of a set is sometimes represented by a dot inside the closed curve. Venn diagrams are mostly useful in depicting operations on sets, as we see in Section A.2.

Definition: A set A is a **subset** of a set B if every element of A is also an element of B.

We denote this as $A \subseteq B$. If B contains at least one element that is not an element of A, then we call A a **proper subset** of B and denote this as $A \subset B$.

---

[1] After John Venn, 1834–1923, an English mathematician.

Alternatively, we call B a **superset** of A, and denote this as $B \supseteq A$, if A is a subset of B. If A is a proper subset of B, we call B a **proper superset** of A and denote this as $B \supset A$. Two sets A, B are called **equal** if and only if $A \subseteq B$ and $B \subseteq A$ (i.e., if they contain exactly the same elements). We write $A = B$ to denote equality of two sets.

Example A.5. The set N of positive integers is a proper subset of the set of real numbers R; and R is a proper subset of the set R^2. The unit square is a proper subset of R^2.

The **universal set** S is a set that contains all elements we are considering in a given study. All sets are then subsets of S. Usually, in a Venn diagram, the universal set is depicted as a rectangle, and generic sets are depicted as circles, squares, or arbitrary-shaped regions within the rectangle.

The **complement** of a set A, denoted as A^c, is the set that contains all elements of the universal set S which are not elements of A. Mathematically,

$$A^c \triangleq \{x \in S : x \notin A\}.$$

The **empty set** contains no elements at all, by definition; it is denoted as $\emptyset$. By default, the empty set is a subset of any set A; moreover, $\emptyset^c = S$.

Consider a set A with $n < \infty$ elements, labeled $e_1, e_2, \ldots, e_n$; how many subsets does A have? Let's count:

- 1 subset with exactly 0 elements, namely, the empty set.

- n subsets with exactly 1 element, namely, the sets $\{e_1\}, \{e_2\}, \ldots, \{e_n\}$.

- $n(n-1)/2$ subsets with exactly 2 elements, namely, the sets $\{e_1, e_2\}$, $\{e_1, e_3\}, \ldots, \{e_1, e_n\}, \{e_2, e_3\}, \{e_2, e_4\}, \ldots, \{e_2, e_n\}, \ldots, \{e_{n-1}, e_n\}$.

- In general, $\binom{n}{k}$ subsets with exactly k elements.

In total, then, there are $1 + n + n(n-1)/2 + \cdots + 1$ subsets, or

$$\sum_{k=0}^{n} \binom{n}{k} = 2^n$$

subsets of any set with n elements.

The set that contains as its elements all subsets of a set A is called the **powerset** of A; it is usually denoted as $2^{|A|}$, and we have to admit that this is a somewhat odd notation.

Of course, the powerset of a set A with an infinite number of elements contains an infinite number of subsets. Moreover, these subsets cannot be enumerated when A has an uncountably infinite number of elements (e.g., when $A = [0, 1]$).

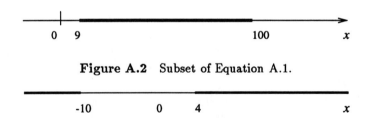

Figure A.2   Subset of Equation A.1.

Figure A.3   Subset of Equation A.2.

A.1.2   Examples of Subsets

In Chapters 3 and 4 we need some expertise and familiarity with subsets of the real line $R = (-\infty, \infty)$ and the two-dimensional space R^2. Here are examples of some subsets of these two sets, described via a relationship and graphically. Make sure you feel comfortable with the relationship description of those sets since we use it a lot in Chapters 3 and 4.

Example A.6.   *Line Segment.*

$$A \triangleq \{x \in R : 9 \leq x \leq 100\} \qquad (A.1)$$

Example A.7.

$$A \triangleq \{x \in R : x^2 + 6x - 40 \geq 0\} \qquad (A.2)$$

Example A.8.   *Vertical Strip.* Consider the set

$$A \triangleq \{(x,y) \in R^2 : 1 \leq x \leq 2\}, \qquad (A.3)$$

shown in Figure A.4. Note that the vertical lines at the points $x = 1$ and $x = 2$ belong to A. Note also that the strip extends from $-\infty$ to $+\infty$ (the same holds true for the sets in Equations A.4, A.5, A.6, A.7, A.9, A.10, and A.11).

Example A.9.   *Horizontal Strip.* See Figure A.5(a).

$$A \triangleq \{(x,y) \in R^2 : -5 \leq y \leq 5\}. \qquad (A.4)$$

Note that the horizontal lines at the points $y = 5$ and $y = -5$ belong to A.

Example A.10.   *Two Vertical Strips.* See Figure A.5(b).

$$A \triangleq \{(x,y) \in R^2 : 1 \leq |x| \leq 2\}. \qquad (A.5)$$

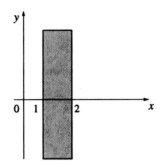

Figure A.4 Subset of Equation A.3.

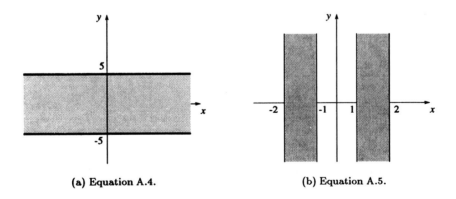

(a) Equation A.4.   (b) Equation A.5.

Figure A.5 The subsets of Equations A.4 and A.5.

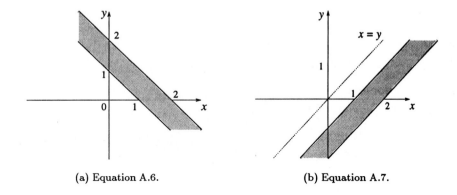

(a) Equation A.6.   (b) Equation A.7.

Figure A.6   Tilted strips.

Example A.11. *"Tilted" Strip.* This set is shown in Figure A.6(a).

$$A \triangleq \{(x,y) \in R^2 : 1 \le x + y \le 2\}. \tag{A.6}$$

Example A.12. *Another "Tilted" Strip.* This set is shown in Figure A.6(b).

$$A \triangleq \{(x,y) \in R^2 : 1 \le x - y \le 2\}. \tag{A.7}$$

Example A.13. *Rectangle.* This set is shown in Figure A.7(a).

$$A \triangleq \{(x,y) \in R^2 : 1 \le x \le 2, \ 10 \le y \le 20\}. \tag{A.8}$$

Example A.14. *"Infinite" Rectangle.* This set is shown in Figure A.7(b).

$$A \triangleq \{(x,y) \in R^2 : x \le 2, \ y \le 20\}. \tag{A.9}$$

Example A.15. *Line.* This set is shown in Figure A.8(a).

$$A \triangleq \{(x,y) \in R^2 : x = 2\}. \tag{A.10}$$

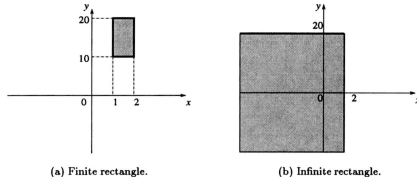

(a) Finite rectangle.　　　　　　(b) Infinite rectangle.

Figure A.7　Rectangles.

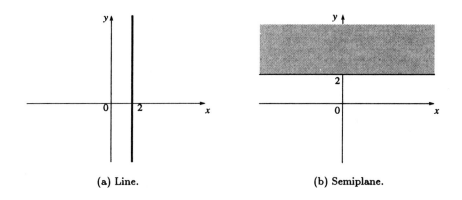

(a) Line.　　　　　　(b) Semiplane.

Figure A.8　A line and a semiplane.

Example A.16. *Semiplane.* This set is shown in Figure A.8(b).

$$A \triangleq \{(x, y) \in R^2 : y \geq 2\}. \tag{A.11}$$

Example A.17. *Circle.*

$$A \triangleq \{(x, y) \in R^2 : x^2 + y^2 = 5\}.$$

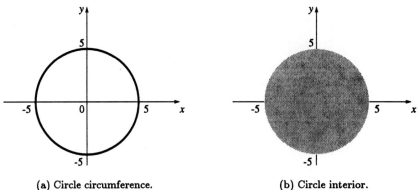

(a) Circle circumference.   (b) Circle interior.

Figure A.9   Circular sets.

This set, shown in Figure A.9(a), contains just the circumference of the circle, while the next one, shown in Figure A.9(b), contains only the interior of the circle.

$$A \triangleq \{(x, y) \in R^2 : x^2 + y^2 < 5\}.$$

A.2   OPERATIONS ON SETS

Given two sets A and B that are subsets of a universal set S, we can create additional sets by the following operations on A and B.

A.2.1   Union of Sets

The **union** of A and B, denoted by $A \cup B$, is defined as

$$A \cup B \triangleq \{x \in S : x \in A \text{ or } x \in B\}.$$

In other words, the union of A and B contains *all* the elements in A and B, regardless of whether these elements are common to both sets. Of course, common elements appear in the union set only once since the elements of a set must be unique.

Example A.18. Let A be the set in Figure A.2 and $B = [0, 10]$. Then

$$A \cup B = [0, 100].$$

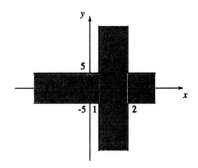

Figure A.10 The union of the sets in Equations A.3 and A.4.

Example A.19. Let A be the set in Figure A.4 and B be the set in Figure A.5(a). Then $A \cup B$ is shown in Figure A.10. The two strips extend, of course, to $-\infty$ and $+\infty$.

Example A.20. Let A be the set in Figure A.9(a) and $B = \{(0,5)\}$. Then $A \cup B = A$.

A.2.2 Intersection of Sets

The **intersection** of A and B, denoted by $A \cap B$, is defined as

$$A \cap B \overset{\triangle}{=} \{x \in S : x \in A \text{ and } x \in B\}.$$

In other words, the intersection of A and B contains only the elements that are common to *both* A and B.

If the intersection of two sets is the empty set, the sets are called **disjoint** (or *mutually exclusive*). In other words, A and B are disjoint if and only if they contain no common elements, as Figure A.11(a) suggests.

Example A.21. Let A be the set in Figure A.2 and $B = [0, 10]$. Then

$$A \cap B = [9, 10].$$

Example A.22. Let A be the set in Figure A.4 and B be the set in Figure A.5(a). Then $A \cap B$ is shown in Figure A.11(b). Note that the boundary lines are included in the intersection.

Example A.23. Let A be the set in Figure A.9(a) and $B = \{(0,5)\}$. Then $A \cap B = B$.

Note that, by definition, intersection and union are symmetric in A and B; therefore, $A \cup B = B \cup A$ and $A \cap B = B \cap A$. The union and intersection operations are easily extended for three or more sets. We leave that as an exercise.

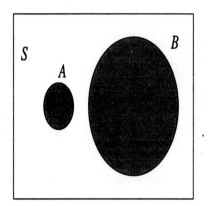

(a) Two disjoint sets.

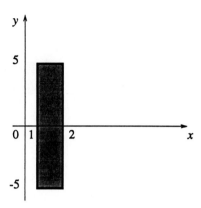

(b) The intersection of the sets in Equations A.3 and A.4.

Figure A.11

A.2.3   Difference of Sets

The **set difference** between A and B, denoted by $A - B$, is defined as

$$A - B \triangleq \{x \in A : x \in A \text{ and } x \notin B\}.$$

In other words, the difference set $A - B$ contains all the elements of A that are not elements of B. It is easy to see that $A - B = A \cap B^c$.

Example A.24. Let A be the set in Figure A.2 and $B = [0, 20]$. Then

$$A - B = (20, 100], \quad B - A = [0, 9).$$

Example A.25. Let A be the set in Figure A.4 and B be the set in Figure A.5(a). Then $A - B$ is shown in Figure A.12(a). The horizontal line segments

$$L_1 \triangleq \{(x,y) : 1 \le x \le 2, \ y = 5\}, \quad L_2 \triangleq \{(x,y) : 1 \le x \le 2, \ y = -5\}$$

do not belong to $A - B$.

Example A.26.   Let A be a circle with radius 10 and center at the origin. Let B be the set in Figure A.5(a). Then $A - B$ is shown in Figure A.12(b).

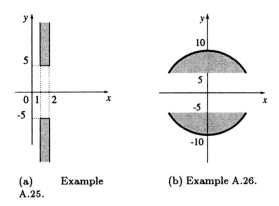

(a)　　Example
A.25.

(b) Example A.26.

Figure A.12　Difference of sets.

A.2.4　Symmetric Difference of Sets

The **symmetric difference** between A and B, denoted by $A\Delta B$, is defined as

$$A\Delta B \triangleq (A - B) \cup (B - A).$$

In other words, the symmetric difference set contains only the elements of $A\cup B$ that are not elements of both A and B. We can easily check that

$$A\Delta B = (A \cap B^c) \cup (B \cap A^c).$$

Example A.27. Let A be the set in Figure A.2 and $B = [0, 10]$. Then

$$A\Delta B = [0, 9) \cup (20, 100].$$

Example A.28. Let A be the set in Figure A.4 and B be the set in Figure A.5(a). Then $A\Delta B$ is shown in Figure A.13.

Example A.29. Let A be the set in Figure A.9(a) and $B = \{(5,0)\}$. Then $A\Delta B$ is the set A *without* the point (5,0).

A.2.5　Properties of Set Operations

The set operations introduced in the previous section have a number of properties, which are easily verified and thus omitted; consult [20] for proofs.

1. Commutative

$$A \cup B = B \cup A, \quad A \cap B = B \cap A.$$

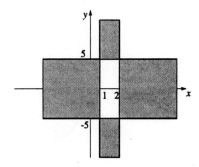

Figure A.13 The symmetric difference of the sets in Equations A.3 and A.4.

2. Associative

$$A \cup (B \cup C) = (A \cup B) \cup C, \quad A \cap (B \cap C) = (A \cap B) \cap C.$$

3. Distributive

$$A \cup (B \cap C) = (A \cup B) \cap (A \cup C), \quad A \cap (B \cup C) = (A \cap B) \cup (A \cap C).$$

4. De Morgan's laws:[2]

$$(A \cap B)^c = A^c \cup B^c, \quad (A \cup B)^c = A^c \cap B^c.$$

A.2.6   Cartesian Products of Sets

These products[3] play a fundamental role in the notion of independence, so we take a close look at them here. Let A and B be two nonempty sets. The **Cartesian product** of A and B is denoted as $A \times B$ and defined as follows:

$$A \times B \triangleq \{(x, y) : x \in A, \ y \in B\}.$$

In other words, the elements of $A \times B$ are *ordered pairs*, with the first element of the pair selected from set A and the second from set B. Observe that $A \times B \neq B \times A$ in general.

Example A.30. An example of a Cartesian product we have seen already is the two-dimensional plane R^2. Here $A = R$ and $B = R$ as well.

Example A.31. Figures A.14 and A.15 show the definition for various sets A, B.

---

[2]Named after A. de Morgan, 1806–1871, an English mathematician.

[3]Named after René Descartes, 1596–1650, a French mathematician.

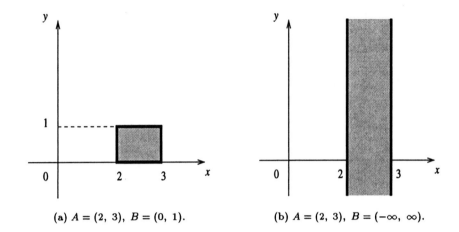

(a) $A = (2, 3)$, $B = (0, 1)$.   (b) $A = (2, 3)$, $B = (-\infty, \infty)$.

Figure A.14 Cartesian products.

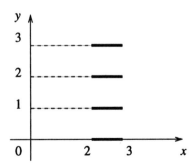

Figure A.15 Cartesian product of $A = [2, 3]$, $B = \{0, 1, 2, 3\}$.

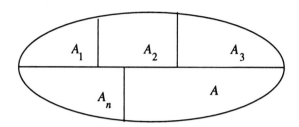

Figure A.16   Partition of A.

A.3   PARTITIONS OF A SET

Partitions are useful in conditioning of events; they are used in the theorem of total probability, in Chapter 2. Consider a nonempty set A and a collection of disjoint sets $\{A_k\}_{k=1}^n$ (where n is a given fixed positive integer) with the property $\cup_{k=1}^n A_k = A$ (see Figure A.16).

Definition: The collection $\{A_k\}_{k=1}^n$ is called a **partition** of A.

A.4   LIMITS OF SEQUENCES OF SETS

In Chapters 6 and 7 we deal with an unbounded number of measurements and events. The notion of limits of sequences of sets is very useful in that respect.

Let $A_1, A_2, \ldots$ denote a sequence of sets that are all elements of a universal set S.

Definition: A sequence of sets is called **monotonic increasing (or monotonic decreasing)** if

$$A_1 \subset A_2 \subset \cdots$$

(or)

$$A_1 \supset A_2 \supset \cdots.$$

Definition: Consider a monotonic increasing sequence of sets $A_1, A_2, \ldots$; the **limit** of the sequence is denoted as $\lim_{n\to\infty} A_n$ (or A_∞) and defined as

$$\lim_{n\to\infty} A_n \triangleq \{x \in S : x \in A_n, \text{ for some } n\}.$$

It is easy to see that

$$\lim_{n\to\infty} A_n = \cup_{n=1}^\infty A_n.$$

Definition: Consider a monotonic decreasing sequence of sets $A_1, A_2, \ldots$; the **limit** of the sequence is defined as

$$\lim_{n \to \infty} A_n \triangleq \{x \in S : x \in A_n, \text{ for all } n = 1, 2, 3, \ldots\}.$$

Again, it is easy to see that

$$\lim_{n \to \infty} A_n = \cap_{n=1}^{\infty} A_n.$$

Definition: Consider an arbitrary sequence of sets $A_1, A_2, \ldots$; the **limit supremum** and **limit infimum** of the sequence are defined as

$$\limsup_{n \to \infty} A_n \triangleq \lim_{n \to \infty} \cup_{i=n}^{\infty} A_i, \quad \liminf_{n \to \infty} A_n \triangleq \lim_{n \to \infty} \cap_{i=n}^{\infty} A_i.$$

Here are some examples of limits of sequences of sets:

Example A.32. *The Singleton* $\{a\}$. Let a be a real number; denote by $A_n = (a - \frac{1}{n}, a + \frac{1}{n}]$ an interval of length $\frac{2}{n}$ around a. The sequence $\{A_n\}$ is monotonic decreasing; then

$$\{a\} = \lim_{n \to \infty} A_n.$$

Example A.33. *The Interval* $[a, b]$. Let $A_n = [a - \frac{1}{n}, b + \frac{1}{n}]$. This sequence is also monotonic decreasing. Then

$$[a, b] = \lim_{n \to \infty} A_n.$$

A.5 ALGEBRAS OF SETS

Consider a universal set S and a collection Q of subsets of S. Let $A, B \in Q$ be two arbitrary elements of this collection. Let A_n, $n = 1, 2, \ldots$, be a sequence of sets with $A_n \in Q$, $\forall n$.

Definition: The collection Q is called a **field** if the following properties are satisfied:

1. $S \in Q$.

2. $A \in Q$ implies $A^c \in Q$.

3. $A, B \in Q$ implies $A \cup B \in Q$.

The idea behind requiring the second and third properties is to make sure that the collection Q is closed under unions and complementations. By induction, a field contains all *finite* unions of its elements. From de Morgan's laws, it contains all finite intersections of its elements as well.

Definition: The field Q is called a σ **field** if the following property is satisfied:

$$A_1, A_2, \ldots \in Q \text{ implies } \cup_{n=1}^{\infty} A_n \in Q.$$

A σ field is therefore guaranteed to contain all (finite or *infinite*) unions, intersections, complements, differences, symmetric differences, and limits of its elements. This is the reason why in Chapter 2 we require that the collection of events be a σ field; only then are we guaranteed that arbitrary set operations on events will create sets on which probability can be defined.

Of particular interest in Chapters 3 and 4 are σ fields associated with the real line R and the plane R^2. Consider the real line R first; intervals of the form

$$[a, b], \quad (a, b], \quad [a, b), \quad (a, b) \tag{A.12}$$

(i.e., all forms of closed/open intervals) are of natural interest in random experiments, since measurements are usually real numbers or integers.

Many σ fields contain the sets in A.12; the "smallest" one is called the **Borel** σ field and is denoted by $\mathcal{B}(R)$. This σ field is intuitively constructed as follows: We start with the intervals in A.12 as elements of a collection σ_1. We then augment this collection with the unions, complements, and limits of every set in σ_1, creating a new collection σ_2. This process is repeated ad infinitum, generating a sequence of collections $\{\sigma_n\}$; the Borel σ field is then σ_∞. Theoretically, the Borel σ field can be thought of as the intersection of all σ fields that contain the sets in A.12.

Consider now the two-dimensional plane. Squares and rectangles are natural events of interest in experiments that involve double measurements (see, for example, Chapter 4). The smallest σ field that contains such subsets of R^2 is called again the Borel σ field on R^2 and is denoted by $\mathcal{B}(R^2)$.

A.6   PROBLEMS

In all problems in this appendix, N is the set of all nonnegative integers, and R is the set of all real numbers.

Set Descriptions

A.1 List all possible subsets of the set $A = \{1, 2, 3, 4\}$.

A.2 Describe the following sets, using English statements; then graph the sets D, E, F, and G.

$A = \{x \in N : x^2 + 10 = 110\}$,
$B = \{x \in R : 3x - 9 \geq 0\}$,
$C = \{x \in R : 3x - 9 = 0\}$,
$D = \{(x,y), x \in R, \; y \in R : 2x + y = 10\}$,
$E = \{(x,y), x \in R, \; y \in R : 2x + y \geq 10\}$,
$F = \{(x,y), x \in R, \; y \in R : 2x + y \leq 10\}$,
$G = \{(x,y), x \in R, \; y \in R : x^2 + y^2 \leq 16\}$.

A.3 State whether the following sets are countable or uncountable.

$A = \{\text{all prime numbers}\}$,
$B = \{x \in R : 1000 > x > 0\}$,
$C = \{x \in N : 1000 > x > 0\}$,
$D = \{\text{all real numbers of the form } m/n, \text{ with } m \in N, n \in N\}$,
$E = \{x \in N : x^2 + 100 = 50\}$.

A.4 Prove that if a set A contains an uncountable subset B, then A is uncountable.

A.5 Consider the set $S = \{\text{SAT scores of all college students in United States}\}$. Provide at least three English statements related to S to describe the empty set.

A.6 Cantor's[4] definition of a set as a collection of elements is too broad; it can lead to paradoxes, such as the following: "A barber in a certain village has declared that he will cut the hair of those villagers and only those villagers who do not cut their own hair."
(a) Identify the universal set S and the set A of villagers whose hair the barber will cut.
(b) Is the barber an element of A?

Set Operations

A.7 Let S be the nonnegative orthant, that is, $S = \{(x,y) \in R^2 : x \geq 0, \; y \geq 0\}$. Let

$$A = \{(x,y) \in S : x + 2y > 10\}, \quad B = \{(x,y) \in S : x + 2y > 100\},$$
$$C = \{(x,y) \in S : x^2 + y^2 > 10\}, \quad D = \{(x,y) \in S : x^2 + y^2 \leq 100\}.$$

Find the algebraic expressions for the following sets and then sketch a few of them in the (x, y) plane.

---

[4] Georg Cantor, 1845–1918, a German mathematician and the father of axiomatic set theory.

(a) A^c, B^c, C^c, D^c,
(b) $A \cap B, A \cap C, A \cap D, B \cap C, A \cap B \cap C \cap D$,
(c) $A \cup B, A \cup C, A \cup D, B \cup C, B \cup D, A \cup B \cup C \cup D$,
(d) $A - B, A - C, B - C, B - D, D - B, C - D, (A - B) - C$.

A.8 Let

$$A = \{n \in N : n \text{ is divisible by } 2\}, \quad B = \{n \in N : n \text{ is divisible by } 3\},$$
$$C = \{n \in N : n \text{ is divisible by } 6\}.$$

Find:
(a) A^c, B^c, C^c,
(b) $A \cap B, A \cap C, B \cap C, A \cap B \cap C$,
(c) $A \cup B, A \cup C, B \cup C, A \cup B \cup C$,
(d) $A - B, A - C, B - C, B - A, C - B, C - A, (A - B) - C$.

A.9 Let $A = \{x \in R : 0 \le x \le 10\}$, $B = \{x \in R : 5 \le x \le 20\}$. Find

$$A^c, \quad B^c, \quad A \cap B, \quad A \cup B, \quad A - B, \quad B - A.$$

A.10 Show that if $A \subseteq B$, then (a) $A \cap B = A$, (b) $A \cup B = B$, (c) $B^c \subseteq A^c$.

A.11 Consider two sets A and B, with $A \cap B = \emptyset$. Show that $A \subseteq B^c$ and $B \subseteq A^c$.

A.12 Show that if $A \subseteq B$ and $B \subseteq C$, then $A \subseteq C$, first using **Venn diagrams** and then using the definition of a subset.

A.13 Let A_1 and A_2 be two sets, not necessarily disjoint. Express $A_1 \cup A_2$ as the union of *two* disjoint sets, using Venn diagrams; then express it as the union of three disjoint sets.

A.14 Let A_k, $k = 1, 2, \ldots, n$, be a sequence of sets, not necessarily disjoint. Let $A_0 = \emptyset$. Show that

$$\cup_{k=1}^n A_k = \cup_{k=1}^n \left(A_k \cap_{j=1}^{k-1} A_j^c \right).$$

A.15 For any $a \in (0, 1]$, let $A_a \triangleq [0, \frac{1}{a})$. Find

$$A \triangleq \cup_{a \in (0,1]} A_a, \quad B \triangleq \cap_{a \in (0,1]} A_a.$$

A.16 Consider the sets of real numbers

$$A_n \triangleq \{r \in R : r > \frac{1}{n}\}, \quad n \ge 1.$$

Calculate $\bigcup_{n=1}^{\infty} A_n$, $\bigcap_{n=1}^{\infty} A_n$.

A.17 Show that $C - A \subseteq (C - B) \cup (B - A)$; in other words, a triangle inequality holds for set differences. When does strict equality hold?

A.18 If $A - B = A - C$, is it true that $B = C$?

A.19 Prove or disprove the statement $(A - B) - C = A - (B - C)$.

A.20 Show that $A \Delta B = A^c \Delta B^c$.

A.21 Show that $A = B \Delta C$ if and only if $C = A \Delta B$.

A.22 Using Venn diagrams or the definitions show that
(a) $(A \cup B)^c \cap C = C - [(A \cap C) \cup (B \cap C)]$,
(b) $[A \cap (B \cup C)]^c = (A^c \cup B^c) \cap (A^c \cup C^c)$,
(c) $(A \cap B \cap C)^c = A^c \cup B^c \cup C^c$,
(d) $(A \cup B) - B = A \cap B^c$,
(e) $(A \cup B) - (A \cap B) = (A \cap B^c) \cup (A^c \cap B)$,
(f) $(A \cup B \cup C) - (A \cap B \cap C) = (A^c \cap B) \cup (B^c \cap C) \cup (C^c \cap A)$,
(g) $[A - (A \cap B)] \cup B = A \cup B$.

A.23 Simplify the following expressions as much as you can:
(a) $(A \cup B) \cup (A \cup B^c)$,
(b) $(A \cup B) \cap (A \cup B^c)$,
(c)$(A \cup B) \cap (B \cup C)$,
(d) $(A \cup B \cup C) \cup B^c$,
(e) $A \cup (A^c \cap B)$.

A.24 State which assertions are true and which are not:
(a) $A \cap B \cap C = A \cap C \cap (B \cup C)$,
(b) $A \cup B \cup C = A \cup (B - A) \cup [C - (A \cup B)]$,
(c) $(A \cup B) - A = B$,
(d) $(A \cup B)^c \cap C = A^c \cap B^c \cap C$,
(e) $(A^c \cup B^c)^c \cup (A^c \cup B)^c = B$,
(f) $(A \cap B)^c \cap (A \cup B) = (A \cap B^c) \cup (A^c \cap B)$.

Set Partitions

A.25 Consider the set $S = \{$SAT scores of all college students$\}$. Construct at least two partitions of this set. Describe them in words.

A.26 Construct a partition of N that contains exactly two sets. Can you generalize to a partition that contains exactly n sets, where n is any given positive integer?

A.27 Consider the plane R^2.
(a) Show graphically a partition that involves sets that are equilateral triangles. Can you derive an algebraic expression for those sets?
(b) Can you construct a partition that involves sets that are circles? Explain.
(c) Can you construct a partition that involves sets that are pentagons? Explain.

A.28 Consider a partition $\{A_i\}$ of a given set S and a nonempty subset B of S. Let $C_i \triangleq B \cap A_i$. Show that $\{C_i\}$ is a partition of B.

A.29 Consider a partition $\{A_i\}$ of a given set S. Let B, C be subsets of S; define $D_i = (B \cap A_i) \cup (C \cap A_i)$. Is $\{D_i\}$ a partition of $B \cup C$?

Cartesian Products

A.30 Find the Cartesian product $A \times B$ when
(a) $A = \{1, 2, 3\}$, $B = \{1, 2, 3\}$,
(b) $A = \{1, 2, 3\}$, $B = \{4, 5, 6\}$,
(c) $A = \{4, 5, 6\}$, $B = \{1, 2, 3, 4\}$,
(d) $A = \{1, 2, 3\}$, $B = R$,
(e) $A = R$, $B = \{1, 2, 3\}$,
(f) $A = [0, 1]$, $B = [1, 2]$,
(g) $A = [0, 1]$, $B = (1, \infty)$,
(h) $A = [0, 1]^2$, $B = R$,
(i) $A = R^n$, $B = R^m$.

Sequences of Sets

A.31 Consider two arbitrary sequences of sets A_i, B_i. Show that

$$(\cup_{i=1}^{\infty} A_i) \triangle (\cup_{i=1}^{\infty} B_i) \; \subseteq \; \cup_{i=1}^{\infty} (A_i \triangle B_i),$$
$$(\cap_{i=1}^{\infty} A_i) \triangle (\cap_{i=1}^{\infty} B_i) \; \subseteq \; \cap_{i=1}^{\infty} (A_i \triangle B_i).$$

A.32 Let I be a nonempty set. Consider a set A and a sequence of sets A_i, $i \in I$. Show that

$$A - \cup_{i \in I} A_i = \cap_{i \in I} (A - A_i), \quad A - \cap_{i \in I} A_i = \cup_{i \in I} (A - A_i).$$

A.33 Consider a sequence of countable sets $A_i, i \in N$. Prove that the union $\cup_{i \in N} A_i$ is also countable.

A.34 Consider two arbitrary sets A, B. Define the sequence

$$A_i = \begin{cases} A, & i = \text{odd}, \\ B, & i = \text{even}. \end{cases}$$

Find $\limsup_{n \to \infty} A_n$, $\liminf_{n \to \infty} A_n$.

A.35 Show that, for any sequence of sets A_i, $\limsup_{n \to \infty} A_n \supseteq \liminf_{n \to \infty} A_n$.

A.36 Show that, for any sequences of sets A_i, B_i,

$$\begin{aligned} \limsup_{n \to \infty}(A_n \cup B_n) &= \limsup_{n \to \infty} A_n \cup \limsup_{n \to \infty} B_n, \\ \liminf_{n \to \infty}(A_n \cap B_n) &= \liminf_{n \to \infty} A_n \cap \liminf_{n \to \infty} B_n. \end{aligned}$$

A.37 Show that for any sequence of sets A_i,
(a) $(\liminf_{n \to \infty} A_n)^c = \limsup_{n \to \infty} A_n^c$,
(b) $(\limsup_{n \to \infty} A_n)^c = \liminf_{n \to \infty} A_n^c$,
(c) If $\lim_{n \to \infty} A_n = A$, then $\lim_{n \to \infty} A_n^c = A^c$,
(d) If $\lim_{n \to \infty} A_n = A$, then $\lim_{n \to \infty} A_n \Delta C = A \Delta C$, for any set C.

Sigma Fields

A.38 Consider a set A with four elements. Construct two fields on A. Construct two σ fields.

A.39 Consider the random experiment in which a die is thrown. Construct the *smallest* field that contains all the outcomes.

A.40 Consider the roulette random experiment in which a number between 1 and 36 is drawn at random (for simplicity, we ignore the 0 and star outcomes of the roulette game). Suppose that the events of interest are only "even" and "odd." Construct the smallest field that contains these events.

A.41 Are Cartesian products of σ fields σ fields?

A.42 Consider an IID sequence of random variables $\{X_i\}_{i=1}^{\infty}$. Let σ_n denote the σ field generated by $\{X_1, \ldots, X_n\}$. Show that $\{\sigma_i\}_{i=1}^{\infty}$ is an increasing sequence of sets.

A.43 Let $S = [0, \infty)$ and Q be the σ field generated by intervals of the form $(n, n+1)$, where n is a positive integer.
(a) Describe some of the sets in Q.
(b) Let $k > m$ be two nonnegative integers. Is $[m, k)$ in Q?
(c) Let $b > a$ be two nonnegative, noninteger real numbers. Are (a, b), $[a, b)$, $(a, b]$, and $[a, b]$ in Q?

A.44 Let $S = (-\infty, \infty)$ and Q be the set

$$Q \triangleq \{A \in R : A \text{ is countable or } A^c \text{ is countable}\}.$$

Show that Q is a σ field.

B

COUNTING METHODS

Ordering of elements of sets arises very naturally in numerous situations in probability theory. In this appendix we review several counting concepts we need to better understand sequential experiments and the binomial model. In the following discussion, we assume that $k < \infty$ *finite* sets $A_1, \ldots, A_k$ are given. The number of elements of set A_i is denoted by n_i, $i = 1, 2, \ldots, k$.

B.1   ORDERING OF ELEMENTS

Suppose we are given k sets $A_1, \ldots, A_k$, and we choose *exactly one* element from each. Let's denote the element we chose from set A_i as x_i. We arrange the chosen elements in a k-tuple $(x_1, x_2, \ldots, x_k)$. How many possible k-tuples can we form?

Theorem B.1 *Let N_o denote the number of possible k-tuples we can form when we choose exactly one element from each one of k sets $A_1, \ldots, A_k$, where set A_i has n_i elements. Then*

$$N_o = n_1 \cdot n_2 \cdots n_k.$$

Proof: The theorem is proved by induction on k. When $k = 1$, the theorem holds true, since when we have only one set A_1 with n_1 elements, there are exactly n_1 1-tuples we can arrange. Assume that the theorem holds true for $k = m$. Therefore, the number of possible m-tuples we can arrange is $n_1 \cdot n_2 \cdots n_m$. Consider now the case $k = m + 1$. We can choose an element from set A_{m+1} in n_{m+1} different ways. For each such choice, the number of $(m + 1)$-tuples we can arrange is equal to $n_1 \cdot n_2 \cdots n_m$. Then the total number of $(m + 1)$-tuples is given by $n_1 \cdot n_2 \cdots n_{m+1}$. QED. □

A very useful corollary is obtained when all sets have the same number of elements.

Corollary B.1 *Let N_o denote the number of possible k-tuples we can form when we choose exactly one element from each one of k sets $A_1, \ldots, A_k$, where each set contains n elements. Then*

$$N_o = n^k.$$

B.2   SAMPLING WITH ORDERING AND WITHOUT REPLACEMENT OF ELEMENTS

Consider now a single set A with n elements; suppose we choose k elements from A *in succession and without replacing the chosen elements.* Again we arrange the chosen elements in a k-tuple $(x_1, x_2, \ldots, x_k)$. How many possible k-tuples can we form?

Theorem B.2 *Let N_o denote the number of possible k-tuples we can form when we choose k elements from a set A of $n \geq k$ elements without replacing the chosen elements. Then*

$$N_o = n \cdot (n - 1) \cdots (n - k + 1).$$

Proof: Define the sequence of sets A_j, $j = 1, \ldots, k$, as follows:

$$A_1 \stackrel{\triangle}{=} A,$$
$$A_j \stackrel{\triangle}{=} A_{j-1} - \{\text{element chosen from } A_{j-1}\}, \quad j = 2, \ldots, k.$$

Sampling with ordering and without replacement of elements of a single set A is equivalent to ordering of elements of the sets $A_1, \ldots, A_k$. The result now follows from Theorem B.1 since the cardinality of the set A_j is equal to $n - (j - 1)$. □

A very useful corollary is obtained when $n = k$. In this case, we talk of a *permutation* of n elements.

Corollary B.2 *Let N_o denote the number of possible n-tuples we can form when we choose all n elements from a set A, in order. Then*

$$N_o = n \cdot (n - 1) \cdots 2 \cdot 1 \stackrel{\triangle}{=} n!. \tag{B.1}$$

Table B.1 Stirling's approximation.

n	Exact Value	Stirling Approximation
1	9.22137008895789e$-$01	1.00000000000000e+00
2	9.59502175744491e$-$01	2.00000000000000e+00
3	1.12317905953687e+00	6.00000000000000e+00
4	1.46913594580583e+00	2.40000000000000e+01
5	2.11119105760909e+00	1.20000000000000e+02
6	3.28739900297308e+00	7.20000000000000e+02
7	5.48808363228209e+00	5.04000000000000e+03
8	9.74179576480877e+00	4.03200000000000e+04
9	1.82663655358405e+01	3.62880000000000e+05
10	3.59869561874103e+01	3.62880000000000e+06
20	2.36600278004017e+05	2.43290200817664e+18
30	1.84346512192855e+10	2.65252859812193e+32
40	7.40526288149906e+15	8.15915283247894e+47
50	1.01882818205865e+22	3.04140932017130e+64
100	9.32484762526934e+57	9.3326215443948e+157

The symbol $n!$ is called the *factorial* of n. (For convenience, we define $0! = 1$.) Equation B.1 is quite often used for large n. The celebrated *Stirling formula*[1] provides a tractable approximation which avoids dealing with factorials:

$$n! \approx \sqrt{2\pi} n^{n+0.5} e^{-n} \qquad \text{(B.2)}$$

The relative accuracy of this formula is shown in Table B.1.

B.3 SAMPLING WITHOUT ORDERING AND WITHOUT REPLACEMENT OF ELEMENTS

Consider again a single set A with n elements; suppose we choose k elements from A, *all at once and without ordering them*. Again we arrange the chosen elements in a k-tuple $(x_1, x_2, \ldots, x_k)$. How many possible k-tuples can we form?

---

[1] After James Stirling, 1692–1770, an English mathematician. This approximation was instrumental in de Moivre's development of the normal pdf and Laplace's proof of the central limit theorem.

Theorem B.3 *Let N_o denote the number of possible k-tuples we can form when we choose k elements from a set A of $n \geq k$ elements, all of them at once and without ordering. Then*

$$N_o = \binom{n}{k}.$$

Proof: From Theorem B.2, there are $n \cdot (n-1) \cdots (n-k+1)$ different ways we can choose a k-tuple, *with* ordering of its elements taken into account. From Theorem B.2, there are $k!$ different ways we can order a k-tuple; all are accounted for as one in our sampling, and thus

$$N_o = \frac{n \cdot (n-1) \cdots (n-k+1)}{k!}.$$

QED   □

The right-hand side of the above equation is usually denoted as $\binom{n}{k}$. The quantity $\binom{n}{k}$ is called a *combinatorial coefficient*. It was discovered and studied by Pascal, in the seventeenth century. An alternative, useful relationship is

$$\binom{n}{k} = \frac{n!}{(n-k)!k!}.$$

A number of useful properties of the combinatorial coefficient are examined as problems at the end of this appendix.

Sampling without ordering and without replacement of elements is in effect equivalent to partitioning the set A into *two* subsets A_1 and A_2, with k and $n-k$ elements each. There is a generalization of this concept; what happens if we partition A into $m > 2$ subsets with l_i elements each? When we choose an element from each subset without ordering and without replacement of elements, how many m-tuples can we form?

Theorem B.4 *Let N_o denote the number of possible m-tuples we can form when we choose (without ordering and without replacement) l_i elements from set A_i such that $l_1 + l_2 + \cdots + l_m = n$. Then*

$$N_o = \frac{n!}{l_1! l_2! l_3! \cdots l_m!}.$$

Proof: There are $\binom{n}{l_1}$ different ways of choosing the element x_1 of the m-tuple. There are $n - l_1$ elements left, when we consider the set A_2, so there are $\binom{n-l_1}{l_2}$ different ways of choosing the element x_2 of the m-tuple. Inductively, there are

$$\binom{n - \sum_{j=1}^{k-1} l_j}{l_k}$$

different ways of choosing the kth element of the m-tuple. Therefore,

$$
N_o = \binom{n}{l_1} \cdot \binom{n - l_1}{l_2} \cdots \binom{n - \sum_{j=1}^{m-1} l_j}{l_m}
$$

$$
= \frac{n!}{(n - l_1)! l_1!} \cdot \frac{(n - l_1)!}{(n - l_1 - l_2)! l_2!} \cdots \frac{n - \sum_{j=1}^{m-1} l_j}{(n - \sum_{j=1}^{m} l_j)! l_m!},
$$

and the desired result follows after the obvious simplifications. $\qquad\square$

Note that N_o is called the *multinomial coefficient*.

B.4 SAMPLING WITHOUT ORDERING AND WITH REPLACEMENT OF ELEMENTS

Consider again a single set A with n elements; suppose we choose k elements from A *in succession and with replacement of the chosen elements.* Again we arrange the chosen elements in a k-tuple $(x_1, x_2, \ldots, x_k)$. How many possible k-tuples can we form? Here $k > n$ is possible, since we replace the elements after we choose them.

Theorem B.5 *Let N_o denote the number of possible k-tuples we can form when we choose k elements from a set A of n elements, in succession and with replacement of the chosen elements. Then*

$$
N_o = \binom{n - 1 + k}{k}.
$$

Proof: Consider sampling a set B with $n + k - 1$ elements without ordering and *without* replacement of elements. This process is equivalent to sampling a set A with n elements without ordering and with replacement of elements. Indeed, when one element of A is replaced, one element of B is "thrown away"; after $k - 1$ samplings, the set B is reduced to n elements. The kth (and last) sampling is the same for both sets, and the fact that we "throw away" an element from B does not matter, since there are no further samplings. Therefore, from Theorem B.3, we get

$$
N_o = \binom{n - 1 + k}{k}.
$$

QED. $\qquad\square$

B.5   PROBLEMS

Apply Concepts

B.1 Consider a color monitor for a quality workstation. Suppose it has 2^{10} pixels horizontally, 2^8 pixels vertically, and 2^3 levels of red, green, and blue per pixel. A color image (sometimes called a *bit-map*) can be displayed on the monitor by selecting appropriate red, green, and blue levels to generate any other color.
(a) How many different colors can we generate per pixel?
(b) How many different images can we display?
(c) How many bytes do we need to store a color image that is the full size of the monitor?

B.2 Suppose that in any given day there are 10 course slots; for simplicity, assume that each class starts on the hour and lasts 1 h. Suppose that you take five classes during a given semester, and each class meets twice a week.
(a) Suppose that on a given day you take three classes. How many schedules are possible for that day?
(b) Suppose that lunchtime is from 12 to 1, and classes can be held during lunchtime as well. How many schedules are there that do not conflict with lunchtimes (first things first...)?
(c) Consider your weekly schedule. In how many ways can you arrange your weekly schedule?
(d) If you want a schedule with no more than three classes per day, how many choices you have?
State any further assumptions you may deem appropriate.

B.3 An airline has 100 first pilots, 150 second pilots, and 1000 flight attendants. A flight needs 1 first pilot, 1 second pilot and 5 flight attendants. How many flights can be scheduled? Do first pilots, second pilots, or flight attendants limit the maximum number of flights?

B.4 Consider a combination lock with three wheels. Each wheel contains 10 positions, numbered 0–9. How many lock combinations are there? If a thief spends 3 s trying out a particular combination, how many minutes will it take (on average) to break your combination, starting from the all-zero combination and trying all combinations in succession?

B.5 A password is a string of, say, eight ASCII characters.
(a) Count the characters on your keyboard; how many passwords can you generate?
(b) Assume that a hacker has a program that tries out passwords by sequentially

going through the ASCII set of characters. If it takes 1 s to try a password, how many days until your password is broken?

B.6 Calculate the odds of winning a lottery in which you must correctly select 6 numbers out of 50. Repeat when a seventh number can be used as a "wild card," replacing any of the six in the winning combination.

B.7 A message to be encrypted contains any arbitrary ASCII character. A particular encryption scheme uses the following *random mapping* idea: A character is mapped randomly onto another one with the restriction that no two characters will be mapped onto the same character. This is necessary for correct decryption. Consider a message of l characters. How many possible encrypted messages can be generated? If it takes an eavesdropper T_l s to break a message of length l, how much time on average is needed to break this encryption scheme by randomly choosing mappings?

Mathematical Skills

B.8 Show that for all integers $0 \leq j \leq m$

$$\binom{m}{j} = \binom{m}{m-j}.$$

Therefore, an equivalent expression for Theorem B.5 is

$$N_o = \binom{n-1+k}{n-1}.$$

B.9 Show that for all integers $0 < j < n$

$$\binom{n}{j} = \binom{n-1}{j-1} + \binom{n-1}{j}.$$

B.10 Show that for all integers $n \geq 0$

$$\sum_{j=0}^{n} \binom{n}{j} \cdot \binom{n}{n-j} = \sum_{j=0}^{n} \binom{n}{j}^2 = \binom{2n}{n}.$$

B.11 Show that for all integers n, $m \geq 0$ and all $k \leq n$, m

$$\sum_{j=0}^{k} \binom{n}{j} \binom{m}{k-j} = \binom{n+m}{k}.$$

B.12 Show that for all integers $n \geq 0$

$$\sum_{j=0}^{n} \binom{n+j}{j} = \binom{2n+1}{n}.$$

(*Hint:* Use the result of Problem B.9.)

B.13 Show that for all integers $n \geq 0$

$$\sum_{j=0}^{n} \binom{n+j}{j} j = n\binom{2n+1}{n} - \binom{2n+1}{n-1}.$$

(*Hint:* Use the result of Problem B.9.)

B.14 Show that for all integers $n \geq 0$

$$\sum_{j=0}^{n} \binom{n}{j} j = n2^{n-1}.$$

B.15 Show that for all integers $n \geq 0$

$$\sum_{j=0}^{n} (-1)^j \binom{n}{j} = 0.$$

B.16 Show that for all integers $n \geq 0$

$$\sum_{j=0}^{n} (-1)^{j+1} \binom{n}{j} j = 0.$$

B.17 Show that for all integers $n \geq 0$

$$\sum_{j=0}^{n} j(j-1) \binom{n}{j} = n(n-1)2^{n-2}.$$

B.18 Show that for all integers $n \geq k \geq 0$ and all $x \in R$

$$\sum_{j=0}^{k} \binom{n}{j} \binom{n-j}{k-j} x^j = \binom{n}{k} (1+x)^k.$$

B.19 Show that for all integers $n \geq 0$

$$\sum_{j=0}^{n} \frac{(2n)!}{(j!)^2(n-j)!(n-j)!} = \binom{2n}{n}^2.$$

(*Hint:* Use the result of Problem B.10.)

B.20 Show that for all integers $n > 0$

$$\sum_{j=1}^{n} (-1)^{j-1} \binom{n}{j} \frac{1}{j} = \sum_{j=1}^{n} \frac{1}{j}.$$

C

HISTORICAL DEVELOPMENT
OF THE THEORY

C.1   A BRIEF HISTORY

The birth of probability theory as a scientific discipline is put by most historians at 1654, when Pascal solved the problem of points (see Section 1.9) at the request of a gambler [62]. In the early years, most of the developments of the discipline of probability were motivated by games of chance (dice throwing, coin flipping, card drawing, and picking balls from urns). The sample spaces for all the problems studied were finite, and all outcomes were taken as equiprobable. All trials of experiments were independent and identically distributed. The tools used in the development were primarily algebraic and mostly combinatorial (combinations, permutations, factorials). A lot of the results in Appendix B were discovered during that period. Calculus (integrals and difference or differential equations) appeared later, in the work of Laplace.

Among the major contributors of the first 250 years one could cite James Bernoulli, Abraham de Moivre, Pierre Simon Laplace, Carl Gauss, and Pafnuti Chebyshev. Bernoulli proved the Weak Law of Large Numbers (the first theorem of the newborn theory) and solved numerous problems. De Moivre gave a complete solution of the duration-of-play problem (see Problem C.2), solved other problems, and derived the normal curve. Laplace solved numerous problems, proved the central limit theorem, and introduced generating functions. Gauss developed mean square estimation techniques and established the importance of the normal pdf in the theory of errors (see Section 5.6). Chebyshev worked on the central limit theorem, moments, and transforms and produced his tail inequality.

The first successful applications of the theory (other than the ones on gaming tables) were in the biological sciences (smallpox vaccinations, mortality tables, and demographics) and astronomy. Gauss was able to discover an asteroid based on least squares estimation techniques!

A number of books soon appeared that prompted new discoveries since they attracted the attention of more mathematicians to the new subject. The first one was Christian Huygens' *De Ratiociniis in Ludo Aleae*, published in 1658 in Latin. This book appeared only 4 years after Pascal's work. It included five problems, which James Bernoulli solved in his *Ars Conjectandi*, published in 1713 in Latin. Pierre Remond de Montmort (1678–1719) gave an account of the developments in *Essai d' Analyse sur les Jeux de Hazards* published in 1708; apparently its success led to a second edition in 1714. De Moivre's *Doctrine of Chances* was published in 1718 and included a list of 53 problems. Thomas Simpson wanted to popularize the subject in a book for the "masses," so he published *The Nature and Laws of Chance*. Laplace's voluminous *Théorie Analytique des Probabilités* included a lot of problems and theoretical results. It was published in 1812 in French, and three more editions followed.

The first crisis in the new science came from the criticisms of Jean le Rond D'Alembert, in the middle of the eighteenth century. This produced the first move toward sample spaces with nonequiprobable outcomes. John Venn published *Logic of Chance* in the later half of the nineteenth century, in an effort to put the theory on a more formal basis. As the theory moved toward adolescence, more and more scientists began attempts to formalize it. Richard von Mises and Bruno de Finetti proposed axiomatizations and interpretations that did not meet with considerable success. We present their efforts briefly in the next section. Andrei Kolmogorov finally put everything in order, in 1933; he elevated the subject to the level of an axiomatic, solid theory, by introducing the familiar three axioms, σ algebras for events and using Lebesgue's measure theory. By proving the Strong Law of Large Numbers, Kolmogorov answered a question that was open for 200 years (Bernoulli's conjecture) and provided a means to unify the theory and applications. His book *Foundations of the Theory of Probability*, published in 1933 in German and in 1950 in English, lay the foundation for the "mature" phase of the theory.

In the twentieth century we have witnessed the transition from finite spaces, where the outcomes are "points," to more complex spaces, where the outcomes are time functions. Einstein introduced diffusion processes, and Markov invented his chains in the first decade of the twentieth century. The theory of random processes (and in particular Markov Chains) really took off in the 1920s to 1940s, inspired now not by games of chance, but by applications in filtering and prediction. Wiener formalized diffusion processes; de Finetti introduced processes with independent increments; Kolmogorov developed a lot of results in Markov Chains; Khinchine introduced stationary processes; and Doob studied martingales.

For a more detailed description of the historical development of the theory, up to the middle of the nineteenth century, see the exciting book [62]. The excellent book by von Plato covers the "modern period" after that, until the

1930s [65]. An account of the developments by the Russian school of St. Petersburg, and in particular the works of Chebyshev, Markov, Lyapunov, and Kolmogorov, can be found in [51].

C.2   ALTERNATIVE AXIOMS

As soon as 1904, 4 years after Hilbert's challenge, efforts to put an axiomatic framework to probability had appeared, some even with measure-theoretic elements in them [65]. These efforts did not, however, lead anywhere. We briefly describe next the two most influential ones, which attracted the most attention. For a more detailed exposition, see [65].

C.2.1   Frequentist Approach

In 1919, Richard von Mises proposed a set of two axioms that were based on his empiricist philosophical viewpoint and that became known as *frequentist probabilities*. His basic idea was to *define* probabilities the way one would calculate them in practice (and that is the way we also suggested calculating them, in terms of relative frequencies and histograms).

Von Mises understood correctly that probability theory applies only to mass phenomena. (A fair number of erroneous attempts had been made until then to apply the theory to phenomena in social sciences that occur even only once.) The basic notion in von Mises' approach is that of a *collective e*, which he defined as an *infinite* sequence of elements (infinite because we are only interested in mass phenomena). So $e = (e_1, e_2, \ldots)$, where e_i are called *labels*. We can interpret that as a sequence of trials of an experiment, so all experiments in his setup are performed infinitely many times.

Let A be an arbitrary set of labels (a *subset* in our terminology). Denote the relative frequency of A in n trials as n_A/n, where n_A denotes the cardinality of A.

Axiom 1. Existence of Limits. For any A, the limit

$$\lim_{n \to \infty} \frac{n_A}{n} \triangleq p(A)$$

exists; $p(A)$ is called the *probability* of A. We can easily see that Axioms 1, 2, and 3a in Chapter 2 can be derived from this axiom. Notice, however, that von Mises does not require the set of events to have any structure, e.g., to be a field or a σ field. Moreover, notice that $p(A)$ can be different if another collective is considered.

Let now A, B be two disjoint sets. Let $p(A)$, $p(B)$ denote their probabilities with respect to a given collective e. Consider the collective e' formed by dropping all labels from e that are not elements of A or B. Let $p'(A)$, $p'(B)$ denote the probabilities of A, B with respect to collective e'.

Axiom 2. Irregularity of the Ordering. Consider a subsequence of labels in collective e' formed without making use of differences in labels. Then limits along those subsequences exist, and, moreover,

$$\frac{p(A)}{p'(A)} = \frac{p(B)}{p'(B)}.$$

The basic idea behind this axiom is to guarantee randomness in performing the experiments.

These axioms have received a number of criticisms since their inception; in response, von Mises had to modify them slightly. Khinchine became one of the most vocal critics, in 1929. The main objections stemmed from the definition of a collective and the existence of limits. It was not so clear how subsequences could be selected, in order to satisfy the requirements of the axioms and guarantee existence of a limit. Entirely intuitive choices for a subsequence were not permissible, based on these axioms, since that could produce two different limits. The fact that probabilities could change, if a different collective were considered, was also a bit disconcerting. Finally, in 1939, Jean Ville dealt a major blow to the frequentist approach, by *proving* that the set of axioms could not handle statements of the form "event A occurs infinitely often" (e.g., "Markov Chain reaches steady state") [64]. Such events would require essential use of countable additivity, as Axiom 3b in Chapter 2 postulates.

C.2.2   Subjectivist Approach

Bruno de Finetti was an Italian prodigy; he derived his axioms when he was in his early twenties and published them in 1930. His basic philosophy was that the probability of an event, no matter how we define it, is a *subjective* quantity. Two different people may assign two different numbers to the same event (based on their ignorance, experience, belief, risk-taking attitude, etc.). Now, as trials are performed (i.e., measurements are taken), these numbers may change since ignorance, experience, etc., may change. He proceeded to calculate what he called the *a posteriori probability* of an event, based on the outcomes of trials. Note that in this approach, there is no need for mass repetitions of an experiment.

Consider a class of events $\mathcal{E}$ that is closed under the operations of logical sums, products, and negations. (These operations correspond to finite unions, intersections, and complements of sets.) On this class, we define a relation $\geq$, read as "at least as probable as"; therefore, given two events A, B, $A \geq B$

means "event A is at least as probable as event B." The events satisfy four axioms:

Axiom 1. For any two events $A, B \in \mathcal{E}$, $A \geq B$ or $B \geq A$.

If $A \geq B$ and $B \geq A$, then we write $A \approx B$, and the two events are defined as identically probable. If $A \geq B$, but not $B \geq A$, then we define A as more probable than B, and we write $A > B$.

Axiom 2. If event A is certain, event B is impossible, and event C is neither certain nor impossible, then $A > C > B$.

Axiom 3. If $A \geq B$ and $B \geq C$, then $A \geq C$.

Axiom 4. If events A, B are incompatible with event C, then

$$A + C \geq A + B, \quad \text{if and only if} \quad A \geq B,$$

where $A + B$ is the logical sum of A and B. (Incompatible events cannot occur at the same time.)

A fifth axiom was added in 1937; however, de Finetti did not publicize his theory adequately. It found some applications only in the 1950s in the field of Bayesian statistics. At the same time, Kolmogorov's measure-theoretic axioms have gained enough ground; since de Finetti strongly opposed the idea of axiomatizing countable additivity (he said we never see it in everyday life), his theory did not take off.

C.3 SOME OF THE EARLY PROBLEMS

C.1 (*Points*; solved by Pascal and Fermat; Pascal's famous triangle was used in the solution of this problem.) Two players A, B play a game; in every round, a point is scored. The first player to score N points wins a stake. The players wish to stop the game *before it is finished*, while their scores are n_A and n_B. How should they divide the stake?

C.2 (*Duration of Play*; solved by de Moivre.) Two players A, B play a game. At the start of the game, A has n counters and B has m counters. After each play, the loser gives the winner one counter. The probability of A's winning a play is $a/(a + b)$; the probability of B's winning a play is $b/(a + b)$. Determine the probability of each player's winning all the opponent's counters.

C.3 (*Combinations of Events*; solved by de Moivre.) Let $A_1, \ldots, A_n$ be n events. Find the probability that exactly k of the events occur; find the probability that at least k of the events occur.

C.4 (*Waldegrave's Problem*; solved in various forms by Montmort, Laplace, and Nicolas Bernoulli.) There are $n + 1$ players in a game. The probability of any player's winning the game is p. The rules of the game are the following:

1. Only two players play at any given time, with players 1 and 2 starting the game.

2. The loser deposits a shilling in a common purse and waits in line after the last player.

3. The winner plays with the next waiting player.

4. The game stops when one player beats all n players in succession.

5. The winner takes the purse.

Calculate the probability that the game will end in k plays.

C.5 (*St. Petersburg Paradox*; solved by many.) Player A throws a fair coin in the air. If heads shows up in the kth throw, player B pays 2^k shillings to A. How many shillings should A pay B to make it a fair game?

C.6 (*Multiple Die Throws*; solved by de Moivre.) A symmetric die has m faces, showing the integers $1, \ldots, m$. And n such dice are thrown simultaneously. Find the probability that the sum of the faces is equal to k.

C.7 (*Le Jeu de Noyaux*; this is the only known contribution by a female, unfortunately anonymous; solved by her and Montmort.) Eight dice, each having only two faces, one black and one white, are thrown at random. Calculate the probability of throwing x black and y white faces.

C.8 (*Runs of Faces*; solved by Laplace.) A symmetric device has m faces, showing the integers $1, \ldots, m$. Find the probability that in $n \geq m$ throws, the faces will show up in the sequence $1, \ldots, m$.

C.9 (*Buffon's Needle*; solved by Buffon and Laplace.) Consider a plane with a grid of parallel lines crossing it as shown in Figure C.1. A needle of length $2l$ is thrown on the plane. Find the probability that the needle will fall across a line.

C.10 (*Croix où Pile*: D'Alembert's paradox.) Consider the game of throwing a fair coin. Find the probability p of throwing one head in two throws. D'Alembert proposed that $p = 2/3$ since when a head shows in the first throw, the game is stopped and thus there are three outcomes only: $\{H, TH, TT\}$. Was he wrong?

C.11 (*Sum of Faces*; solved by James Bernoulli. Do you see any generating functions used here?) Consider the experiment of throwing n fair dice simultaneously. Let $A_m = \{\text{sum of faces equals } m\}$. Let C_{mn} be the coefficient of the term x^m in the expansion of the polynomial

$$(x + x^2 + x^3 + x^4 + x^5 + x^6)^n$$

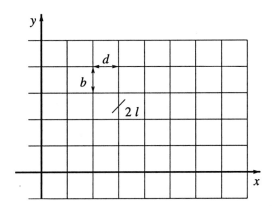

Figure C.1 Buffon's needle.

in a series of powers of x. Show that

$$P[A_m] = \frac{C_{mn}}{6^n}.$$

C.12 (*Game of Tennis*; solved by James Bernoulli.) Two players A, B play a game of tennis. Assume that the probability that player A wins a point is p and that player B wins a point is $1 - p$. Moreover, assume that successive points are won independently. Find $p(i, j)$, the probability that player A wins the game given that player A has at present i points and player B has j.

MODELING OF RANDOMNESS IN ENGINEERING SYSTEMS: A SUMMARY

In this appendix, we provide a brief summary of the main concepts discussed throughout the book, in an effort to highlight the structure of the theory and its application to a realistic experiment.

D.1 ELEMENTARY CONCEPTS

The fundamental notions in studying any (engineering) system that involves randomness are those of an *experiment*, its *outcomes*, and *events*. *Probabilities* are assigned to events. Probabilities (in the viewpoint adopted in this book) can be associated with *mass* phenomena only. The set of all outcomes of an experiment is the sample space S; the set of events is denoted as Q. The probability assignment is a function from Q on the set $[0, 1]$, denoted as $P[A]$.

D.2 PROBABILISTIC MODELS

What is a probabilistic model of an engineering experiment? From the mathematical point of view, it is the triplet (S, Q, P). So, how does one determine it?

1. Sample space S is not unique for a given experiment.

2. When S is discrete, the most convenient choice for Q is the powerset of S; when S is continuous, Q is tough to define. The point is that Q must be closed under (finite or infinite) set operations, in order to avoid logical inconsistencies.

3. Probability P can be determined via:

- *Measurements.* This is an accurate approach, but also time-consuming and expensive.

- *Hypothesis.* This may not be as accurate, but it is inexpensive, can be done on paper, and sometimes is the only choice.

Our choice for P should satisfy the three axioms. The trouble comes from Axiom 3. Moreover, P is a function defined on sets, so it is not so easy to manipulate. That leads to the artifact of a random variable X. Random variable X transforms the triplet (S, Q, P) into $(R, \mathcal{B}(R), F_X)$ or $(R, \mathcal{B}(R), f_X)$, where R is the real line, $\mathcal{B}(R)$ is the set of all Borel subsets of the real line, F_X is the cdf, and f_X is the pdf. Since R, $\mathcal{B}(R)$ are the same in all models, we usually drop them from any modeling discussions. We then end up with

The function $F_X(x)$ or $f_X(x)$ is the probabilistic model of an experiment.

So now, how is $F_X(x)$ or $f_X(x)$ determined for an experiment? We can follow one of three approaches:

1. *Measurements.* The fundamental notion here is that of a *histogram*, which approximates the pdf.

2. *Hypothesis.* We can assume any real-valued function for the pdf, as long as it satisfies the properties of Lemma 3.3 (i.e., it is nonnegative and integrates to 1).

3. *Transformations of another random variable with a known pdf*, via Theorem 3.1.

More often than not, more than one random variable is involved in a given experiment. When a finite number are involved, we speak of a *random vector*. The model then is a multidimensional, joint *pdf* or joint *cdf*.

   When an infinite number of random variables are involved, we speak of a *random process* or a *sequence of random variables*, depending on whether the number is uncountably or countably infinite. The probabilistic model of the system in such cases must be given in terms of all finite, n-dimensional jpdf's or jcdf's. This requires gathering a lot of information; multidimensional histograms to approximate a jpdf are not fun. Even if one follows the hypothesis approach, still an enormous amount of information must be supplied. Therefore, by necessity, the models for random processes or sequences (experimental or theoretical) have been centered on the notions of

1. *Independent, identical distributions.* This property essentially requires only a single pdf to be determined. It applies to random processes in both discrete time (IID sequences) and continuous time (processes with stationary, independent increments).

2. *Stationarity.* This property reduces somewhat the amount of information to be gathered since jpdf's do not depend on the time origin. Still an enormous amount of information is required, and this leads to Item 3.

3. *Wide-sense stationarity.* This property focuses on the mean and correlation of any two of the random variables involved in a random process. The characterization of the random process is reduced to determining a number (the mean) and a function of a single argument (the *autocorrelation* or the *power spectral density*).

4. *Limited dependence.* This property reduces again the amount of information needed to describe the random process. *Markov Chains* in discrete space or *Markov processes* in continuous space need only a *transition probability* or *transition rate matrix* to be defined.

D.3 WHAT MODELS HAVE WE DEVELOPED?

For a single random variable, some of the most widely used models are presented in Section 3.6. For a vector random variable, once independence is assumed, any of the models in Section 3.6 applies. If dependence is present, the Gaussian model in Section 4.10 is the most widely used one. Since in general it is difficult to obtain the full N-dimensional pdf, we often resort to *partial information* about the model. Then the *mean, variance,* and *covariance matrix* of the vector suffice.

For the case of a sequence of random variables, the most widely used models are those of an IID sequence and a Markov Chain. In the case of an IID sequence, any model in Section 3.6 applies. In the case of a Markov Chain, the most widely used model involves *homogeneous* transition probabilities.

For the case of a random process in continuous time, the most widely used models are those of the Gaussian and the Poisson process in Chapter 8. Markov process models are also used.

D.4 WHAT TOOLS HAVE WE DEVELOPED?

We have developed a series of tools to use in proofs or to calculate measures of interest in a real life-experiment, such as probabilities, means, and cdf's or pdf's.

1. Conditional probabilities and the total probability theorem (Equation 2.10):

$$P[A] = \sum_B P[A|B]P[B],$$

$$F_Y(y) = \sum_B P[Y \le y|B]P[B].$$

2. Bayes' rule (Equation 2.12)

3. Tail inequalities (Inequalities 3.83, 3.84, 3.87)

4. Transformations of a single or multiple random variables (Theorems 3.1, 3.2, 4.6, 4.7, 4.8, 4.9, and 4.10)

5. Transforms of a single or multiple random variables, namely, Laplace, Fourier, and Z transforms

6. Strong Law of Large Numbers (Theorem 6.5)

7. Central limit theorem (Theorem 6.6)

8. The orthogonality principle (Equations 5.17 and 9.17) and minimum mean square estimation (Theorems 5.1, 5.2, 9.4, 9.6, and the Kalman filter).

D.5   MATHEMATICAL SUBTLETIES

We should be aware of some purely mathematical subtleties, *even though we will never see them as part of a model.*

The collection of events Q should contain not only "events of interest" but also their unions, intersections, complements, and limits. (In math lingo, Q must be closed under these set operations, or Q must be a σ field.) Why? Only then do symbols such as $P[A \cup B]$ and $P[\cup_{n=1}^{\infty} A_n]$ and theorems such as $P[A] + P[A^c] = 1$ make sense. The set $(-\infty, x]$ and its associated Borel σ field were instrumental here.

Characteristic functions, Laplace transforms, and generating functions are mathematically equivalent descriptions of a random variable, and therefore of an experiment, since there is a one-to-one correspondence between $F_X(x), \Phi_X(\omega)$, $G_N(z), L_X(s)$, etc.

When we work with the initial model (S, Q, P), the real difficulty lies in proving that the chosen P satisfies Axiom 3. Notice that $f_X(x)$-based models satisfy the three axioms *automatically.* Axiom 3 is guaranteed through the additive properties of the integral

$$\int_{A \cup B} f_X(x)\, dx = \int_A f_X(x)\, dx + \int_B f_X(x)\, dx,$$

when $A \cap B = \emptyset$.

Why these particular three axioms? Because experience (accumulated over 300 years of hard work by mathematicians) has shown that these three axioms provide the most elegant, economical foundation of a *rigorous* and *complete* theory. Bear in mind that conditional probabilities, densities, theorems, and transforms have been put to use with success since the seventeenth century. The three axioms were finally proposed in the form you see them in most books, in 1933. Only then was a rigorous theory, without logical gaps, developed.

D.6 WHAT CAN WE DO WITH A MODEL?

Once we define $N \leq \infty$ random variables on an experiment, we have a probabilistic model. We can characterize this model through a pdf, means, correlations, power spectral densities, or transition probabilities. In a nutshell, this is what we can do:

1. Calculate probabilities of various events (e.g., through simple integration, using conditional probabilities, Bayes' rule).

2. Bound probabilities, through Markov, Chebyshev, or Chernoff inequalities.

3. Calculate averages (e.g., through simple integration, or via the Strong Law of Large Numbers and ergodic theorems).

4. Calculate variances, covariances, correlations (e.g., through simple integration).

5. Filter, predict, or smooth a random process associated with the experiment at hand.

REFERENCES

[1] M. Abramowitz and I. Stegun. *Handbook of Mathematical Functions.* National Bureau of Standards, Washington, 1964.

[2] G. Adomian. *Stochastic Systems.* Academic Press, New York, 1983.

[3] T. M. Apostol. *Mathematical Analysis*, 2d ed. Addison Wesley, Reading, MA, 1974.

[4] L. Arnold. *Stochastic Differential Equations: Theory and Applications.* John Wiley & Sons, New York, 1974.

[5] R. B. Ash. *Real Analysis and Probability.* Academic Press, New York, 1972.

[6] S. N. Bernstein. Principes de la théorie des équations differentielles stochastiques. *Trudy Steklov Fiz.-Mat. Inst.*, 5:95–125, 1934.

[7] R. N. Bhattacharya and E. C. Waymire. *Stochastic Processes with Applications.* Wiley Series in Probability and Mathematical Statistics. John Wiley & Sons, New York, 1990.

[8] P. Billingsley. *Probability and Measure*, 2d ed. John Wiley & Sons, New York, 1986.

[9] Emile Borel. Les probabilités dénombrables et leurs applications arithmétiques. *Oeuvres de Emile Borel*, 2:1055–1078. (Original in *Rendiconti del Circolo Matematico di Palermo*, 27:247–270, 1909.)

[10] H. Bruneel and B. G. Kim. *Discrete Time Models for Communication Systems Including ATM.* Kluwer Academic Publishers, Boston, 1993.

[11] Y. S. Chow and H. Teicher. *Probability Theory: Independence, Interchangeability, Martingales*, 2d ed. Springer Texts in Statistics. Springer-Verlag, New York, 1988.

[12] K. L. Chung. *Markov Chains with Stationary Transition Probabilities.* Springer-Verlag, New York, 1960.

[13] K. L. Chung. *A Course in Probability Theory*. Academic Press, New York, 1974.

[14] H. Cramér. On the theory of stationary processes. *Ann. Math.*, 41:215–230, 1940.

[15] B. de Finetti. *Probability, Induction and Statistics*. John Wiley & Sons, New York, 1972.

[16] J. L. Doob. Markoff chains—denumerable case. *Trans. Amer. Math. Soc.*, 58:455–473, 1945.

[17] J. L. Doob. *Stochastic Processes*. John Wiley & Sons, New York, 1953.

[18] E. B. Dynkin. *Markov Processes*. Academic Press, New York, 1965 (two volumes).

[19] E. Eberlein and M. S. Taqqu, eds. *Dependence in Probability and Statistics: A Survey of Recent Results*. Progress in Probability and Statistics. Birkhauser, Boston, 1986.

[20] H. B. Enderton. *Elements of Set Theory*. Academic Press, New York, 1977.

[21] W. Fischer and K. Meier-Hellstern. The Markov modulated Poisson process (MMPP) cookbook. *Performance Eval.*, 18:149–171, 1992.

[22] J. E. Freund and R. E. Walpole. *Mathematical Statistics*, 4th ed. Prentice-Hall, Englewood Cliffs, NJ, 1987.

[23] I. I. Gihman and A. V. Skorohod. *The Theory of Stochastic Processes*. Springer-Verlag, New York, 1974–1979 (three volumes).

[24] I. I. Gihman and A. V. Skorohod. *The Theory of Stochastic Processes II*. Springer-Verlag, New York, 1975.

[25] R. M. Gray and L. D. Davisson. *Random Processes, A Mathematical Approach for Engineers*. Prentice-Hall, Englewood Cliffs, NJ, 1986.

[26] P. R. Halmos. *Measure Theory*. Van Nostrand Reinhold, New York, 1950.

[27] P. R. Halmos. *Naive Set Theory*. Undergraduate Texts in Mathematics. Springer-Verlag, New York, 1974.

[28] H. Heffes and D. M. Lucantoni. A Markov modulated characterization of packetized voice and data traffic and related statistical multiplexer performance. *IEEE J. Selected Areas in Commun.*, SAC–4, September 1986.

[29] E. Hopf. *Ergodentheorie.* Chelsea Publishing Co., New York, 1948. Reprinted from German.

[30] K. Ito. Stochastic integral. *Proc. Japanese Acad. Tokyo*, 20:519–524, 1944.

[31] K. Ito. *Foundations of Stochastic Differential Equations in Infinite Dimensional Spaces.* CBMS-NSF regional conference series in applied mathematics; 47.

[32] G. M. Jenkins and D. G. Watts. *Spectral Analysis and Its Applications.* Holden-Day, San Francisco, 1968.

[33] R. E. Kalman. A new approach to linear filtering and prediction problems. *J. Basic Eng.*, 82:35–45, March 1960.

[34] R. E. Kalman and R. S. Bucy. New results in linear filtering and prediction theory. *J. Basic Eng.*, 83:95–107, December 1961.

[35] M. Kaplan. A sufficient condition for nonergodicity of a Markov Chain. *IEEE Trans. Infor. Theory*, 25:470–471, July 1979.

[36] K. Karhunen. Über lineare methoden in der wahrscheinlichkeitsrechnung. *Ann. Acad. Sci. Fennicae, Ser. A. Math. Phys.*, 37:3–79, 1947 (in German).

[37] K. Karhunen. Über die struktur stationärer zufälliger funktionen. *Ark. Math.*, 1:141–160, 1950 (in German).

[38] S. Karlin and H. M. Taylor. *A First Course in Stochastic Processes*, 2d ed. Academic Press, New York, 1975.

[39] A. Khinchine. Correlation theory of stationary random processes. *Usp. Mat. Nauk*, 5:42–51, 1938.

[40] A. C. King and C. B. Read. *Pathways to Probability.* Holt, Rinehart and Winston, New York, 1963.

[41] L. Kleinrock. *Queuing Systems*, vol. I: *Theory.* John Wiley & Sons, New York, 1975.

[42] A. N. Kolmogorov. Über die analytischen methoden in wahrscheinlichkeitsrechnung. *Math. Ann.*, 104:1–16, 1931 (in German).

[43] A. N. Kolmogorov. Anfangsgründe der theorie der Markoffschen Ketten mit unendlich vielen möglichen Zuständen. 1:607–610, 1936 (in German).

[44] A. N. Kolmogorov. Markov Chains with a countable number of possible states. *Bull. Math. Univ. Moscow*, 2:1–16, 1937 (in Russian).

[45] A. N. Kolmogorov. *Foundations of the Theory of Probability*. Chelsea Publishing Company, New York, 1950.

[46] A. M. Law and W. D. Kelton. *Simulation Modeling and Analysis*, 2d ed. McGraw-Hill, New York, 1991.

[47] N. Levinson. The Wiener RMS error criterion in filter design and prediction. *J. Math. Physics*, 25, 1947.

[48] S. Lin and D. J. Costello Jr. *Error Control Coding: Fundamentals and Applications*. Prentice-Hall, Englewood Cliffs, NJ, 1983.

[49] J. Lindeberg. Eine neue Herleitung des Exponentialgesetzes in der Wahrscheinlichkeitsrechnung. *Mathematische Zeitschrift*, 15:211–225, 1922.

[50] M. Loève. *Probability Theory*, vol. 2, 4th ed. Springer-Verlag, New York, 1978.

[51] L. E. Maistrov. *Probability Theory: A Historical Sketch*. Academic Press, New York, 1974.

[52] A. A. Markov. Extension of the law of large numbers to dependent events. *Bull. Soc. Phys. Math. Kazan*, 15(2):155–156, 1906 (in Russian).

[53] A. V. Oppenheim and R. W. Schaffer. *Digital Signal Processing*. Prentice-Hall, Englewood Cliffs, NJ, 1975.

[54] A. V. Oppenheim and A. S. Willsky. *Signals and Systems*. Prentice-Hall, Englewood Cliffs, NJ, 1983.

[55] A. Papoulis. *Probability, Random Variables and Stochastic Processes*, 3rd ed. McGraw-Hill, New York, 1991.

[56] The Rand Corporation. *A Million Random Digits with 100,000 Normal Deviates*. The Free Press of Glencoe, IL, 1955.

[57] H. G. Romig. *50–100 Binomial Tables*. John Wiley & Sons, New York, 1953.

[58] K. S. Shanmugan and A. M. Breipohl. *Random Signals: Detection, Estimation and Data Analysis*. John Wiley & Sons, New York, 1988.

[59] Ernst Stadlober and Franz Niederl. *Generation of Non-Uniform Random Variates with C-Rand*. Institute of Statistics, Technical University Graz, Research Report 15, Austria, 2nd Update with WIN-Rand, September 1995.

[60] W. Stallings. *Data and Computer Communications*, 4th ed. MacMillan, New York, 1994.

[61] H. Stark and J. W. Woods. *Probability, Random Processes and Estimation Theory for Engineers*, 2d ed. Prentice-Hall, Englewood Cliffs, NJ, 1994.

[62] I. Todhunter. *A History of the Mathematical Theory of Probability.* Chelsea Publishing Company, New York, 1949.

[63] H. L. Van Trees. *Detection, Estimation and Modulation Theory.* John Wiley & Sons, New York, 1968.

[64] J. Ville. *Etude Critique de la Notion de Collectif.* Gauthier-Villars, Paris, 1939.

[65] J. von Plato. *Creating Modern Probability.* Cambridge University Press, Cambridge, 1994.

[66] N. Wiener and R. E. A. C. Paley. *Fourier Transforms in the Complex Domain.* American Mathematical Society, New York, 1934.

[67] R. W. Wolff. Poisson arrivals see time averages. *Oper. Res.*, 30:223–231, 1982.

[68] E. Wong. *Introduction to Random Processes.* Springer-Verlag, New York, 1983.

INDEX

A

Absorbing barrier, 408
Absorption time, 408
Algebras of sets, 625
Almost-sure convergence, 357,
    363–367, 454, 512
Alternative axioms, 645
Amplitude modulation, 491
Anti-PASTA property, 447
Aperiodic Markov Chain, 545
ARMA process, 409
Autocorrelation function of
    random process, 398
  and derivatives, 411
  and integral, 412
  and mean square continuity, 410
  and periodicity, 418
  and power spectral density, 479
  band-limited white noise,
    482–483
  definition, 398
  delayed random process, 481
  first-order filter, 488
  output of linear system, 487
  Poisson random process, 442
  relation to autocovariance, 398
  signal plus noise, 502
  sinusoid process, 399–400, 481
  white noise, 483
  WSS random process, 404
Autocovariance function of
    random process, 398, 486
  Poisson random process, 442

relation to autocorrelation, 398
Wiener process, 459
Autoregressive moving-average
    process, 409, 489, 504, 532
Average:
  estimation of, 347
  lifetime, 130
  of estimation error, 492
  of function of random variable,
    128
  of function of two random
    variables, 256–257
  of Poisson process, 442
  of random process, 398
  of random sum, 341
  of random variable, 124
  of transformation, 128
  properties, 124
  square error, 306, 309
  table of, 127
Axiom of continuity, 53, 77
Axioms, 30

B

Bandpass filter, 490
Bayes' rule, 38, 156, 324
Bayes, Thomas, 37
Bernoulli random variable, 96–97,
    99, 138, 146–147
  sequence of, 339, 345, 348, 448,
    456, 572, 600
Bernoulli splitting, 456
Bernoulli, Daniel, 129, 306

Bernoulli, James, 39, 96, 98, 138,
    344–345, 347, 643–644,
    648–649
Bernoulli, John, 109
Bernoulli, Nicolas, 647
Bernstein, S. N., 32
Binary symmetric channel, 323
Binomial coefficient, 97, 139
Binomial random process, 453
Binomial random variable, 97, 112,
    114, 139, 141, 143, 147, 241,
    349, 353, 448, 453
Birkoff, G., 413
Birth-death process, 565, 567
    limiting pmf of, 569
Boltzmann's kinetic theory, 528
Borel field, 28
Borel-Cantelli lemmas, 366
Borel, Emil, 70, 347, 366
Brownian motion, 459
Buffon's needle, 648
Bus-based switch, 596
Busy cycle, 343, 602

C

Cantelli, F. P., 366
Cantor, G., 627
Cartesian product, 27, 198, 233,
    622
Cascaded filters, 518
Cauchy criterion
    almost-sure convergence, 358
    convergence in probability, 360
    mean square convergence, 361,
        411, 413
    sure convergence, 357
Cauchy random variable, 102, 124,
    149, 255
Cauchy, A. L., 102
Causal filtering, 493–494
Causal system, 485
Central limit theorem:

IID random variables, 350
    in practice, 352
    Lindeberg, 351
    random processes, 462
Central moment, 126
Certain event, 29
CFR distribution, 131
Chain rule, 57, 300
Chapman-Kolmogorov equations,
    539, 567
Characteristic function:
    of pdf, 135
    of random sum, 341
    of sum of random variables, 336
    table of, 140
Chebyshev inequality, 141, 345
Chebyshev, Pafnuti, 123, 141, 351,
    643, 645
Chernoff inequality, 144
Chevalier de Méré, 13
Chi-squared random variable, 105,
    111
Chi-squared test, 112
Cholesky factorization, 271
Classes of states in Markov Chain,
    543
Clipping transformation, 119
Coefficient of variation, 450, 566,
    603
Collection of events, 28
Communicating classes in Markov
    Chain, 543
Complement of set, 613
Conditional cdf
    n random variables, 266
    two random variables, 224
Conditional expectation, 257
Conditional pdf, 85, 266
Conditional probabilities, 35
Consumer price index, 122
Continuity of random process, 410

Continuous random variable, 100–107

Continuous sample space, 27

Continuous-time linear system, 484

Continuous-time Markov Chain, 529, 561

Continuous-time random process, 388

Convergence:
  almost-sure, 357
  Cauchy criterion, 357–358, 360–361
  in distribution, 362
  in probability, 358
  mean square, 360
  relationship between modes, 363
  sequences of random variables, 355
  sure, 355

Convolution integral, 239

Corollaries of axioms, 33–34

Correlation coefficient
  random process, 398
  two random variables, 258

Cosine transformation, 118

Counting process, 454

Covariance matrix, 268

Covariance:
  of Poisson process, 442
  of random process, 398
  of two random variables, 258

CPU scheduling, 49, 591

Cross-correlation of random process, 398, 482
  and cross-power spectral density, 479
  output of a linear system, 487

Cross-power spectral density of random process, 479, 482

Cross-power spectral density
  of output of linear system, 487

Cyclostationary random processes, 404

D

D'Alembert, Jean le Rond, 42, 648

De Finetti, Bruno, 387, 400, 644, 646–647

De Moivre, Abraham, 35, 101, 635, 643–644, 647

De Montmort, Pierre Remond, 98, 644

De Morgan's laws, 622, 626

De Morgan, A., 622

Decimator, 471

Derivative of a random process, 411
  sinusoid process, 412

Descartes, René, 622

Deutsche mark banknote, 101

Deviates of random variable, 145

DFR distribution, 131

Diagonalization, 271

Difference of sets, 620

Dirac, Paul, 82

Discrete random variable, 96–99

Discrete sample space, 27

Discrete uniform random variable, 99, 353

Discrete-time linear system, 485

Discrete-time Markov Chain, 529, 535

Discrete-time random process, 388

Disjoint sets, 619

Dow Jones industrial average, 12, 36, 86, 218, 225, 241, 318

Drifts of MC, 557, 572

E

Ehrenfest model, 14, 528, 586, 593

Eigenfunctions, 419, 461

Eigenvalues, 271–272, 419–420, 461

Eigenvectors, 271–272

Einstein, Albert, 387, 460, 479, 644
Elementary event, 29
Empty set, 613
Equivalent events:
  linear transformation, 250
  nonlinear transformation, 252
  two random variables, 115–116
Ergodic Markov Chains, 549
Ergodic theorems, 347, 415, 447,
    454, 552
Ergodicity, 347, 413, 447, 454, 549,
    557, 560, 574, 602
  criteria, 415
Erlang random variable, 103, 138,
    240, 270, 444
Erlang, A. K., 19, 103
Estimation criteria, 304, 492
Estimation error, 304
Estimation of single random
    variable:
  with constant, 308
  with linear function, 309
  with nonlinear function, 313
Estimation:
  causal filtering of processes, 494
  causal filtering, 493
  criteria, 304
  ideal filters, 502
  Kalman filter, 503
  Levinson's algorithm, 499
  of mean, 347–349
  of periodogram, 510, 512
  of probability, 348
  of rate, 454
  of spectral density, 511
  orthogonality principle, 310, 494
  prediction of processes, 497
  signal plus noise, 500
  smoothing of processes, 499
  Wiener filter, 502
Estimators, unbiased, 310
Ethernet, 17, 571

Events, 28
  algebra of, 198
  Cartesian product, 198
  certain, 29
  disjoint, 30
  elementary, 29, 41
  equivalent, 115–117, 238, 250,
    252, 443
  impossible, 29
  independence of, 40–42
  null, 33
Expectation, 124, 398
  conditional, 257
Experiment design, 110
Exponential random variable, 102,
    136, 138, 141, 143, 148, 209,
    213, 215, 240, 243, 270, 338,
    342, 348, 443, 450–451, 457,
    562

F

F random variable, 105
Factorial, 635
Failure rate function, 131
  table of, 131
Fast Fourier transforms, 239
Fault-tolerant systems, 134
Field, 625
Filtering of random processes, 493
  Kalman, 508
  optimal, 494
First moment, 124
First-order filter, 488
Fisher, Sir Ronald, 105
FM signals, 389, 475
Fourier series of random process,
    417
Fourier transform:
  and autocorrelation function, 479
  and cross-correlation function,
    479
  and pdf, 135

and transfer function, 485
inverse, 342
of signal, 477–478
table of, 140
Fourier, Jean Baptiste Joseph, 135
Frequency modulation, 389
Frequentist approach, 645
FTP, 2, 4, 15, 36, 87, 111, 197,
    220, 225, 259, 311, 314, 347,
    391
Function of single random variable:
  cdf of, 116
  examples, 118–119, 121
  mean of, 128
  pdf of, 116
Function of two random variables;
  linear:
    cdf of, 237
    examples, 239–241, 250–251
    independent case, 239
    pdf of, 237, 249
  nonlinear:
    cdf of, 242
    examples, 243–244, 246, 248,
      253–255
    pdf of, 253

G

Galton, Sir Francis, 258
Gamma function, 103
Gamma random variable, 103, 137,
    241, 270, 354
Gaussian random process, 458
  Brownian motion, 459
  CLT, 462
  generation of variates, 463
  linear transformations, 462
  Ornstein-Uhlenbeck process, 461
  sum of, 462
  sums, 462
  white noise, 483
  Wiener process, 459

Gaussian random variable, 101,
    136, 354, 461
  independent, 261
Gauss, Carl, 101, 316, 325–326,
    335, 643
Generating function, 138
  table of, 140
Generation of variates:
  Bernoulli, 147
  binomial, 148
  Cauchy, 149
  exponential, 148
  Gaussian process, 463
  IID sequences, 420
  independent-increment process,
    421
  Laplace, 148
  Markov Chain:
    continuous-time, 575
    discrete-time, 574
  n-dimensional random vector,
    271
  Poisson process, 457
  rejection method, 149
  single random variable:
    via cdf, 147
    via pdf, 149
  transformation method, 147
  vector random variable, 270
  white Gaussian noise, 484
  Wiener process, 460
  WSS process, 421, 488
Geometric random variable, 98,
    342, 550
Global balance equations, 569
Goodness-of-fit test, 112
Gossett, W. S., 106

H

Handel's *Messiah*, 91
High-pass filter, 490

Higher moments of random
    variable, 126
Hilbert, David, 30
Histogram bins, 110–114, 218, 220
Histogram:
    experimental, 86–87, 91, 93, 218,
      220–221
    one random variable, 86, 109
    two random variables, 217
Historical problems:
    Buffon's needle, 648
    combinations of events, 647
    croix ou pile, 648
    D'Alembert's paradox, 648
    duration of play, 647
    game of tennis, 649
    le jeu de noyaux, 648
    multiple die throws, 648
    points, 13, 647
    runs of faces, 648
    St. Petersburg paradox, 648
    sum of faces, 648
    Waldegrave's problem, 647
Holder's Inequality, 288
Homogeneous Markov Chain, 529
Huygens, Christian, 124

I

IBP, 600
Ideal filters, 490, 502
IFR distribution, 131
Impossible event, 29, 33
Impulse response:
    continuous-time, 484
    discrete-time, 485
    shot noise, 449
Independence:
    of events, 40, 42
    of functions of random variables,
      235
    of two random variables, 231
    tests for, 231, 234

Independent experiments, 233
Independent increment process,
    399, 402, 421, 439, 448,
    452–453, 459, 531, 535
Independent increments, 400
Independent, identically
    distributed random variables,
    334–335, 344, 347, 394,
    401–402, 420, 531
Indicator function, 44, 96, 348
Innovation sequence, 508
Integral of a random process, 412
Interarrival times, 443
Interrupted Bernoulli process, 600
Interrupted Poisson process, 449
Intersection of sets, 619
Inverse transform formula, 135
IPP, 449
Irreducible Markov Chain, 543

J

Jacobian of a transformation, 252
Jensen's Inequality, 182
Joint cdf:
    n random variables, 265
    properties, 205, 265
    random process, 394, 397
    two random variables, 204
Joint central moment, 258
Joint characteristic function, 293
Joint moment, 258
Joint pdf:
    n Gaussian random variables,
      268
    n random variables, 265
    properties, 212
    two dependent Gaussian random
      variables, 262
    two independent Gaussian
      random variables, 261
    two random variables, 212
Joint pmf:

dependent Poisson random
  variables, 261
independent Poisson random
  variables, 260
n random variables, 265
Joint stationarity, 402
Jointly stationary random
  processes, 402
Jointly wide-sense stationary
  random processes, 403

K

Kalman filter, 503, 508
Kalman, R. E., 504
Karhunen-Loève expansion, 417,
  461
Khinchine, A., 401, 403, 479–480,
  644, 646
Kolmogorov's 0-1 law, 368
Kolmogorov's existence theorem,
  394
Kolmogorov's Inequality for sums,
  373
Kolmogorov, A. N., 30, 32, 70, 76,
  333, 347, 387, 393–394, 483,
  492, 494, 528, 644

L

Lagrange, Joseph Louis, 35, 82,
  335
Laplace random variable, 102, 148,
  325, 354
Laplace transform, 137
  table of, 140
Laplace, Marquis Pierre Simon de,
  258, 70, 82, 102, 124,
  137–138, 325, 350–351,
  643–644, 647–648
Law of Large Numbers:
  Strong, 347
  Weak, 344
Least squares method, 325

Legendre, A. M., 324
Level of significance, 111–113
Levinson's algorithm, 499
Life expectancy table, 185
Lifetime, 129
  average, 130
  residual, 130
Limit in the mean (l.i.m.), 360
Limit infimum, 625
Limit supremum, 625
Limiting probabilities of Markov
  Chain
  continuous-time, 569
  discrete-time, 548
Limits of sequences of sets, 624
Linear prediction:
  multiple measurements, 316
  two measurements, 315
Linear system:
  causal, 484
  cross-correlation of input and
    output, 487
  mean of output, 486
  psd of output, 487
  random output, 486
Linear transformation:
  of Gaussian random process, 462
  of random process, 484
  of random variable, 118
Log-normal random variable, 181
Logarithmic transformation, 121
Low-pass filter, 490

M

Marginal cdf:
  n random variables, 266
  two random variables, 207
Marginal pdf, 215
Markov Chain, 405
  accessible states, 543
  aperiodic, 545

Chapman-Kolmogorov
   equations, 539, 567
communicating states, 543
conditional pmf, 539
continuous-time, 561
discrete-time, 535
drifts, 557
ergodicity of, 549
generation of values, 574
global balance equations, 569
irreducible, 543
limiting pmf, 540, 569
marginal pmf, 538, 567
multidimensional pmf, 538
n-step transition probability, 537
one-step transition matrix, 536
period of, 545
periodic, 545
recurrent, 546
reducible, 543
return time to state, 549
sojourn times, 549
state classification, 543
state transition diagram, 536
steady-state, 548
transient, 546
transition matrix, 561
transition rate matrix, 564
Markov inequality, 141, 363
Markov modulated Bernoulli
   process, 600
Markov modulated Poisson
   process, 451, 565
Markov process:
  classification, 529
  definition of, 528
  multidimensional pmf, 530
Markov, A. A., 141, 351, 527–528,
   644–645
Mass phenomena, 23
Match of length k, 300

Maximum-Likelihood estimation,
   322
Maximum-Likelihood estimator,
   323
Mean ergodic random process, 415
Mean square continuity, 410
Mean square convergence,
   360–361, 363, 365, 412
Mean square estimation
  error, 306, 492
  examples, 311, 313–314, 316–322,
    495–497
  filtering, 494
  Levinson's algorithm, 499
  linear prediction, 316
  prediction, 497
  smoothing, 499
  Wiener filter, 502
  with constant, 308
  with linear function, 309
  with nonlinear function, 313
Mean square:
  continuity of random process,
    410
  derivative of random process, 411
  integral of random process, 412
  periodic random process, 417
Mean time to failure, 130
Measurable function, 71
Median, 181
Memoryless property, 102, 169,
   446, 563
Mercer equation, 419
Minkowski's Inequality, 288
Mixed random variable, 108
MMPP, 451, 565
MMSE, 307
Moment around mean, 126
Moment generating function, 139
  table of, 140
Moment theorem, 136
Montmort, Pierre Remond de, 647

Multinomial coefficient, 637

N

N-dimensional Poisson model, 268
N-dimensional random vector:
  conditional cdf, 266
  conditional pdf, 266
  definition, 265
  functions of, 267
  Gaussian, 268
  generation of variates, 271
  independence, 267
  joint cdf, 265
  joint pdf, 265
  joint pmf, 265
  marginal cdf, 266
  Poisson, 268
  properties of joint cdf, 265
  transformations of, 268
NBA, 14
Negative binomial random
  variable, 98
Negative information, 56
Netstat experiment, 86–87,
  113–114, 196, 205, 218, 225,
  259
NFL, 14
Nielsen ratings, 192
Noise, 4–5, 101, 107, 235, 305, 322,
  389, 448–449, 463, 482–484,
  488–489, 499–501
Nonseparable random processes,
  394
Normal random variable, 101
Null event, 33

O

Ordering of set elements, 633
Ornstein-Uhlenbeck process, 461
Orthogonal random variables, 259
Orthogonality principle, 310, 494

Outcome, 26

P

Pakes' lemma, 558
Paley-Wiener condition, 496
Paley, R. E. A. C., 496
Parallel filters, 518
Pareto random variable, 107, 124
Partitions of set, 624
Pascal random variable, 98
Pascal, Blaise, 10, 13, 98, 636,
  643–644, 647
PASTA property, 447
PDF, 82
Pearson, Karl, 105
Percentile of a random variable,
  156, 160, 166–169
Periodic Markov Chain, 545
Periodogram estimate:
  almost-sure convergence,
    512–513
  definition, 510
  mean square convergence, 512
  mean, 511, 513
  smoothing, 512
  variance, 511, 513
Periodograms, 510
Permutation, 634
Phase-shift key system, 521
PMF, 77
Poisson process, 554, 568
  alternative definitions, 448
  anti-PASTA property, 447
  arrival instants, 440
  autocorrelation, 442
  autocovariance, 442
  average, 442
  Bernoulli splitting, 456
  binomial, 453
  compound, 451
  conditional arrival time, 445
  definition, 439

distributions, 441
ergodicity, 454
generating values of, 457
interarrival times, 443
interrupted, 449
Markov modulated, 451, 565
Markovian property, 554
memoryless property, 446
multidimensional cdf, 441
PASTA property, 447
rate, 454
shot noise, 448
sum of, 455
thinned, 456
time of nth arrival, 444
Poisson random variable, 99, 114,
    139, 144, 209, 214, 216, 240,
    260–261, 268–269, 323–324,
    337, 353, 440
Poisson, S. D., 99
Polar coordinates, 297
Power spectral density:
    band-limited white noise, 482
    definition, 479
    estimation, 510
    of output of linear system, 487
    properties, 480–481
Powerset, 33, 71, 613, 651
Prediction of random processes,
    493, 497
Probabilistic model, 44
    revisited, 95, 221
Probability density function:
    calculation as histogram, 86
    conditional, 85
    Gaussian process, 459
    Markov Chain, 529
    n random variables, 265
    of function of single random
        variable, 116
    of function of two random
        variables, 237, 239

of normalized sums of IID
    random variables, 336
of sums of IID random variables,
    336
of two functions of two random
    variables, 253
one random variable, 82
properties, 83
two random variables, 212
Probability distribution function
    classification, 77
    conditional, 81
    IID random variables, 334
    Markov process, 528
    n random variables, 265
    one random variable, 76
    properties, 76
    random process, 394
    two random variables, 204
Probability generating function,
    138
Probability mass function:
    Markov Chain, 529–530
    n random variables, 265
    one random variable, 77
    Poisson process, 440, 442
Probability:
    and relative frequencies, 42, 102,
        109–110, 112, 121, 218, 645
    assignment, 42–46, 68, 76, 93,
        198
    axioms, 30
    model, 44
    revisited model, 95, 221
    space, 26
Product form sets:
    and circles, 201, 203
    and infinite strips, 200
    and lines, 200–201
    and points, 201
    and rectangles, 201–202
    and semi-infinite strips, 199–200

and semiplanes, 199
and triangles, 202–203
Pólya, G., 350, 406

Q

Q function, 101, 167–168
Quantizer transformation, 119

R

Random experiment:
  definition, 25
  examples, 25–26
  unrelated, 233
Random number generation, 145
Random process:
  autocorrelation, 398
  autocovariance, 398
  classification, 389
  continuity of, 410
  continuous-time, 388
  correlation coefficient, 398
  cross-correlation, 398
  cyclostationary, 404
  derivative of, 411
  discrete-time, 388
  ergodicity, 413
  examples of, 389–391, 394, 399,
    401, 417
  generation of variates, 420
  independent-increment, 400
  integral of, 412
  joint distribution function, 394
  Karhunen-Loève expansion, 419
  mean ergodic, 415
  mean of, 398
  mean square continuity of, 410
  mean square periodic, 417
  multiple, 397
  stationary, 401
  wide-sense stationary, 403
Random sum, 340
Random variable:

average of, 124
cdf, 76
characteristic function of, 135
classification, 77
conditional cdf, 81
continuous, 100
definition, 70
discrete, 96
generating function of, 138
generation of values, 147, 149
higher moments of, 126
histograms, 109
independence, 231
Laplace transform of, 137
mixed, 108
orthogonal, 259
pdf, 82
standard deviation of, 124
transformation of, 116
uncorrelated, 259
variance of, 124
Random walk:
  absorbing barriers, 408
  classification, 406
  definition, 406
  drifts, 560
  mean, 407
  period of, 556
  pmf, 407
  recurrence, 557
  reflecting barriers, 408
  restricted, 408
  state transition diagram, 554
  variance, 407
Range of random variable, 70
Rayleigh random variable, 107,
  248, 254–255
Rayleigh, Sir J. W., 107
Realization of random process, 388
Record times, 587
Recurrence of Markov Chain, 546
Reflecting barrier, 408

Rejection method, 149
Relative frequency, 109
Reliability, 129
   table of, 131
Residual lifetime, 130
Return times of Markov Chain,
   549
Rice random variable, 106, 248
Riemann, G. B., 412

S

Sample path of random process,
   388
Sample space, 26
Sampling:
   with ordering and without
      replacement, 634
   without ordering and with
      replacement, 637
   without ordering and without
      replacement, 635
Schwarz' inequality, 288, 403
Sequence of IID random variables,
   334
   convergence modes, 363
Set function, 68
Set:
   σ field, 626
   algebras of, 625
   Cartesian product, 622
   complement of, 613
   De Morgan's laws, 622
   definition, 611
   difference, 620
   disjoint, 619
   empty, 613
   intersection, 619
   limits of sequences, 624
   mutually exclusive, 619
   ordering of elements, 633
   partitions of, 624
   powerset of, 613

   proper superset, 613
   properties of operations, 621
   subset of, 612
   superset of, 613
   symmetric difference, 621
   union, 618
   Venn diagrams, 612
Shot noise, 448–449
Sigma field, 626
Signal plus noise, 500
Significance level of χ^2 test, 111
Signum, 290
Simpson, T., 325
Simulation, 46, 93, 146, 347, 454,
   595
Sinusoid random process:
   autocorrelation function of,
      399–400
   autocovariance function of, 399
   definition of, 389
   derivative of, 412
   mean of, 399
   mean square continuity of, 410
   one-dimensional cdf of, 395–396
   power spectral density of,
      481–482
   sample paths of, 390
   stationarity of, 402
   two-dimensional cdf of, 395, 397
   wide-sense cyclostationarity of,
      405
   wide-sense stationarity of, 404
Smoothing of random processes,
   493, 499
Snedecor, G. W., 105
Sojourn times of Markov Chain,
   549
Standard deviation, 124
State transition diagram, 536–537
Stationary increments, 400
Stationary random processes, 401